Second Edition

AN INTRODUCTION TO BIOSTATISTICS

Thomas Glover
Hobart and William Smith Colleges

Kevin Mitchell
Hobart and William Smith Colleges

D1373451

WAVELAND

PRESS, INC.

Long Grove, Illinois

For information about this book, contact:
Waveland Press, Inc.
4180 IL Route 83, Suite 101
Long Grove, IL 60047-9580
(847) 634-0081
info@waveland.com
www.waveland.com

Cover photo: Kevin Mitchell. *Linckia laevigata*,
Lady Elliot Island (Great Barrier Reef), Queensland, Australia.
See Example 11.17.

10-digit ISBN 1-57766-580-5
13-digit ISBN 978-1-57766-580-9

Printed in the United States of America

7   6   5   4   3   2

To our families:

Emily Glover

Ellen, Dorothy, and Dianne Mitchell

CONTENTS

Preface   ix

1 Introduction to Data Analysis   1

1.1 Introduction   1
1.2 Populations and Samples   3
1.3 Variables or Data Types   4
1.4 Measures of Central Tendency: Mean, Median, and Mode   5
1.5 Measures of Dispersion and Variability   8
1.6 Descriptive Statistics for Frequency Tables or Grouped Data   12
1.7 The Effect of Coding Data   14
1.8 Tables and Graphs   16
1.9 Quartiles and Box Plots   20
1.10 Accuracy, Precision, and the 30–300 Rule   23
1.11 Problems   24

2 Introduction to Probability   33

2.1 Definitions 33
2.2 Use of Permutations and Combinations   36
2.3 Introduction to Set Theory and Venn Diagrams   40
2.4 Axioms and Rules of Probability   43
2.5 Probability Rules and Mendelian Genetics (Optional)   51
2.6 Problems   56

3 Probability Distributions   65

3.1 Discrete Random Variables   66
3.2 The Binomial Distribution   71
3.3 The Poisson Distribution   77
3.4 Continuous Random Variables   81
3.5 The Normal Distribution   83
3.6 The Standard Normal Distribution   86
3.7 Problems   91

4 Sampling Distributions   **101**
- 4.1 Definitions 101
- 4.2 Distribution of the Sample Mean 103
- 4.3 Confidence Intervals for the Population Mean 109
- 4.4 Confidence Intervals for the Population Variance 115
- 4.5 Confidence Intervals for a Population Proportion 118
- 4.6 Problems 122

5 Introduction to Hypothesis Testing   **127**
- 5.1 An Overview: The Famous Cornflakes Example 127
- 5.2 Typical Steps in a Statistical Test of Hypothesis 132
- 5.3 Type I versus Type II Errors in Hypothesis Testing 134
- 5.4 Binomial Example of Hypothesis Testing (Optional) 139
- 5.5 Problems 141

6 One-Sample Tests of Hypothesis   **143**
- 6.1 Hypotheses Involving the Mean (μ) 143
- 6.2 Hypotheses Involving the Variance (σ^2) 151
- 6.3 Nonparametric Statistics and Hypothesis Testing 154
- 6.4 The One-Sample Sign Test 155
- 6.5 Confidence Intervals Based on the Sign Test 158
- 6.6 The One-Sample Wilcoxon Signed-Rank Test 159
- 6.7 Problems 165

7 Tests of Hypothesis Involving Two Samples   **173**
- 7.1 Comparing Two Variances 173
- 7.2 Testing the Difference Between Two Means of Independent Samples . 177
- 7.3 Confidence Intervals for $\mu_1 - \mu_2$ 182
- 7.4 The Difference Between Two Means with Paired Data 184
- 7.5 The Wilcoxon Rank-Sum (Mann-Whitney U) Test 187
- 7.6 The Sign Test and Paired Data 190
- 7.7 The Wilcoxon Signed-Rank Test for Paired Data 192
- 7.8 Problems 194

8 k-Sample Tests of Hypothesis: The Analysis of Variance   **211**
- 8.1 Model I ANOVA 213
- 8.2 Mean Separation Techniques for Model I ANOVAs 221
- 8.3 Model II ANOVA 226
- 8.4 The Kruskal-Wallis Test 228
- 8.5 Problems 234

9 Two-Factor Analysis   **245**
- 9.1 Randomized Complete Block Design ANOVA 245
- 9.2 Factorial Design Two-Way ANOVA 254
- 9.3 The Friedman k-Sample Test: Matched Data 264
- 9.4 Problems 270

10 Linear Regression and Correlation   **279**
   10.1 Simple Linear Regression 282
   10.2 Simple Linear Correlation Analysis 294
   10.3 Correlation Analysis Based on Ranks 300
   10.4 Problems 308

11 Goodness of Fit Tests for Categorical Data   **319**
   11.1 The Binomial Test 320
   11.2 Comparing Two Population Proportions 324
   11.3 The Chi-Square Test for Goodness of Fit 326
   11.4 The Chi-Square Test for $r \times k$ Contingency Tables 330
   11.5 The Kolmogorov-Smirnov Test 338
   11.6 The Lilliefors Test 344
   11.7 Problems 347

A Proofs of Selected Results   **363**
   A.1 Summation Notation and Properties 363
   A.2 Expected Values 367
   A.3 The Formula for SS_{Treat} in a One-Way ANOVA 373
   A.4 ANOVA Expected Values 373
   A.5 Calculating H in the Kruskal-Wallis Test 376
   A.6 The Method of Least Squares for Linear Regression 377

B Answers to Even-Numbered Problems   **379**
   Answers for Chapter 1 379
   Answers for Chapter 2 382
   Answers for Chapter 3 384
   Answers for Chapter 4 386
   Answers for Chapter 5 389
   Answers for Chapter 6 390
   Answers for Chapter 7 393
   Answers for Chapter 8 399
   Answers for Chapter 9 407
   Answers for Chapter 10 414
   Answers for Chapter 11 421

C Tables of Distributions and Critical Values   **429**
   C.1 Cumulative Binomial Distribution 430
   C.2 Cumulative Poisson Distribution 436
   C.3 Cumulative Standard Normal Distribution 438
   C.4 Student's t Distribution 440
   C.5 Cumulative Chi-Square Distribution 443
   C.6 Wilcoxon Signed-Rank Test Cumulative Distribution 445
   C.7 Cumulative F Distribution 448
   C.8 Critical Values for the Wilcoxon Rank-Sum Test 457
   C.9 Critical Values of the q Statistic for the Student-Newman-Keuls Test . 461
   C.10 Fisher's Z Transformation of Correlation Coefficient r 463
   C.11 Correlation Coefficient r Corresponding to Fisher's Z Transformation 466
   C.12 Cumulative Distribution for Kendall's Test (τ) 469

C.13 Critical Values for the Spearman Rank Correlation Coefficient, r_s . . .   472
C.14 Critical Values for the Kolmogorov-Smirnov Test   473
C.15 Critical Values for the Lilliefors Test   475

References   **477**

Index   **479**

Guide to Hypothesis Testing   **485**

Our goal in writing this book was to generate an accessible and relatively complete introduction for undergraduates to the use of statistics in the biological sciences. The text is designed for a one quarter or one semester class in introductory statistics for the life sciences. The target audience is sophomore and junior biology, environmental studies, biochemistry, and health sciences majors. The assumed background is some coursework in biology as well as a foundation in algebra but not calculus. Examples are taken from many areas in the life sciences including genetics, physiology, ecology, agriculture, and medicine.

This text emphasizes the relationships among probability, probability distributions, and hypothesis testing. We highlight the expected value of various test statistics under the null and research hypotheses as a way to understand the methodology of hypothesis testing. In addition, we have incorporated nonparametric alternatives to many situations along with the standard parametric analysis. These nonparametric techniques are included because undergraduate student projects often have small sample sizes that preclude parametric analysis and because the development of the nonparametric tests is readily understandable for students with modest math backgrounds. The nonparametrics can be skipped or skimmed without loss of continuity.

We have tried to include interesting and easily understandable examples with each concept. The problems at the end of each chapter have a range of difficulty and come from a variety of disciplines. Most are not real life examples but are realistic in their design and data values. The end-of-chapter problems are randomized within each chapter to require the student to choose the appropriate analysis. Many undergraduate texts present a concept or test and immediately give all the problems that can be solved by that technique. This approach prevents students from having to make the real life decision about the appropriate analysis. We believe this decision making is a critical skill in the introduction of statistical analysis and have provided a large number of opportunities to practice and develop this skill.

The material for this text derives principally from a required biostatistics course one of us (Glover) has taught to undergraduates for more than twenty years and from a second course in nonparametric statistics and field data analysis that the other of us (Mitchell) has taught more recently during several term abroad programs to Queensland, Australia. Recent shifts in undergraduate curricula have de-emphasized

calculus for biology students and are now highlighting statistical analysis as a fundamental quantitative skill. We hope that our text will make teaching and learning that
skill somewhat less arduous.

Supplemental Materials

The material in this textbook can be supported by a wide variety of statistical
packages and ancillary materials. The selection of these support materials is dic-
tated by personal interests and cost considerations. Presently we use SPSS software
(http://www.spss.com/) in the laboratory sessions of our course. This software is
easy to use, relatively flexible, and can complete nearly all the statistical techniques
presented in our text. Another similarly useful and accessible statistical package is
Minitab (www.minitab.com). A number of free online statistical tools are also available. Among the best and most complete is The R Project for Statistical Computing
which is available at http://www.r-project.org/.

For student purchase we have used Texas Instrument calculators ranging from the
TI-30 to the TI-83, TI-84, and TI-89 models. The price range for these calculators is
considerable and might be a factor in choosing a required calculator for a particular
course. Although calculators such as the TI-30 do less automatically, they sometimes
give the student clearer insights into the statistical tests by requiring a few more
computational steps. The ease of computation afforded by computer programs or
sophisticated calculators sometimes leads to a "black box" mentality about statistics
and their calculation.

For both students and instructors we recommend D. J. Hand et al., editors, 1994,
A Handbook of Small Data Sets, Chapman & Hall, London. This book contains 510
small data sets ranging from the numbers of Prussian military personel killed by horse
kicks from 1875–1894 (data set #283) to the shape of bead work on leather goods
of Shoshoni Indians (data set #150). The data sets are interesting, manageable,
and amenable to statistical analysis using techniques presented in our text. While a
number of the data sets from the Handbook were utilized as examples or problems
in our text, there are many others that could serve as engaging and useful practice
problems.

An Answer Manual for instructors is available free on CD from the publisher. Instructors can access a printable version with both questions and answers that correlate
to this text, or a printable version with answers only. The CD also includes additional
exercises and answers not found in the text.

Acknowledgments

For the preparation of this second edition, thanks are due to the following people: Don
Rosso at Waveland Press for his support and guidance; Topa (Sophie) Petit, *University
of South Australia*, for her many suggestions; Ann Warner, *Hobart and William Smith
Colleges*, for her meticulous word processing of the various drafts of this manuscript;
and the students of Hobart and William Smith Colleges for their many comments and
suggestions, particularly Aline Gadue for her careful scrutiny of the first edition.

Thomas J. Glover
Kevin J. Mitchell
Geneva, NY

Introduction to Data Analysis

Concepts in Chapter 1:

- Scientific Method and Statistical Analysis
- Parameters: Descriptive Characteristics of Populations
- Statistics: Descriptive Characteristics of Samples
- Variable Types: Continuous, Discrete, Ranked, and Categorical
- Measures of Central Tendency: Mean, Median, and Mode
- Measures of Dispersion: Range, Variance, and Standard Deviation
- Descriptive Statistics for Frequency Data
- Effects of Coding on Descriptive Statistics
- Tables and Graphs
- Quartiles and Box Plots
- Accuracy, Precision, and the 30–300 Rule

1.1 Introduction

The modern study of the life sciences includes experimentation, data gathering, and interpretation. This text offers an introduction to the methods used to perform these fundamental activities.

The design and evaluation of experiments, known as the **scientific method,** is utilized in all scientific fields and is often implied rather than explicitly outlined in many investigations. The components of the scientific method include observation, formulation of a potential question or problem, construction of a hypothesis, followed by a prediction, and the design of an experiment to test the prediction. Let's consider these components briefly.

Observation of a Particular Event

Generally an observation can be classified as either quantitative or qualitative. Quantitative observations are based on some sort of measurement, e.g., length, weight, temperature, and pH. Qualitative observations are based on categories reflecting a quality or characteristic of the observed event, e.g., male versus female, diseased versus healthy, and mutant versus wild type.

Statement of the Problem

A series of observations often leads to the formulation of a particular problem or unanswered question. This usually takes the form of a "why" question and implies a cause and effect relationship. For example, suppose upon investigating a remote

Fijian island community you realized that the vast majority of the adults suffer from hypertension (abnormally elevated blood pressures with the systolic over 165 mmHg and the diastolic over 95 mmHg). Note that the individual observations here are quantitative while the percentage that are hypertensive is based on a qualitative evaluation of the sample. From these preliminary observations one might formulate the question: Why are so many adults in this population hypertensive?

Formulation of a Hypothesis

A hypothesis is a tentative explanation for the observations made. A good hypothesis suggests a cause and effect relationship and is testable.

The Fijian community may demonstrate hypertension because of diet, life style, genetic makeup, or combinations of these factors. Because we've noticed extraordinary consumption of octopi in their diet and knowing octopods have a very high cholesterol content, we might hypothesize that *the high level of hypertension is caused by diet.*

Making a Prediction

If the hypothesis is properly constructed, it can and should be used to make predictions. Predictions are based on deductive reasoning and take the form of an "if-then" statement. For example, a good prediction based on the hypothesis above would be: *If the hypertension is caused by a high cholesterol diet, then changing the diet to a low cholesterol one should lower the incidence of hypertension.*

The criteria for a valid (properly stated) prediction are:

1. An "if" clause stating the hypothesis.

2. A "then" clause that

   (a) suggests altering a causative factor in the hypothesis (change of diet);

   (b) predicts the outcome (lower level of hypertension);

   (c) provides the basis for an experiment.

Design of the Experiment

The entire purpose and design of an experiment is to accomplish one goal, that is, to test the hypothesis. An experiment tests the hypothesis by testing the correctness or incorrectness of the predictions that came from it. Theoretically, an experiment should alter or test only the factor suggested by the prediction, while all other factors remain constant.

How would you design an experiment to test the diet hypothesis in the hypertensive population?

The best way to test the hypothesis above is by setting up a controlled experiment. This might involve using two randomly chosen groups of adults from the community and treating both identically with the exception of the one factor being tested. The control group represents the "normal" situation, has all factors present, and is used as a standard or basis for comparison. The experimental group represents the "test" situation and includes all factors except the variable that has been altered, in this case the diet. If the group with the low cholesterol diet exhibits *significantly* lower levels of hypertension, the hypothesis is supported by the data. On the other hand, if the change in diet has no effect on hypertension, then a new or revised hypothesis should be formulated and the experimental procedure redesigned. Finally, the generalizations that are drawn by relating the data to the hypothesis can be stated as conclusions.

While these steps outlined above may seem straightforward, they often require considerable insight and sophistication to apply properly.

In our example how the groups are chosen is not a trivial problem. They must be constructed without bias and must be large enough to give the researcher an acceptable level of confidence in the results. Further, how large a change is significant enough to support the hypothesis? What is *statistically significant* may not be *biologically significant.*

A foundation in statistical methods will help you design and interpret experiments properly. The field of statistics is broadly defined as the methods and procedures for collecting, classifying, summarizing, and analyzing data, and utilizing the data to test scientific hypotheses. The term *statistics* is derived from the Latin for state, and originally referred to information gathered in various censuses that could be numerically summarized to describe aspects of the state, e.g., bushels of wheat per year, or number of military-aged men. Over time statistics has come to mean the scientific study of numerical data based on natural phenomena. Statistics applied to the life sciences is often called **biostatistics** or **biometry**. The foundations of biostatistics go back several hundred years, but statistical analysis of biological systems began in earnest in the late nineteenth century as biology became more quantitative and experimental.

1.2 Populations and Samples

Today we use statistics as a means of informing the decision-making processes in the face of the uncertainties that most real world problems present. Often we wish to make generalizations about populations that are too large or too difficult to survey completely. In these cases we sample the population and use characteristics of the sample to extrapolate to characteristics of the larger population. See Figure 1.1.

Real-world problems concern large groups or **populations** about which inferences must be made. (Is there a size difference between two color morphs of the same species of sea star? Are the offspring of a certain cross of fruit flies in a 3 : 1 ratio of normal to eyeless?) Certain characteristics of the population are of particular interest (systolic blood pressure, weight in grams, resting body temperature). The values of these characteristics will vary from individual to individual within the population. These characteristics are called **random variables** because they vary in an unpredictable way or in a way that appears or is assumed to depend on chance. The different types of variables are described in Section 1.3.

A descriptive measure associated with a random variable when it is considered over the *entire population* is called a **parameter**. Examples are the mean weight of all green turtles, *Chelonia mydas*, or the variance in clutch size of all tiger snakes, *Notechis scutatus*. In general, such parameters are difficult, if not impossible, to determine because the population is too large or expensive to study in its entirety. Consequently, one is forced to examine a subset or **sample** of the population and make inferences about the entire population based on this sample. A descriptive measure associated with a random variable of a *sample* is called a **statistic**. The mean weight of 25 female green turtles laying eggs on Heron Island or the variability in clutch size

Population(s) have traits called random variables.
Summary characteristics of the population random variables
are called parameters: μ, σ^2, N.

$\downarrow$

Random samples of size n
of the population(s) generate numerical data: X_i's.

$\downarrow$

These data can be organized into
summary statistics: $\overline{X}$, s^2, n,
graphs, and figures (Chapter 1).

$\downarrow$

The data can be analyzed using
an understanding of basic probability (Chapters 2–4)
and various tests of hypotheses (Chapters 5–11).

$\downarrow$

The analyses lead to conclusions or inferences
about the population(s) of interest.

FIGURE 1.1.   The general approach to statistical analysis.

of 50 clutches of tiger snake eggs collected in southeastern Queensland are examples of statistics.

While such statistics are not equal to the population parameters, it is hoped that they are sufficiently close to the population parameters to be useful or that the potential error involved can be quantified. Sample statistics along with an understanding of probability form the foundation for inferences about population parameters. See Figure 1.1 for review.

Chapter 1 provides techniques for organizing sample data. Chapters 2 through 4 present the necessary probability concepts, and the remaining chapters outline various techniques to test a wide range of predictions from hypotheses.

1.3   Variables or Data Types

There are several data types that arise in statistics. Each statistical test requires that the data analyzed be of a specified type. Here are the most common types of variables.

1. **Quantitative variables** fall into two major categories:

    (a) **Continuous variables** or **interval data** can assume any value in some (possibly unbounded) interval of real numbers. Common examples include length, weight, temperature, volume, and height. They arise from measurement.

    (b) **Discrete variables** assume only isolated values. Examples include clutch size, trees per hectare, arms per sea star, or items per quadrat. They arise from counting.

2. **Ranked (ordinal) variables** are not measured but nonetheless have a natural ordering. For example, candidates for political office can be ranked by individual voters. Or students can be arranged by height from shortest to tallest and correspondingly ranked without ever being measured. The rank values have no inherent meaning outside the "order" that they provide. That is, a candidate ranked 2 is not twice as preferable as the person ranked 1. (Compare this with measurement variables where a plant 2 feet tall *is* twice as tall as a plant 1 foot tall. With measurement variables such ratios are meaningful, while with ordinal variables they are not.)

3. **Categorical data** are qualitative data. Some examples are species, gender, genotype, phenotype, healthy/diseased, and marital status. Unlike with ranked data, there is no "natural" ordering that can be assigned to these categories.

When measurement variables are collected for either a population or a sample, the numerical values have to be abstracted or summarized in some way. The summary descriptive characteristics of a population of objects are called **population parameters** or just **parameters**. The calculation of a parameter requires knowledge of the measurement variables value for *every* member of the population. These parameters are usually denoted by Greek letters and do not vary within a population. The summary descriptive characteristics of a sample of objects, that is, a subset of the population, are called **statistics**. Sample statistics can have different values, depending on how the sample of the population was chosen. Statistics are denoted by various symbols, but (almost) never by Greek letters.

Let's consider some important summary characteristics of populations and samples now.

1.4 Measures of Central Tendency: Mean, Median, and Mode

Mean

There are several commonly used measures to describe the location or center of a population or sample. The most widely utilized measure of central tendency is the *arithmetic mean* or *average*.

The **population mean** is the sum of the values of the variable under study divided by the total number of objects in the population. It is denoted by a lowercase μ ("mu"). Each value is algebraically denoted by an X with a subscript denotation i. For example, a small theoretical population whose objects had values 1, 6, 4, 5, 6, 3, 8, 7 would be denoted

$$X_1 = 1, X_2 = 6, X_3 = 4, X_4 = 5, X_5 = 6, X_6 = 3, X_7 = 8, X_8 = 7. \qquad (1.1)$$

We would denote the population size with a capital N. In our theoretical population $N = 8$.

The population mean μ would be

$$\frac{1 + 6 + 4 + 5 + 6 + 3 + 8 + 7}{8} = 5.$$

FORMULA 1.1. The algebraic shorthand formula for a population mean is

$$\mu = \frac{\sum_{i=1}^{N} X_i}{N}.$$

The Greek letter Σ ("sigma") indicates summation. The subscript $i = 1$ indicates to start with the first observation and the superscript N means to continue until and including the Nth observation. The subscript and superscript may represent other starting and stopping points for the summation within the population or sample. For the example above,

$$\sum_{i=2}^{5} X_i$$

would indicate the sum of $X_2 + X_3 + X_4 + X_5$ or $6 + 4 + 5 + 6 = 21$.

Notice also that $\sum_{i=1}^{N} X_i$ is written $\sum_{i=i}^{N} X_i$ when the summation symbol is embedded in a sentence. In fact, to further reduce clutter, the summation sign may not indexed at all, for example $\sum X_i$. It is implied that the operation of addition begins with the first observation and continues through the last observation in a population, that is,

$$\sum X_i = \sum_{i=1}^{N} X_i.$$

If sigma notation is new to you or if you wish a quick review of its properties, read Appendix A.1 before continuing.

FORMULA 1.2. The sample mean is defined by

$$\overline{X} = \frac{\sum_{i=1}^{n} X_i}{n},$$

where n is the sample size. *The sample mean is usually reported to one more decimal place than the data and always has appropriate units associated with it.*

The symbol $\overline{X}$ (read "X bar") indicates that the observations of a subset of size n from a population have been averaged. $\overline{X}$ is fundamentally different from μ because samples from a population can have different values for their sample mean, that is, they can vary from sample to sample within the population. The population mean, however, is constant for a given population.

Again consider the small theoretical population $1, 6, 4, 5, 6, 3, 8, 7$. A sample of size 3 may consist of $5, 3, 4$ with $\overline{X} = 4$ or $6, 8, 4$ with $\overline{X} = 6$.

Actually there are 56 possible samples of size 3 that could be drawn from the population in (1.1). Only four samples have a *sample* mean the same as the population mean, i.e., $\overline{X} = \mu$:

Sample	Sum	$\overline{X}$
X_3, X_6, X_7	$4 + 3 + 8$	5
X_2, X_3, X_4	$6 + 4 + 5$	5
X_5, X_3, X_4	$6 + 4 + 5$	5
X_8, X_6, X_4	$7 + 3 + 5$	5

Each sample mean $\overline{X}$ is an unbiased estimate of μ but depends on the values included in the sample and sample size for its actual value. We would expect the average of all possible $\overline{X}$'s to be equal to the population parameter, μ. This is, in fact, the definition of an **unbiased estimator** of the population mean.

If you calculate the sample mean for each of the 56 possible samples with $n = 3$ and then average these sample means, they will give an average value of 5, that is, the population mean, μ. Remember that most real populations are too large or too difficult to census completely, so we must rely on using a single sample to estimate or approximate the population characteristics.

Median

The second measure of central tendency is the median. The **median** is the "middle" value of an **ordered** list of observations. Though this idea is simple enough, it will prove useful to define it in terms of an even simpler notion. The **depth** of a value is its position relative to the nearest extreme (end) when the data are listed in order from smallest to largest.

EXAMPLE 1.1. The table below gives the circumferences at chest height (CCH) (in cm) and their corresponding depths for 15 sugar maples, *Acer saccharum*, measured in a forest in southeastern Ohio.

CCH	18	21	22	29	29	36	37	38	56	59	66	70	88	93	120
Depth	1	2	3	4	5	6	7	8	7	6	5	4	3	2	1

The **population median** M is the observation whose depth is $d = \frac{N+1}{2}$, where N is the population size.

Note that this parameter is not a Greek letter and is seldom computed in practice. Rather a sample median $\tilde{X}$ (read "X tilde") is the statistic used to approximate or estimate the population median. $\tilde{X}$ is defined as the observation whose depth is $d = \frac{n+1}{2}$, where n is the sample size. In Example 1.1, the sample size is $n = 15$, so the depth of the sample median is $d = 8$. The sample median $\tilde{X} = X_{\frac{n+1}{2}} = X_8 = 38$ cm.

EXAMPLE 1.2. The table below gives CCH (in cm) for 12 cypress pines, *Callitris preissii*, measured near Brown Lake on North Stradbroke Island.

CCH	17	19	31	39	48	56	68	73	73	75	80	122
Depth	1	2	3	4	5	6	6	5	4	3	2	1

Since $n = 12$, the depth of the median is $\frac{12+1}{2} = 6.5$. Obviously no observation has depth 6.5, so this is interpreted as the average of both observations whose depth is 6 in the list above. So $\tilde{X} = \frac{56+68}{2} = 62$ cm.

Mode

The **mode** is defined as the most frequently occurring value in a data set. The mode of Example 1.2 would be 73 cm, while Example 1.1 would have a mode of 29 cm. In symmetrical distributions the mean, median, and mode are coincident. Bimodal distributions may indicate a mixture of samples from two populations (e.g., weights of males and females). While the mode is not often used in biological research, reporting the number of modes, if more than one, can be informative.

Each measure of central tendency has different features. The mean is a purposeful measure only for a quantitative variable, whether it is continuous (e.g., height) or discrete (e.g., clutch size). The median can be calculated whenever a variable can be ranked (including when the variable is quantitative). Finally, the mode can be calculated for categorical variables, as well as for quantitative and ranked variables.

The sample median expresses less information than the sample mean because it utilizes only the ranks and not the actual values of each measurement. The median, however, is resistant to the effects of **outliers**. Extreme values or outliers in a sample can drastically affect the sample mean, while having little effect on the median. Consider Example 1.2 with $\overline{X} = 58.4$ cm and $\hat{X} = 62$ cm. Suppose X_{12} had been mistakenly recorded as 1220 cm instead of 122 cm. The mean $\overline{X}$ would become 149.9 cm while the median $\tilde{X}$ would remain 62 cm.

1.5 Measures of Dispersion and Variability: Variance, Standard Deviation, and Range

EXAMPLE 1.3. The table that follows gives the weights of two samples of albacore tuna, *Thunnus alaluga* (in kg). How would you characterize the differences in the samples?

Sample 1	Sample 2
8.9	3.1
9.6	17.0
11.2	9.9
9.4	5.1
9.9	18.0
10.9	3.8
10.4	10.0
11.0	2.9
9.7	21.2

SOLUTION. Upon investigation we see that both samples are the same size and have the same mean, $\overline{X}_1 = \overline{X}_2 = 10.11$ kg:

Sample 1	Sample 2
8.9	3.1
9.6	17.0
11.2	9.9
9.4	5.1
9.9	18.0
10.9	3.8
10.4	10.0
11.0	2.9
9.7	21.2
$n_1 = 9$	$n_2 = 9$
$\overline{X}_1 = 10.11$ kg	$\overline{X}_2 = 10.11$ kg

In fact, both samples have the same median. To see this, arrange the data sets in rank order as in Table 1.1. We have $n = 9$, so $\hat{X} = X_{\frac{n+1}{2}} = X_5$, which is 9.9 kg for both samples.

Neither of the samples has a mode. So by all the descriptors in Section 1.4 these samples appear to be identical. Clearly they are not. The difference in the samples is reflected in the scatter or spread of the observations. Sample 1 is much more

TABLE 1.1. The ordered samples of *Thunnus alaluga*

Depth	Sample 1	Sample 2
1	8.9	2.9
2	9.4	3.1
3	9.6	3.8
4	9.7	5.1
5	9.9	9.9
4	10.4	10.0
3	10.9	17.0
2	11.0	18.0
1	11.2	21.2
	$\tilde{X}_1 = 9.9$ kg	$\tilde{X}_2 = 9.9$ kg

uniform than Sample 2, that is, the observations tend to cluster much nearer the mean in Sample 1 than in Sample 2. We need descriptive measures of this scatter or dispersion that will reflect these differences.

Range

The simplest measure of dispersion or "spread" of the data is the range.

FORMULAS 1.3. The difference between the largest and smallest observations in a group of data is called the **range**:

$$\text{Sample range} = X_n - X_1$$
$$\text{Population range} = X_N - X_1$$

When the data are ordered from smallest to largest, the values X_n and X_1 are called the **sample range limits**.

In Example 1.3 we have from Table 1.1

$$\text{Sample 1: range} = X_9 - X_1 = 11.2 - 8.9 = 2.3 \text{ kg}$$
$$\text{Sample 2: range} = X_9 - X_1 = 21.2 - 2.9 = 18.3 \text{ kg}$$

The range for each of these two samples reflects some differences in dispersion, but the range is a rather crude estimator of dispersion because it uses only two of the data points and is somewhat dependent on sample size. As sample size increases, we expect largest and smallest observations to become more extreme and, therefore, the sample range to increase even though the population range remains unchanged. It is unlikely that the sample will include the largest and smallest values from the population, so the sample range usually underestimates the population range and is, therefore, a biased estimator.

Variance

To develop a measure that uses all the data to form an index of dispersion consider the following. Suppose we express each observation as a distance from the mean

$x_i = X_i - \overline{X}$. These differences are called **deviates** and will be sometimes positive (X_i is above the mean) and sometimes negative (X_i is below the mean).

If we try to average the deviates, they always sum to 0. Because the mean is the central tendency or location, the negative deviates will exactly cancel out the positive deviates. Consider a simple numerical example

$$X_1 = 2 \qquad X_2 = 3 \qquad X_3 = 1 \qquad X_4 = 8 \qquad X_5 = 6.$$

The mean $\overline{X} = 4$, and the deviates are

$$x_1 = -2 \qquad x_2 = -1 \quad x_3 = -3 \qquad x_4 = 4 \qquad x_5 = 2.$$

Notice that the negative deviates cancel the positive ones so that $\sum(X_i - \overline{X}) = 0$.

Algebraically one can demonstrate the same result more generally,

$$\sum_{i=1}^{n}(X_i - \overline{X}) = \sum_{i=1}^{n} X_i - \sum_{i=1}^{n} \overline{X}.$$

Since $\overline{X}$ is a constant for any sample,

$$\sum_{i=1}^{n}(X_i - \overline{X}) = \sum_{i=1}^{n} X_i - n\overline{X}.$$

Since $\overline{X} = \frac{\sum X_i}{n}$, then $n\overline{X} = \sum X_i$, so

$$\sum_{i=1}^{n}(X_i - \overline{X}) = \sum_{i=1}^{n} X_i - \sum_{i=1}^{n} X_i = 0. \qquad (1.2)$$

To circumvent this unfortunate property, the widely used measure of dispersion called the **sample variance** utilizes the squares of the deviates. The quantity

$$\sum_{i=1}^{n}(X_i - \overline{X})^2$$

is the sum of these squared deviates and is referred to as the **corrected sum of squares,** denoted by CSS. Each observation is corrected or adjusted for its distance from the mean.

FORMULA 1.4. The corrected sum of squares is utilized in the formula for the sample variance,

$$s^2 = \frac{\sum_{i=1}^{n}(X_i - \overline{X})^2}{n - 1}.$$

The sample variance is usually reported to two more decimal places than the data and has units that are the square of the measurement units.

This calculation is not as intuitive as the mean or median, but it is a very good indicator of scatter or dispersion. If the above formula had n instead of $n-1$ in the denominator, it would be exactly the average squared distance from the mean. Returning to Example 1.3, the variance of Sample 1 is 0.641 kg^2 and the variance of Sample 2 is 49.851 kg^2, reflecting the larger "spread" in Sample 2.

A sample variance is an unbiased estimator of a parameter called the **population variance.**

FORMULA 1.5. A population variance is denoted by σ^2 ("sigma squared") and is defined by

$$\sigma^2 = \frac{\sum_{i=1}^{N}(X_i - \mu)^2}{N}.$$

It really *is* the average squared deviation from the mean for the population. The $n-1$ in Formula 1.4 makes it an unbiased estimate of the population parameter. (See Appendix A.2 for a proof.) Remember that "unbiased" means that the average of all possible values of s^2 for a certain size sample will be equal to the population value σ^2.

Formulas 1.4 and 1.5 are theoretical formulas and are rather tedious to apply directly. Computational formulas utilize the fact that most calculators with statistical registers simultaneously calculate n, $\sum X_i$, and $\sum X_i^2$.

FORMULA 1.6. The corrected sum of squares $\sum(X_i - \overline{X})^2$ may be computed more simply as

$$\text{CSS} = \sum X_i^2 - \frac{(\sum X_i)^2}{n}.$$

$\sum X_i^2$ is the uncorrected sum of squares and $\frac{(\sum X_i)^2}{n}$ is the correction term.

To verify Formula 1.6, using the properties in Appendix A.1 notice that

$$\sum(X_i - \overline{X})^2 = \sum(X_i^2 - 2X_i\overline{X} + \overline{X}^2) = \sum X_i^2 - 2\overline{X}\sum X_i + \sum \overline{X}^2.$$

Remember that $\overline{X} = \frac{\sum X_i}{n}$, so $n\overline{X} = \sum X_i$; hence

$$\sum(X_i - \overline{X})^2 = \sum X_i^2 - 2\overline{X}(n\overline{X}) + \sum \overline{X}^2 = \sum X_i^2 - 2n\overline{X}^2 + n\overline{X}^2 = \sum X_i^2 - n\overline{X}^2.$$

Substituting $\frac{\sum X_i}{n}$ for $\overline{X}$ yields

$$\sum(X_i - \overline{X})^2 = \sum X_i^2 - n\left(\frac{\sum X_i}{n}\right)^2 = \sum X_i^2 - \frac{n(\sum X_i)^2}{n^2} = \sum X_i^2 - \frac{(\sum X_i)^2}{n}.$$

FORMULA 1.7. Use of the computational formula for the corrected sum of squares gives the computational formula for the sample variance

$$s^2 = \frac{\sum X_i^2 - \frac{(\sum X_i)^2}{n}}{n-1}.$$

Returning to Example 1.3, Sample 2,

$$\sum X_i = 91, \qquad \sum X_i^2 = 1318.92, \qquad n = 9,$$

so

$$s^2 = \frac{1318.92 - \frac{(91)^2}{9}}{9-1} = \frac{1318.92 - 920.11}{8} = \frac{398.81}{8} = 49.851 \text{ kg}^2.$$

Remember, the numerator must always be a positive number because it's a sum of squared deviations. Because the variance has units that are the square of the measurement units, such as squared kilograms above, they have no physical interpretation.

With a similar derivation, the population variance computational formula can be shown to be

$$\sigma^2 = \frac{\sum X_i^2 - \frac{(\sum X_i)^2}{N}}{N}.$$

Again, this formula is rarely used since most populations are too large to census directly.

Standard Deviation

FORMULAS 1.8. A more "natural" calculation is the **standard deviation**, which is the positive square root of the population or sample variance, respectively.

$$\sigma = \sqrt{\frac{\sum X_i^2 - \frac{(\sum X_i)^2}{N}}{N}} \quad \text{and} \quad s = \sqrt{\frac{\sum X_i^2 - \frac{(\sum X_i)^2}{n}}{n-1}}.$$

These descriptions have the same units as the original observations and are, in a sense, the average deviation of observations from their mean.

Again, consider Example 1.3.

$$\text{For Sample 1: } s_1^2 = 0.641 \text{ kg}^2, \quad \text{so } s_1 = 0.80 \text{ kg}.$$
$$\text{For Sample 2: } s_2^2 = 49.851 \text{ kg}^2, \text{ so } s_2 = 7.06 \text{ kg}.$$

The standard deviation of a sample is relatively easy to interpret and clearly reflects the greater variability in Sample 2 compared to Sample 1. *Like the mean, the standard deviation is usually reported to one more decimal place than the data and always has appropriate units associated with it.* Both the variance and standard deviation can be used to demonstrate differences in scatter between samples or populations.

1.6 Descriptive Statistics for Frequency Tables or Grouped Data

When large data sets are organized into frequency tables or presented as grouped data, there are shortcut methods to calculate the sample statistics: $\overline{X}$, s^2, and s.

EXAMPLE 1.4. The following table shows the number of sedge plants, *Carex flacca*, found in 800 sample quadrats in an ecological study of grasses. Each quadrat was 1 m^2.

Plants/quadrat (X_i)	Frequency (f_i)
0	268
1	316
2	135
3	61
4	15
5	3
6	1
7	1

To calculate the sample descriptive statistics using Formulas 1.2, 1.7, and 1.8 would be quite arduous, involving sums and sums of squares of 800 numbers. Fortunately, the following formulas limit the drudgery for these calculations.

It is clear that $X_1 = 0$ occurs $f_1 = 268$ times, $X_2 = 1$ occurs $f_2 = 316$ times, etc., and that the sum of observations in the first category is $f_1 X_1$, the sum in the second category is $f_2 X_2$, etc. The sum of all observations is, therefore,

$$f_1 X_1 + f_2 X_2 + \cdots + f_c X_c = \sum_{i=1}^{c} f_i X_i,$$

where c denotes the number of categories. The total number of observations is $n = \sum_{i=1}^{c} f_i$, and as a result:

FORMULA 1.9. The sample mean for a grouped data set is given by

$$\overline{X} = \frac{\sum_{i=1}^{c} f_i X_i}{\sum_{i=1}^{c} f_i}.$$

Similarly, the computational formula for the sample variance for a grouped data set can be derived directly from

$$s^2 = \frac{\sum_{i=1}^{c} f_i (X_i - \overline{X})^2}{n - 1}.$$

FORMULA 1.10. The sample variance for a grouped data set is given by

$$s^2 = \frac{\sum_{i=1}^{c} f_i X_i^2 - \frac{(\sum f_i X_i)^2}{n}}{n - 1},$$

where $n = \sum_{i=1}^{c} f_i$.

To apply Formulas 1.9 and 1.10, we need to calculate only three sums: $n = \sum f_i$; the sum of observations $\sum f_i X_i$; and the uncorrected sum of squared observations $\sum f_i X_i^2$.

Returning to Example 1.4, it is now straightforward to calculate $\overline{X}$, s^2, and s.

Plants/quadrat (X_i)	f_i	$f_i X_i$	$f_i X_i^2$
0	268	0	0
1	316	316	316
2	135	270	540
3	61	183	549
4	15	60	240
5	3	15	75
6	1	6	36
7	1	7	49
Sum	800	857	1805

Note that column 4 in the table above is generated by first squaring X_i and then multiplying by f_i, not by squaring the values in column 3. In other words, $f_i X_i^2 \neq (f_i X_i)^2$.

$$\overline{X} = \frac{\sum_{i=1}^{c} f_i X_i}{\sum_{i=1}^{c} f_i} = \frac{857}{800} = 1.1 \text{ plants/quadrat},$$

$$s^2 = \frac{\sum_{i=1}^{c} f_i (X_i^2) - \frac{\left(\sum_{i=1}^{c} f_i X_i\right)^2}{n}}{n - 1} = \frac{1805 - \frac{(857)^2}{800}}{800 - 1} = 1.11 \text{ (plants/quadrat)}^2,$$

and

$$s = \sqrt{1.11} = 1.1 \text{ plants/quadrat}.$$

Example 1.4 summarized data for a discrete variable taking on whole number values from 0 to 7. Continuous variables can also be presented as grouped data in frequency tables.

EXAMPLE 1.5. The following data were collected by randomly sampling a large population of rainbow trout, *Salmo gairdnerii*. The variable of interest is weight in pounds.

X_i (lb)	f_i	$f_i X_i$	$f_i X_i^2$
1	2	2	2
2	1	2	4
3	4	12	36
4	7	28	112
5	13	65	325
6	15	90	540
7	20	140	980
8	24	192	1536
9	7	63	567
10	9	90	900
11	2	22	242
12	4	48	576
13	2	26	338
Sum	110	780	6158

Rainbow trout have weights that can range from almost 0 to 20 lb or more. Moreover their weights can take on any value in that interval. For example, a particular trout may weigh 7.3541 lb. When data are grouped as in Example 1.5 intervals are implied for each class. A fish in the 3-lb class weighs somewhere between 2.50 and 3.49 lb and a fish in the 9-lb class weighs between 8.50 and 9.49 lb. Fish were weighed to the nearest pound allowing analysis of grouped data for a continuous measurement variable. In Example 1.5,

$$\overline{X} = \frac{\sum_{i=1}^{c} f_i X_i}{\sum_{i=1}^{c} f_i} = \frac{780}{110} = 7.1 \text{ lb}$$

and

$$s^2 = \frac{\sum_{i=1}^{c} f_i (X_i)^2 - \frac{\left(\sum_{i=1}^{c} f_i X_i\right)^2}{n}}{n-1} = \frac{6158 - \frac{(780)^2}{110}}{110 - 1} = 5.75 \text{ (lb)}^2.$$

Therefore,

$$s = \sqrt{5.75} = 2.4 \text{ lb.}$$

Again, consider that calculation time is saved by working with 13 classes instead of 110 individual observations. Whether measuring the rainbow trout to the nearest pound was appropriate will be considered in Section 1.10.

1.7 The Effect of Coding Data

While grouping data can save considerable time and effort, coding data may also offer similar savings. Coding involves conversion of measurements or statistics into easier to work with values by simple arithmetic operations. It is sometimes used to change units or to investigate experimental effects.

Additive Coding

Additive coding involves the addition or subtraction of a constant from each observation in a data set. Suppose the data gathered in Example 1.5 were collected using a scale that weighed the fish 2 lb too low. We could go back to the data and add 2 lb to each observation and recalculate the descriptive statistics. A more efficient tack would be to realize that *if a fixed amount c is added or subtracted from each observation in a data set, the sample mean will be increased or decreased by that amount, but the variance will be unchanged.*

To see why, if $\overline{X}_c$ is the coded mean, then

$$\overline{X}_c = \frac{\sum(X_i + c)}{n} = \frac{\sum X_i + \sum c}{n} = \frac{\sum X_i + nc}{n} = \frac{\sum X_i}{n} + c = \overline{X} + c.$$

If s_c^2 is the coded sample variance, then

$$s_c^2 = \frac{\sum[(X_i + c) - (\overline{X} + c)]^2}{n-1} = \frac{\sum(X_i + c - \overline{X} - c)^2}{n-1} = \frac{\sum(X_i - \overline{X})^2}{n-1} = s^2,$$

therefore, $s_c = s$.

If the scale weighed 2 lb light in Example 1.5 the new, corrected statistics would be $\overline{X}_c = 7.1 + 2.0 = 9.1$ lb, and $s_c^2 = 5.75$ $(\text{lb})^2$, and $s_c = 2.4$ lb.

Multiplicative Coding

Multiplicative coding involves multiplying or dividing each observation in a data set by a constant. Suppose the data in Example 1.5 were to be presented at an international conference and, therefore, had to be presented in metric units (kilograms) rather than English units (pounds). Since 1 kg equals 2.20 lb, we could convert the observations to kilograms by multiplying each observation by $1/2.20$ or 0.45 kg/lb. Again, the more efficient approach would be to realize the following.

If each of the observations in a data set is multiplied by a fixed quantity c, the new mean is c times the old mean because

$$\overline{X}_c = \frac{\sum cX_i}{n} = \frac{c\sum X_i}{n} = c\overline{X}.$$

Further the new variance is c^2 times the old variance because

$$s_c^2 = \frac{\sum(cX_i - c\overline{X})^2}{n-1} = \frac{\sum[c(X_i - \overline{X})]^2}{n-1} = \frac{\sum c^2(X_i - \overline{X})^2}{n-1} = c^2\frac{\sum(X_i - \overline{X})^2}{n-1} = c^2s^2$$

and from this it follows that the new standard deviation is c times the old standard deviation, $s_c = cs$. (Remember, too, that division is just multiplication by a fraction.)

To convert the summary statistics of Example 1.5 to metric we simply utilize the formulas above with $c = 0.45$ kg/lb.

$$\overline{X}_c = c\overline{X} = 0.45 \text{ kg/lb } (7.1 \text{ lb}) = 3.20 \text{ kg.}$$

$$s_c^2 = c^2s^2 = (0.45 \text{ kg/lb})^2(5.75 \text{ lb}^2) = 1.164 \text{ kg}^2.$$

$$s_c = cs = 0.45 \text{ kg/lb } (2.4 \text{ lb}) = 1.08 \text{ kg.}$$

Our understanding of the effects of coding on descriptive statistics can sometimes help determine the nature of experimental manipulations of variables.

EXAMPLE 1.6. Suppose that a particular variety of strawberry yields an average 50 g of fruit per plant in field conditions without fertilizer. With a high nitrogen fertilizer this variety yields an average of 100 g of fruit per plant. A new "high yield" variety of strawberry yields 150 g of fruit per plant without fertilizer. How much would the yield be expected to increase with the high nitrogen fertilizer?

SOLUTION. We have two choices here: The effect of the fertilizer could be additive, increasing each value by 50 g ($X_i + 50$) or the effect of the fertilizer could be multiplicative, doubling each value ($2X_i$). In the first case we expect the yield of the new variety with fertilizer to be 150 g + 50 g = 200 g. In the second case we expect the yield of the new variety with fertilizer to be 2×150 g = 300 g. To differentiate between these possibilities we must look at the variance in yield of the original variety with and without fertilizer. If the effect of fertilizer is additive, the variances with and without fertilizer should be similar because additive coding doesn't effect the variance: $X_i + 50$ yields s^2, the original sample variance. If the effect is to double the yield, the variance of yields with fertilizer should be four times the variance without fertilizer because multiplicative coding increases the variance by the square of the constant used in coding. $2X_i$ yields $4s^2$, doubling the yield increases the sample variance four fold.

1.8   Tables and Graphs

The data collected in a sample are often organized into a table or graph as a summary representation. The data presented in Example 1.4 were arranged into a frequency table and could be further organized into a **relative frequency table** by expressing each row as a percentage of the total observations or into a **cumulative frequency distribution** by accumulating all observations up to and including each row. The cumulative frequency distribution could be manipulated further into a **relative cumulative frequency distribution** by expressing each row of the cumulative frequency distribution as a percentage of the total. See columns 3–5 in Table 1.2 for the relative frequency, cumulative frequency, and relative cumulative frequency distributions for Example 1.4. (Here $n = \sum f_i$ and r is the row number.)

TABLE 1.2. The relative frequencies, cumulative frequencies, and relative cumulative frequencies for Example 1.4

X_i Plants/quadrat	f_i Frequency	$\frac{f_i}{n}(100)$ Relative frequency	$\sum_{i=1}^{r} f_i$ Cumulative frequency	$\frac{\sum_{i=1}^{r} f_i}{n}(100)$ Relative cumulative frequency
0	268	33.500	268	33.500
1	316	39.500	584	73.000
2	135	16.875	719	89.875
3	61	7.625	780	97.500
4	15	1.875	795	99.375
5	3	0.375	798	99.750
6	1	0.125	799	99.875
7	1	0.125	800	100.000
$\sum$	800	100.0		

Discrete data, as in Example 1.4, are sometimes expressed as a **bar graph** of relative frequencies. See Figure 1.2. In a bar graph the bar *heights* are the relative frequencies. The bars are of equal width and spaced equidistantly along the horizontal axis. Because these data are discrete, that is, because they can only take certain values along the horizontal axis, the bars do not touch each other.

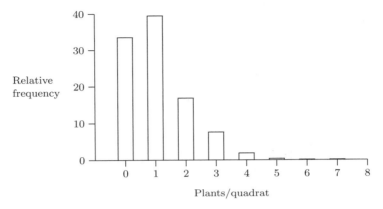

FIGURE 1.2. A bar graph of relative frequencies for Example 1.4.

The data in Example 1.5 can be summarized in a similar fashion with relative frequency, cumulative frequency, and relative cumulative frequency columns. See Table 1.3.

TABLE 1.3. The relative frequencies, cumulative frequencies, and relative cumulative frequencies for Example 1.5

X_i	f_i	$\frac{f_i}{n}(100)$	$\sum_{i=1}^{r} f_i$	$\frac{\sum_{i=1}^{r} f_i}{n}(100)$
1	2	1.82	2	1.82
2	1	0.91	3	2.73
3	4	3.64	7	6.36
4	7	6.36	14	12.73
5	13	11.82	27	24.55
6	15	13.64	42	38.18
7	20	18.18	62	56.36
8	24	21.82	86	78.18
9	7	6.36	93	84.55
10	9	8.18	102	92.73
11	2	1.82	104	94.55
12	4	3.64	108	98.18
13	2	1.82	110	100.00
$\sum$	110	100.00		

Because the data in Example 1.5 are continuous measurement data with each class implying a range of possible values for X_i, e.g., $X_i = 3$ implies each fish weighed between 2.50 lb and 3.49 lb, the pictorial representation of the data set is a **histogram** not a bar graph. Histograms have the observation classes along the horizontal axis. The *area* of the strip represents the relative frequency. (If the classes of the histogram are of equal width, as they often are, then the heights of the strips will represent the relative frequency, as in a bar graph.) See Figure 1.3. The strips in this case touch each other because each X value corresponds to a range of possible values.

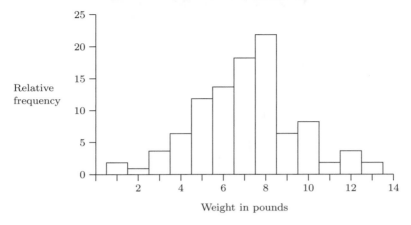

FIGURE 1.3. A histogram for the relative frequencies for Example 1.5.

While the categories in a bar graph are predetermined because the data are discrete, the classes representing ranges of continuous data values must be selected by the investigator. In fact, it is sometimes revealing to create more than one histogram of the same data by employing classes of different widths.

EXAMPLE 1.7. The list below gives snowfall measurements for 50 consecutive years (1951–2000) in Syracuse, NY (in inches per year). The data have been rearranged in order of increasing annual snowfall. Create a histogram using classes of width 30 inches and then create a histogram using narrower classes of width 15 inches. (Source: http://neisa.unh.edu/Climate/IndicatorExcelFiles.zip)

71.7	73.4	77.8	81.6	84.1	84.1	84.3	86.7	91.3	93.8
93.9	94.4	97.5	97.6	98.1	99.1	99.9	100.7	101.0	101.9
102.1	102.2	104.8	108.3	108.5	110.2	111.0	113.3	114.2	114.3
116.2	119.2	119.5	122.9	124.0	125.7	126.6	130.1	131.7	133.1
135.3	145.9	148.1	149.2	153.8	160.9	162.6	166.1	172.9	198.7

SOLUTION. Use the same scale for the horizontal axis (inches of annual snowfall) in both histograms. Remember that the *area* of a strip represents the relative frequency of the associated class. Since the snowfall classes of the second histogram (15 in) are one-half those of the first histogram (30 in), then the vertical scale must be multiplied by a factor of 2 so that equal areas in each histogram will represent the same relative frequencies. Thus, a single year in the second histogram will be represented by a strip half as wide but twice as tall as in the first histogram, as indicated in the key in the upper left corner of each diagram.

In this case, the narrower classes of the second histogram provide more information. For example, nearly one-third of all recent winters in Syracuse have produced snowfalls in the 90–105 inch range. There was one year with a very large amount of snowfall of approximately 200 in. While one could garner this same information from the data itself, normally one would use a (single) histogram to summarize data and not list the entire data set.

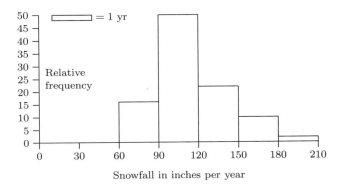

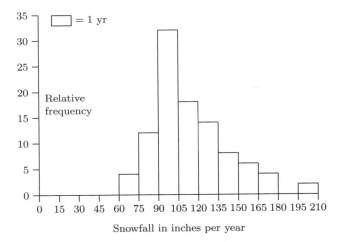

FIGURE 1.4. Two histograms for the data in Example 1.7. The areas of the strips represent the relative frequencies. The same area represents the same relative frequency in both graphs.

It is worth emphasizing that to make valid comparisons between two histograms, equal areas must represent equal relative frequencies. Since the relative frequencies of all the classes in a histogram sum to 1, this means that *the total area under each of the histograms being compared must be the same.* Look at Figure 1.4 for an application of this idea.

Histograms are often used as graphical tests of the shape of samples usually testing whether the data are approximately "bell-shaped" or not. We will discuss the importance of this consideration in future chapters.

1.9   Quartiles and Box Plots

In the previous sections we have used sample variance, standard deviation, and range to obtain measures of the spread or variability. Another quick and useful way to visualize the spread of a data set is by constructing a box plot that makes use of quartiles and the sample range.

Quartiles and Five-Number Summaries

As the name suggests, quartiles divide a distribution in quarters. More precisely, the pth **percentile** of a distribution is the value such that p percent of the observations fall at or below it. For example, the median is just the 50th percentile. Similarly, the **lower** or **first quartile** is the 25th percentile and the **upper** or **third quartile** is the 75th percentile. Because the second quartile is the same as the median, quartiles are appropriate ways to measure the spread of a distribution when the median is used to measure its center.

Because sample sizes are not always evenly divisible by 4 to form quartiles, we need to agree on how to break a data set up into approximate quarters. Other texts, computer programs, and calculators may use slightly different rules which produce slightly different quartiles.

FORMULA 1.11. To calculate the first and third quartiles, first order the list of observations and locate the median. The **first quartile** Q_1 is the median of the observations falling below the median of the entire sample and the **third quartile** Q_3 is the median of the observations falling above the median of the entire sample. The **interquartile range** is defined as

$$\text{IQR} = Q_3 - Q_1.$$

The sample IQR describes the spread of the middle 50% of the sample, i.e., the difference between the first and third quartiles. As such, it is a measure of variability and is commonly reported with the median.

> **EXAMPLE 1.8.** Find the first and third quartiles and the IQR for the cypress pine data in Example 1.2.
>
> | CCH | 17 | 19 | 31 | 39 | 48 | 56 | 68 | 73 | 73 | 75 | 80 | 122 |
> |-------|----|----|----|----|----|----|----|----|----|----|----|-----|
> | Depth | 1 | 2 | 3 | 4 | 5 | 6 | 6 | 5 | 4 | 3 | 2 | 1 |
>
> **SOLUTION.** The median depth is $\frac{12+1}{2} = 6.5$. So there are six observations below the median. The quartile depth is the median depth of these six observations: $\frac{6+1}{2} = 3.5$. So the first quartile is $Q_1 = \frac{31+39}{2} = 35$ cm. Similarly, the depth for the third quartile is also 3.5 (from the right), so $Q_3 = \frac{73+75}{2} = 74$ cm. Finally, the IQR $= Q_3 - Q_1 = 74 - 35 = 39$ cm.

A compact way to report the descriptive information involving the quartiles and the range is with a **five-number summary** of the data. It consists of the median, the two quartiles, and two extremes.

EXAMPLE 1.9. Provide the five-number summary for this sample of 15 weights (in lb) of lake trout caught in Geneva's Lake Trout Derby in 1994.

Weight		
2.26	3.57	7.86
2.45	1.85	3.88
4.60	4.90	3.60
3.89	2.14	1.52
2.83	1.84	2.12

SOLUTION. The sample size is $n = 15$. The median depth is $d(\tilde{X}) = \frac{15+1}{2} = 8$. The first quartile is determined by the seven observations *below* the median; hence the quartile depth is $\frac{7+1}{2} = 4$. The ordered data set and depths are

Weight	Depth	Weight	Depth	Weight	Depth
1.52	1	2.26	6	3.88	5
1.84	2	2.45	7	3.89	4
1.85	3	2.83	8	4.60	3
2.12	4	3.57	7	4.90	2
2.14	5	3.60	6	7.86	1

So $\tilde{X} = 2.83$ lb, $Q_1 = 2.12$ lb, and $Q_3 = 3.89$ lb. The extremes are 1.52 lb and 7.86 lb. The five-number summary is usually presented in the form of a chart:

Median:		2.83	
Quartiles:	2.12		3.89
Extremes:	1.52		7.86

Two other measures of variability are readily computed from the five-number summary. The IQR is the difference in the quartiles, IQR $= 3.89 - 2.12 = 1.77$ lb. The range is the difference in the extremes, $7.86 - 1.52 = 6.34$ lb.

Box Plots

The visual counterpart to a five-number summary is a **box plot**. Box plots can contain more or less detail, depending on the patience of the person constructing them. Below are instructions for a moderately detailed version that contains all the essentials.

1. Draw a horizontal or vertical reference scale based on the range of the data set.

2. Calculate the median, the quartiles, and the IQR.

3. Determine the **fences** f_1 and f_3 using the formulas below. Points lying outside these fences will be considered **outliers** and may warrant further investigation.

$$f_1 = Q_1 - 1.5(\text{IQR})$$

$$f_3 = Q_3 + 1.5(\text{IQR})$$

When an outlier is detected, one should consider its source. Is it a misrecorded data point? If it is legitimate, is it special in some way or other?

4. Locate the two "adjacent values." These are the smallest and largest data values inside the fences.

5. Lightly mark the median, quartiles, and adjacent values on the scale. Choose a scale to spread these points out sufficiently.

6. Beside the scale, construct a box with ends at the quartiles and a dashed interior line drawn at the median. Generally this will *not* be at the middle of the box!

7. Draw a "whisker" (line segment) from the quartiles to the adjacent values that are marked with crosses "×." Mark any outliers beyond the fences (equivalently, beyond the adjacent values) with open circles "○."

EXAMPLE 1.10. Construct a box plot for the lake trout data in Example 1.9.

SOLUTION. We have already computed the quartiles and median. The fences are

$$f_1 = Q_1 - 1.5(\text{IQR}) = 2.12 - 1.5(1.77) = -0.535 \text{ lb}$$

$$f_3 = Q_3 + 1.5(\text{IQR}) = 3.89 + 1.5(1.77) = 6.545 \text{ lb}$$

Only 7.86 lb is an outlier, and this is visually obvious in the box plot shown in Figure 1.5. The adjacent values are 1.52 lb and 4.90 lb.

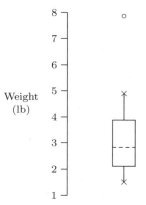

FIGURE 1.5. A box plot for the lake trout data in Example 1.9. Notice that the range can be determined as the difference in the most extreme values, and the IQR can be calculated as the difference in the quartiles at the top and bottom edges of the box.

When a box plot, such as the one in Figure 1.5, is not symmetric about the dashed median line, this is an indication that the data are not symmetrically distributed. We will discuss the importance of this later in the text.

Some other graphic representations used in preliminary data analysis include stem-and-leaf diagrams, polygons, ogives, and pictographs. Virtually all statistical packages available for computers offer some of these techniques to rapidly investigate the shape of a data set. Most of these manipulations are tedious to do by hand and are less useful than the bar graphs and histograms previously presented. We leave the configuration of these techniques and their interpretation for other authors and your instructor.

1.10 Accuracy, Precision, and the 30–300 Rule

All biologists are aware of the importance of accuracy and precision in data collection and recording. While these two terms are used synonymously in everyday speech, they have different meanings in statistics. **Accuracy** is the closeness of a measured or computed value to its true value, while **precision** is the closeness of repeated measurements of the same quantity to each other. A biased but sensitive instrument may yield inaccurate but precise readings. On the other hand, an insensitive instrument might result in an accurate reading, but the reading would be imprecise, since another reading of the same object would be unlikely to yield an equally accurate value. Unless there is bias in a measuring instrument, precision will lead to accuracy.

Some measurements are by their nature precise. When we count eggs in a monitor lizard's nest and record the number as 9 or 13, these are exact numbers and, therefore, precise variates. Most continuous variables, however, are approximate with the exact value unknown and unknowable. Recordings of continuous variable data imply a level of precision by the number of digits used. For example, if the length of an adult female monitor lizard is recorded as 97.2 cm, the implied true value of the length is between 97.15 and 97.25 cm. In other words, the last digit recorded defines an interval in which the exact value of the variable resides. A measurement of 97 cm implies a length between 96.5 and 97.5 cm.

In most studies too much precision can slow down data collection while not contributing significantly to the resolution of scientific questions. While it doesn't make sense to measure large eucalyptus trees to the nearest millimeter or to weigh sperm whales to the nearest gram, what level of precision should be recorded? To how many significant figures should we record measurements? Many biologists use the **thirty–three-hundred (30–300) rule** to determine precision for data sets. This rule is easy to apply and will save a great deal of time and effort. Array the sample by order of magnitude from largest to smallest measurement. The number of unit steps between the largest and smallest value should be between 30 and 300. For example, if you were collecting small shells in the intertidal zone of a beach and the largest was 9 mm and the smallest was 5 mm, the number of units steps would be 4 (a unit step is a millimeter in this example). If you recorded the lengths to the nearest tenth of a millimeter with the largest being 9.2 mm and the smallest 5.1 mm in length, the unit step is now 0.1 mm and there are 41 unit steps ($9.2 - 5.1 = 4.1$ mm or 41 tenths of mm) in the data array. The data set will now give you enough precision for most statistical analyses and allow for a reasonable error in recording, i.e., a mistake of 1 in the last digit recorded is now less than 2.5% as opposed to 25% when the data were recorded to the nearest millimeter.

If sedge plant heights were measured to the nearest tenth of a centimeter with the tallest being 194.3 cm and the shortest being 27.1 cm, the unit step would be tenths of centimeters and the data array would have 1672 unit steps ($194.3 - 27.1 = 167.2$ or 1672 tenths of cm). Clearly there is more precision in this data set than is needed. Recording these plant heights to the nearest centimeter would yield 167 unit steps ($194 - 27 = 167$ cm) and would give enough precision for analysis while saving time and effort in data collection.

As stated earlier, statistics such as the sample mean and standard deviation should have one more decimal place than the data, while the sample variance should carry two more. For example, the mean for the sedge plant height might be 141.3 cm for data collected to the nearest centimeter. After calculation the sample statistics often have to be rounded to the appropriate number of significant figures. The rules for rounding are very simple. A digit to be rounded is not changed if it is followed by a digit less than 5. If the digit to be rounded is followed by a digit greater than 5 or by 5 followed by other nonzero digits, it is increased by one. When the digit to be rounded is followed by a 5 standing alone or followed by zeros, it is unchanged if it is even but increased by one if it is odd. So a mean for the sedge data of 141.35 cm would be rounded to 141.4 cm, while a mean of 141.25 cm would be rounded to 141.2 cm. Similar rounding should be done for the standard deviation and variance.

1.11   Problems

1. If $X_1 = 9$, $X_2 = 8$, $X_3 = 13$, $X_4 = 6$, and $X_5 = 9$, evaluate the following:

(a) $\sum X_i$ (b) $\sum_{i=2}^{4} X_i$ (c) $\sum (X_i - \overline{X})$

(d) $\sum (X_i - \overline{X})^2$ (e) $\sum X_i^2$ (f) $\left(\sum X_i \right)^2$

(g) $\sum X_i^2 - \dfrac{\left(\sum X_i \right)^2}{n}$ (h) $\sum X_i - 9$ (i) $\sum_{i=1}^{3} (X_i + 2)$

(j) $\sum_{i=1}^{n} 5$ (k) $\sum 2X_i$ (l) $2 \sum X_i$

2. The following data are the carapace (shell) lengths in centimeters of a sample of adult female green turtles, *Chelonia mydas*, measured while nesting at Heron Island in Australia's Great Barrier Reef. Calculate the following descriptive statistics for this sample: sample mean, corrected sum of squares, sample variance, standard deviation, and range. Remember to use the appropriate number of decimal places in these descriptive statistics and to include the correct units with all statistics.

110   105   117   113   95   115   98   97   93   120

3. The red-tailed tropic bird, *Phaethon rubricauda*, is an extremely rare sea bird that nests on several islands of the Queensland coast of Australia. As part of a conservation effort to manage these endangered birds, every nesting pair was measured and weighed. Below are the body weights of these birds (in kg).

Female	2.45	2.57	2.81	2.37	2.01	2.50	2.32
Male	2.86	2.65	2.75	2.60	2.30	2.49	2.84

(a) Determine the following descriptive characteristics for the weights of the females: mean, variance, and standard deviation. Is this a sample or population? Again, pay attention to number of decimal places and appropriate units.

(b) Determine the mean, variance, and standard deviation for the male weights.

(c) Comment on the differences or similarities between the two data sets.

4. As part of a larger study of the effects of strenuous exercise on human fertility and fecundity, the ages (in years) of menarche (the beginning of menstruation) for 10 Olympic female endurance athletes (runners and swimmers) who had vigorously trained for at least 18 months prior to menarche were recorded.

$$13.6 \quad 13.9 \quad 14.0 \quad 14.2 \quad 14.9 \quad 15.0 \quad 15.0 \quad 15.1 \quad 15.4 \quad 16.4$$

(a) Calculate the following descriptive statistics: sample mean, variance, standard deviation, and median.

(b) Do you feel that the sample mean is significantly higher than the overall population mean for nonathletes of 12.5 years? Provide a rationale for your answer.

5. In a study to determine the distribution of *Cepaea nemoralis* snails at the Hanley Wildlife Preserve in upstate New York, 100 sampling quadrats were surveyed, with the results in the table below. Calculate the mean, median, and variance of the number of snails per quadrat in this study. (*Note:* The corrected sum of squares is 345.)

No. of snails (X_i)	f_i	$f_i X_i$	$f_i X_i^2$
0	69	0	0
1	18	18	18
2	7	14	28
3	2	6	18
4	1	4	16
5	1	5	25
8	1	8	64
15	1	15	225
	100	70	394

6. In a new reality television program called "Solitary Survivor," 15 volunteers were placed in isolation chambers without clocks, electronic devices, or any other means of measuring time. After a week their individual daily sleep cycles (day lengths) were measured and are recorded below.

Day lengths (hr)	26.0	25.5	26.5	24.3	24.2	26.5	27.4	26.6
	25.3	26.1	25.9	25.4	26.2	25.1	27.1	

(a) Calculate the mean and variance for this sample.

(b) Do you feel that the average day length for these people is significantly longer than 24 hr? Explain.

7. Larvae of the insect family, Myrmeleontidae, live in sandy soil and dig cone-shaped pits that trap ants and other prey that fall into them. These so-called ant lions stay in the soil with their heads just below the bottom of the pit and wait for unwary prey. An ecology class determined the number of ant lion pits in a sample of 100 randomly selected 1-m-square quadrats.

Pits/quadrat (X_i)	f_i	$f_i X_i$	$f_i X_i^2$
0	5	0	0
1	15	15	15
2	23	46	92
3	21	63	189
4	17	68	272
5	11	55	275
6	5	30	180
7	2	14	98
8	1	8	64
	100	299	1185

(a) Is the variable measured here discrete or continuous? Explain.

(b) Calculate the sample mean and variance for these data.

(c) Calculate the median and range for these data.

(d) Which pair of calculations (b) or (c) is more informative? Why?

8. An undergraduate ecology student doing research on niche dimensions decided to repeat part of R. H. MacArthur's famous study of foraging behavior of warblers in northeastern coniferous forests. She marked the heights of various branches in several conifers with colored tape and observed two similar species of warbler with binoculars and recorded the average foraging height for each bird. The heights in feet for the individuals observed were:

Bay-breasted warbler	17	10	13	12	13	11	13	16	17	19		
Blackburnian warbler	15	17	17	18	15	16	17	24	20	16	24	15

(a) Calculate the mean and standard deviation of the foraging heights for each species. Comment on the results.

(b) Determine the median and range for each species. Which of the two statistics, the standard deviation or the range, is a better reflection of the variability in the foraging height? Explain.

9. From a certain population of the freshwater sculpin, *Cottus rotheus*, a sample of 100 fish were dissected to determine the number of tail vertebrae in each. The data collected are summarized below. Find the mean, median, and standard deviation for this data set.

No. of vertebrae (X_i)	f_i	$f_i X_i$	$f_i X_i^2$
20	3	60	1,200
21	51	1,071	22,491
22	40	880	19,360
23	6	138	3,174
	100	2,149	46,225

10. Descriptive statistics for an extensive study of human morphometrics has been completed in the United States. The measurements on height have to be converted to centimeters from inches for publication in a British journal. The mean in the study was 68 inches and the standard deviation was 10 inches. What should the reported mean and variance be? (1 inch = 2.54 cm.)

11. (a) Invent a sample of size 5 for which the mean is 20 and the median is 15.

    (b) Invent a sample of size 2 for which the mean is 20 and the variance is 50.

12. Find the median and quartile depths for samples of size $n = 22, 23, 24$, and 25.

13. (a) Complete a five-number summary for each of the samples in Example 1.3.

    (b) Construct parallel box plots for these same data. Interpret these plots.

14. From a large population of tiger salamanders, *Ambystoma tigrinum*, a random sample of 81 were collected and their lengths recorded to the nearest centimeter. Calculate the mean, median, variance, and standard deviation for this sample.

X_i	f_i
10	2
11	8
12	17
13	22
14	14
15	10
16	7
17	1
	81

15. In Problems 2–4, 8, and 14, which data sets satisfy the 30–300 rule? Explain.

16. Why are Problems 5 and 7 exempt from the 30–300 rule?

17. For Problem 7 make a table including the relative frequency, cumulative frequency, and relative cumulative frequencies. Make a pictorial representation of this data set. Should you use a bar graph or histogram here?

18. For Problem 14 make a table including the relative frequency, cumulative frequency, and relative cumulative frequencies. Represent the data with a histogram.

19. In a study of lead concentration in breast milk of nursing mothers, the following data were reported for mothers aged 21 to 25 living in Riyadh, Saudi Arabia. (Based on data in: Younes, B. et al. 1995. Lead concentrations in breast milk of nursing mothers living in Riyadh. *Annals of Saudi Medicine*, **15**(3): 249–251.)

n	$\overline{X} \pm s$ (μg/dl)
19	0.777 ± 0.410

   (a) Determine the sample variance.

   (b) Determine $\sum X_i$.

   (c) Determine $(\sum X_i)^2$.

   (d) Determine $\sum X_i^2$.

20. The Sumatran orangutan, *Pongo pygmaeus abelii*, is an endangered species. When a small number of these orangutans was discovered on a remote island of Sumatra, zoologists were sent to evaluate this newfound group. Using tranquilizer darts they managed to capture and weigh every adult member of the group. Their data (in kilograms) are given below. Summarize these data with the appropriate indices. Are the data samples?

Weight (kg)	
Females	Males
49.7	90.1
48.0	88.0
55.0	79.0
46.3	85.5
44.9	92.3
49.0	90.6
51.3	88.3
	89.5
	77.3

21. You have been asked to prepare an information pamphlet on the water quality of the public water supply for a town in in upstate New York. You need to present data on various water characteristics in a meaningful and easily understood format. In addition to the pamphlet each consumer will be given a kit to test their own tap water for contaminants. Supposing the average lead contaminant concentration for the water supply is 8.0 μg/L, what additional information should be supplied to the consumer to allow him to make an assessment of the quality of the water coming from his tap?

22. From a site believed to be 30,000 years old in Irian Jaya near the border with Papua New Guinea archeologists recovered 12 adzes (double-edged stone blades for scythe-like tools). Given below are the lengths and diameters (in cm) at the base of the adzes.

Length	Diameter
20	8.0
23	7.5
19	10.3
26	9.9
25	10.1
28	11.0
29	11.3
20	11.2
22	10.0
24	9.8
19	9.5
24	11.0

Discuss these measurements in light of the 30–300 rule. Summarize the data appropriately. Do the data represent a sample or a population? Explain.

23. The timber rattlesnake, *Crotalus horridus*, is a top predator in the forest ecosystems of eastern North America. It is found nowhere else in the world and has been the subject of much controversy and myth. The males sexually mature in 5 years, but the females require 7 to 11 years to mature. Their fecundity is quite low with females giving birth to 4 to 14 young every 3 to 5 years. Their long development time and small brood size put these animals at significant risk of extirpation or extinction due to habitat destruction and indiscriminate hunting. Healthy populations of *C. horridus* have adult individuals that range in length from 91 to 137 cm with a mean length of 115 cm. At a recent "snake rodeo" adult timber rattlers were captured from a region in the Southern Tier of New York. These snakes' lengths were crudely measured to the nearest 5 cm and are recorded below.

Length (cm)	Frequency
70	5
75	7
80	8
85	10
90	10
95	13
100	15
105	10
110	14
115	11
120	9
125	3
130	0
135	1
140	4
	120

Calculate the mean, standard deviation, and range for the rodeo sample. Do the rodeo data indicate a population that is significantly shorter than the so-called healthy population? Discuss briefly. Why do you think the 30–300 rule was violated here?

24. The list below provides snowfall data for 50 consecutive years in Buffalo, NY (in inches per year). The data have been rearranged in order of increasing annual snowfall. Create two histograms for these data: The first should use classes of width 20 in, the second should use classes of width 10 in. *(HSDS, #278)*

25.0	38.8	39.9	40.1	46.7	49.1	49.6	51.1	51.6
53.5	54.7	55.5	55.9	58.8	60.3	63.6	65.4	66.1
69.3	70.9	71.4	71.5	71.8	72.9	74.4	76.2	77.8
78.1	78.4	79.0	79.3	79.7	80.7	82.4	82.4	83.0
83.6	83.6	84.8	85.5	87.4	88.7	89.6	89.8	89.9
90.9	97.0	98.3	101.4	102.4	103.9	104.5	105.2	110.0
110.5	110.5	113.7	114.5	115.6	120.5	120.7	124.7	126.4

Do the two histograms yield different impressions of the data set?

25. Match each histogram with its corresponding box plot.

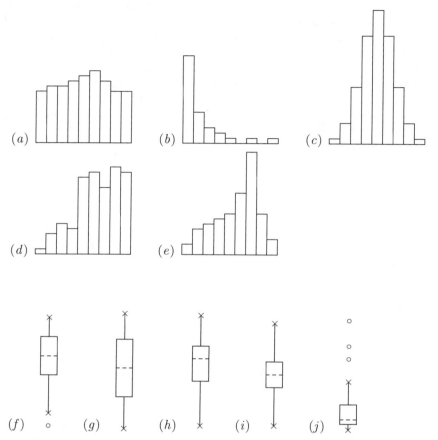

26. A preliminary study of white-tailed deer, *Odocoileus virginianus*, and a closely related mule deer, *O. hemionus*, generated the following weights (in kg) for mature males.

O. virginianus		O. hemionus	
90	77	99	84
71	90	80	104
83	83	120	104
85	84	91	101
86	91	106	109
92	70	95	80
75	71	100	91
76	75	109	73
83	89	101	98
85	92	93	102

Calculate appropriate descriptive statistics for each sample. Also calculate five-number summaries and box plots for each sample. Are there significant differences between the two samples? Are there any outliers?

27. As we have seen in this chapter, the variance and standard deviation are measures of dispersion or scatter of the values of the X's in a sample or population. We also learned that the most important statistic of central tendency, the sample mean, varies from sample to sample (see page 6). Because means utilize a number of X's in their calculation, they tend to be less variable than the individual X's. An extreme value of X (large or small) contributes only one nth to the mean value and is, therefore, somewhat dampened out.

A measure of the variability in $\overline{X}$'s then depends on two factors, the variability in the X's and the number of X's averaged to generate $\overline{X}$. We utilize two statistics to estimate this variability. The first is $\frac{s^2}{n}$, which is called the variance of the sample mean. The second is $\frac{s}{\sqrt{n}}$, which is called the standard deviation of the sample mean or more commonly the **standard error**. The importance and utility of these statistics will be amplified in Chapter 4 when the Central Limit Theorem is outlined. Calculate the variance of the sample mean and the standard error for the data sets presented in Problems 2 and 4.

28. *Rafflesia arnoldii* is one of the strangest and rarest plants on Earth. It lives as a parasite on the *Tetrostigma* vine, which grows only in primary undisturbed rainforests of Sumatra and Borneo in the Indonesia Archipelago. Its real claim to fame is that it produces the world's largest flowers. Individual blossoms can be nearly a meter in diameter and can weigh up to 11 kilograms. It doesn't produce leaves, stems, or roots. Other than the gigantic blooms the plant exists only as filaments within its unfortunate host.

Because this plant is so rare it is considered endangered. During a recent expedition fourteen flowers were discovered and measured. Below are the summary data for this sample.

Flower diameter	Flower weight
$\sum X_i = 1286$ cm	$\overline{X} = 9.5$ kg
$\sum X_i^2 = 118,876$ cm^2	$\sum X_i^2 = 1379$ kg^2

(*a*) Calculate the mean and standard deviation for the flower diameter measured in centimeters.

(*b*) For the flower weights, measured to the nearest kilogram, calculate the variance, standard deviation, and standard error.

29. For years we've had a remarkable clematis growing by our backyard deck. This plant grows about ten feet tall each year and has hundreds of purple blooms. The blooms have 3, 4, 5 or 6 petals. In a fit of abject boredom one day I counted the petals on 100 blossoms obtaining the results below. Calculate the mean, median, standard deviation, and sample variance. Discuss the data set relative to the 30–300 rule.

# of petals	# of blossoms
3	6
4	30
5	55
6	9

30. QCR: Quick Concept Review. At the end of the problems in each chapter will be a small number of true and false questions designed as a rapid check of your conceptual mastery of the chapter's content. Answers to these questions and page numbers for appropriate review are presented with the answers to the even-numbered problems at the back of the text.

(*a*) Parameters are descriptive measures of populations and are usually denoted by Greek letters, while descriptive measures of samples are termed statistics and are almost never denoted by Greek letters.

(*b*) When the average of all possible values of a statistic equals the population parameter that the statistic is estimating, we say the statistic is unbiased.

(*c*) The corrected sum of squares in a sample variance formula is the sum of squared deviations or squared distances of each observation from the population mean.

(*d*) The sample mean and variance for continuous variables should be reported to two more decimal places than the original data and the standard deviation to one more decimal place.

(*e*) The 30–300 Rule normally doesn't apply to discrete variables.

(*f*) Histograms are appropriate for discrete variables while bar diagrams are a more suitable representation of continuous variables.

2

Introduction to Probability

Concepts in Chapter 2:

- Definition of Classical and Empirical Probabilities
- Use of Permutations and Combinations
- Set Theory Definitions and Venn Diagrams
- Axioms and Rules of Probability
  - The General Addition Rule
  - Conditional Probability
  - Independence of Events
  - The General Multiplication Rule
- Application of Probability Rules to Mendelian Genetics (Optional)

Introduction

The concepts outlined in Sections 1.3 through 1.10 allow an investigator to describe a sample of observations in a summary fashion through either descriptive statistics or graphic representation. They are usually sufficient if sample description is the sole aim of the data collection. Except in the most preliminary of studies description of the data set is not the ultimate goal. In the vast majority of cases data are collected to test hypotheses (see Figure 1.1). In order to test hypotheses adequately, we must have at least a rudimentary knowledge of probability, so we take the next few chapters to introduce this fundamental material. This knowledge of probability can be extraordinarily useful in solving problems in Mendelian genetics as well.

2.1 Definitions

The probability of an event is a numerical measure of the likelihood or degree of predictability that the event will occur. The events whose probabilities we wish to calculate all occur as the outcomes of various experiments. By an **experiment** we simply mean an activity with an observable outcome. Examples of experiments include:

1. Flipping a coin and observing whether it lands heads or tails;

2. Rolling a red die and a green die and observing the sum of the dots on each of their uppermost faces;

3. Selecting a playing card from a full deck;

4. Selecting a year and tabulating the annual snowfall x (in cm) in Buffalo, NY for that year.

It is helpful to use the language of sets to describe an experiment. The set consisting of all possible outcomes of an experiment is called the **sample space** of the experiment. If S_1, S_2, S_3, and S_4 denote the sample spaces of the four experiments listed above, then

- $S_1 = \{\text{heads, tails}\}$;

- $S_2 = \{2, 3, 4, 5, 6, 7, 8, 9, 10, 11, 12\}$;

- $S_3 = \{A\heartsuit, 2\heartsuit, \ldots, K\heartsuit, A\spadesuit, 2\spadesuit, \ldots, K\spadesuit, A\diamondsuit, 2\diamondsuit, \ldots, K\diamondsuit, A\clubsuit, 2\clubsuit, \ldots, K\clubsuit\}$;

- $S_4 = \{x \geq 0\}$.

An **event** A is a subset of the sample space. We say that an event occurs when the outcome of the experiment is an element of A. For example, in experiment 2 above, if the event is "an odd number occurs," then this event occurs precisely when the outcome is an element of the set $A = \{3, 5, 7, 9, 11\}$. Similarly, in experiment 3, if the event is "the suit is black," then this event occurs precisely when the outcome is an element of the set $A = \{A\spadesuit, 2\spadesuit, \ldots, K\spadesuit, A\clubsuit, 2\clubsuit, \ldots, K\clubsuit\}$. A **simple event** is an event that consists of a single element, that is, a set consisting of a single outcome of an experiment. For example, the sample space of experiment 1 consists of the simple events {heads} and {tails}. Notice that in experiment 2, the event $A = \{5\}$ is *not* a simple event. Keeping track of the color of the dice (red first, say), the event $A = \{5\}$ is the consolidation or union of four simple events: $1 + 4$, $2 + 3$, $3 + 2$, and $4 + 1$.

The two approaches to probability used most extensively in biostatistics are termed (1) empirical or relative frequency and (2) classical or *a priori*. The **relative frequency** approach is especially useful with certain empirical questions. For example, suppose one is interested in the sex ratio in populations of European starlings, *Sternus vulgaris*. Using mist nets, samples from various flocks can be collected. What is the probability that a randomly chosen bird will be a female? This probability, denoted by P, can be estimated by taking a sample of starlings and calculating the ratio of the number of females caught to the number of starlings caught:

$$P(\text{female}) = \frac{\text{number of females}}{\text{number of starlings}}.$$

That is,

FORMULA 2.1. The empirical probability of an event A is defined as

$$P(A) = \frac{n_A}{n} = \frac{\text{number of times } A \text{ occurred}}{\text{number of trials run}}.$$

The empirical probability obtained is approximate since different samples might produce different estimates of this probability. However, if the number of trials or sample size is large, the change in the relative frequencies is usually slight, so values obtained this way can be quite accurate. Other examples of relative frequency probabilities include the probability of a randomly collected fruit fly having white eyes instead of the normal wild-type red color, the probability that a randomly caught fox will be rabid, and the probability that a patient will recover from bypass surgery.

The second approach to probability, termed **classical** or *a priori*, forms the theoretical basis for several statistical tests that we will develop in later chapters and is immediately useful for answering questions involving Mendelian genetics. Let S be the sample space of an experiment. Then the following three axioms form the basis for this approach. (Axioms are general truths that are offered without proof and are intuitively obvious.)

1. For every event A, $0 \leq P(A) \leq 1$, that is, the probability of any event is a real number between 0 and 1, inclusive.

2. $P(S) = 1$ and $P(\emptyset) = 0$.

3. If $A_1, A_2, \ldots, A_m$ are mutually exclusive events (i.e., mutually disjoint subsets), then

$$P(A_1 \text{ or } A_2 \text{ or } A_3 \text{ or } \ldots \text{ or } A_m) = \sum_{i=1}^{m} P(A_i) = P(A_1) + P(A_2) + \cdots + P(A_m).$$

This method is especially easy to employ when the sample space, S, of an experiment can be partitioned into a finite number n of mutually exclusive simple events, $\{s_1, s_2, \ldots, s_n\}$, that are **equally likely** or **equiprobable**. Using the axioms 2 and 3 above, since

$$1 = P(S) = P(s_1 \text{ or } s_2 \text{ or } \ldots \text{ or } s_n) = \sum_{i=1}^{n} P(s_i)$$

and since by assumption $P(s_1) = P(s_2) = \cdots = P(s_n)$, then $P(s_i) = \frac{1}{n}$ for each i. Under the assumption of equiprobability, to calculate the classical probability of a particular event A, count the number of simple events in A and multiply by $\frac{1}{n}$. In other words, count the number of elements in A and divide by the total number elements in the sample space.

FORMULA 2.2. The classical probability of an event A under the assumption of equiprobability of simple events is

$$P(A) = \frac{n(A)}{n(S)} = \frac{\text{number of elements in } A}{\text{number of elements in } S}.$$

Notice that this analysis was done without actually running any trials.

For example, in experiment 3, the sample space consists of 52 simple events,

$$\{A\heartsuit, 2\heartsuit, \ldots, K\heartsuit, A\spadesuit, 2\spadesuit, \ldots, K\spadesuit, A\diamondsuit, 2\diamondsuit, \ldots, K\diamondsuit, A\clubsuit, 2\clubsuit, \ldots, K\clubsuit\},$$

each of which is equally likely. So to determine the probability of the event A, "The suit is black," we count the number of elements of A and divide by the number of elements in the sample space: $P(A) = \frac{26}{52} = \frac{1}{2}$.

In experiment 2, we saw the event $A = \{5\}$ was the union of four simple events: $1 + 4$, $2 + 3$, $3 + 2$, and $4 + 1$. Since each die can land in 6 different ways, the entire sample space consists of $6 \times 6 = 36$ possible simple events. So $P(A) = \frac{4}{36} = \frac{1}{9}$.

In the example above, if the dice had been loaded, i.e., biased in some way to make certain outcomes more likely than others, the probability of a sum of 5 on the two dice could only be determined empirically. A large number of rolls would be required to estimate the appropriate probability.

When possible, classical probabilities are preferred because they are exact and don't require data collection for their determination. Some classical probabilities require sophisticated methods of counting outcomes that derive from the area of mathematics termed combinatorics. Some basic combinatorial methods are presented in the following section.

2.2 Use of Permutations and Combinations

Two types of counting problems often occur. The first asks in how many ways can n different objects be arranged or ordered. Such rearrangements are called **permutations**. For example, how many different ways can n students be ranked using their performance on an exam? There are n possible ranks for the first student, $n-1$ ranks remain for the second, $n-2$ ranks for the third, and so on. The total number is called **n factorial**.

FORMULA 2.3. The number n factorial is denoted by $n!$ and is defined by

$$_nP_n = n! = n \cdot (n-1) \cdot (n-2) \cdots 2 \cdot 1.$$

Notice that factorial expressions get large very quickly. For example, there are

$$10! = 10 \cdot 9 \cdots 1 = 3,628,800$$

ways to order 10 different objects. Nonparametric statistics depend heavily on rankings and orderings, so permutations will play an important role in the underlying theory of these statistical tests. (Note that $0!$ is defined to be 1. This makes certain formulas work nicely.)

Permutations that arrange only a portion of n objects, say, k objects, have n choices for the first position, $n-1$ choices for the second position, etc., through $n-k+1$ choices for the last position. So $_nP_k = n(n-1)(n-2)\cdots(n-k+1)$. This formula can be written more compactly using factorials:

$$n! = n(n-1)(n-2)\cdots(n-k+1)(n-k)(n-k-1)\cdots 2 \cdot 1$$
$$= [n(n-1)\cdots(n-k+1)](n-k)!.$$

Therefore,

$$\frac{n!}{(n-k)!} = n(n-1)(n-2)\cdots(n-k+1) = {_nP_k}.$$

So we have

FORMULA 2.4. The number of permutations that arrange only k of n objects is given by

$$_nP_k = n(n-1)(n-2)\ldots(n-k+1) = \frac{n!}{(n-k)!}.$$

For example, how many ways can a geneticist utilize 5 distinct stocks of fruit flies, *Drosophila melanogaster*, in 3 different experiments, if each stock can be used only once and the order of their use is important? Here $n = 5$ and $k = 3$, so

$$_5P_3 = \frac{5!}{(5-3)!} = \frac{120}{2} = 60.$$

There are 60 different ways to set up the three experiments using the 5 stocks.

A second type of counting problem involves determining the number of ways in which k objects can be chosen from a larger group of n objects. For example, how many different teams of three students can be formed from a group of 10 students? We can think of this as a modified permutation problem. There are 10 choices for the first student, 9 for the second, and 8 for the third, or $10 \cdot 9 \cdot 8$ total ways that three students can be chosen in order. But as far as the team is concerned, the order in which the students are selected is irrelevant. We know that there are 3! permutations of three students, so the total number of different ways to select 3 students from 10 without regard for order is $\frac{10 \cdot 9 \cdot 8}{3!}$. This number is usually written as $\binom{10}{3}$. More generally, we have

FORMULA 2.5. The number of ways of choosing k objects from n without regard to order is

$$_nC_k = \binom{n}{k} = \frac{n(n-1)(n-2)\cdots(n-k+1)}{k!}. = \frac{n!}{k!(n-k)!}$$

The symbol $\binom{n}{k}$ is read as "n choose k." It represents the number of **combinations** of n distinct objects chosen k at a time. The formula also can be written as a ratio of factorials:

$$\binom{n}{k} = \frac{n(n-1)\cdots(n-k+1)}{k!}$$

$$= \frac{n(n-1)\cdots(n-k+1)[(n-k)\cdots 2 \cdot 1]}{k![(n-k)\cdots 2 \cdot 1]}$$

$$= \frac{n!}{k!(n-k)!}.$$

For large values of n and k, permutations, combinations, and factorials can be arduous to calculate. Most scientific calculators have function keys which permit you to do these calculations with relative ease.

For example, suppose the geneticist above decides to do a single experiment that will utilize 3 stocks of *Drosophila* at once, and, therefore, order is not important. How many experimental protocols are possible from his 5 available stocks? Then

$$_5C_3 = \frac{5!}{3!(5-3)!} = 10.$$

is less when order isn't important

There are only 10 ways now to design this new experiment because order is not important.

Permutation and combination problems assume that there is no repetition in the selection. That is, once an item has been selected it cannot be selected again in that particular grouping. But there are problems where such repeated selection or repetition occurs. For example, how many different two-digit numbers are there? There are 10 choices for both the first and second digits, so there are a total of $10 \cdot 10 = 10^2$ possible numbers, namely, 00 to 99. More generally, there are n^k ways to choose k items from n with repetition, keeping track of order.

EXAMPLE 2.1. Eight consumers are asked to rank three brands of Australian beer using 1 for the most preferable and 3 for the least preferable. How many different rankings can occur keeping track of the individual consumers?

SOLUTION. This problem combines a couple of the ideas discussed. First each consumer has to rank the three products. That's a permutation question, so there are $3! = 6$ possible rankings by each consumer. This ranking process is repeated 8 times, once by each consumer, so there are 6^8 different rankings.

This example was solved using the **Multiplication Principle** of **staged experiments** or processes. Each step or stage of the experiment or process was enumerated separately and the total number of ways the experiment can be run was the product of the individual step values. In this example there were 8 stages or steps each with 3! or 6 possible outcomes, so the total number of possible outcomes is $6 \times 6 \times 6 \times 6 \times 6 \times 6 \times 6 \times 6 = 1,679,616$.

When using the Multiplication Principle to enumerate a staged experiment or process:

- Consider carefully whether repetition is allowed or not.

- Sometimes it is easier to enumerate the complement of the desired event rather than the event itself and subtract the complement from the total number of possibilities.

- If a particular stage in the experiment has a special restriction, you should consider the restriction first.

The aids above will become more clear as they are applied to a few practice problems.

EXAMPLE 2.2. Suppose a die is rolled 3 times and the outcome is recorded for each roll. How many different results can be recorded for the set of three rolls?

SOLUTION. A three-stage experiment with repetition allowed: $6 \times 6 \times 6 = 216$.

How many sets of results will have a different side up for each roll?

SOLUTION. This is a three-stage experiment without repetition: $_6P_3 = 6 \times 5 \times 4 = 120$.

How many sets of results will have at least 2 rolls the same?

SOLUTION. The problem of all sides different was enumerated in the previous calculation and the total possible in the first part above, so there are $216 - 120 = 96$ ways that at least two of the rolls will be the same.

We can use the counting techniques developed earlier in this section to solve classical probability problems.

What is the probability of a different side up on each of 3 rolls?

SOLUTION. This is a classical probability problem (see Formula 2.2). Let event A be a different side up in each roll. Then $n(S) = 216$ and $n(A) = 120$. So

$$P(A) = \frac{n(A)}{n(S)} = \frac{120}{216} = 0.556.$$

What is the probability that the first roll is a one, the second is even (2, 4, or 6), and the third is odd (1, 3, or 5)?

SOLUTION. Again use classical probability. In how many ways can the required events occur? $1 \times 3 \times 3 = 9 = n(A)$.

$$P(A) = \frac{n(A)}{n(S)} = \frac{9}{216} = 0.042.$$

Note that the restrictions are all met in the enumeration for the required events.

EXAMPLE 2.3. An insect toxicologist would like to test the effectiveness of three new pyrethroid insecticides on populations of European corn borer moths, *Ostrinia nubilalis*, but she has 7 different geographically isolated strains available. If each strain could be used more than once, how many different tests could she perform?

SOLUTION. A three-stage experiment with repetition permitted: $7 \times 7 \times 7 = 343$, so there are 343 ways to test each pyrethroid once.

If each strain can be used only once, how many ways could she perform her tests?

SOLUTION. The answer is a permutation,

$$_7P_3 = \frac{7!}{(7-3)!} = 7 \cdot 6 \cdot 5 = 210.$$

If the experiment is redesigned to test a single pyrethroid with 3 of the strains, how many different experiments can be done?

SOLUTION. This reduces to 7 strains choose 3. The answer is a combination.

$$_7C_3 = \frac{7!}{3!(7-3)!} = 35. \qquad \binom{7}{3}$$

EXAMPLE 2.4. In a project to investigate the water quality of Lake Erie, 20 samples were taken. Unknown to the investigators, 5 of the sample containers were contaminated before use. If 10 of the samples are selected at random, what is the probability that 2 of these selected samples will be contaminated?

SOLUTION. We use the counting techniques developed earlier to solve a classical probability problem. The denominator of the classical probability is $\binom{20}{10}$, the number of ways to select 10 samples at random from a group of 20. The numerator can now be considered a staged experiment with restrictions. First 2 of the 5 contaminated samples must be chosen, then 8 of the 15 uncontaminated samples must be chosen. Stage 1: $\binom{5}{2} = 10$ and Stage 2: $\binom{15}{8} = 6,435$. By the Multiplication Principle: number of ways for Stage 1 × number of ways for Stage 2 is $10 \times 6,435 = 64,350$. So

$$P(\text{2 of the chosen 10 contaminated}) = \frac{\binom{5}{2}\binom{15}{8}}{\binom{20}{10}} = \frac{\frac{5!}{3!2!} \cdot \frac{15!}{7!8!}}{\frac{20!}{10!10!}} = \frac{64,350}{184,756} = 0.348.$$

If 10 of the samples are selected at random, what is the probability that none will be contaminated?

SOLUTION. Again the denominator of the classical probability is $\binom{20}{10}$, the number of ways to select 10 samples at random from the complete group of 20. We can still consider the numerator a two-staged experiment with restrictions, the first stage being $\binom{5}{0} = 1$, the number of ways of selecting 0 contaminated samples from 5. The second stage is $\binom{15}{10}$, the number of ways of selecting 10 uncontaminated samples. Therefore, the probability that none of the samples chosen are contaminated is

$$P(\text{none contaminated}) = \frac{\binom{5}{0}\binom{15}{10}}{\binom{20}{10}} = \frac{\frac{5!}{5!0!} \cdot \frac{15!}{5!10!}}{\frac{20!}{10!10!}} = \frac{15!10!}{20!5!} \qquad \frac{15! \, 10! \, 10!}{20! \, 10! \, 5!}$$

$$= \frac{10 \cdot 9 \cdot 8 \cdot 7 \cdot 6}{20 \cdot 19 \cdot 18 \cdot 17 \cdot 16} = 0.0163.$$

Notice that all combinatoral probability problems reduce to the quotient of two enumerations.

2.3 Introduction to Set Theory and Venn Diagrams

The use of probability to solve problems quickly and efficiently in biology is often a stumbling block for many beginning students. The development of set theory along with some axioms of probability will lead to several very useful methods to attack basic probability problems.

Set theory, developed by George Cantor (1845–1918) in the late nineteenth century, and Venn diagrams, named for the English logician John Venn (1834–1923), are ways of organizing experimental situations in order to more easily solve probability problems. Consider the following definitions and pictorial representations.

Definitions

1. A **set** is a collection of definite, distinct objects of interest.

2. The objects are called **elements** or members of the set.

3. The set of all elements for which there is interest in a given discussion is called the **universal set** or **sample space** and is denoted by S.

4. Any subset A of the sample space or universal set, S, is called an **event**.

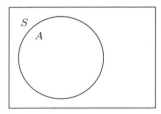

5. If every element of the set B is an element of the set A, then B is said to be a **subset** of A. This is denoted by $B \subset A$. Notice that we also have $A \subset S$ and $B \subset S$ here.

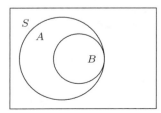

6. Two sets, A and B, are said to be **conjoint** when they have at least one element in common.

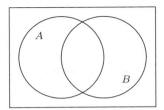

7. Two sets, A and B, are said to be **disjoint** when they have no elements in common. Disjoint sets are sometimes said to be **mutually exclusive**.

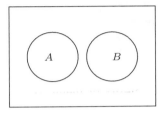

8. Given a universal set, S, and a subset $A \subset S$, then the **complement** of A (written A') is the set of elements of S that are not elements of A. See the shaded area of the Venn diagram below.

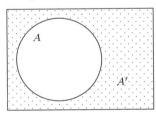

9. The **union** of A and B, written $A \cup B$, is the set of elements that belong to either A or B or to both A and B. See the shaded area of the next Venn diagram.

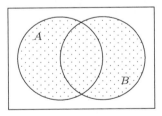

10. The **intersection** of A and B, written $A \cap B$, is the set of elements that belong to both A and B. See the shaded area of the Venn diagram below.

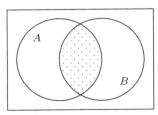

With a bit of practice these rather formal definitions are easy to apply.

EXAMPLE 2.5. In a study to investigate the side effects of an experimental cancer drug, 500 patients are given the drug as part of their cancer treatment. S is the universal set of all cancer patients treated with the experimental drug.

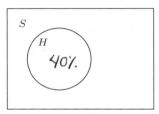

Suppose 40% of the patients receiving the experimental drug experience hypertension. H is the set of those experiencing hypertension. $H \subset S$.

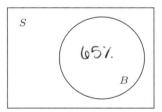

Sixty-five percent of the patients receiving the experimental drug experience blurred vision. B is the set of those experiencing blurred vision. $B \subset S$.

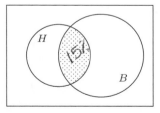

Fifteen percent of the patients receiving the experimental drug experience both hypertension and blurred vision. So the set H and the set B are conjoint, and the 15% represents the intersection of the two sets.

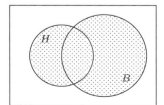

A patient experiencing either symptom or both symptoms is represented by union of H and B, remembering that union means H or B or both H and B.

$$65 + 40 - 15 = 90\%$$

A patient with no symptoms is represented by $(H \cup B)'$.

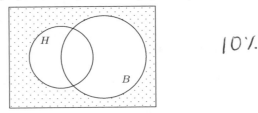

10%.

2.4   Axioms and Rules of Probability

Let's now consider the axioms of probability that we will use to develop certain general rules of probability. The set of all possibilities for an experiment or objects in a set of interest is called S, as stated earlier, and is referred to as the **sample space.** The statements below just reiterate the axioms for probability introduced in Section 2.1.

1. Let S denote the sample space for an experiment. Then $P(S) = 1$.

2. $P(A_i) \geq 0$ for all events A_i.

3. Let $A_1, A_2, A_3, \ldots$ be a collection of mutually exclusive events. Then
$$P(A_1 \text{ or } A_2 \text{ or } A_3 \text{ or } \ldots) = P(A_1) + P(A_2) + P(A_3) + \cdots .$$

4. $P(A') = 1 - P(A)$.

5. The **empty set**, $\emptyset$, is a subset of every set. It corresponds to a physical event that cannot occur and, therefore, is considered an impossible event. It is assigned a probability of zero. $P(\emptyset) = 0$.

EXAMPLE 2.6. Consider the outcome when rolling a single unbiased die. The set of possible outcomes is $\{1, 2, 3, 4, 5, 6\} = S$.

$$P(A_1) = \tfrac{1}{6} \qquad \text{for } A = \text{``1''}$$
$$P(A_2) = \tfrac{1}{6} \qquad \text{for } A = \text{``2''}$$
$$P(A_3) = \tfrac{1}{6} \qquad \text{for } A = \text{``3''}$$
$$P(A_4) = \tfrac{1}{6} \qquad \text{for } A = \text{``4''}$$
$$P(A_5) = \tfrac{1}{6} \qquad \text{for } A = \text{``5''}$$
$$P(A_6) = \tfrac{1}{6} \qquad \text{for } A = \text{``6''}$$
$$\sum_{\text{all } i} P(A_i) = 1.$$

Each outcome in S is **mutually exclusive**, that is, when one outcome occurs another cannot simultaneously occur. For example, one can't get a three and a two *simultaneously* on a single roll of a die. The probability of a three or two is, therefore,

$$P(3 \text{ or } 2) = P(3) + P(2) = \tfrac{1}{6} + \tfrac{1}{6} = \tfrac{1}{3} = 0.333$$

because of the mutually exclusive nature of these outcomes. The probability of something other than a three, $P(\text{not } 3)$, is

$$P(3') = 1 - P(3) = 1 - \tfrac{1}{6} = \tfrac{5}{6} = 0.833.$$

The probability of 3.5, since 3.5 is an impossible event, is zero. $P(3.5) = 0$ [remember $P(\emptyset) = 0$ for all impossible events].

44   CHAPTER 2: Introduction to Probability

The General Addition Rule

Returning to Example 2.5, the study of the side effects of an experimental drug treatment for cancer, determine the probability that a patient will have some side effects, i.e., hypertension, blurred vision, or both hypertension and blurred vision.

Clearly the probability required is not just the sum of the individual probabilities. $P(H) + P(B) = 0.40 + 0.65 = 1.05$ and no probability can exceed 1.0 or certainty. The problem here is that these events, hypertension and blurred vision, are not mutually exclusive. A patient can experience both.

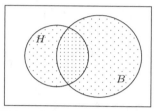

When we add the probabilities of H and B together we are actually counting the intersection $H \cap B$ or overlap twice. An accurate determination of the probability of either H or B or both, i.e., $H \cup B$, requires that the intersection be subtracted from the sum of the individual probabilities to account for this double counting.

$$P(H \cup B) = P(H) + P(B) - P(H \cap B).$$

$P(H \cup B) = 0.40 + 0.65 - 0.15 = 0.90$, so 90% of the patients experienced at least one side effect or the probability that a randomly chosen patient experiences some side effect is 0.90. The formula above is a specific application of the general addition formula.

FORMULA 2.6. The General Addition Rule Of Probability. If A_1 and A_2 are any events, then

$$P(A_1 \cup A_2) = P(A_1) + P(A_2) - P(A_1 \cap A_2).$$

The probability of either A_1 or A_2 or both equals the probability of A_1 plus the probability of A_2 minus the probability of both. This general addition formula is sometimes called the "either/or" rule.

EXAMPLE 2.7. The maidenhair tree, *Ginkgo biloba*, was thought to be extinct until early in the twentieth century when a population was found in eastern China. Now this ornamental tree is cultivated all over the world. Thirty-five percent of the specimens have variegated leaves, while the rest have normal green leaves. Seventy percent of the trees have white flowers and the remainder have pink flowers. Only 20% of the trees have variegated leaves and white flowers. What is the probability that a randomly collected specimen will have variegated leaves or white flowers?

SOLUTION. First rewrite the given information in set notation and draw the Venn diagram.

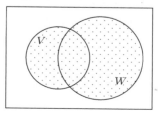

$P(V) = 0.35$ $P(V') = 0.65 = P(\text{green})$

$P(W) = 0.70$ $P(W') = 0.30 = P(\text{pink})$

$P(V \cap W) = 0.20.$

From the diagram, clearly what is required to answer the question above is the union of V and W, $P(V \cup W)$. Applying the General Addition Rule,

$$P(V \cup W) = P(V) + P(W) - P(V \cap W) = 0.35 + 0.70 - 0.20 = 0.85.$$

So a randomly chosen specimen has an 85% chance of being either variegated, or white-flowered, or both.

What is the probability that a randomly chosen specimen has white flowers and normal green leaves?

SOLUTION. Again refer to the Venn diagram. The probability of white-flowered and not variegated is the probability of white-flowered minus the probability of white-flowered *and* variegated.

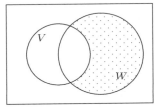

$$P(W \cap V') = P(W) - P(W \cap V)$$
$$= 0.70 - 0.20$$
$$= 0.50.$$

What is the probability that a randomly collected specimen has normal green leaves and pink flowers?

SOLUTION. Once again utilize the Venn diagram. Clearly the shaded area represents the event *not* variegated and *not* white,

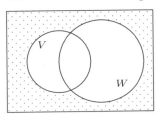

$$P(V' \cap W') = 1 - P(V \cup W) = 1 - 0.85 = 0.15.$$

We expect 15% of the specimens to have normal green leaves and pink flowers.

From Example 2.7 it's clear that proper use of Venn diagrams can clarify this type of problem. Note that there are four kinds of trees and exactly four areas in S with each area representing a single type of tree. Use the negation of a subset, for example, variegated, V, and not variegated, V', rather than variegated V and green G because these two forms are mutually exclusive and all inclusive in this problem:

$$P(V) + P(V') = 1.$$

It should be further noted that the General Addition Rule can be applied to mutually exclusive events or events that are disjoint, as well as to conjoined sets. When applied to mutually exclusive events, $P(A_1 \cap A_2) = 0$ by definition and the General Addition Rule becomes Axiom 3 presented earlier. If A_1 and A_2 are mutually exclusive, then

$$P(A_1 \cup A_2) = P(A_1) + P(A_2) - P(A_1 \cap A_2) = P(A_1) + P(A_2).$$

Conditional Probability

Sometimes information about one characteristic modifies the probabilities for a second characteristic. Return to Example 2.7. What is the probability of selecting at random a white-flowered specimen? Stated in the example the percentage of white-flowered trees is 70, so the probability of a randomly selected tree being white-flowered is 0.70. What if a tree has variegated leaves but has not bloomed yet? Is the probability of white flowers still 0.70? To investigate this question we write the required probability as $P(W|V)$, which is read as "the probability of white flowers *given* variegated leaves," and return to the Venn diagram.

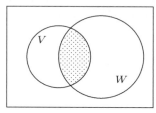

Given variegated leaves directs us to consider only the circle representing plants with such leaves, V. We wish to know what percentage of *this circle* is covered by the overlap with W. In other words, we want to express the intersection, $V \cap W$, as a percentage of V. So

$$P(W|V) = \frac{P(V \cap W)}{P(V)} = \frac{0.20}{0.35} = 0.57.$$

If a tree has variegated leaves, there is a 57% chance it will have white flowers. To put it another way, 70% of all ginkgoes are white-flowered but only 57% of the variegated-leaved ginkgoes are white-flowered.

FORMULA 2.7. Let A_1 and A_2 be events such that the $P(A_1) \neq 0$. The conditional probability of A_2 given that A_1 has already occurred is denoted $P(A_2|A_1)$ and is defined as

$$P(A_2|A_1) = \frac{P(A_2 \cap A_1)}{P(A_1)}.$$

EXAMPLE 2.8. Suppose that on a field trip to Guatemala you decide to study handsome fungus beetles, *Stenotarsus rotundus*. The population you investigate is composed of 60% females and 40% males. In addition, it has two color morphs, dull brown (70%) and bronze (30%). Half of all the insects are dull brown females. What is the probability that a randomly collected individual is either dull brown or female?

SOLUTION. Use a Venn diagram.

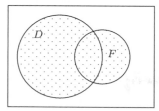

$$P(F) = 0.60 \qquad\qquad P(D) = 0.70$$

$$P(F') = P(M) = 0.40 \quad P(D') = P(B) = 0.30$$

$$P(F \cap D) = 0.50.$$

Using the General Addition Rule,

$$P(F \cup D) = P(F) + P(D) - P(F \cap D) = 0.70 + 0.60 - 0.50 = 0.80.$$

So 80% of the beetles express one or both of the required characteristics, i.e., dull brown color or female.

If a beetle is female, what is the probability that it is dull brown? Bronze?

SOLUTION.
$$P(D|F) = \frac{P(D \cap F)}{P(F)} = \frac{0.50}{0.60} = 0.833.$$

Eighty-three percent of females are dull brown. This means that $100 - 83.3 = 16.7\%$ of females must be bronze. We can also find this using conditional probability:

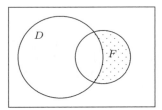

$$P(D'|F) = \frac{P(D' \cap F)}{P(F)} = \frac{0.10}{0.60} = 0.167.$$

Here you can find $P(D' \cap F)$ from the Venn diagram as

$$P(D' \cap F) = P(F) - P(F \cap D) = 0.60 - 0.50 = 0.10.$$

Note that all females must be either dull brown or bronze, so $P(D|F) + P(D'|F) = 1.0$.

If a beetle is male, what is the probability that it is dull brown? Bronze?

SOLUTION. To determine what fraction of males are dull brown,

$$P(D|M) = P(D|F') = \frac{P(D \cap F')}{P(F')} = \frac{0.20}{0.40} = 0.50,$$

where

$$P(D \cap M) = P(D \cap F') = P(D) - P(D \cap F) = 0.70 - 0.50 = 0.20.$$

To determine what fraction of males are bronze,

$$P(B|M) = P(D'|F') = \frac{P(D' \cap F')}{P(F')} = \frac{0.20}{0.40} = 0.50,$$

where

$$P(D' \cap F') = 1 - P(D \cup F) = 1 - 0.80 = 0.20.$$

Again, note that $P(D|M) + P(D'|M) = 1$ because all males must be either dull brown or bronze.

So knowing the sex of the beetle changes the probability of the color morph in this example:

$$P(D|F) = 0.83 \qquad P(B|F) = 0.17 \qquad\qquad (2.1)$$
$$P(D|M) = 0.50 \qquad P(B|M) = 0.50. \qquad\qquad (2.2)$$

Because the percentages of the two color morphs are different in the two sexes, we consider the two characteristics, color morph and gender, not to be independent of each other. Knowing one trait changes the probabilities for the other trait.

Independence of Events

When events exhibit independence from one another, the probabilities associated with these events simplify. Consider a simple experiment involving rolling a die and flipping a coin. The two aspects of this experiment are independent of each other.

The result on the die is not affected by the result on the coin and vise versa. If the die turns up a 6, the probability of a head on the coin is the same as if the die turned up a 5 or any other number (1 through 4).

DEFINITION 2.1. Two events A_1 and A_2 such that $P(A_1) \neq 0$ are **independent events** if $P(A_2|A_1) = P(A_2)$.

In other words, knowing that A_1 has occurred doesn't affect the outcome or probability of A_2 in any way. Contrast this to equations (2.1) and (2.2) where knowing the gender does change the probability of the color morph of the beetle. From the definition of independent events we can design a test for independence. Formula 2.7 says

$$P(A_2|A_1) = \frac{P(A_2 \cap A_1)}{P(A_1)}$$

as long as $P(A_1) \neq 0$. But by definition, if A_1 and A_2 are independent, then $P(A_2|A_1) = P(A_2)$. So if A_1 and A_2 are independent, then

$$P(A_2|A_1) = \frac{P(A_2 \cap A_1)}{P(A_1)} = P(A_2).$$

Multiplying by $P(A_1)$ we obtain

FORMULA 2.8. Events A_1 and A_2 are independent if and only if

$$P(A_2 \cap A_1) = P(A_1) \cdot P(A_2).$$

It follows that Formula 2.8 can be used as a test of independence or to calculate the probability of both A_1 and A_2 occurring if they are known to be independent.

EXAMPLE 2.9. *Cepaea nemoralis*, a widely studied species of land snail, exhibits several forms or morphs within any given population. In one extensively studied population, 45% have pink background coloring while 55% have yellow background coloring. In addition 30% of this population are striped with 20% of the total being pink and striped. Is the presence or absence of striping independent of background color?

SOLUTION. Begin with set notation and the Venn diagram.

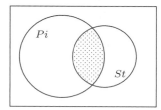

$$P(Pi) = 0.45 \qquad P(Y) = P(Pi') = 0.55$$

$$P(St) = 0.30 \quad P(St') = 0.70$$

$$P(Pi \cap St) = 0.20.$$

The test of independence is whether $P(Pi \cap St) = P(Pi) \cdot P(St)$. But $0.20 \neq 0.45 \times 0.30 = 0.135$. Since the probability of the intersection of Pi and St is *not* equal to the product of the individual probabilities the two events, pink background and striping, are *not* independent.

Given that a snail is pink, what is the probability it will have stripes?

SOLUTION. This time

$$P(St|Pi) = \frac{P(Pi \cap St)}{P(Pi)} = \frac{0.20}{0.45} = 0.44.$$

Again note $P(St|Pi) \neq P(St)$ because $0.44 \neq 0.30$. The probability of being striped is higher in the pink morph than in the population as a whole, indicating lack of independence.

If the presence of a black lip in the opening of the shell (found in 40% of the individuals) is controlled by genes independent of the genes controlling background color and striping, what is the probability that a randomly chosen individual is pink with a black lip?

SOLUTION. Since Pi and B are independent,

$$P(Pi \cap B) = P(Pi) \cdot P(B) = 0.45 \cdot 0.40 = 0.18.$$

Here we utilize the independent nature of the two traits to calculate the intersection directly.

While the multiplication rule above derives directly from earlier definitions, it must only be applied to situations where the individual events are *known* to be independent of each other.

The General Multiplication Rule

From Formula 2.7 for conditional probability,

$$P(A_2|A_1) = \frac{P(A_1 \cap A_2)}{P(A_1)},$$

we obtain the following relationship:

FORMULA 2.9. The General Multiplication Rule. Let A_1 and A_2 be any events (not necessarily independent). Then

$$P(A_1 \cap A_2) = P(A_2|A_1) \cdot P(A_1).$$

EXAMPLE 2.10. In a large genetics study utilizing guinea pigs, *Cavia porcellus*, 30% of the offspring produced had white fur and 40% had pink eyes. Two-thirds of the guinea pigs with white fur had pink eyes. What is the probability of a randomly selected offspring having *both* white fur and pink eyes?

SOLUTION. $P(W) = 0.30$, $P(Pi) = 0.40$, and $P(Pi|W) = 0.67$. By Formula 2.9,

$$P(Pi \cap W) = P(Pi|W) \cdot P(W) = 0.67 \cdot 0.30 = 0.20.$$

Twenty percent of all offspring are expected to have *both* white fur and pink eyes.

Application: Honest Answers to Sensitive Questions

An interesting and important application of the general multiplication rule in Formula 2.9 is the design and analysis of surveys that involve sensitive questions. For example, suppose that the Dean's Office at your university is interested in determining the extent of plagiarism on campus. While the dean might ask a random sample of students to obtain the information, the dean knows that at least some students may not respond honestly or at all, because some answers are incriminating or socially unacceptable. Stanley Warner developed the **randomized response technique** in 1965 to deal with such issues. A variation of it is described in Example 2.11. The primary assumption of this method is that if a respondent's privacy can be absolutely guaranteed, the tendency to give non-incriminating or socially acceptable answers that are untrue will decrease and, therefore, increase the honesty (validity) of the responses.

> **EXAMPLE 2.11.** The dean wishes to estimate the number of students that plagiarize material in term papers. Instead of asking only this question, she asks a pair of "Yes" or "No" questions, one of them sensitive and the other not. The probability of answering "Yes" to the non-sensitive question must be known in advance. For example, the dean knows that 52% of the students at her campus are either sophomores or juniors. She asks:
>
> **Sensitive:**   Have you ever plagiarized a term paper while at this university?
>
> **Non-sensitive:** Are you either a sophomore or junior?
>
> The student is asked to flip a coin (privately) to determine which of the two questions to answer. If the coin toss is heads, the student is asked to answer the first question (S) and if tails, to answer the second one (NS). Since the dean (or other interviewer) does not know which side the coin landed on, the dean does not know which question the student is answering. Thus, there is no stigma or guilt attached to answering "Yes" in this situation, since the interviewer does not know whether the student is answering, "Yes, I have plagiarized," or "Yes, I am a sophomore or junior." For this reason, we expect that students will answer honestly.
>
> When the results were tallied, 32% of the respondents answered "Yes." Determine the rate of plagiarism on campus.
>
> **SOLUTION.** The coin flip is a way of randomly assigning half the students to answer each of the questions. Since 52% of the students on campus are either sophomores or juniors, then we expect that 52% of the students who get tails will answer "Yes." More formally, whether a student is a sophomore or junior and the outcome of the coin toss are independent events. This means that
>
> $$P(\text{Replied Yes} \mid \text{Asked NS}) = P(\text{Replied Yes} \mid \text{Tails})$$
> $$= P(\text{Sophomore or junior} \mid \text{Tails})$$
> $$= P(\text{Sophomore or junior}) \qquad (2.3)$$
> $$= 0.52,$$
>
> where we have used independence (Definition 2.1) at the next to last step. What the dean wants to know is $P(\text{Student is a plagiarizer})$. Since the coin toss and whether the student has plagiarized are independent events, we can use the definition of independence in reverse to obtain
>
> $$P(\text{Student is a plagiarizer}) = P(\text{Student is a plagiarizer} \mid \text{Heads})$$
> $$= P(\text{Replied Yes} \mid \text{Heads})$$
> $$= \frac{P(\text{Replied Yes} \cap \text{Heads})}{P(\text{Heads})}. \qquad (2.4)$$

Since $P(\text{Replied Yes}) = P(\text{Replied Yes} \cap \text{Heads}) + P(\text{Replied Yes} \cap \text{Tails})$, then

$$P(\text{Replied Yes} \cap \text{Heads}) = P(\text{Replied Yes}) - P(\text{Replied Yes} \cap \text{Tails})$$
$$= P(\text{Replied Yes}) - P(\text{Replied Yes} \,|\, \text{Tails}) \cdot P(\text{Tails})$$
$$= P(\text{Replied Yes}) - P(\text{Sophomore or junior}) \cdot P(\text{Tails})$$
$$(2.5)$$

where we used the general multiplication rule (Formula 2.9) and then (2.3) in the last two steps. Substituting (2.5) into (2.4),

$$P(\text{Student is a plagiarizer}) = \frac{P(\text{Replied Yes}) - P(\text{Sophomore or junior}) \cdot P(\text{Tails})}{P(\text{Heads})}.$$
$$(2.6)$$

But the dean knows the values of all the terms on the right side of this last equation. In this case,

$$P(\text{Student is a plagiarizer}) = \frac{0.32 - 0.52(0.50)}{0.50} = 0.12.$$

It appears that 12% of the students have plagiarized on this campus.

More generally with this type of random response survey, if we are interested in whether the respondent has some sensitive property X and we know in advance the probability of having non-sensitive attribute Y, then we can interpret (2.6) as saying

$$P(\text{Has } X) = \frac{P(\text{Replied Yes}) - P(\text{Has } Y) \cdot P(\text{Asked NS})}{P(\text{Asked S})}.$$
$$(2.7)$$

$P(\text{Replied Yes})$ will be known from the responses and $P(\text{Asked S})$ and $P(\text{Asked NS})$ represent known probabilities of the outcome of the random event (like a coin flip) used to assign respondents to answer one or the other of the questions. Consequently, $P(\text{Has } X)$ can be determined.

2.5 Probability Rules and Mendelian Genetics (Optional)

Traditionally, biology students have struggled with genetics problems, often employing tedious methods such as Punnett squares to solve most transmission genetics problems. Utilizing variations of the rules and techniques developed earlier in this chapter, we can address many problems quickly and easily.

Consider a typical problem involving the cross of two pure breeding strains of the garden pea, *Pisum sativum*. One parent is tall with yellow seeds while the other is dwarf with green seeds. These physical characteristics are termed the **phenotypes** of the plants.

Tall, Yellow × dwarf, green

The offspring of this cross or mating are called the **first filial generation**, or F_1. The phenotype of the F_1 was all tall with yellow seeds. When one phenotype (e.g., dwarf) disappears in the F_1, it is said to be **recessive** to the phenotype that is expressed (e.g., tall), which is said to be **dominant**. In the cross above,

Tall, Yellow × dwarf, green

↓

F_1 all Tall, Yellow

so both Tall and Yellow are considered dominant to the traits dwarf and green.

Through a myriad of crosses involving a wide variety of organisms, geneticists have determined that most higher organisms are diploid, i.e., each organism contains two copies of each gene, one from each of its parents. Furthermore, when diploid organisms produce gametes, sperm or eggs (or in the case of higher plants, ova and pollen grains), they contribute only one copy of each gene to the resulting gamete. This is known as **Gregor Mendel's Law of Segregation**, or separation. So a pure-breeding dwarf plant has two copies of a gene influencing height. These are similar in nature and the plant is said to be **homozygous** for the gene causing dwarfing.

$$\text{Phenotype} = \text{dwarf}$$
$$\text{Genotype} = \text{dd}$$

We conventionally use a lowercase letter to denote the form of the gene that causes the recessive condition. Different forms or states of a gene are called alleles (from the Greek root *allo-*, meaning "other"). A homozygous tall plant would have a different allele and would be denoted as

$$\text{Phenotype} = \text{Tall}$$
$$\text{Genotype} = \text{DD}.$$

Notice that alleles of the same gene are represented by the same letter with the dominant allele usually the uppercase and the recessive the lowercase letter. Conventionally, the letter used is the first letter of the recessive phenotype. Do not use different letters for each allele, e.g., d for dwarf and T for tall because that implies different genes (gene loci) rather than different alleles are influencing those phenotypes.

Since tall *pure-breeding plants* produce gametes that all carry the D allele and dwarf pure-breeding plants produce gametes that all carry the d allele, F_1 plants will have both alleles Dd and are said to be **heterozygous:**

$$\text{Tall} \times \text{dwarf}$$
$$\text{DD} \times \text{dd}$$
$$\downarrow$$
$$F_1 \text{ Tall}$$
$$\text{Dd}$$

The F_1 individuals all express the dominant **phenotype** (phenotype is the outward appearance), but have a fundamentally different **genotype** (genetic makeup) from their tall parents. An F_1 individual can make two kinds of gametes, those carrying a D allele and those carrying a d allele. The fact that these alleles are carried on a pair of chromosomes that always separate during gamete formation allows us to predict the resulting offspring of a cross using classical probability. For example, the heterozygote can be viewed as analogous to a coin with the probability of a D gamete equal to $\frac{1}{2}$ (the head on a coin) and the probability of a d gamete equal to $\frac{1}{2}$ (the tail on a coin).

The cross between two heterozygotes then becomes a two-staged experiment, with the first stage being the contribution of the left parent and the second stage being the contribution of the right parent:

$$
\begin{array}{ccc}
 & \text{Tall} & \times & \text{Tall} \\
F_1 & \text{Dd} & \times & \text{Dd} \\
 & & \downarrow & \\
\text{Stages:} & \text{1st} & & \text{2nd}
\end{array}
$$

With 2 possibilities for the first stage and 2 for the second stage, according to the Multiplication Principle (see Example 2.1) there are 4 possible, equally likely outcomes.

$$
\begin{array}{c}
\text{DD} \\
\text{Dd} \\
\text{dD} \\
\text{dd}
\end{array}
$$

These outcomes collapse into three genotypes: 1DD, 2Dd (since Dd = dD), and 1dd and two phenotypes: 3 Tall (1DD; 2Dd) and 1 dwarf (1dd). A short hand for the tall phenotype is D_, since a single dose of the D allele will result in a tall individual. The second position can be either D or d, and the resulting phenotype will be the same. A cross between a heterozygote and a dwarf individual will also be approached as a two-staged experiment with classical probability:

$$
\begin{array}{ccc}
\text{Tall} & \times & \text{dwarf} \\
\text{Dd} & \times & \text{dd} \\
 & \downarrow & \\
 & \text{Dd} & \\
 & \text{Dd} & \\
 & \text{dd} & \\
 & \text{dd} &
\end{array}
$$

These outcomes collapse into two genotypes 2Dd and 2dd and two phenotypes Tall (2D_) and dwarf (2dd).

We say the ratio of heterozygous to homozygous recessive offspring is 1:1 and the tall to dwarf phenotypes are 1:1. For a single gene locus with two alleles there are a limited number of crosses with easily determined outcomes. To generalize for a locus with alleles, A and a, and dominance of allele A, see Table 2.1.

TABLE 2.1. Possible matings considering a single gene locus with two alleles

Mating Type	Parents	Offspring Genotype(s)	Offspring Phenotype(s)
1	AA × AA	All AA	All A_
2	AA × Aa	$\frac{1}{2}$AA; $\frac{1}{2}$Aa	All A_
3	AA × aa	All Aa	All A_
4	Aa × Aa	$\frac{1}{4}$AA; $\frac{2}{4}$Aa; $\frac{1}{4}$aa	$\frac{3}{4}$A_; $\frac{1}{4}$aa
5	Aa × aa	$\frac{1}{2}$Aa; $\frac{1}{2}$aa	$\frac{1}{2}$A_; $\frac{1}{2}$aa
6	aa × aa	All aa	All aa

Since there are 3 genotypes possible for each parent, there are 3×3 possible matings. With matings 2, 3, and 5 in Table 2.1 occurring two ways each, AA × Aa or Aa × AA, for example, are equivalent. These outcomes follow from the Mendelian principles that genes are on chromosomes and individual chromosomes pair and separate during the cell division leading to gamete formation. They lead to a classical probability solution to any problem involving a single gene locus and two alleles.

Now consider two gene loci, one controlling height and the other seed color. With pure-breeding (homozygous) parents we have the following

$$\begin{matrix} \text{Tall, Yellow} & \times & \text{dwarf, green} \\ \text{DDGG} & & \text{ddgg} \end{matrix}$$

$$\downarrow$$

$$F_1 \text{ Tall, Yellow}$$
$$\text{DdGg}$$

If the resulting F_1 is crossed together, what kind of offspring are expected? From our earlier discussion the results for the "D" gene locus will be a 1DD:2Dd:1dd genotypic ratio and a 3D_:1dd phenotypic ratio. We can use the general predictions above to infer a 1GG:2Gg:1gg genotypic ratio and a 3G_:1gg phenotypic ratio for the "G" gene locus. If the "D" and "G" gene loci are on different pairs of chromosomes, they will behave *independently* of each other. We know then that the probabilities for the first locus can be *multiplied* times the probabilities for the second locus to generate the combined probabilities (see Formula 2.8):

$$\begin{matrix} \text{Tall, Yellow} & \times & \text{Tall, Yellow} \\ \text{DdGg} & & \text{DdGg} \end{matrix}$$

$$\downarrow$$

Genotypes: $\left(\frac{1}{4}DD + \frac{2}{4}Dd + \frac{1}{4}dd\right)\left(\frac{1}{4}GG + \frac{2}{4}Gg + \frac{1}{4}gg\right) = 1$

$$\begin{matrix} \frac{1}{16}DDGG & \frac{2}{16}DdGG & \frac{1}{16}ddGG \\ \frac{2}{16}DDGg & \frac{4}{16}DdGg & \frac{2}{16}ddGg \\ \frac{1}{16}DDgg & \frac{2}{16}Ddgg & \frac{1}{16}ddgg \end{matrix}$$

This yields a 1:2:1:2:4:2:1:2:1 genotypic ratio. Note that there are 3×3 or 9 distinct genotypes with their probabilities summing to one.

Phenotypes: $\left(\frac{3}{4}D_- + \frac{1}{4}dd\right)\left(\frac{3}{4}G_- + \frac{1}{4}gg\right) = 1$

$$\frac{9}{16}D_-G_- \quad \text{Tall, Yellow}$$
$$\frac{3}{16}D_-gg \quad \text{Tall, green}$$
$$\frac{3}{16}ddG_- \quad \text{dwarf, Yellow}$$
$$\frac{1}{16}ddgg \quad \text{dwarf, green}$$

This yields a 9:3:3:1 phenotypic ratio.

The 9 genotypes collapse into 4 distinct phenotypes whose probabilities again sum to one. The approach for two gene loci follows from our understanding of independence and independent events. As long as the gene loci under consideration are all on different pairs of chromosomes, they will behave independently and the results can be predicted using basic probability rules. Gregor Mendel referred to those independent loci as exhibiting **independent assortment** (his second law of gene behavior).

Consider a third gene locus in peas with two alleles (W, w). Plants with the ww genotype have wrinkled seed, while plants with Ww or WW genotypes have round seeds (W$_-$). If this gene locus is on a third pair of chromosomes, we can easily extend to three characteristics via Formula 2.8:

$$\text{Tall, Yellow, Smooth} \quad \times \quad \text{Tall, Yellow, Smooth}$$
$$\text{DdGgWw} \qquad\qquad \text{DdGgWw}$$
$$\downarrow$$

Genotypes: $\left(\frac{1}{4}\text{DD} + \frac{2}{4}\text{Dd} + \frac{1}{4}\text{dd}\right)\left(\frac{1}{4}\text{GG} + \frac{2}{4}\text{Gg} + \frac{1}{4}\text{gg}\right)\left(\frac{1}{4}\text{WW} + \frac{2}{4}\text{Ww} + \frac{1}{4}\text{ww}\right) = 1$

Offspring are the result of a three-stage experiment with $3 \times 3 \times 3 = 27$ possible genotypes from DDGGWW to ddggww.

Phenotypes: $\left(\frac{3}{4}\text{D}_- + \frac{1}{4}\text{dd}\right)\left(\frac{3}{4}\text{G}_- + \frac{1}{4}\text{gg}\right)\left(\frac{3}{4}\text{W}_- + \frac{1}{4}\text{ww}\right) = 1$

There are 8 phenotypes generated with the following probabilities.

$\frac{27}{64}$D$_-$G$_-$W$_-$	Tall, Yellow, Round
$\frac{9}{64}$D$_-$G$_-$ww	Tall, Yellow, wrinkled
$\frac{9}{64}$D$_-$ggW$_-$	Tall, green, Round
$\frac{3}{64}$D$_-$ggww	Tall, green, wrinkled
$\frac{9}{64}$ddG$_-$W$_-$	dwarf, Yellow, Round
$\frac{3}{64}$ddG$_-$ww	dwarf, Yellow, wrinkled
$\frac{3}{64}$ddggW$_-$	dwarf, green, Round
$\frac{1}{64}$ddggww	dwarf, green, wrinkled

In most situations the entire array isn't required, but certain phenotypes or genotypes are of interest and can be calculated by inspection of the proper genotypic or phenotypic arrays.

EXAMPLE 2.12. Consider three gene loci in tomato, the first locus affects fruit shape with the oo genotype causing oblate or flattened fruit and OO or Oo normal round fruit. The second locus affects fruit color with yy having yellow fruit and YY or Yy red fruit. The final locus affects leaf shape with pp having potato or smooth leaves and PP or Pp having the more typical cut leaves. Each of these loci is located on a different pair of chromosomes and, therefore, acts independently of the other loci. In the following cross

$$\text{OoYyPp} \times \text{OoYypp},$$

what is the probability that an offspring will have the dominant phenotype for each trait? What is the probability that it will be heterozygous for all three genes? What is the probability that it will have round, yellow fruit and potato leaves?

SOLUTION. Genotypic array:

$$\left(\tfrac{1}{4}\text{OO} + \tfrac{2}{4}\text{Oo} + \tfrac{1}{4}\text{oo}\right)\left(\tfrac{1}{4}\text{YY} + \tfrac{2}{4}\text{Yy} + \tfrac{1}{4}\text{yy}\right)\left(\tfrac{1}{2}\text{Pp} + \tfrac{1}{2}\text{pp}\right)$$

Phenotypic array:

$$\left(\tfrac{3}{4}\text{O}_- + \tfrac{1}{4}\text{oo}\right)\left(\tfrac{3}{4}\text{Y}_- + \tfrac{1}{4}\text{yy}\right)\left(\tfrac{1}{2}\text{P}_- + \tfrac{1}{2}\text{pp}\right)$$

The probability of dominant phenotype for each trait from the phenotypic array above is

$$P(\text{O}_-\text{Y}_-\text{P}_-) = P(\text{O}_-) \times P(\text{Y}_-) \times P(\text{P}_-) = \tfrac{3}{4} \times \tfrac{3}{4} \times \tfrac{1}{2} = \tfrac{9}{32}.$$

The probability of heterozygous for all three genes from the genotypic array above is

$$P(OoYyPp) = P(Oo) \times P(Yy) \times P(Pp) = \tfrac{2}{4} \times \tfrac{2}{4} \times \tfrac{1}{2} = \tfrac{4}{32} = \tfrac{1}{8}.$$

The probability of a round, yellow-fruited plant with potato leaves from the phenotypic array above is

$$P(O_-yypp) = P(O_-) \times P(yy) \times P(pp) = \tfrac{3}{4} \times \tfrac{1}{4} \times \tfrac{1}{2} = \tfrac{3}{32}.$$

Each answer applies the probability rules for independent events to the separate gene loci.

Suppose an offspring has round, red fruit and cut leaves (all dominant traits), what is the probability that it is heterozygous for all three gene loci?

SOLUTION. This is not a trivial problem, but it can be easily solved using conditional probability, the genotypic and phenotypic arrays, and a Venn diagram.

The probability in question is $P(OoYyPp|O_-Y_-P_-)$, that is, probability of heterozygous for each locus given that it expresses the dominant phenotype for each locus. According to Formula 2.7,

$$P(OoYyPp|O_-Y_-P_-) = \frac{P(OoYyPp \cap O_-Y_-P_-)}{P(O_-Y_-P_-)}.$$

The probability of all dominant is $P(O_-Y_-P_-) = \tfrac{9}{32}$ as given above. All OoYyPp plants are also $O_-Y_-P_-$, so the OoYyPp plants are a subset of the dominant phenotype plants. This makes the intersection of the two sets, $O_-Y_-P_-$ and OoYyPp, equal to the subset OoYyPp.

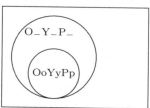

$P(OoYyPp \cap O_-Y_-P_-) = P(OoYyPp) = \tfrac{4}{32}$ as given earlier. So

$$P(OoYyPp|O_-Y_-P_-) = \frac{\tfrac{4}{32}}{\tfrac{9}{32}} = \tfrac{4}{9}.$$

We expect $\tfrac{4}{9}$ of all the round, red-fruited plants with cut leaves to be completely heterozygous.

Appropriate application of the probability rules can break down most elementary genetics problems into manageable pieces that can be confidently manipulated to generate the required answers.

2.6   Problems

1. Evaluate the following expressions:

(a) $5!$ (b) $\binom{7}{3}$ (c) $\binom{6}{0}$ (d) $\binom{4}{4}$ (e) $\binom{9}{1}$ (f) $\binom{8}{5}$

2. (*a*) Show that $\binom{5}{3}$ is the same as $\binom{5}{2}$.

    (*b*) Explain why this makes sense from the viewpoint of the selection process.

    (*c*) Explain why $\binom{n}{k}$ should be the same as $\binom{n}{n-k}$.

3. Suppose you are planning to study the respiratory physiology of the common toado, *Tetracteros hamiltoni*, a small fish found in tidal pools. Unknown to you, the pool from which you decide to collect specimens has 25 individuals, 5 of which have a fungal infection in their gills that may confound the experiment you plan to run.

    (*a*) What is the probability that you collect 10 fish from this pool and none have infected gills?

    (*b*) What is the probability that you collect 10 fish and half have infected gills?

4. An oceanic island has 15 species of birds on it. Nearby, a new island is formed by volcanic activity. Since this new island is somewhat smaller than the original island, zoogeographers theorize that it can support 8 species of birds. If the colonizing birds must migrate from the oceanic island to the newly formed island, how many different communities of 8 species could be formed on the new island?

5. (*a*) Each year the density of 7 species of Odonata (dragonflies and damselflies) is monitored in a wetland preserve. If the density of each species is to be compared with the density of every other species, how many comparisons must be made?

    (*b*) Now generalize: Assume that n different populations are sampled. If the populations are to be compared or tested in pairs, how many different comparisons must be made?

6. From a group of 30 patients receiving a standard chemotherapy treatment for lymphoma, 5 can be chosen for an additional experimental treatment that may ameliorate most side effects. How many different groups of 5 can be chosen from the initial group of 30?

7. It is often said that your chances of winning the lottery if you buy a ticket are just slightly higher than if you don't buy one! Suppose a Lotto game consists of picking 6 of 48 numbers. What is the probability of winning with the very first Lotto ticket you purchase?

8. Frost damage to apple blossoms can severely reduce apple yield in commercial orchards. It has been determined that the probability of a late spring frost causing blossom damage to Empire apple trees in the Hudson Valley of New York State is 0.6. In a season when two frosts occur, what is the probability of an apple tree being injured in this period? (Assume the two frosts are independent in their ability to cause damage.)

9. If a zoologist has 6 male guinea pigs and 9 female guinea pigs, and randomly selects 2 of them for an experiment, what are the probabilities that

    (*a*) both will be males?

    (*b*) both will be females?

    (*c*) there will be one of each sex?

10. In probability the concepts of independent events and mutually exclusive events are often confused. Clearly define each type of event and give a good example for each.

11. Let A, B, and C be any three events defined over sample space S. From the Venn diagrams that follow, describe the shaded areas using set notation.

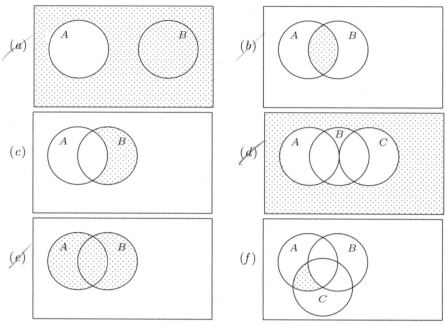

12. How would you respond to a student who claimed that the probability of getting at least 1 head in 2 tosses of a fair coin is $\frac{2}{3}$, his argument being that there are 3 possible outcomes, H, TH, and TT, two of which, H and TH, satisfy the event "at least one head."

13. Suppose you are planning to study a species of crayfish in the ponds at a wildlife preserve. Unknown to you 15 of the 40 ponds available lack this species. Because of time constraints you feel you can survey only 12 ponds. What is the probability that you choose 8 ponds with crayfish and 4 ponds without crayfish?

14. In a study of the effects of acid rain on fish populations in Adirondack mountain lakes, samples of yellow perch, *Perca flavescens*, were collected. Forty percent of the fish had gill filament deformities and 70% were stunted. Twenty percent exhibited both abnormalities.

    (*a*) Find the probability that a randomly sampled fish will be free of both symptoms.

    (*b*) If a fish has a gill filament deformity, what is the probability it will be stunted?

    (*c*) Are the two symptoms independent of each other? Explain.

15. The blood bank of the local hospital has a supply of plasma in which some units have been contaminated with a virus causing hepatitis. If 4 of the 20 units in the supply are contaminated, what is the probability that a severely injured accident victim given 5 units of plasma will not be exposed to the hepatitis virus?

16. A plant pathologist studying fire blight (a bacterial disease of apple trees) surveyed a large population of commercial apple trees of a variety called Empire. She noted that 60% of the trees had pink flowers and the remainder had white flowers. Thirty percent of the trees had some evidence of fire blight infection and 10% were pink-flowered and infected.

   (a) Draw a Venn diagram representing this situation.

   (b) What percentage of the trees were white-flowered and infected?

   (c) What is the probability that a white-flowered tree will be infected?

   (d) What is the probability that a pink-flowered tree will be infected?

   (e) Are flower color and disease resistance independent? Explain.

   (f) What kind of trees should she recommend to growers? Explain.

17. Two important indicators of stream pollution are high biological oxygen demand (BOD) and low pH. Of the more than 250 streams draining into a large lake, 30% have high BOD and 20% have low pH levels, with 10% having both characteristics.

   (a) Draw an appropriate Venn diagram.

   (b) Demonstrate that the two indicators, high BOD and low pH, are not independent of each other.

   (c) If a stream has high BOD, what is the probability it will also have low pH?

   (d) If a stream has normal levels of BOD, what is the probability it will also have low pH.

   (e) What is the probability that a stream will not exhibit either pollution indicator, i.e., will have normal BOD and pH levels?

18. A large fruit-eating bat called the black flying fox, *Pleropus alceto*, occupies a large mangrove swamp on Indooroopilly Island. Assume that about 80% of these bats are infected with an ectoparasitic mite and 30% have larger tick parasites. Twenty percent are infected with both.

   (a) Find the probability that a randomly chosen bat will have some parasites.

   (b) If a randomly chosen bat has mites, what is the probability that it will not have ticks?

   (c) Are the presence of the two types of ectoparasites independent of each other?

19. In the cross below, assume each gene locus is assorting independently from every other gene locus and each capital letter allele is dominant to its lowercase allele.

Female   AaBbCc × AabbCc   Male

Find the probabilities that the following events occur:

   (a) The female parent produces a gamete carrying abc.

   (b) The male parent produces a gamete carrying abc.

   (c) The first offspring is aabbcc.

   (d) An offspring has the same genotype as its maternal parent.

   (e) An offspring has the same phenotype as its paternal parent.

   (f) The first two offspring are homozygous recessive.

20. Tay-Sachs disease is a rare autosomal recessive disease in humans. Homozygous recessive individuals (aa) lack an enzyme called hexosaminidase A and, therefore, accumulate gangliosides in their nervous system leading to paralysis, epilepsy, blindness, and eventual death. Heterozygotes (Aa) are phenotypically normal. A second gene independently assorting from the A locus exhibits autosomal dominant expression in the form of dramatically shortened fingers, a condition known as brachydactyly. B_ individuals are brachydactylous and bb are normal. In a marriage between the following two individuals

$$\text{AaBb} \times \text{Aabb},$$

  (a) find the probability that the first child has Tay-Sachs disease;

  (b) find the probability that the first child has Tay-Sachs and brachydactyly;

  (c) find the probability that the first child is a carrier (Aa) for Tay-Sachs *given* that he appears normal;

  (d) find the probability that the first two children are normal.

21. In sweet pea, *Lathyrus odoratus*, flower color is controlled by two independently assorting gene loci (A, a and B, b). In order to produce purple-colored flowers a plant must have at least one capital allele at each locus (i.e., must be A_B_ to have purple-colored flowers). Homozygous recessive genotypes for either aa or bb or both result in white-flowered plants. Suppose the following cross is made: AaBb × AaBb.

  (a) What is the probability that the first offspring produced has purple flowers?

  (b) What is the probability that the first two plants produced are white-flowered?

  (c) Given that the first plant is purple-flowered, what is the probability that it has the same genotype as its parents?

22. In 1992 the Mattel Toy Company marketed a Teen Talk Barbie doll with a computer chip programmed with four sayings chosen randomly from a pool of 270 developed from interviews with thousands of young girls. One of the phrases was, "Math class is tough." Several organizations objected to the phrase saying it imparted a negative attitude towards mathematics in girls. A spokeswoman for Mattel claimed that there was less than a 1% chance of getting a doll that says, "Math class is tough." Using a classical probability approach and combinatorics demonstrate that her assessment of this possibility is a bit low.

23. In humans the blood type for the ABO blood group is determined by a locus with three alleles. The possible genotypes and resulting phenotypes are listed below.

Genotypes	Blood type
$I^A I^A$ or $I^A I^O$	A
$I^B I^B$ or $I^B I^O$	B
$I^A I^B$	AB
$I^O I^O$	O

The six genotypes yield four blood types with both the I^A and I^B alleles exhibiting dominance to the I^O allele, but showing codominance in the $I^A I^B$ genotype.

An independent gene locus controls the rh blood group with RR or Rr being rh positive and rr being rh negative. Two individuals with the following blood types plan to start a family.

$$I^A I^O Rr \times I^B I^O Rr$$

Find the following probabilities for their offspring:

(a) The first child has type O blood.

(b) The first child has type A positive blood.

(c) The first child has RR genotype.

(d) The first child is RR genotype given that he is rh positive.

(e) The first two children are Type O.

(f) None of the first five have Type O blood.

(g) Given that the first two children are B negative, the third is A positive.

(h) The first child is an AB negative boy.

24. The gene causing Duchenne's Muscular Dystrophy (DMD) is very large and contains 55 exons. Data from many studies indicate that there are 76 different point mutations, 114 different deletions, and 8 different duplications which cause DMD. Suppose these genetic defects occur equally frequently (they don't really); find the following:

(a) the probability a randomly chosen DMD patient has a point mutant.

(b) the probability a randomly chosen DMD patient has a duplication or deletion.

(c) the probability a randomly chosen DMD patient doesn't have a deletion.

25. In studies linking obesity to Non-Insulin Dependent Diabetes (NIDD) it is thought that obesity genes interact with diabetes susceptibility genes. Consider the following:

• Obesity reaches 30% in industrialized countries.

• Seven percent of obese patients develop NIDD, i.e., $P(\text{NIDD}|\text{obese}) = 0.07$.

• Ninety percent of patients with NIDD are obese, i.e., $P(\text{obese}|\text{NIDD}) = 0.90$.

(a) What proportion of the population in the industrialized world are obese and have NIDD?

(b) What proportion of the population have NIDD?

26. While DDT (dichlorodiphenyltrichloroethane) has been banned for use as an insecticide in the United States since 1972 because of damage to non-target species, it is still used in some countries and persists in the environment. The following table characterizes 1000 cores taken from an arctic ice sheet. Each sample was tested for levels of DDT and its breakdown products DDE (dichlorodiphenyldichloroethylene) and DDD (dichlorodiphenyldichloroethane). In addition each core was also tested for nitrates. For each type of contaminant the core was scored as high, moderate, or low.

		DDT, DDE, DDD			
		High	Moderate	Low	Total
	High	105	80	55	240
Nitrates	Moderate	70	175	145	390
	Low	25	45	300	370
	Total	200	300	500	1000

Calculate the following probabilities for an ice core picked at random from this group.

(a) It has high levels of nitrate.

(b) It has high levels of both nitrates and DDT.

(c) It has high levels of nitrates given low levels of DDT.

(d) It has moderate levels of nitrates given low levels of DDT.

(e) Find the probability that two cores picked at random will both have moderate levels of DDT. Think about sampling without replacement here.

(f) Are high levels of nitrates and low levels of DDT independent of each other?

27. Mad cow disease (bovine spongiform encephalopathy), caused by twisted brain proteins called prions, is a serious health concern in cattle from Alberta, Canada. If five cattle in a herd of 40 are infected and 10 cattle are chosen at random for testing, find the following probabilities:

(a) None of the sample cattle have mad cow disease.

(b) At least one of the sample cattle has mad cow disease.

(c) Two of the sample cattle have mad cow disease.

28. As the flood waters from Hurricane Katrina (2005) receded there was considerable concern about the presence of pathogens and environmental toxins in the remaining ponds, pools, and puddles. A team of microbiologists sampled some of these water sources. They found that 65% of the samples had high levels of coliform bacteria indicating fecal contamination. Forty percent had significant levels of heavy metal contamination (lead, mercury, arsenic, and cadmium). Thirty percent had high levels of both coliforms and heavy metals.

(a) Draw a suitable Venn diagram for this situation.

(b) Are the two types of contaminants independent of each other?

(c) What is the probability that a randomly selected sample will have high coliform levels, but not high heavy metal levels?

(d) What is the probability that a randomly selected sample will have high levels of at least one type of contaminant?

(e) If a sample has a high coliform count, what is the probability that it will also contain high levels of heavy metals?

29. Sociologists profiling the refugees from Hurricane Katrina over the age of sixty in the New Orleans Superdome found the following:

- 70% had annual incomes of \$8,000 or less (low income).

- 35% had diabetes and/or asthma.

- 30% had low income and at least one of the chronic ailments.

(*a*) Draw a Venn diagram of these results.

(*b*) What percentage of the elderly refugees were low income or chronically ill?

(*c*) Are low income and diabetes/asthma independent?

(*d*) What is the probability that a low income refugee is chronically ill?

(*e*) What is the probability that a refugee earning more than \$8,000 annually is chronically ill?

30. Suppose a physical education class is made up of 25 students, 10 of whom are classified as cigarette smokers. A random sample of 6 students is to be chosen for an exercise physiology experiment. What is the probability that exactly half the sample will be smokers?

31. An amateur horticulturist has been growing tulips for many years. In one variety he notices that 40% have deep red flowers while the remainder are bright red. He also notes that 50% have fringed or crenulated petals, 20% have ruffled, curled petals referred to as parrot beak petals, and the rest have traditional, smooth petals. Half of the deep red flowers are fringed and 20% of them are parrot beaked.

(*a*) Draw a Venn diagram of these characteristics for this variety. How many morphs are there? How many areas are there in your Venn diagram?

(*b*) Are petal shape and color independent characteristics?

(*c*) If a tulip is deep red, what is the probability that it is crenulate?

(*d*) If a tulip is bright red, what is the probability that it is crenulate?

(*e*) Comment on these last two calculations.

32. A large study of life style effects on heart disease yielded the following statistics:

Smokers	40%
Obese	30%
Heart Disease	10%
Smokers and Obese	20%
Smokers and Heart Disease	6%
Obese and Heart Disease	5%
Smoker, Obese, and Heart Disease	4%

(*a*) Draw a Venn diagram to model these statistics. Label each area with its probability.

(*b*) What percentage of the study have heart disease but are not smokers or obese?

(*c*) What percentage of the study will be either smokers or obese?

(*d*) What is the probability that an obese person will develop heart disease?

(*e*) What is the probability that a smoker will develop heart disease?

(*f*) Does this study indicate life style affects the incident of heart disease? Check for independence here.

33. A large food processing company uses a pool of food tasters to test the palatability of various foods. This pool is made up of 30 people of different ages and ethnic backgrounds. Suppose in a study to "restructure" scraps of high-value meat into products that resemble fresh steak, casein is used as the "glue" or binder. Casein is a protein derived from milk. If 4 of the 30 people in the tasting panel have severe allergies to milk and milk proteins and 10 are chosen to evaluate the restructured steaks, find the following using combinatorics and classical probability:

(*a*) The probability that none of the 10 is allergic to milk;

(*b*) The probability that at least 1 of the 10 is allergic to milk;

(*c*) The probability that 3 of the 10 will be allergic to milk.

34. Here's a variation on the randomized response method called Boruch's forced response method. A respondent is confronted with the (sensitive) question "Have you ever cheated on your federal income taxes?" The respondent is asked to reply to this question dependent on the roll of a die. If the roll is a 1 or 2, the respondent is forced to answer "Yes." When the die roll is a 6, the respondent is forced to answer "No." If the roll is a 2, 3, or 4, the respondent is asked to answer the tax question truthfully. Since the interviewer does not know what the roll of the die was, the privacy of the respondent is preserved. Answering "Yes" could either mean the die roll was 1 or 2 or that the person is a tax cheat. Suppose 40% of respondents answer "Yes." What is the probability that a person is a tax cheat?

35. Suppose you know $P(B|A)$ but you want to find $P(A|B)$. Prove that

$$P(A|B) = \frac{P(B|A) \cdot P(A)}{P(B)}.$$

36. QCR. True or False. Justify your answer.

(*a*) In general, classical probabilities are more exact than empirical probabilities, but they always require sophisticated combinatoric solutions.

(*b*) In combinatorics, permutations are used to determine orders or arrangements of objects, while combinations partition objects into groups or classes without regard to order.

(*c*) The intersection of two mutually exclusive subsets is the empty set, $\emptyset$.

(*d*) The General Addition Rule cannot be applied to mutually exclusive events.

(*e*) For independent events $P(A_2 \mid A_1) = P(A_2)$.

(*f*) For independent events $P(A_2 \cap A_1) = P(A_2) \cdot P(A_1)$.

(*g*) The General Multiplication Rule can be applied only to independent events.

3

Probability Distributions

Concepts in Chapter 3:

- Definitions: Discrete versus Continuous Variables
- Probability Density Functions for Discrete Variables
- Expected Values from Probability Density Functions
- Cumulative Distribution Functions
- The Binomial Distributions
- The Poisson Distributions
- General Discussion of Continuous Random Variables
- The Normal Distributions
- The Standard Normal Distribution

Introduction

With the basic definitions and rules of probability in hand from Chapter 2, we now turn to the techniques needed to predict the actions of various random variables that are fundamental in the study of life sciences. But first we need a few definitions and some general background.

A **random variable** is a variable whose actual value is determined by chance operations. The outcome when you roll a die (number of dots) or flip a coin (head or tail) is a random variable. Intuitively we realize this, but the probability of an albino offspring from two heterozygous parents or the probability of an individual being less than 6 feet tall are also determined as outcomes of random variables.

In general we consider two classes of random variables, **discrete** and **continuous**. These classes are handled somewhat differently and, therefore, will be discussed separately. A discrete random variable can assume only certain values, either finite or countably infinite. Examples include the outcomes when rolling a die (the only outcomes possible are 1, 2, 3, 4, 5, and 6) or the number of patients who are HIV positive (only integers from 1 to N). A continuous random variable can assume any value within an interval or intervals of real numbers, and the probability that it assumes any specific value is 0. For example, a rainbow trout may be any length from 0 to 100 cm. The probability that a randomly measured trout is exactly 50 cm long is considered to be zero.

> **NOTATION.** We will use uppercase letters such as X and Y to denote random variables. If we have several such variables, we will often use subscripts with the variables, e.g., X_1, X_2, X_3, or more generally X_i. To indicate a particular (but unknown) *value* of a random variable X or Y we will use the corresponding lowercase letters, x or y, respectively. For example, if X is the discrete random variable representing the roll of a die, then the probability that the outcome has the particular value x would be

denoted $P(X = x)$. If Y is the continuous random variable representing the height of a woman in the U.S., then the probability that such a woman has height greater than some value y would be denoted $P(Y > y)$.

We first examine the behavior and predictability of discrete random variables.

3.1 Discrete Random Variables

The pattern of behavior of a discrete random variable is described by a mathematical function called a **density function** or **probability distribution**. The density function or distribution is a record of how often each value of the discrete random variable occurs. In other words, the density function associates a probability to each possible outcome of the random variable. More formally,

DEFINITION 3.1. Let X be a discrete random variable. The **probability density function** or **probability distribution** f for X is

$$f(x) = P(X = x),$$

where x is any real number. (In words, this says that the value of $f(x)$ is the probability that the random variable X is equal to the value x.) Note that

- f is defined for all real numbers;
- $f(x) \geq 0$ since it is a probability;
- $f(x) = 0$ for most real numbers because X is discrete and cannot assume most real values;
- summing f over all possible values of X produces 1, i.e.,

$$\sum_{\text{all } x} f(x) = 1.$$

A probability density function is also referred to as a **pdf**.

EXAMPLE 3.1. A fair 6-sided die is rolled with the discrete random variable X representing the number obtained per roll. Give the density function for this variable.

SOLUTION. Since the die is fair, each of the six outcomes, 1 through 6, is equally likely to occur (classical probability from Chapter 2). So each of these possible outcomes has a probability of $\frac{1}{6}$. In cases when all possible outcomes are equally likely, we say the variable X is **uniformly distributed**. When the number of outcomes is small, as it is here, it is convenient to give the density function in the form of a table that lists only the possible values.

Random variable $X = x$	1	2	3	4	5	6
Density $f(x)$	$\frac{1}{6}$	$\frac{1}{6}$	$\frac{1}{6}$	$\frac{1}{6}$	$\frac{1}{6}$	$\frac{1}{6}$

Values of X not listed in the table are presumed to be impossible and their corresponding values of f are 0. For example $f(0.1) = 0$, $f(7.9) = 0$, and $f(-3) = 0$. Notice that the values of f sum to 1, as they should.

EXAMPLE 3.2. A fair 6-sided die is rolled twice with the discrete random variable X representing the sum of the numbers obtained on both rolls. Give the density function of this variable.

SOLUTION. There are 6×6 different outcomes possible, so each has a probability of $\frac{1}{36}$. The possible values range from 2 (a 1 on each die or "snake eyes" in dice vernacular) to 12 (a 6 on each die or "box cars"). While 2 and 12 can occur only one way, all other values occur at least two ways, e.g., a 3 is either a 1 then a 2 or a 2 then a 1; a 4 is either a 1 then a 3, a 3 then a 1, or two 2's. The complete density function is

Random variable $X = x$	2	3	4	5	6	7	8	9	10	11	12
Density $f(x)$	$\frac{1}{36}$	$\frac{2}{36}$	$\frac{3}{36}$	$\frac{4}{36}$	$\frac{5}{36}$	$\frac{6}{36}$	$\frac{5}{36}$	$\frac{4}{36}$	$\frac{3}{36}$	$\frac{2}{36}$	$\frac{1}{36}$

We can also represent this distribution graphically. Technically we should use a bar graph, but for reasons that will become clear when we discuss continuous random variables, we will use a histogram. In Figure 3.1 the horizontal axis corresponds to the possible values or sample space of the random variable X. Remember that in a histogram the *area* of the bar corresponds to the relative frequency of the outcome. Since the width of each bar is 1, this means that the height and area of each bar equals $f(x)$, the value of the density function, which is just the probability of the corresponding outcome. For example, the bar corresponding to $X = 5$ (a sum of 5 on the two rolls) has both height and *area* equal to $\frac{4}{36}$. If the areas under all the bars are added together, the total will be 1 since this corresponds to summing the probabilities of all possible outcomes in the sample space.

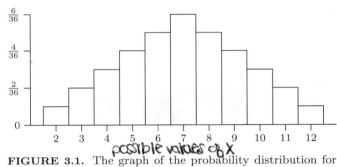

FIGURE 3.1. The graph of the probability distribution for sum of two rolls of a fair die.

The shape of the graph of this density function is symmetric about 7. Moreover, sums in the middle of the range are more common than the more extreme values. From this density function you can see why "7" is such an important outcome in many dice games.

One can use either the density function table or bar graph to find the probability of various outcomes. For example,

$$P(X = 10) = \tfrac{3}{36} = \tfrac{1}{12}.$$

To determine $P(X = 7 \text{ or } 11)$, since 7 and 11 are mutually exclusive events, the General Addition Law applies, so

$$P(X = 7 \text{ or } 11) = \tfrac{6}{36} + \tfrac{2}{36} = \tfrac{2}{9}.$$

Suppose you rolled a die 1000 times and recorded the outcomes, what would be the long-run average value? Suppose you rolled a pair of dice 50,000 times and recorded the outcomes, what would you expect the mean to be? Actually doing these processes would be quite time-consuming, but density functions give us an *a priori* way to determine these means rather quickly.

DEFINITION 3.2. The long-term **expected value** or **mean** for a discrete random variable X with density function f is given by

$$\mu = E(X) = \sum_{\text{all } x} x f(x).$$

In other words, in the sum each value x is weighted by its density or probability.

> **EXAMPLE 3.3.** Find the expected values for the random variable X in Examples 3.1 and 3.2.
>
> **SOLUTION.** For Example 3.1,
>
> $$\mu = E(X) = \sum_{x=1}^{6} x f(x) = \sum_{x=1}^{6} x \cdot \frac{1}{6} = \frac{1+2+3+4+5+6}{6} = 3.5. \quad \text{\textit{— mean}}$$
>
> So the long-term average value or the expected value when you roll a single fair die is 3.5. Note this value is not a possible actual outcome.
> For rolling a die twice as in Example 3.2,
>
> $$\mu = E(X) = \sum_{x=2}^{12} x f(x) = 2 \cdot \tfrac{1}{36} + 3 \cdot \tfrac{2}{36} + 4 \cdot \tfrac{3}{36} + \cdots + 11 \cdot \tfrac{2}{36} + 12 \cdot \tfrac{1}{36} = 7.$$
>
> Intuitively we expect the mean for 2 rolls of a die to be twice the mean for a single die.

This notion of expectation can be generalized.

DEFINITION 3.3. Let $H(X)$ denote a function of the discrete random variable X. Then its **expected value** is

$$E[H(X)] = \sum_{\text{all } x} H(x) f(x).$$

When $H(X) = X$, the formula gives the expected value or mean of X, $E(X) = \mu$, again.

> **EXAMPLE 3.4.** Find the expected value of X^2 in Examples 3.1 and 3.2.
>
> **SOLUTION.** For Example 3.1,
>
> $$\begin{aligned}
E(X^2) &= \sum_{x=1}^{6} x^2 f(x) \\
&= (1)^2 \cdot \tfrac{1}{6} + (2)^2 \cdot \tfrac{1}{6} + (3)^2 \cdot \tfrac{1}{6} + (4)^2 \cdot \tfrac{1}{6} + (5)^2 \cdot \tfrac{1}{6} + (6)^2 \cdot \tfrac{1}{6} \\
&= \tfrac{91}{6} \\
&= 15.167.
\end{aligned}$$
>
> For Example 3.2,
>
> $$\begin{aligned}
E(X^2) &= \sum_{x=2}^{12} x^2 f(x) \\
&= (2)^2 \cdot \tfrac{1}{36} + (3)^2 \cdot \tfrac{2}{36} + \cdots + (11)^2 \cdot \tfrac{2}{36} + (12)^2 \cdot \tfrac{1}{36} \\
&= \tfrac{1974}{36} \\
&= 54.833.
\end{aligned}$$

If X is a discrete random variable so that all N values of the sample space are equally likely, as in Example 3.1, then we can express the variance of X as an expected value. If all values x_i are equally likely, then $f(x_i) = \frac{1}{N}$. Starting with the theoretical formula for the variance of a population, we have

$$\sigma^2 = \frac{\sum(x_i - \mu)^2}{N} = \sum_{i=1}^{N}(x_i - \mu)^2 \left(\tfrac{1}{N}\right) = \sum_{i=1}^{N}(x_i - \mu)^2 f(x_i) = E\left[(X - \mu)^2\right].$$

This motivates the following definition.

DEFINITION 3.4. Let X be any discrete random variable. Then the **variance** of X, denoted by σ^2, σ_X^2, or $\text{Var}(X)$, is defined by

$$\sigma^2 = E\left[(X - \mu)^2\right].$$

In the third part of the theorem below, we give a convenient way to calculate this quantity. There are a few other simple properties of expected values that we will also exploit.

THEOREM 3.1. Let X denote a finite discrete random variable with N possible values $x_1, x_2, \ldots, x_N$ and mean $\mu = E(X)$.

1. If c is any constant, then $E(cX) = cE(X)$ and $E(X + c) = E(X) + c$.
2. If G and H are functions of X, then $E[G(X) + H(X)] = E[G(X)] + E[H(X)]$. *distribution*
3. $\sigma^2 = E(X^2) - \mu^2 = E(X^2) - [E(X)]^2$.

PROOF. For the first part of (1), if c is any constant, then using Definition 3.3

$$E(cX) = \sum_{i=1}^{N} cx_i f(x_i) = c\sum_{i=1}^{N} x_i f(x_i) = cE(X).$$

The second part of (1) is left as an exercise. For (2)

$$E[G(X) + H(x)] = \sum_{i=1}^{N}[G(x_i) + H(x_i)]f(x_i)$$

$$= \sum_{i=1}^{N}[G(x_i)f(x_i) + H(x_i)f(x_i)]$$

$$= \sum_{i=1}^{N} G(x_i)f(x_i) + \sum_{i=1}^{N} H(x_i)f(x_i)$$

$$= E[G(X)] + E[H(X)].$$

For (3), starting with Definition 3.4 we have

$$\sigma^2 = E\left[(X - \mu)^2\right] = E\left[(X^2 - 2\mu X + \mu^2)\right] = E(X^2) - E(2\mu X) + E(\mu^2), \qquad (3.1)$$

where the last equality follows from part (2). Now using part (1), since μ is a constant, $E(2\mu X) = 2\mu E(X) = 2\mu \cdot \mu = 2\mu^2$ and $E(\mu^2) = \mu^2$, so (3.1) becomes

$$\sigma^2 = E\left[(X - \mu)^2\right] = E(X^2) - 2\mu^2 + \mu^2 = E(X^2) - \mu^2 = E(X^2) - [E(X)]^2.$$

This is a handy computational formula for the population variance of a random variable.

For additional properties of expected value see Appendix A.2. Note that Definition 3.4 and Theorem 3.1 carry over to continuous random variables.

EXAMPLE 3.5. Find the variances for X in Examples 3.1 and 3.2.

SOLUTION. For Example 3.1,

$$\sigma^2 = E(X^2) - [E(X)]^2 = 15.167 - (3.5)^2 = 2.917.$$

For Example 3.2,

$$\sigma^2 = E(X^2) - [E(X)]^2 = 54.833 - (7)^2 = 5.833.$$

Why is σ^2 in Example 3.2 greater than σ^2 in Example 3.1? Consider the shapes of the density functions here to rationalize your answer.

Now let's apply these ideas about expectation and density functions to a biological example.

EXAMPLE 3.6. Breeding pairs of the common robin, *Turdus migratorius*, typically produce a clutch of 3 to 6 eggs. Assume the estimated density function for the random variable X (eggs per clutch) is given in the table below. Based on this information, what is $E(X)$ and what is σ_X?

No. of eggs: x	3	4	5	6
Density: $f(x)$	0.35	0.45	0.16	0.04

SOLUTION. The mean or expected value is

$$\mu = E(X) = \sum_{x=3}^{6} x f(x) = 3(0.35) + 4(0.45) + 5(0.16) + 6(0.04) = 3.89.$$

The variance is

$$\sigma^2 = E(X^2) - [E(X)]^2 = [3^2(0.35) + 4^2(0.45) + 5^2(0.16) + 6^2(0.04)] - (3.89)^2$$
$$= 15.79 - 15.13$$
$$= 0.66.$$

So the mean number of eggs per clutch is 3.89 with a standard deviation of $\sigma = \sqrt{.66} = 0.81$ eggs per clutch.

The density function for a random discrete variable tells us the probability of particular values of the variable. However, many statistical problems require knowing the probability of observing a value of a random variable *at least* as large (or small) as a given value of X. In other words, instead of using $P(X = x)$, which defines the density function, we need $P(X \le x)$. For example, what is the probability of a roll of 5 or less on a pair of dice? From Example 3.2

$$P(X \le 5) = [P(X = 2) + P(X = 3) + P(X = 4) + P(X = 5)]$$
$$= \tfrac{1}{36} + \tfrac{2}{36} + \tfrac{3}{36} + \tfrac{4}{36}$$
$$= \tfrac{10}{36}$$
$$= 0.28.$$

Here we answered the question by adding all the appropriate individual densities together. To answer this type of question in general, we define the following concept.

DEFINITION 3.5. Let X be a discrete random variable with density f. The **cumulative distribution function (CDF)** for X is denoted by F and is defined by

$$F(x) = P(X \leq x),$$

for all real x.

> **EXAMPLE 3.7.** Find the CDF for Example 3.2, where X was the sum on two dice rolled together.

> **SOLUTION.** The cumulative distribution function is easily tabulated from the density function. Simply add all values of f for the values of X less than or equal to x.

Random variable $X = x$	2	3	4	5	6	7	8	9	10	11	12
Density $f(x)$	$\frac{1}{36}$	$\frac{2}{36}$	$\frac{3}{36}$	$\frac{4}{36}$	$\frac{5}{36}$	$\frac{6}{36}$	$\frac{5}{36}$	$\frac{4}{36}$	$\frac{3}{36}$	$\frac{2}{36}$	$\frac{1}{36}$
CDF: $F(x)$	$\frac{1}{36}$	$\frac{3}{36}$	$\frac{6}{36}$	$\frac{10}{36}$	$\frac{15}{36}$	$\frac{21}{36}$	$\frac{26}{36}$	$\frac{30}{36}$	$\frac{33}{36}$	$\frac{35}{36}$	$\frac{36}{36}$

> *What is the probability of an 8 or less on a roll of a pair of dice?*

> **SOLUTION.** $P(X \leq 8) = F(8) = \frac{26}{36}$.

> *What is the probability of between 4 and 10 on a roll of a pair of dice?*

> **SOLUTION.** $P(4 < X < 10) = F(9) - F(4) = \frac{30}{36} - \frac{6}{36} = \frac{2}{3}$. Note here that less than 10 means $F(9)$ because the CDF is up to and including the values in parentheses. The utility of the CDF may not be obvious from this rather elementary example, but it should be clear after we examine the binomial and Poisson distributions.

3.2 The Binomial Distribution

The most important discrete variable distribution in biology is the **binomial distribution**. This distribution arises in a context referred to as the independent trials model, which has four assumptions:

1. A fixed number n of trials are carried out.

2. The outcome of each trial can be classified in precisely one of two mutually exclusive ways termed "success" and "failure." The term "binomial" literally means two names.

3. The probability of a success, denoted by p, remains constant from trial to trial. The probability of a failure is $1 - p$.

4. The trials are independent; that is, the outcome of any particular trial is not affected by the outcome of any other trial.

The random variable of interest in this situation is the number of successes within the n trials.

The mathematics that we have already developed allows us to determine any binomial density function. Let X be the discrete random variable which represents the number of successes in the n trials. Obviously, X can have values ranging from 0 to

n. The value of the density function $f(x)$ is the probability of exactly x successes in n trials. The number of ways that x successes can occur in n trials is

$$\binom{n}{x} = \frac{n!}{x!(n-x)!}.$$

Because the trials are independent, the probabilities of success and failure in successive trials multiply. The probability of any one combination of x successes and $n-x$ failures is $p^x(1-p)^{n-x}$. Since there are $\frac{n!}{x!(n-x)!}$ such combinations of successes and failures, we have

DEFINITION 3.6. The probability density function of a **binomial random variable** with n trials and probability p of success on any trial is

$$f(x) = \frac{n!}{x!(n-x)!}p^x(1-p)^{n-x}.$$

The binomial random variable probability density function is characterized by the two parameters n, the number of trials or sample size, and p, the probability of success on a trial.

> **EXAMPLE 3.8.** Assuming that sex determination in human babies follows a binomial distribution, find the density function and CDF for the number of females in families of size 5.
>
> **SOLUTION.** $P(\text{female}) = P(\text{success}) = p = 0.5$ and $P(\text{male}) = P(\text{failure}) = 1-p = 0.5$. No value judgement is implied in applying the terms success and failure here! From Definition 3.6 with $n = 5$,
>
> $$f(x) = \binom{5}{x}(0.5)^x(0.5)^{5-x} = \binom{5}{x}(0.5)^5$$
>
> is the density function of interest. Thus,

$$f(0) = \frac{5!}{0!(5-0)!}(0.5)^0(1-0.5)^5 = 0.03125,$$

$$f(1) = \frac{5!}{1!(5-1)!}(0.5)^1(1-0.5)^4 = 0.15625,$$

$$f(2) = \frac{5!}{2!(5-2)!}(0.5)^2(1-0.5)^3 = 0.31250,$$

$$f(3) = \frac{5!}{3!(5-3)!}(0.5)^3(1-0.5)^2 = 0.31250,$$

$$f(4) = \frac{5!}{4!(5-4)!}(0.5)^4(1-0.5)^1 = 0.15625,$$

$$f(5) = \frac{5!}{5!(5-5)!}(0.5)^5(1-0.5)^0 = 0.03125.$$

Both the pdf and the CDF are given in the table below.

Random variable $X = x$	0	1	2	3	4	5
Density $f(x)$	0.03125	0.15625	0.3125	0.3125	0.15625	0.03125
CDF $F(x)$	0.03125	0.1875	0.5000	0.8125	0.96875	1.00000

$P(X \leq 3)$

The probability of at most 3 females in 5 births is $F(3) = 0.8125$. The probability of exactly 3 females in 5 births is $f(3) = 0.3125$. The probability of at least 3 females in 5 births is $1 - F(2) = 1 - 0.5000 = 0.5000$. $P(X \geq 3)$

The graph of the pdf for the binomial distribution with $n = 5$ and $p = 0.5$ is shown in Figure 3.2. It shows that the distribution is symmetric about 2.5.

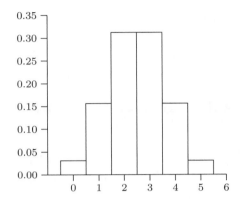

FIGURE 3.2. The binomial distribution with $n = 5$ and $p = 0.5$.

We could determine the mean and variance for this particular binomial directly by using expected values and the definitions of these terms. In fact, by the symmetry in $f(x)$, we expect the mean to be 2.5. However, the mean and variance for any binomial random variable can be calculated by making use of the following facts.

- From part (2) of Theorem 3.1, if X and Y are *any* random variables, then $E(X + Y) = E(X) + E(Y)$.

- If X and Y are *independent* random variables, then $\mathrm{Var}(X + Y) = \mathrm{Var}(X) + \mathrm{Var}(Y)$, the variance of $X + Y$ is equal to the sum of the individual variances of X and Y.

By definition of the binomial model each of the n trials is independent, so both facts apply. Let $X_1, X_2, \ldots, X_n$ be the results of the individual trials (1 for a success and 0 for a failure). The random variable X, which is the total number of successes, is just the sum of these n random variables. So

$$X = X_1 + X_2 + \cdots + X_n = \sum_{i=1}^{n} X_i.$$

Now because each X_i has only two values 1 and 0 with probabilities p and $1 - p$, respectively,

$$E(X_i) = 1 \cdot p + 0 \cdot (1 - p) = p$$

and from Theorem 3.1

$$\mathrm{Var}(X_i) = E(X_i^2) - [E(X_i)]^2 = [1^2 \cdot p + 0^2 \cdot (1 - p)] - p^2 = p - p^2 = p(1 - p).$$

Therefore, by the two rules above for the mean and variance, the total number of successes, X, in n individual trials has the following parameters:

$$\mu = E(X) = E(X_1 + \cdots + X_n) = \sum_{i=1}^{n} E(X_i) = \sum_{i=1}^{n} p = np$$

and

$$\sigma^2 = \text{Var}(X) = \text{Var}(X_1 + \cdots + X_n) = \sum_{i=1}^{n} \text{Var}(X_i) = \sum_{i=1}^{n} p(1 - p) = np(1 - p).$$

We have proved the following theorem:

THEOREM 3.2. Let X be a binomial random variable with parameters n and p. Then $\mu = E(X) = np$ and $\sigma_X^2 = \text{Var}(X) = np(1 - p)$.

Notice that "the" binomial distribution is really an infinite family of distributions, each defined by its particular values of n and p.

The formula for the mean might have been anticipated. If there are n trials each with probability of success p, then we should have expected np successes "on average." Notice that in the case where $n = 5$ and $p = 0.5$, we get $\mu = 5(0.5) = 2.5$, as we suspected at the end of Example 3.8.

Table C.1 provides the CDFs for values of $n = 5$ to 20 for selected values of p. The pdf for $n = 20$ and $p = 0.5$ is illustrated in Figure 3.3. Notice that it has the same general shape as the pdf for the binomial with $n = 5$ and $p = 0.5$ (see Figure 3.2). It, too, is symmetric about $E(X) = np = 20 \times 0.5 = 10$.

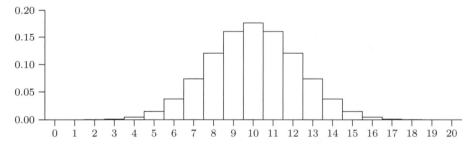

FIGURE 3.3. The binomial distribution with $n = 20$ and $p = 0.5$ is symmetric about its mean of 10.

Not all binomials are symmetric, just those with $p = 0.5$. Figure 3.4 illustrates the binomial with $n = 10$ and $p = 0.25$. While the graph of the binomial with $n = 10$ and $p = 0.25$ is not symmetric about its expected value of $E(X) = np = 2.5$; nonetheless this point has geometric significance. If we imagined the distribution as being located along a seesaw, then $\mu = E(X)$ is always the balance point for the distribution. For example, in Figures 3.1, 3.2, and 3.3, μ was the balance point because the distributions were symmetric about μ. In Figure 3.4 there is somewhat more area under the bars to the left of $\mu = 2.5$ than to the right. However, some of the area to the right of 2.5 is further away than that to the left. As anyone who has played on a seesaw knows, a small weight (area) at a distance from the fulcrum will compensate for a larger weight (area) closer to the fulcrum.

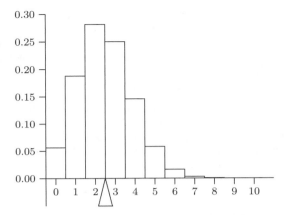

FIGURE 3.4. The binomial distribution with $n = 10$ and $p = 0.25$ is not symmetric, but the mean or expected value of $np = 2.5$ acts as a balance point for the graph distribution.

Even though binomial distributions are not symmetric if $p \neq 0.5$, as n becomes larger, they do tend to approach symmetry about their mean. Figure 3.5 illustrates the pdf for the binomial with $n = 20$ and $p = 0.25$. The expected value is $\mu = np = 5$. While the graph is not exactly symmetric about 5, it is nearly so.

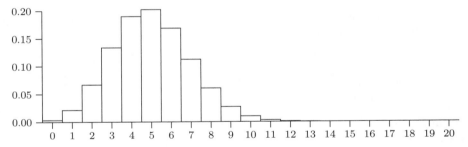

FIGURE 3.5. The binomial distribution with $n = 20$ and $p = 0.25$ is close to, but not exactly, symmetric about $\mu = np = 5$.

Binomial distributions have wide applicability, as the next two examples show.

EXAMPLE 3.9. A particular strain of inbred mice has a form of muscular dystrophy that has a clear genetic basis. In this strain the probability of appearance of muscular dystrophy in any one mouse born of specified parents is $\frac{1}{4}$. If 20 offspring are raised from these parents, find the following probabilities.

(*a*) Fewer than 5 will have muscular dystrophy;

(*b*) Five will have muscular dystrophy;

(*c*) Fewer than 8 and more than 2 will have muscular dystrophy.

SOLUTION. This problem could be approached using the density function directly

$$f(x) = \frac{20!}{x!(20-x)!} \left(\frac{1}{4}\right)^x \left(\frac{3}{4}\right)^{20-x} ,$$

of binomial random variable

but this approach would require an extraordinary amount of calculation. Fortunately Table C.1 of the Appendices and similar tables allow easy solution to selected binomial problems. From Theorem 3.2 it is clear that the sample size n and the probability of one of the two kinds of events p completely define a particular binomial distribution. For values of n and p not found in Table C.1 the CDFs can be generated rather easily with statistical programs on most computers.

(a) To determine the probability that fewer than 5 will have muscular dystrophy, use $n = 20$ and $p = 0.25$. Then $P(X < 5) = F(4)$. In Table C.1 we look up the column corresponding to $n = 20$ and $p = 0.25$. The row denoted by the 4 in the margin gives the CDF for 4 successes in 20 trials, i.e., the probability of 4 or fewer successes in 20 trials:

$$P(X < 5) = F(4) = 0.4148.$$

This number could be obtained by determining the values of $f(0)$, $f(1)$, $f(2)$, $f(3)$, and $f(4)$ and adding the results.

(b) To determine the probability that 5 will have muscular dystrophy, use

$$P(X = 5) = f(5) = F(5) - F(4) = 0.6172 - 0.4148 = 0.2024.$$

Here again we could evaluate the density function directly, but a simple subtraction of adjacent CDF values is easier and quicker than calculating

$$f(5) = \frac{20!}{5!(20-5)!} \left(\frac{1}{4}\right)^5 \left(\frac{3}{4}\right)^{20-5} = 0.2024.$$

(c) To determine the probability that fewer than 8 and more than 2 will have muscular dystrophy, use

$$P(2 < X < 8) = F(7) - F(2) = 0.8982 - 0.0913 = 0.8069.$$

Again, direct computation of $f(7) + f(6) + f(5) + f(4) + f(3)$ would eventually yield the same result.

From Example 3.9 the utility of CDF tables should be apparent. Let's look at one final example.

EXAMPLE 3.10. Suppose it is known that the probability of recovery from infection with the Ebola virus is 0.10. If 16 unrelated people are infected, find the following:

(a) The expected number who will recover;
(b) The probability that 5 or fewer recover;
(c) The probability that at least 5 will recover;
(d) The probability that exactly 5 will recover.

SOLUTION. Use Table C.1 with $n = 16$ and $p = 0.10$.

(a) The expected number who will recover is

$$E(X) = \mu = np = 16(0.10) = 1.6.$$

(b) The probability that 5 or fewer recover is

$$P(X \leq 5) = F(5) = 0.9967.$$

(c)   The probability that at least 5 will recover is

$$P(X \geq 5) = 1 - F(4) = 1 - 0.9830 = 0.0170.$$

(d)   The probability that exactly 5 will recover is

$$P(X = 5) = f(5) = F(5) - F(4) = 0.9967 - 0.9830 = 0.0137.$$

Note that 5 is a long way from the expected value of 1.6, and therefore, the probability is small.

3.3   The Poisson Distribution

A distribution related to the binomial that has utility for description of some discrete random variables is the Poisson distribution. The distribution is named for the French mathematician Siméon Denis Poisson (1781–1840). The discrete random variable of the number of occurrences of an event in a continuous interval of time or space, sometimes called rare, random events in time or space, arises from what are known as Poisson processes. The number of radioactive decays in a time interval or the number of sedge plants per sampling quadrat may be examples of Poissonally distributed discrete random variables. The random variable X in a Poisson process is the number of occurrences of the event in an interval of time or space. The underlying assumptions of this model are that:

1. Events occur one at a time; two or more events do not occur precisely at the same moment or location.

2. The occurrence of an event in a given period is independent of the occurrence of an event in any previous or later nonoverlapping period.

3. The expected number of events during any one period is the same as during any other period. This expected number of events is denoted, again, by μ.

Using calculus, one can show the following:

DEFINITION 3.7. The probability density function for the **Poisson random variable** X is given by

$$f(x) = \frac{e^{-\mu}(\mu)^x}{x!} \qquad x = 0, 1, 2, \ldots,$$

where e is the natural exponential $2.71828\ldots$. The expected value is $E(X) = \mu$ and so, too, is the variance, $\sigma^2 = \mu$.

This definition shows that a Poisson random variable depends on the single parameter μ, unlike a binomial distribution that depends on two parameters n and p.

Note that the CDFs for selected Poisson random variables are given in Table C.2. To be able to use a Poisson distribution the numerical value of μ must first be determined.

EXAMPLE 3.11. A radioactive source emits decay particles at an average rate of 4 particles per second. Find the probability that 2 particles are emitted in a 1-second interval.

SOLUTION. Here $\mu = 4$ particles per second. We want

$$P(X = 2) = f(2) = \frac{e^{-4}(4)^2}{2!} = 0.1465.$$

Using Table C.2,

$$P(X = 2) = F(2) - F(1) = 0.2381 - 0.0916 = 0.1465.$$

Table C.2 catalogues the cumulative distribution function for Poisson distributions with means equal to 0.5 and the integers 1 through 15. As with the binomial, the Poisson is an infinite family of distributions. If μ is not listed in Table C.2, the densities can be calculated directly from Definition 3.7 and the CDF with most statistical programs.

EXAMPLE 3.12. An ichthyologist studying the spoonhead sculpin, *Cottus ricei*, catches specimens in a large bag seine that she trolls through the lake. She knows from many years experience that on average she will catch 2 fish per trolling run. Find the probabilities of catching

(*a*)   No fish on a particular run;

(*b*)   Fewer than 6 fish on a particular run;

(*c*)   Between 3 and 6 fish on a particular run.

SOLUTION. $\mu = 2$ fish/run.

(*a*)   For no fish, use column 3 in Table C.2,

$$P(X = 0) = f(0) = F(0) = 0.1353.$$

(*b*)   For fewer than 6 fish, $P(X < 6) = F(5) = 0.9834.$

(*c*)   For between 3 and 6 fish,

$$P(3 < X < 6) = F(5) - F(3) = 0.9834 - 0.8571 = 0.1263.$$

Alternatively,

$$P(3 < X < 6) = f(4) + f(5) = \frac{e^{-2}2^4}{4!} + \frac{e^{-2}2^5}{5!} = 0.0902 + 0.0361 = 0.1263.$$

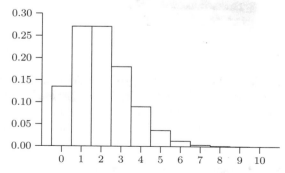

FIGURE 3.6. The graph of the Poisson density function for $\mu = 2$ for Example 3.12.

The graph of the pdf for the Poisson distribution with $\mu = 2$ for Example 3.12 is given in Figure 3.6. Notice that it is *not* symmetric. Note, too, that even though it appears that both $f(9)$ and $f(10)$ are 0, they are not. Both are simply very, very small (less than 0.0009). In fact, $f(x)$ is never 0.

As μ increases, however, the graph of the pdf for the Poisson distribution becomes more nearly symmetric about $x = \mu$. Figure 3.7 illustrates this with $\mu = 10$.

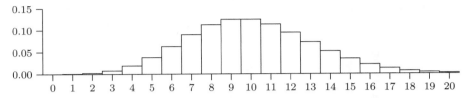

FIGURE 3.7. The Poisson distribution with $\mu = 10$ is almost, but not quite, symmetric about $x = 10$.

EXAMPLE 3.13. A group of forest ecology students survey several 10-m × 10-m plots in a subtropical rainforest. They find a mean of 30 trees per plot. Under the assumption that the trees are randomly distributed, what is the probability of finding no more than 3 trees in a 1-m^2 plot? Of finding exactly 1 tree in a 1-m^2 plot? At least 2 trees?

SOLUTION. If there were a mean of 30 trees in 100 m^2, the expected number in a 1-m^2 observation plot is

$$\mu = \frac{30}{100} = 0.3 \text{ tree/m}^2.$$

The density function is

$$f(x) = \frac{e^{-0.3}(0.3)^x}{x!}.$$

The probability of finding at most 3 trees in a 1-m^2 plot is

$$P(X \leq 3) = f(0) + f(1) + f(2) + f(3).$$

Table C.2 does not list $\mu = 0.3$, so we must do the calculations by hand.

$$
\begin{aligned}
P(X \leq 3) &= \sum_{x=0}^{3} \frac{e^{-0.3}(0.3)^x}{x!} \\
&= \frac{e^{-0.3}(0.3)^0}{0!} + \frac{e^{-0.3}(0.3)^1}{1!} + \frac{e^{-0.3}(0.3)^2}{2!} + \frac{e^{-0.3}(0.3)^3}{3!} \\
&= \frac{0.7408}{1} + \frac{0.2222}{1} + \frac{0.0667}{2} + \frac{0.0200}{6} \\
&= 0.9996.
\end{aligned}
$$

The probability of finding exactly one tree is

$$f(1) = \frac{e^{-0.3}(0.3)^1}{1!} = 0.2222.$$

Using complements, the probability of finding at least two trees is

$$1 - [f(0) + f(1)] = 1 - (0.7408 + 0.2222) = 0.0370.$$

Poisson Approximation to the Binomial Distribution

Another important application of the Poisson distribution is that under certain circumstances it is a very good approximation for the binomial distribution. Generally, for a good approximation, the binomial parameter n must be large (at least 20) and p must be small (no greater than 0.05). See Figure 3.8. The approximation is very good if $n \geq 100$ and $np \leq 10$. Recall that for a binomial, $\mu = np$. So np is the value of μ in the Poisson approximation.

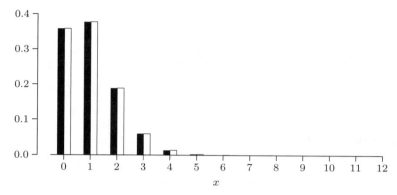

FIGURE 3.8. The binomial distribution for $n = 20$ trials with probability of success $p = 0.05$ (shaded bars) and its Poisson approximation with $\mu = np = 20(0.05) = 1$ (unshaded bars). The binomial probabilities are given by $\binom{20}{x}(0.2)^x(0.8)^{20-x}$, while the corresponding Poisson probabilities are given by $\frac{e^{-1}1^x}{x!}$.

EXAMPLE 3.14. A certain birth defect occurs with probability $p = 0.0001$. Assume that $n = 5000$ babies are born at a particular large, urban hospital in a given year.

(a) What is the approximate probability that there is at least 1 baby born with the defect?

(b) What is the probability that there will be no more than 2 babies born with the defect?

SOLUTION. This is a binomial problem with $n = 5000$ and $p = 0.0001$. Obviously Table C.1 does not help here. One could work out the answers by hand using the binomial density function. However, since n is very large and $np \leq 10$, a Poisson approximation is possible. Here $\mu = np = 0.5$. So from Table C.2,

(a) the probability that there is at least 1 baby with the defect is

$$P(X \geq 1) = 1 - F(0) = 1 - 0.6065 = 0.3935,$$

(b) and the probability that there are no more than 2 with the defect is

$$P(X \leq 2) = F(2) = 0.9856.$$

Using the binomial distribution, the answers would have been the same to four decimal places, 0.3935 and 0.9856, respectively.

3.4 Continuous Random Variables

We now turn to the second major class of random variables, the *continuous* variables. Unlike counts that are common discrete random variables, continuous random variables can assume any values over an entire interval. Measurements such as the lengths of rainbow trout or weights of manatees would be examples of continuous random variables.

DEFINITION 3.8. The **probability density function for a continuous random variable** X is a function f defined for all real numbers x such that

- $f(x) \geq 0$;

- the region under the graph of f and above the x-axis has an area of 1;

- for any real numbers a and b, $P(a \leq X \leq b)$ is given by the area bounded by the graph of f, the lines $x = a$ and $x = b$, and the x-axis.

The definition above may appear complicated and is not immediately intuitive but some examples will help our understanding.

EXAMPLE 3.15. Each of the following is a graph of a density function of a continuous variable. Let's look at their characteristics.

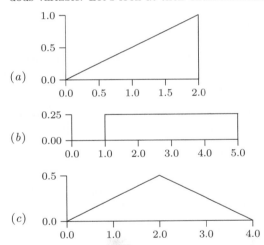

Each graph gives the values of a density function f over an interval of the number line with the convention that the values of f are zero elsewhere. These graphs are the analogues of the histogram probability density functions that were used for discrete random variables. As with the earlier graphs, the total area under the curves and above the x-axis represents the sum of the probabilities of all possible events in the sample space and is, therefore, 1.

In (a) the area of the triangle is $\frac{1}{2}bh = \frac{1}{2} \times 1.0 \times 2.0 = 1.0$; in (b) the area of rectangle is $bh = 0.25 \times 4.0 = 1.0$; in (c) the area is again the area of a triangle, $\frac{1}{2}bh = \frac{1}{2} \times 4.0 \times 0.5 = 1.0$. Probabilities of events can, therefore, be determined as areas under these curves, as we will see below. For complex curves the calculation of areas (probabilities) would require integration using calculus. For the examples above, probabilities can be determined using plane geometry.

For density (a), find $P(X < 1.0)$ and $P(X \leq 1.0)$.

SOLUTION. See the figure on the left below.

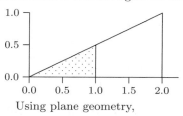 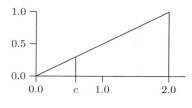

Using plane geometry,

$$P(X < 1.0) = \text{area of shaded triangle} = \tfrac{1}{2} \times 0.5 \times 1.0 = 0.25.$$

Note, too, that $P(X \leq 1.0) =$ area of shaded triangle again $= 0.25$. With continuous variables $P(X < c) = P(X \leq c)$ because the probability that X exactly equals any particular value is zero. Another way to think about it is that $P(X = c) = 0$ because it represents the area of the line segment under $f(c)$. (See the figure on the right above.) *This is a fundamental difference between continuous and discrete variables.* Remember in the binomial or Poisson distributions $P(X \leq c) \neq P(X < c)$ because $P(X = c)$ was a probability greater than zero.

For density (b), assume the graph represents the density for body weights (in lb) of cane toads, Bufo marinus. Find $P(X < -1.0)$ and $P(X = 2.0)$.

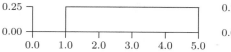 

SOLUTION. Since -1.0 is left of the graph (on the left above), by definition the probability is zero. A toad that weighs less than -1.0 lb is clearly an impossible event and therefore has, by definition, a probability of zero.

$P(X = 2.0)$ is the probablity of collecting a toad that is exactly 2 lb, not the smallest bit more or less than 2 lb. You can think of it as $P(X = 2.0000000\ldots)$. This probability is represented by the line (in the figure on the right above) and the area of that line is $P(X = 2.0) = 0$.

Find $P(X > 2.0)$ and $P(2.0 \leq X \leq 4.0)$.

SOLUTION. $P(X > 2.0)$ is the area of the shaded rectangle in the figure on the left below. The probability that a randomly collected cane toad is more than 2 lb in weight is

$$bh = 3.0 \times 0.25 = 0.75.$$

In other words, 75% of the cane toads weigh more than 2.0 lb.

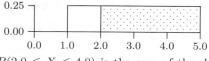 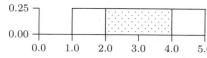

$P(2.0 \leq X \leq 4.0)$ is the area of the shaded rectangle of the figure on the right above, so

$$bh = 2.0 \times 0.25 = 0.50.$$

So there is a 50% probability that a cane toad weighs between 2.0 lb and 4.0 lb, inclusive.

For density (c), assume the graph represents the time in minutes for the onset of anaphylactic shock following bee stings in susceptible children. What is the probability that shock will occur in less than 2.0 minutes? In less than or equal to 1.0 minute?

SOLUTION. $P(X < 2.0)$ is the area of the shaded triangle in the graph on the left below, so

$$P(X < 2.0) = \tfrac{1}{2}bh = \tfrac{1}{2} \times 2.0 \times 0.5 = 0.50.$$

Or by symmetry we see that half of the graph is shaded.

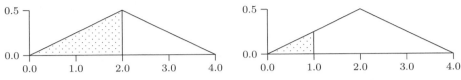

$P(X \leq 1.0)$ is the shaded area in the preceding figure on the right.

$$P(X \leq 1.0) = \tfrac{1}{2}bh = \tfrac{1}{2} \times 1.0 \times 0.25 = 0.125.$$

What is the probability that shock will occur in more than 1 minute and less than 3 minutes?

SOLUTION. We must find $P(1 < X < 3)$. See area III in the figure below.

From the previous question, area I equals is 0.125 and because of symmetry area II is also 0.125. So

$$P(1 < X < 3) = 1 - (\text{areas I and II}) = 1 - (0.125 + 0.125) = 0.75.$$

Seventy-five percent of the susceptible children experience anaphylactic shock in between one and three minutes after being stung.

Note for all three densities above we used areas of the graph to determine probabilities. These areas can be described in terms of the cumulative distribution function F, which gives the probability that a random variable will assume a value less than or equal to the value specified. For density (c), $P(X < 2.0) = F(2.0)$ and $P(X \leq 1.0) = F(1.0)$, so

$$P(1.0 < X < 3.0) = F(3.0) - F(1.0).$$

Again note $P(X < 1.0) = P(X \leq 1.0) = F(1.0)$ because $f(1) = 0$, the probability of exactly 1.0 is zero.

3.5 The Normal Distribution

The vast majority of continuous variable distributions in biology take shapes other than triangles and rectangles and are, therefore, more difficult to manipulate. Fortunately a great many of the continuous variables of interest in biology form a bell-shaped curve or can be transformed to form such a curve. See Figure 3.9.

The characteristics of this family of curves were developed by Abraham de Moivre (1667–1754), Pierre Simon, Marquis de Laplace (1749–1827), and Karl Friedrich Gauss (1777–1855). In fact, this distribution is sometimes called the **Gaussian distribution,** although the adjective "**normal,**" first coined by Sir Francis Galton in 1877, is more routinely used.

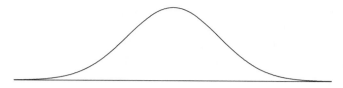

FIGURE 3.9. A typical bell-shaped curve.

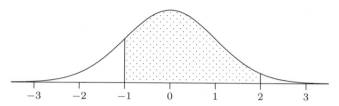

FIGURE 3.10. $P(-1 \leq X \leq 2)$ for a continuous density function.

A continuous density function definition (Definition 3.8) is illustrated in Figure 3.10. The area under the entire curve from $-\infty$ to $+\infty$ is equal to 1. The area of the shaded region represents $P(-1 \leq X \leq 2)$. If you have taken calculus, you may recognize that this area is given by the definite integral $\int_{-1}^{2} f(x)\, dx$.

As with discrete random variables, the cumulative distribution function or CDF for a continuous random variable whose density is f is denoted by F, so that $F(a) = P(X \leq a)$. Geometrically, this probability is represented by the area under the density function f at and to the left of the vertical line $x = a$. (This is the "left tail" of the distribution.) For example, in Figure 3.10 $F(-1) = P(X \leq -1)$ is the area of the *unshaded* region starting at the vertical line $x = -1$ and going left. In general, finding such areas requires integral calculus. In fact, to find the values of the cumulative distribution generally requires the use of improper integrals over infinite intervals. These take the form $\int_{-\infty}^{a} f(x)\, dx$. Fortunately, the distributions that we will be working with have been carefully tabulated and their CDFs are readily available.

The area to the right of a vertical line $x = b$ is the "right tail" of the distribution. It represents $P(X > b)$, which is the same as $1 - P(X \leq b)$ or $1 - F(b)$. In Figure 3.10 the small unshaded region to the right of the vertical line at $x = 2$ is $1 - F(2)$.

One consequence of the definition of a continuous density function is that $P(X = a)$ is 0 for every value of a. This follows from the fact that the *area* of the vertical line segment that represents $P(X = a)$ is 0. In calculus terms,

$$P(X = a) = \int_{a}^{a} f(x)\, dx = 0$$

because both endpoints of the interval are the same.

The most important family of continuous random variables are the normal random variables. Like the binomial distributions, normal distributions depend on two parameters.

DEFINITION 3.9. The **probability density function for a normal random variable** has the form

$$f(x) = \frac{1}{\sigma\sqrt{2\pi}} e^{-(x-\mu)^2/2\sigma^2},$$

where σ is the standard deviation of the random variable and μ is its mean.

The density formula certainly is not simple! And there's more bad news. To find the cumulative distribution, f needs to be integrated. But no closed-form antiderivative exists for f. That means one can't use the Fundamental Theorem of Calculus to evaluate the required integrals. So the tables for the normal distribution were created using some clever approximations to the actual areas under the curve.

Figure 3.11 gives the graph of the normal density function with mean μ and standard deviation σ. There are several features to note.

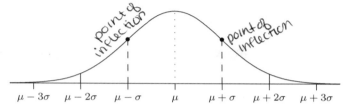

FIGURE 3.11. The normal density function with mean μ and standard deviation σ.

The graph of a normal density function f is symmetric about its mean $x = \mu$ (the dotted vertical line). This is a consequence of the $(x - \mu)^2$ term in the definition of f. For this reason μ is called a **location** parameter because it indicates where the graph is centered or positioned. See Figure 3.12.

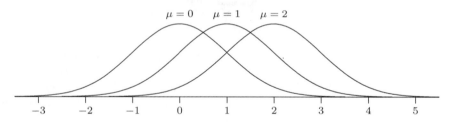

FIGURE 3.12. The shift of the graph of the normal density function for various values of μ, the location parameter.

The points of inflection (where the curve changes from bending up to bending down) on the curve f occur at $\mu \pm \sigma$. This can be verified by taking the second derivative of f. Thus σ determines the shape of f. Smaller values of σ lead to a "narrow" steeper curve while larger values of σ spread the curve out and make it "flatter." See Figure 3.13.

Approximately two-thirds (68%) of the area under the curve lies between $\mu - \sigma$ and $\mu + \sigma$, that is, within one standard deviation of the mean μ. This is the area between the dashed vertical lines in Figure 3.11. Approximately 95% of the area under the curve lies within two standard deviations of μ, that is, between $\mu - 2\sigma$ and $\mu + 2\sigma$. (This is the area between the solid vertical lines in Figure 3.11.) Approximately 99% of the area under the curve lies between $\mu - 3\sigma$ and $\mu + 3\sigma$, that is, within three standard deviations of the mean μ.

As mentioned earlier, the normal distribution is really an infinite family of distributions. Just as the binomial distribution was defined by n and p and the Poisson

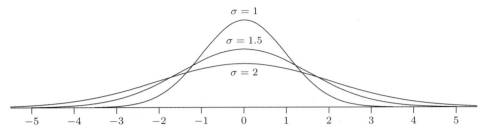

FIGURE 3.13. The effect on the graph of the normal density function of various values of σ, the "spread" of the distribution.

distribution by a particular μ, the normal distribution is defined by μ and σ, the only parameters required for the normal density function (Definition 3.9).

Fortunately, this family of curves can be collapsed into one curve called the **standard normal** curve whose mean is 0 and whose standard deviation is 1. Areas under this curve have been tabulated and are usually called **Z tables**. They can be used to determine the CDF of any normal distribution. In the following section we develop and apply the standard normal curve.

3.6   The Standard Normal Distribution

DEFINITION 3.10. Let X be a normal random variable with mean μ, and standard deviation σ. The transformation

$$Z = \frac{X - \mu}{\sigma}$$

expresses X as the **standard normal random variable** Z with $\mu = 0$ and $\sigma = 1$.

This process transforms or expresses X as a number of standard deviations above or below the mean $\mu = 0$ and is called **normalizing** or **standardizing** X. Using expected values we can show that the mean of Z is $\mu_Z = 0$ and the standard deviation of Z is $\sigma_Z = 1$. For the mean, using Theorem 3.1, which holds for continuous random variables, $\mu_Z = E(Z)$. So by Definition 3.10

$$\mu_Z = E\left[\frac{X - \mu_X}{\sigma_X}\right] = \frac{1}{\sigma_X}E[X - \mu_X] = \frac{1}{\sigma_X}\left(E[X] - E[\mu_X]\right) = \frac{1}{\sigma_X}(\mu_X - \mu_X) = 0.$$

For the variance, using the fact that $\mu_Z = 0$, Theorem 3.1, and Definition 3.4,

$$\sigma_Z^2 = E[(Z - \mu_Z)^2] = E[Z^2] = E\left[\left(\frac{X - \mu_X}{\sigma_X}\right)^2\right] = \frac{1}{\sigma_X^2}E[(X - \mu_X)^2] = \frac{1}{\sigma_X^2}\cdot\sigma_X^2 = 1.$$

Therefore, the standard deviation is $\sigma_Z = 1$.

Since any normal random variable can be standardized, the cumulative distribution function of the standard normal can be used to find probabilities (areas under the normal curve). See Table C.3 in the Appendices.

EXAMPLE 3.16. Suppose that the scores on an aptitude test are normally distributed with a mean of 100 and a standard deviation of 10. (Some of the original IQ tests were purported to have these parameters.) What is the probability that a randomly selected score is below 90?

SOLUTION. We must find $P(X < 90) = F(90)$. The scores form a distribution represented in the left panel of Figure 3.14. Without some very sophisticated mathematics we cannot calculate $F(90)$.

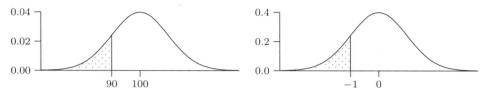

FIGURE 3.14. On the left, $\mu = 100$, $\sigma = 10$, and $X < 90$. On the right, $\mu = 0$, $\sigma = 1$, and $Z < -1.0$. Note the different scales.

Transform X into a standard normal variable. $\mu = 100$ and $\sigma = 10$, so

$$Z = \frac{X - \mu}{\sigma} = \frac{90 - 100}{10} = -1.0.$$

Thus a score of 90 can be represented as 1 standard deviation below the mean,

$$P(X < 90) = P(Z < -1.0).$$

See the right panel in Figure 3.14.

Table C.3 catalogues the CDF for the standard normal distribution from $Z = -3.99$ to $Z = 3.99$ in increments of 0.01. The probabilities are found by looking up the Z value to one decimal along the left margin of the table and then finding the second decimal across the top of the table. Where the appropriate row and column intersect in the body of the table lies the probability for up to and including that particular Z value.

For $P(X < 90) = P(Z < -1.0)$ from Table C.3 we use the row marked -1.0 and the column marked 0.00. The probability from this intersection is 0.1587, so the probability of a score less than 90 is 0.1587.

What is the probability of a score between 90 and 115?

SOLUTION. We wish to find the shaded area of Figure 3.15 on the left.

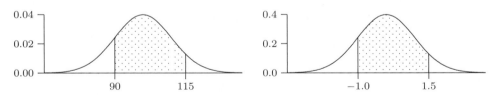

FIGURE 3.15. On the left, $\mu = 100$, $\sigma = 10$, and $90 < X < 115$. On the right, $\mu = 0$, $\sigma = 1$, and $-1.0 < Z < 1.5$.

$$P(90 < X < 115) = P\left(\frac{90 - 100}{10} < Z < \frac{115 - 100}{10}\right)$$
$$= P(-1.0 < Z < 1.5)$$
$$= F(1.5) - F(-1.0).$$

$F(1.5)$ from Table C.3 is 0.9332 and $F(-1.0)$ given above is 0.1587. So $P(90 < X < 115) = 0.9332 - 0.1587 = 0.7745$. The probability of a score between 90 and 115 is 0.7745 or 77.45% of the scores are between those two values.

What is the probability of a score of 125 or higher?

SOLUTION. We want $P(X \geq 125)$. See Figure 3.16.

$$P(X \geq 125) = 1 - P(X < 125) = 1 - P\left(Z < \frac{125 - 100}{10}\right) = 1 - F(2.5).$$

From Table C.3 $F(2.5) = 0.9938$. Remember that Table C.3 is cumulative from the left, so
$$P(Z \geq 2.5) = 1 - F(2.5) = 1 - 0.9938 = 0.0062.$$

Only 0.62% of the scores will be 125 or higher.

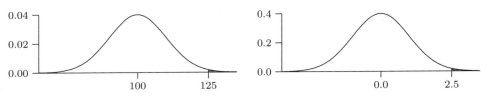

FIGURE 3.16. On the left, $\mu = 100$, $\sigma = 10$, and $X \geq 125$. On the right, $\mu = 0$, $\sigma = 1$, and $Z \geq 2.5$.

EXAMPLE 3.17. Suppose that diastolic blood pressure X in hypertensive women centers about 100 mmHg and has a standard deviation of 16 mmHg and is normally distributed. Find $P(X < 90)$, $P(X > 124)$, $P(96 < X < 104)$. Then find x so that $P(X \leq x) = 0.95$.

SOLUTION. Simply transform the values of the variable X to the standard normal by using $Z = \frac{X - \mu}{\sigma} = \frac{X - 100}{16}$. In particular, when $X = 90$,

$$Z = \frac{90 - 100}{16} = -\frac{10}{16} = -0.625.$$

So using Table C.3

$$P(X < 90) = P(Z < -0.625) = F(-0.625) \approx 0.2660.$$

Since Table C.3 is constructed to give probabilities for Z values to the nearest hundredth, the value -0.625 cannot be evaluated precisely using the table. In this case you can interpolate between the probabilities for the two bracketing Z values (-0.62 and -0.63) as we have done or you may choose the table value that gives you the more conservative answer.

When $X = 124$,
$$Z = \frac{124 - 100}{16} = \frac{24}{16} = 1.5.$$

So,

$$P(X > 124) = P(Z > 1.5) = 1 - F(1.5) = 1 - 0.9332 = 0.0668.$$

To find $P(96 < X < 104)$, perform both transformations simultaneously,

$$P(96 < X < 104) = P\left(\frac{96 - 100}{16} < Z < \frac{104 - 100}{16}\right)$$
$$= P(-0.25 < Z < 0.25)$$
$$= F(0.25) - F(-0.25)$$
$$= 0.5987 - 0.4013$$
$$= 0.1974.$$

Finally, to find x so that $P(X \leq x) = 0.95$, we simply work backward. First we find the value of Z so that $P(Z \leq z) = 0.95$. From Table C.3, $z = 1.645$. Now use the transformation equation to solve for x.

$$z = 1.645 = \frac{x - 100}{16}$$

or

$$x = 100 + 1.645(16) = 126.32.$$

This means that approximately 95% of these hypertensive women will have diastolic blood pressures less than 126.32 mmHg.

Normal Approximation to the Binomial Distribution

An additional property of the normal distributions is that they provide good approximations for certain binomial distributions. Let X be a binomial random variable with parameters n and p. For large values of n, X is approximately normal with mean $\mu = np$ and variance $\sigma^2 = np(1-p)$.

A simple rule of thumb is that this approximation is acceptable for values n and p such that $np > 5$ and $n(1-p) > 5$. For example, if $p = 0.5$, then n should be at least 11 before the approximation may be used, since $11(0.5) > 5$ and $11(1-0.5) > 5$. For values of p other than 0.5, the sample size must be larger before the normal approximation is acceptable. For example, if $p = 0.1$, then n should be at least 51. Remember that when p is very small, the Poisson distribution may also be used to approximate the binomial.

EXAMPLE 3.18. Return to Example 3.9 with a genetic form of muscular dystrophy in mice. This example was solved using binomial probabilities with $p = 0.25$ and $n = 20$. Suppose a larger sample of progeny are generated and we would like to know the probability of fewer than 15 with muscular dystrophy in a sample of 60.

SOLUTION. Use the normal approximation to the binomial here with $p = 0.25$ and $n = 60$, so $np = \mu = 60(0.25) = 15$ and $n(1-p) = 60(0.75) = 45$. So the normal approximation will be acceptable here. Since $\sigma^2 = np(1-p) = 60(0.25)(0.75) = 11.25$, then $\sigma = \sqrt{11.25} = 3.35$. Let F_B and F_N denote the binomial and normal cumulative distribution functions. If we were to do the calculation using the binomial density function directly, we would find that the probability of less than 15 of 60 mice with muscular dystrophy is $P(X < 15) = F_B(14) = 0.4506$. This exact calculation requires summing 15 terms,

$$\sum_{k=0}^{14} \binom{60}{k} (0.25)^k (0.75)^{60-k}.$$

On the other hand, the normal approximation can be done much more quickly as

$$P(X < 15) = F_B(14) \approx F_N\left(\frac{14.5 - \mu}{\sigma}\right) = F_N\left(\frac{14.5 - 15}{3.35}\right)$$

$$= F_N(-0.15) = 0.4404$$

using Table C.3. So with relatively little effort, the actual CDF for a binomial with $n = 60$ and $p = 0.25$ is reasonably well-approximated by the normal distribution using $\mu = 15$ and $\sigma = 3.35$. But why was 14.5 instead of 14 used in this approximation? This 0.5 correction term was used because we were estimating a discrete distribution using a continuous one. For this reason, it is called a **continuity correction**.

To see why this correction is necessary, consider the pdf for a binomial with $n = 20$ and $p = 0.5$. The probabilities for 0 through 20 successes in 20 trials are closely approximated by the normal distribution in Figure 3.17.

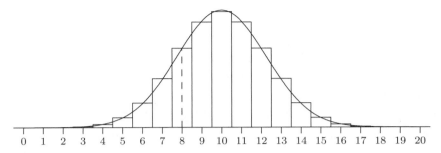

FIGURE 3.17. The binomial distribution with $n = 20$ and $p = 0.5$ is closely approximated by a normal distribution with $\mu = np = 10$ and $\sigma = \sqrt{np(1-p)} = \sqrt{5}$.

The height of each bar of the binomial density function (e.g., see the dashed line in Figure 3.17) is the probability of the corresponding discrete outcome. But the width of each bar is 1, so the *area* of the bar also represents the corresponding probability of the same discrete outcome. (This area property is the reason we used histograms rather than bar graphs to illustrate the binomial and Poisson distributions.)

$F_B(8)$ is the sum of the bars representing 0 through 8 successes in 20 trials. In Figure 3.17, the bar representing 8 successes starts at 7.5 and goes to 8.5 on the x-axis. The normal approximation of $F(8)$ would be more accurate if all of the "8"-bar were included in the area under the curve, so we use 8.5 as the cutoff point to approximate $F_B(8)$ rather than 8.0.

To see the effect of this 0.5 correction term, check that $F_B(8) = 0.2517$ from Table C.1. For the normal approximation, use $\mu = np = 20(0.5) = 10$ and $\sigma^2 = 20(0.5)(0.5) = 5$. So $\sigma = 2.24$. Then from Table C.3,

$$\text{Using } 8: \qquad F\left(\frac{8 - 10}{2.24}\right) = F(-0.89) = 0.1867,$$

which is not a very close approximation!

$$\text{Using } 8.5: \qquad F\left(\frac{8.5 - 10}{2.24}\right) = F(-0.67) = 0.2514,$$

which is a very good approximation!

This correction for continuity improves the accuracy of the approximation, particularly when evaluating CDFs for values near the mean because there the bars (and half-bars) are larger. In the tails of the distribution where the bars are small, the correction has less effect.

FORMULA 3.1. Let X be a binomial random variable with parameters n and p. For large values of n, X is approximately normal with mean $\mu = np$ and variance $\sigma^2 = np(1-p)$. The approximation is acceptable for values n and p such that $np > 5$ and $n(1-p) > 5$. In this

situation,

$$F_B(X) \approx F_N \left(\frac{X + 0.5 - np}{\sqrt{np(1-p)}} \right),$$

where F_B and F_N are the cumulative binomial and normal distributions, respectively.

3.7   Problems

1. Some games use tetrahedral dice that have four triangular faces that are numbered 1 through 4. Assume each face is equally likely to land face down.

(a) Determine the probability density function f for the total number (X) on the down faces of 2 such tetrahedron-shaped dice thrown together.

(b) Calculate the cumulative distribution function of X.

(c) Using expected values, find the mean and variance of X.

(d) Find $f(7)$.

(e) Find $F(7)$.

(f) Describe in complete sentences the meaning of $f(7)$ and $F(7)$.

(g) Find $P(3 < X \leq 7)$.

(h) Find $P(3 \leq X \leq 7)$.

(i) Find $P(X > 6)$.

2. In a certain population of koala, *Phascolarctos cinereus*, the heights of individuals are distributed as indicated by the density curve shown below. Areas under the curve are shown in the figure. Let X represent the length of an individual koala chosen at random from the population. Find the following.

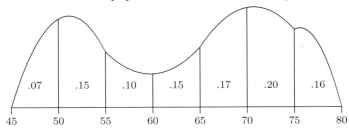

(a) $P(X < 55)$

(b) $P(55 < X \leq 65)$

(c) $P(X = 55)$

(d) $F(65)$

(e) $P(X > 65)$

(f) Why do you think the density is bimodal for this population?

✓ **3.** Calculate the density and cumulative distribution functions for a binomial random variable with parameters $n = 6$ and $p = 0.2$. Compare your answers to those in Table C.1.

4. A population of marine gastropods has shell lengths that are normally distributed with mean $\mu = 7$ mm and variance $\sigma^2 = 2.25$ mm^2. What proportion of the population will have a shell length between 5.5 mm and 8.5 mm?

5. Cystic fibrosis is a rare but severe genetic disorder. It is caused by an autosomal recessive allele. Homozygotes for this allele (cc) have difficulty breathing due to a defect in sodium transport. Heterozygotes (Cc) manifest the same physiology as homozygous (CC) normal individuals. A second gene independently assorting from the C locus exhibits autosomal dominant expression in the form of dramatically shortened fingers, a condition known as brachydactyly. B_ individuals are brachydactylous and bb individuals are normal. In a marriage between the following two individuals

optional

$$\text{Ccbb} \times \text{CcBb},$$

find the probability that

(*a*) the first child has cystic fibrosis;

(*b*) the first child has cystic fibrosis and brachydactyly;

(*c*) the first three children are normal;

(*d*) the first child is a carrier of the cystic fibrosis gene given that the child appears normal;

(*e*) in a family of four children exactly two will have normal respiration and two will have cystic fibrosis.

6. In an area of heavy infestation, gypsy moths lay egg clusters on trees with a mean number of clusters per tree equal to 3. Calculate the probability that a randomly chosen tree has

(*a*) no egg clusters;

(*b*) at least 1 egg cluster.

7. In the plant *Capsella bursa-pastoris* (Shepard's purse) consider an epistatic genetic system consisting of two independently assorting gene loci (A and B) with each locus having two alleles. A plant with either a capital A or capital B allele has triangular seed capsules and only a completely homozygous recessive (aabb) individual will have ovoid seed capsules. In an AaBb × AaBb cross, we have the following genotypic array:

$$\left(\tfrac{1}{4}\text{AA} + \tfrac{2}{4}\text{Aa} + \tfrac{1}{4}\text{aa}\right)\left(\tfrac{1}{4}\text{BB} + \tfrac{2}{4}\text{Bb} + \tfrac{1}{4}\text{bb}\right).$$

(*a*) What is the probability of an offspring with triangular seed capsules? Ovoid seed capsules?

(*b*) What is the probability of an offspring with the AaBb genotype?

(*c*) If the offspring has triangular seed capsules, what is the probability that it is AaBb?

(*d*) If two offspring with triangular seed capsules from the above cross are themselves crossed, what is the probability they will both be AaBb and produce an offspring with ovoid seed capsules?

(*e*) From the original cross, what is the probability that 2 of 6 progeny will have ovoid seed capsules?

(*f*) From the original cross, what is the probability of at least 1 progeny with ovoid seed capsules in a sample of 7 offspring?

8. Suppose that you have read that the leopard frog, *Rana pipiens*, has a sex ratio in most populations of 60% females and 40% males. If this is true, what is the probability that

(*a*) in a random sample of 20 individuals collected from Odell's Pond fewer than 10 will be female?

(*b*) in a random sample of 13 exactly 8 will be female?

9. In a large population, 16% of the people are left-handed. In a random sample of 10 find

(*a*) the probability that exactly 2 will be left-handed;

(*b*) the probability that fewer than 2 will be left-handed.

10. In a study of the effectiveness of an insecticide against gypsy moths, *Lymantria dispar*, a large area of land was sprayed. Later the area was examined for live adult insects by randomly selecting and surveying 10-m × 10-m squares. Past experience has shown the average number of live adult insects per square after spraying to be 3. If these insects are Poissonally distributed, find

(*a*) the probability that a square will contain exactly 3 live adult insects;

(*b*) $f(3)$;

(*c*) the probability that a square will have more than 3 live adult insects.

11. If the capacities of the cranial cavity of a certain population of humans are approximately normally distributed with a mean of 1400 cc and a standard deviation of 125 cc, find the probability that a person randomly picked from this population will have a cranial cavity capacity

(*a*) greater than 1450 cc;

(*b*) less than 1350 cc;

(*c*) between 1300 cc and 1500 cc.

12. The "Summary Data for the April 2000 MCAT" (Medical College Admission Test) published by the Association of American Medical Colleges indicated that the mean score on the biological sciences section was 8.4 (the scale is 1 to 15) with a standard deviation of 2.4. Assume that MCAT scores are approximately normally distributed.

(*a*) Because individual scores are given only as whole numbers, a score of 9 must be considered 8.5 to 9.5 on a continuous scale. What percent of the applicants would you expect to score 9 on the biological sciences section?

(*b*) If a highly selective medical school looks at applicants only in the 90th percentile or above, what minimum score would an applicant need on this section to be considered by this school?

13. A commercial plant breeder knows from experience that when she crosses two varieties of tomato, *Lycopersicom esculentum*, about 10% of the resulting offspring will be resistant to pith necrosis caused by the bacterium *Pseudomonas corrugata* and about 20% will be resistant to early blight caused by the fungus *Alternaria solani*.

(*a*) If she wishes to breed plants that are resistant to both pith necrosis and early blight and the resistance factors are independent of one another, what proportion of the offspring will be of use to her?

(*b*) What is the probability that in 100 such offspring none will be worthy of further study?

(*c*) What is the average number of useful offspring per 100.

14. (*a*) Architects often work with physical anthropologists to design work areas that comfortably fit the largest number of people. In the United States a standard door frame has an 80-inch clearance. Assuming that male American's heights are normally distributed with a mean of 70 inches and a standard deviation of 4 inches, what percentage of the population is at risk of hitting the door frame?

(*b*) If a pneumatic door closer is placed within the door frame that lowers the clearance 3 inches, what percentage of the male population is now at risk of not clearing the door opening?

15. In a certain population of the herring, *Pomolobus aestivalis*, the lengths of the individual fish are normally distributed. The mean length of a fish is 54 mm and the standard deviation is 4 mm.

(*a*) What is the probability that the first fish captured is less than 62 mm long?

(*b*) What percentage of the fish are longer than 59 mm?

(*c*) If the first 2 fish that you caught were less than 44 mm long, would you be suspicious of the claim of a mean of 54 mm? Explain with a probability argument.

16. An instructor is administering a final examination. She tells her class that she will give an A grade to the 10% of the students who earn the highest marks. Past experience with the same examination has yielded grades that are normally distributed with a mean of 70 and a standard deviation of 10. If the present class runs true to form, what numerical score would a student need to earn an A grade?

17. The time for symptoms of fire blight (caused by *Erwinia amylovora*) to develop in apple seedlings maintained at a temperature of 25°C is normally distributed with a mean of 7 days and a variance of 4 days^2. At the same temperature the development time for crown gall disease (caused by *Agrobacterium tumifaciens*) is normally distributed with a mean of 12 days and variance of 4 days^2. If a seedling inoculated with both bacteria is kept at 25°C for 9 days, what is the probability that

(*a*) fire blight will be evident;

(*b*) crown gall disease will be evident;

(*c*) both fire blight and crown gall disease will be evident, assuming that there is no interaction between the two species of fungi?

18. In a study of the *Protoreaster linckii* sea star it was found that the population had two color morphs—red and green. Researchers at a particular site collected a random sample of 67 *P. linckii* sea stars. Of these 12 were red. Assuming that 25% of the population were red color morphs, find the probability that

    (*a*) a random sample of 67 will have no more than 12 red morphs;

    (*b*) a random sample of 67 will have at least 12 red color morphs. (Use a normal approximation of the binomial.)

19. Suppose a variable X is normally distributed with a standard deviation of 10. Given that 30.85% of the values of X are greater than 70, what is the mean of X?

20. Suppose that 60% of a population has been found to have blood type O.

    (*a*) In a randomly chosen sample of 100 people from this population, how many are expected to have type O blood?

    (*b*) Using the normal approximation calculate the probability that between 65 and 70 inclusive are blood type O in a random sample of 100.

21. Given a binomial variable with a mean of 20 and a variance of 16, find n and p.

22. Given that $\mu = 25$ and $\sigma = 10$, evaluate the following two expressions.

    (*a*) $E\left(\frac{X-\mu}{\sigma}\right)$

    (*b*) $\text{Var}(4X - 3)$

23. Among American women aged 18–24 years, 10% are less than 61.2 inches tall, 80% are between 61.2 and 67.4 inches tall, and 10% are more than 67.4 inches tall. Assuming that the distribution of heights can be adequately approximated by the normal curve, find the mean and standard deviation of the distribution.

24. Morton's foot is characterized by the second toe (metatarsal) being longer (at least 0.3 mm) than the big toe. The problem is that the bone behind the big toe (first metatarsal) is too short. This inherited trait occurs in about 30% of the population and can lead to bursitis or an irritation of the tendons attached to the second toe. Suppose you examine the feet of 19 classmates. Find the probabilities that:

    (*a*) None of them have Morton's foot.

    (*b*) At least 7 have Morton's foot.

    (*c*) Between 4 and 7 have Morton's foot (inclusive).

    (*d*) At most 12 clasmates have Morton's foot.

    (*e*) Suppose that at a family picnic at the beach you notice that 11 of 19 people have this condition. Should this be considered unusual?

25. It is well known that taste in humans includes sweet, sour, salty, and bitter. A lesser known fifth taste called umami was identified almost 100 years ago in Japanese seaweed broth. Umami is not easy to describe but is often characterized as full bodied, savory, or rich. Umami is caused by the free form of the amino acid glutamine. Scientists have identified both the specific receptors on the tongue for the umami taste and the particular nerves that transmit the signal to the brain and, therefore, have accepted umami as a legitimate taste. Not all people can identify the umami taste and most who cannot confuse it with saltiness.

Suppose a population consists of 80% umami tasters. Find the probabilities of the following events.

(*a*) A random sample of 17 people has exactly 14 tasters.

(*b*) A random sample of 17 people has 14 or more tasters.

(*c*) A random sample of 17 people has less than 14 tasters.

(*d*) A random sample of 20 people has at least 16 tasters.

(*e*) A random sample of 25 people has 20 tasters.

26. LDL-cholesterol levels in men in industrialized societies tend to be normally distributed with a mean of 150 mg/dl and a standard deviation of 25 mg/dl. Individuals with Familial Hypercholesterolemia (FH), a dominant autosomal trait, have LDL-cholesterol levels that are normally distributed with a mean of 350 mg/dl and a standard deviation of 50 mg/dl. A patient is examined and found to have an LDL-cholesterol level of 225 mg/dl. What is the probability of a level at least this high given a normal genotype? What is the probability of a level no higher than this given the patient has FH? Do you think the patient has FH? Discuss. (Consider FH is a rare disease.)

27. The 17-year cicada, *Magicicada septendecim*, is a very unusual Homopteran insect that spends around 17 years as a subterranean nymph. It then emerges (most recently in 2004) as a winged, breeding adult and mates. The females then lay their eggs on twigs of deciduous trees. The development time is not uniform for the members of any brood, but, in fact, has the following probability density function.

x (years)	14	15	16	17	18
$f(x)$	.05	.05	.20	.50	.20

(*a*) Construct the cumulative distribution function, $F(x)$, for $f(x)$.

(*b*) Find $F(17)$.

(*c*) Find $P(X < 17)$.

(*d*) Find the expected value for X, $E(X)$.

(*e*) Find the mean development time.

(*f*) Find the variance of development time. Hint: $E(X^2) = 281.55$.

28. In New York State turbidity levels (a measure of the cloudiness of water) are measured in nephelometric turbidity units (NTU) using a nephelometer to estimate the light scattering from suspended particulate material. For high quality water, regulations require that 95% of the turbidity samples collected have measurements below 0.5 NTU. If for a particular water source the measurements are normally distributed with a mean of 0.4 NTU and a standard deviation of 0.15 NTU, does this water source meet the turbidity criterion for high quality water?

29. The American coot, *Fulica americana*, lays clutches of eggs that depend on habitat quality and population density. In a large wetland the distribution of clutch sizes for this species is characterized by the following probability density function:

x (# of eggs)	7	8	9	10	11	12
$f(x)$	0.10	0.10	0.15	0.35	0.25	0.05

(a) Construct the cumulative density function, $F(x)$, for the pdf above.

(b) Using the expected value approach, find the mean and variance for the clutch size in this population.

(c) Find $F(10)$.

(d) Find $P(X = 12)$.

(e) Find $P(X > 9)$.

(f) Find $P(X = 12 \,|\, X > 9)$.

30. Amazingly, the West Nile virus has been found in more than 200 species of wild vertebrates in North America. Monitoring the spread of this pathogen is difficult but essential. Suppose 10% of a population of American crows in eastern Oklahoma are infected with the virus. How many crows would have to be sampled (analyzed) to be at least 50% certain of finding one infected individual? How many to be 90% certain?

31. If the infected and non-infected birds in the preceding problem are binomially distributed, find the following:

(a) In a sample of 15 less than 3 will be infected.

(b) In a sample of 15 exactly 3 will be infected.

(c) In a sample of 15 more than 3 will be infected.

(d) If you found 8 in a sample of 15 to be infected, would you continue to believe that infected birds follow a binomial distribution with $p = 0.10$? Explain.

32. Environmental agencies around the world regulate dust pollutants on the basis of mass, not chemistry, and most governments focus on the particles easiest to catch and quantify; those that are 10 micrometers across (the PM-10 fraction). Federal law prohibits PM-10 concentrations in air from exceeding 150 micrograms per cubic meter (μg/m^3) over any 24-hour period. If the managers of an industrial park know that they are creating PM-10 concentrations that average 120 μg/m^3 with a standard deviation of 15 μg/m^3, assuming the PM-10 concentrations are normally distributed, find the following probabilities:

(a) That a day's PM-10 concentration is below 140 μg/m^3.

(b) That a day's PM-10 concentration is between 100 and 140 μg/m^3.

(c) That a day's PM-10 concentration is above the federal acceptable limit of 150 μg/m^3.

(d) On average how many days a year would the industrial park be expected to be in violation of the federal law?

33. A common childhood language disorder called specific language impairment (SLI) stems from a brain-based difficulty in discerning the acoustic building blocks of spoken words, especially in noisy settings such as classrooms. Researchers estimate that 10% of U.S. elementary school students experience substantial problems understanding what others say and in speaking comprehensibly, despite good health, normal hearing, and average or above intelligence. As a district school psychologist you decide to test elementary school students for SLI. In a class of 20:

(*a*) How many would you expect to have SLI?

(*b*) What is the probability of none with SLI?

(*c*) What is the probability of at least 2 with SLI?

(*d*) If you found 5 or more with SLI, would you feel that there is a problem with that classroom environment?

(*e*) If the probability of SLI is really 0.07, find the probability of 2 in 20 with this language disorder.

34. Blood pressure is an important health indicator in humans. Blood pressure results from two forces. One, the systolic pressure, is created by the heart as it pumps blood into the arteries and through the circulatory system. The other, the diastolic pressure, is the force of the arteries as they resist the blood flow. In a particular population the systolic pressures are normally distributed with a mean of 110 mmHg and a standard deviation of 8 mmHg.

(*a*) Systolic pressure less than 120 mmHg are considered normal. What percentage of the population has normal systolic blood pressures?

(*b*) Systolic pressures above 140 mmHg are considered hypertensive and in need of medical intervention. What portion of the population will have these values?

(*c*) Systolic pressures of 120 to 139 mmHg are considered "prehypertensive." What percentage of this population is prehypertensive?

35. Two persons are on a reducing diet. The first person weighs 178 lb and belongs to an age and height group for which the mean weight is 144 lb with a standard deviation of 12 lb. The second person weighs 194 lb and belongs to an age and height group for which the mean weight is 160 lb with a standard deviation of 16 lb. Assuming weight to be normally distributed within each age and height group, determine which person is most overweight. Explain your rationale.

36. In Kentucky if you shoot a deer that is in the 95% percentile in weight you are eligible for membership in the mythical Davy Crockett Big Buck Club. If male white tails in Kentucky have weights that are normally distributed with a mean of 81.3 kg and a standard deviation of 6.7 kg, how big a deer must you kill to be able to apply for membership in this prestigious organization?

37. White-tailed does upon reaching sexual maturity at age two give birth to between one and four offspring every year. Below is the probability density function for live births in mature white-tailed does in optimal habitat.

x	0	1	2	3	4
$f(x)$	0.07	0.25	0.55	0.10	0.03

Find the cumulative density function, mean, and variance for this probability density function.

38. The corpse flower of Sumatra, *Amorphophallus titanium*, is one of the strangest and rarest plants known. This giant exotic plant produces a flower that is bright red and resembles rotting meat. In addition it smells like a road kill, producing some of the same chemicals that dead bodies produce. The characteristic phenotype and smell attract flies and carrion beetles as pollinators. If this plant follows a Poisson distribution over its home range with a mean of seven plants per hectare, find the following:

   (a) The probability that a hectare has 3 or fewer specimens.

   (b) The probability that a randomly chosen hectare has exactly 4 specimens.

   (c) The probability that a randomly chosen hectare has more than 5 specimens.

   (d) The probability that two randomly chosen hectares each have less than 4 specimens.

39. The bacterium *Staphylococcus aureus* is commonly found on the skin and rarely causes major problems. If it gets under the skin or into the lungs it can cause boils or pneumonia. If the strain of bacterium causing these problems is resistant to methicillin, the infections can be life threatening. Methicillin resistant *Staphylococcus aureus* or MRSA is, therefore, a major concern in hospitals and clinics. Supposing the death rate from MRSA lung infections is 20%, find the following probabilities:

   (a) Three of 20 patients with lung infections die.

   (b) Less than 3 die.

   (c) At least 3 die.

   (d) Between 6 and 3 die.

   (e) How many would you expect to die in a random sample of 20? of 15?

40. The waters around Crystal River, Florida, support a large population of manatees or sea cows, *Trichechus manatus*. These docile, slow moving mammals are often injured by careless power boaters and sport fishers. A local conservation group claims that 30% of the adult manatees exhibit propeller scars.

   (a) If during a survey cruise 20 adult manatees are clearly sighted, how many would you expect to exhibit propeller scars?

   (b) In samples of 20, what is the probability a random sample has less than 5 with scars?

   (c) In samples of 20, what is the probability a random sample has 5 or less with scars?

   (d) In samples of 20, what is the probability a random sample has exactly 5 with scars?

   (e) In samples of 20, what is the probability that a sample has more than 8 with scars?

   (f) If your random sample of 20 had 12 with scars, would you say the 30% claim is too low? (Use a probability argument here.)

41. The American Kennel Club has very precise breed standards for pure bred dogs. For instance, Golden Retriever males should be 23 to 24 inches in height at the withers. Dogs that deviate more than 1 inch from this range are disqualified as show dogs. Assume that height at the withers in Goldens is normally distributed with a mean of 23.5 in and a standard deviation of 1 in. Determine the following:

(a) What percentage of the dogs meet the breed standard?

(b) What percentage of the dogs would be disqualified, that is, are outside the 22 to 25 in range?

(c) What percentage would be disqualified for being too tall?

(d) What is the probability that a randomly chosen dog is taller than 26 in?

(e) A beautiful Golden female named Sydney is 23 in at the withers. The breed standard for females is 21.5 to 22.5 in. If for females the mean is 22 in and standard deviation is 0.75 in, what percentile is Sydney's height?

42. QCR. True or False. Justify your answer.

(a) The density function, $f(x)$, for a discrete distribution can be thought of as the probability that the variable X equals x.

(b) The binomial distribution is a family of distributions with each individual distribution characterized by two parameters, n and p.

(c) The total number of successes in n individual binomial trials is expected to be $E(X) = np = \mu$.

(d) The cumulative distribution function for various binomial distributions are found in Table C.1. Individual probability density functions must be calculated directly from the pdf formula.

(e) Like the binomial, the Poisson is a family of distributions that require knowledge of both the mean and sample size to determine individual probabilities.

(f) When a binomial distribution has a large n and small p value, the Poisson distribution can be used to approximate the binomial distribution.

(g) In continuous distributions $P(X = x) = f(x) = 0$, while $f(x)$ is usually not zero in discrete distributions.

(h) The standard normal distribution has a mean of zero and a standard deviation of one. Standardizing or normalizing various normal distributions allows one to arrange CDF values for an infinite number of curves into one table, for example, Table C.3.

4

Sampling Distributions

Concepts in Chapter 4:

- Random Samples and Sampling Distributions
- Binomial Sampling Distributions
- Continuous Sampling Distributions
- Central Limit Theorem
- Confidence Intervals for μ
- Confidence Intervals for σ^2
- Confidence Intervals for a Population Proportion

Sampling Distributions

The previous chapter described the behavior of both discrete and continuous random variables with particular emphasis on binomial, Poisson, and normal variables. In this chapter we show the relationship of probability density functions to statistical inference through the concept of sampling distributions.

4.1   Definitions

A **simple random sample** is a sample of size n drawn from a population of size N in such a way that every possible sample of size n has the same probability of being selected. Variability among simple random samples drawn from the same population is called **sampling variability**, and the probability distribution that characterizes some aspect of the sampling variability, usually the mean but not always, is called a **sampling distribution**. These sampling distributions allow us to make objective statements about population parameters without measuring every object in the population.

> **EXAMPLE 4.1.** Suppose a population of *Peripatus juanensis*, ancient wormlike animals of the Phylum Onychophora that live beneath stones, logs, and leaves in the tropics and subtropics, has a mean length of 60 mm and a standard deviation of 5 mm. These are population parameters, i.e., $\mu = 60$ mm and $\sigma = 5$ mm. If a random sample of 25 *Peripatus* is collected and the average length determined for this sample, it will likely not be exactly 60 mm. It may be 56.5 mm or 61.7 mm or a myriad of other values. This one sample will generate a single $\overline{X}$ that is a guess at μ. How good this guess is depends on the nature of the sampling process (randomness), how variable the attribute is (σ), and how much effort is put into the sampling process (sample size).
>
> Now suppose we return our sample to the population (sampling with replacement) and collect a second random sample of size 25 and calculate the mean from this sample. The new $\overline{X}$ will most likely not be identical to the first $\overline{X}$ because of sampling

variability. If we return our sample to the population and sample a third time we generate a third independent estimate of μ in the form of $\overline{X}$. Repeating this sampling process over and over indefinitely will generate hundreds then thousands of $\overline{X}$'s, and as long as the samples are random and have equal size n, their means will form a theoretical construct that we call the **sampling distribution for the mean**.

In real life we sample the population only once, but we realize that our sample comes from a theoretical sampling distribution of all possible samples of a particular size. This sampling distribution concept provides a link between sampling variability and probability. Choosing a random sample is a chance operation and generating the sampling distribution consists of many repetitions of this chance operation, so probabilities concerning a random sample can be interpreted as relative frequencies in the sampling distribution.

The concept of a sampling distribution and its utility can be demonstrated with an example of a binomial variable.

EXAMPLE 4.2. Suppose that in a genetics experiment with fruit flies, *Drosophila melanogaster*, a very large progeny is generated from parents heterozygous for a recessive allele causing vestigial wings. Recall from your study of elementary genetics that a cross between heterozygous parents should produce offspring in a 3:1 ratio of dominant to recessive phenotypes:

$$\text{Vv} \times \text{Vv} \longrightarrow \tfrac{3}{4}\text{V}_- \ (\tfrac{2}{4}\text{Vv}; \ \tfrac{1}{4}\text{VV: normal wings}), \quad \tfrac{1}{4}(\text{vv: vestigial wings}).$$

Mendelian principles lead to the expectation of 25% vestigial-winged offspring. How many of a random sample of 20 of the progeny would you expect to be vestigial-winged?

SOLUTION. For the binomial with $p = 0.25$ and $n = 20$,

$$E(X) = np = 20(0.25) = 5,$$

so the long-term theoretical average expected in samples of size 20 is 5. But the probability of exactly 5 in a random sample of 20 is

$$f(5) = \frac{20!}{5!(20-5)!}(0.25)^5(0.75)^{15} = 0.2024.$$

If we let $\hat{p}$ represent the sample proportion of individuals with vestigial wings, then a sample that contains 5 individuals with vestigial wings has $\hat{p} = \frac{5}{20} = 0.25$ and will occur in 20.24% of the sampling distribution. *Each* sample generates $\hat{p}$, an estimate of p, the population value, but only 20.24% of the estimates match the true population parameter. Table 4.1 shows the sampling distribution of $\hat{p}$.

From Table 4.1 we see that only 20.24% of the estimates $\hat{p}$ are the same as the population parameter p. $\hat{p}$ is demonstrating sampling variability. Notice also that as $\hat{p}$ gets farther from p, its frequency diminishes. Estimates of p ranging from 0.20 to 0.30 have a high probability. $P(0.20 \leq \hat{p} \leq 0.30) = 0.1896 + 0.2024 + 0.1686 = 0.5606$. This indicates that even small samples such as 20 can estimate p with a certain degree of accuracy. Stated another way, random samples with between 4 and 6 flies with vestigial wings in 20 flies will occur 56% of the time when the value of p is actually 0.25.

Further investigation of the sampling distribution of $\hat{p}$ when $n = 20$ and $p = 0.25$ shows that $\hat{p} < 0.10$ and $\hat{p} > 0.40$ are very rare values, $P(\hat{p} < 0.10) = 0.0032 + 0.0211 =$

TABLE 4.1. The sampling distribution of $\hat{p}$ with $n = 20$ and $p = 0.25$. As $\hat{p}$ gets farther from p, its probability diminishes.

$\hat{p}$	Probability	$\hat{p}$	Probability	$\hat{p}$	Probability
0.00	0.0032	0.35	0.1124	0.70	0.0000
0.05	0.0211	0.40	0.0609	0.75	0.0000
0.10	0.0670	0.45	0.0270	0.80	0.0000
0.15	0.1339	0.50	0.0100	0.85	0.0000
0.20	0.1896	0.55	0.0030	0.90	0.0000
0.25	0.2024	0.60	0.0007	0.95	0.0000
0.30	0.1686	0.65	0.0002	1.00	0.0000

0.0243 and $P(\hat{p} > 0.40) = 0.0270 + 0.0100 + 0.0030 + 0.0007 + 0.0002 = 0.0409$. Less than 2 vestigial-winged flies or more than 8 vestigial-winged flies would not occur often in samples drawn at random from a population with $p = 0.25$. So if you expected $\frac{1}{4}$ vestigial-winged flies and sampled 20 and found 13 with vestigial wings ($\hat{p} = 0.65$), one of three things must have happened: Maybe your sample wasn't random; maybe you were just unlucky and got a rare result by chance $P(\hat{p} \geq 0.65) = 0.0002$; or maybe your hypothesis that the population had $\frac{1}{4}$ of the flies with vestigial wings is wrong.

Generally, the less likely an event is to occur according to sampling distribution probabilities, the more likely the underlying premise is incorrect. This is especially true if the sampling process generated a simple random sample. We will discuss these different causes for unexpected results in more detail when hypothesis testing is formally considered.

Before turning to another example let's consider the effect of sample size on the shape of the sampling distribution of $\hat{p}$ for $p = 0.25$. See Figure 4.1.

The *area* of each bar in Figure 4.1 represents the probability of the corresponding value of $\hat{p}$ for a given value of n. As you can see, all the sampling distributions cluster around $p = 0.25$, the true value of $\hat{p}$. However, the spread is much more compressed in the distributions with larger values of n (c and d) than in the distributions with smaller values of n (a and b), as the following table shows:

n	$P(0.20 \leq \hat{p} \leq 0.30)$
20	0.5606
40	0.6389
80	0.7554
160	0.8799

Larger sample sizes (more work!) tend to improve the probability that $\hat{p}$ is close to p. However, the probability that a sample generates a $\hat{p}$ *exactly equal* to p actually gets smaller as the sample size increases, e.g., $P(\hat{p} = 0.25 \mid n = 20) = 0.2024$, while $P(\hat{p} = 0.25 \mid n = 80) = 0.1444$. This effect can be seen clearly in Figure 4.1 in the areas of the bars over 0.25 in (a) and (c).

4.2 Distribution of the Sample Mean

Perhaps the most important sampling distribution in applied statistics is the distribution of the **sample mean**. To construct a sampling distribution for the sample mean,

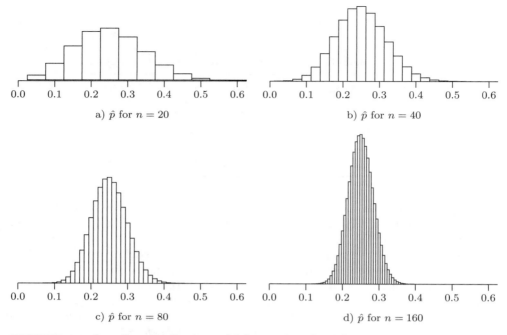

FIGURE 4.1. Sampling distributions of $\hat{p}$ for $p = 0.25$ for different size samples. For each n, the *area* of each bar is proportional to the probability of the corresponding value of $\hat{p}$.

we do the following:

1. From a finite population of size N, randomly draw all possible samples of size n.

2. For each sample compute $\overline{X}$.

3. List all the distinct observed values of $\overline{X}$ and the frequency of each of these distinct values.

Construction of a sampling distribution for the sample mean or any other statistic is a very arduous task for samples of any appreciable size. We do examples here for their heuristic value and not to demonstrate acceptable testing procedures.

EXAMPLE 4.3. Consider a population of size $N = 6$ consisting of the 6 sides of a fair die. The sample space is $\{1, 2, 3, 4, 5, 6\}$. As we demonstrated in Chapter 3, this population has the following parameters:

$$\mu = \frac{\sum X_i}{N} = \frac{21}{6} = 3.5, \qquad \sigma^2 = \frac{\sum (X_i - \mu)^2}{N} = \frac{17.5}{6}.$$

Note the population distribution is a uniform distribution.

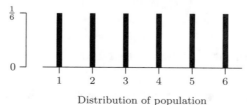

Distribution of population

As an exercise in sampling distribution construction, let's draw all possible samples of size $n = 2$ from this population that correspond to rolling the die twice. For each sample drawn calculate the sample mean $\overline{X}$.

SOLUTION. Any of the 6 faces can occur on either roll, so this process corresponds to sampling with replacement. Therefore, the total number of possible samples is $6 \times 6 = 36$. (In general, sampling with replacement n times from a population of size N generates N^n possible samples.) Table 4.2 lists all possible samples of size $n = 2$.

TABLE 4.2. The possible samples of size 2 (and the sample mean) corresponding to two rolls of a die.

Roll 1	Roll 2					
	⚀	⚁	⚂	⚃	⚄	⚅
⚀	1,1 (1.0)	1,2 (1.5)	1,3 (2.0)	1,4 (2.5)	1,5 (3.0)	1,6 (3.5)
⚁	2,1 (1.5)	2,2 (2.0)	2,3 (2.5)	2,4 (3.0)	2,5 (3.5)	2,6 (4.0)
⚂	3,1 (2.0)	3,2 (2.5)	3,3 (3.0)	3,4 (3.5)	3,5 (4.0)	3,6 (4.5)
⚃	4,1 (2.5)	4,2 (3.0)	4,3 (3.5)	4,4 (4.0)	4,5 (4.5)	4,6 (5.0)
⚄	5,1 (3.0)	5,2 (3.5)	5,3 (4.0)	5,4 (4.5)	5,5 (5.0)	5,6 (5.5)
⚅	6,1 (3.5)	6,2 (4.0)	6,3 (4.5)	6,4 (5.0)	6,5 (5.5)	6,6 (6.0)

From Table 4.2, we see that for any sample, the mean ranges from 1 to 6. Notice that some samples generate identical means, e.g., $(2,2)$ has $\overline{X} = 2$ as do $(1,3)$ and $(3,1)$. Table 4.3 lists all the distinct observed values of $\overline{X}$ and their relative frequencies. In other words, Table 4.3 gives the sampling distribution of $\overline{X}$'s generated by the samples in Table 4.2.

What are the characteristics of the sampling distribution in Table 4.3?

$$\mu_{\overline{X}} = \frac{\sum f_i \overline{X}_i}{\sum f_i} = \frac{1 \cdot 1.0 + 2 \cdot 1.5 + 3 \cdot 2.0 + \cdots + 1 \cdot 6.0}{36} = \frac{126}{36} = 3.5.$$

So the mean of the $\overline{X}$'s equals the mean of the original X's, $\mu_X = \mu_{\overline{X}}$. However,

$$\sigma_{\overline{X}}^2 = \frac{\sum f_i (\overline{X}_i - \mu_{\overline{X}})^2}{\sum f_i}$$

$$= \frac{1 \cdot (1.0 - 3.5)^2 + 2 \cdot (1.5 - 3.5)^2 + 3 \cdot (2.0 - 3.5)^2 + \cdots + 1 \cdot (6.0 - 3.5)^2}{36}$$

$$= \frac{52.5}{36}$$

$$= \frac{17.5}{12}.$$

TABLE 4.3. The sampling distribution of $\overline{X}$

$\overline{X}_i$	Frequency f_i	Relative frequency
1.0	1	0.028
1.5	2	0.056
2.0	3	0.083
2.5	4	0.111
3.0	5	0.139
3.5	6	0.167
4.0	5	0.139
4.5	4	0.111
5.0	3	0.083
5.5	2	0.056
6.0	1	0.028
Total	36	1.000

The variance of the sampling distribution of the mean is *not* equal to the population variance. However, the variance of the sampling distribution of the mean, $\frac{17.5}{12}$, is exactly the population variance, $\frac{17.5}{6}$, divided by the sample size, 2. In general, for a sample of size n,

$$\sigma_{\overline{X}}^2 = \frac{\sigma_X^2}{n}.$$

The square root of $\sigma_{\overline{X}}^2$ is called the **standard error of the mean**, or usually just the **standard error** to differentiate it from the standard deviation,

$$\text{SE} = \sigma_{\overline{X}} = \frac{\sigma_X}{\sqrt{n}}.$$

In this example,

$$\text{SE} = \sigma_{\overline{X}} = \sqrt{\sigma_{\overline{X}}^2} = \sqrt{\frac{17.5}{12}} = 1.21.$$

Figure 4.2 is a graphical representation of the sampling distribution of $\overline{X}$'s. Notice how the $\overline{X}$'s cluster in the middle of this distribution and, further, how different it is from the original uniform population distribution at the beginning of this example. None of the above results is coincidence and, in fact, can be generalized for two situations, Case I: sampling from a normally distributed population, and Case II: sampling from a nonnormally distributed population.

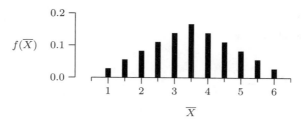

FIGURE 4.2. The sampling distribution of $\overline{X}$.

THEOREM 4.1.

Case I. When sampling from a *normally* distributed population with mean μ_X and variance σ_X^2, the distribution of the sample mean (sampling distribution) will have the following attributes:

1. The distribution of $\overline{X}$'s will be normal.

2. The mean $\mu_{\overline{X}}$ of the distribution of $\overline{X}$'s will be equal to μ_X, the mean of the population from which the samples were drawn.

3. The variance $\sigma_{\overline{X}}^2$ of the distribution of $\overline{X}$'s will be equal to $\frac{\sigma_X^2}{n}$, the variance of the original population of X's divided by the sample size. $\sigma_{\overline{X}}$ is the population standard error.

Case II. When sampling from a *nonnormally* distributed population with mean μ_X and variance σ_X^2, the distribution of the sample mean (sampling distribution) will have the following attributes:

1. The distribution of $\overline{X}$'s will be *approximately normal*, with the approach to normality becoming better as the sample size increases. Generally, the sample size required for the sampling distribution of $\overline{X}$ to approach normality depends on the shape of the original distribution. Samples of 30 or more give very good normal approximations for this sampling distribution of $\overline{X}$ in nearly all situations. (This property of means is known as the **Central Limit Theorem.**)

2. The mean $\mu_{\overline{X}}$ of the distribution of $\overline{X}$'s will be equal to μ_X, the mean of the population from which the samples were drawn.

3. The variance $\sigma_{\overline{X}}^2$ of the distribution of $\overline{X}$'s will be equal to $\frac{\sigma_X^2}{n}$, the variance of the original population of X's divided by the sample size. $\sigma_{\overline{X}}$ is the population standard error.

Parts 2 and 3 of both cases above can be proved using rules for expectation and variance. See Appendix A.2 for formal proofs (Facts 10 and 13).

Part 1 of both cases (the normality or approximate normality of the sampling distribution of the mean) will allow us to develop powerful tools for statistical inference in later chapters. Part 3 of both cases shows that means of samples are less variable than individual observations because the standard error $\left(\frac{\sigma}{\sqrt{n}}\right)$ is always smaller than the standard deviation (σ).

Let's apply Theorem 4.1 to a couple of practical examples before continuing.

> **EXAMPLE 4.4.** The mean blood cholesterol concentration of a large population of adult males (50–60 years old) is 200 mg/dl with a standard deviation of 20 mg/dl. Assume that blood cholesterol measurements are normally distributed. What is the probability that a randomly selected individual from this age group will have a blood cholesterol level below 250 mg/dl?
>
> **SOLUTION.** Apply the standard normal transformation as in Chapter 3.
>
> $$P(X < 250) = P\left(Z < \frac{250 - 200}{20}\right) = P(Z < 2.5) = F(2.5).$$
>
> From Table C.3, $F(2.5) = 0.9938$. So we expect over 99% of the men in the population to have cholesterol levels below 250 mg/dl.
>
> *What is the probability that a randomly selected individual from this age group will have a blood cholesterol level above 225 mg/dl?*

SOLUTION. Again utilize the standard normal transformation and Table C.3.

$$P(X > 225) = P\left(Z > \frac{225 - 200}{20}\right)$$
$$= P(Z > 1.25)$$
$$= 1 - F(1.25)$$
$$= 1 - 0.8944$$
$$= 0.1056.$$

Remember that Table C.3 is a cumulative distribution function from the left, so to find the right-hand tail one must subtract the left side from 1.

What is the probability that the mean of a sample of 100 men from this age group will have a value below 204 mg/dl?

SOLUTION. This question requires the understanding and application of Theorem 4.1. Sample means have a sampling distribution with $\mu_{\overline{X}} = \mu_X = 200$ mg/dl and

$$\sigma_{\overline{X}} = \frac{\sigma_X}{\sqrt{n}} = \frac{20}{\sqrt{100}} = 2.0 \text{ mg/dl},$$

and because the X's were normally distributed, the $\overline{X}$'s will also be normally distributed. The Z transformation for a mean is then

$$Z = \frac{\overline{X} - \mu_{\overline{X}}}{\sigma_{\overline{X}}} = \frac{\overline{X} - \mu_{\overline{X}}}{\frac{\sigma_X}{\sqrt{n}}} = \frac{204 - 200}{2.0}.$$

So

$$P(\overline{X} < 204) = P\left(Z < \frac{204 - 200}{2.0}\right) = P(Z < 2.0) = F(2.0) = 0.9772.$$

Because means are less variable than individual observations, the standard error is always smaller than the standard deviation and reflects the sample size: $\sigma = 20$ mg/dl, while $\sigma_{\overline{X}} = 2.0$ mg/dl for a sample of size $n = 100$.

 If a group of 25 older men who are strict vegetarians have a mean blood cholesterol level of 188 mg/dl, would you say that vegetarianism significantly lowers blood cholesterol levels? Explain.

SOLUTION. First consider the probability of a mean as low as 188 mg/dl or lower for a randomly chosen sample of 25 from the baseline population. Using Table C.3,

$$P(\overline{X} < 188) = P\left(Z \le \frac{188 - 200}{\frac{20}{\sqrt{25}}}\right) = P(Z \le -3.0) = F(-3.0) = 0.0013. \quad \frac{3}{10,000}$$

By chance alone, the probability of getting a mean as low as 188 mg/dl or lower in the baseline population is 0.0013 or about 1 in 769 samples. This is a very low probability if the sample had been drawn at random from the baseline population. Remember we used $\mu_{\overline{X}} = 200$ mg/dl and $\sigma_{\overline{X}} = \frac{\sigma}{\sqrt{25}} = 4.0$ mg/dl to determine this probability. More likely the vegetarians are from a different distribution with a significantly lower mean cholesterol level than the general population. Although other factors might be involved, from the data given it appears that diet may affect blood cholesterol levels.

EXAMPLE 4.5. Portions of prepared luncheon meats should have pH values with a mean of 5.6 and a standard deviation of 1.0. The usual quality control procedure is to randomly sample each consignment of meat by testing the pH value of 25 portions. The consignment is rejected if the pH value of the sample mean exceeds 6.0. What is the probability of a consignment being rejected?

SOLUTION. Note here there is no mention of the shape of the distribution of pH values for individual portions. They may or may not be normally distributed. The question, however, involves a sample mean, and according to Theorem 4.1 we can assume $\mu_{\overline{X}} = \mu_X = 5.6$ and $\sigma_{\overline{X}} = \frac{\sigma_x}{\sqrt{n}} = \frac{1.0}{\sqrt{25}} = 0.2$. Also because of the Central Limit Theorem, these $\overline{X}$'s are at least approximately normally distributed. So,

$$P(\overline{X} > 6.0) = P\left(Z > \frac{6.0 - 5.6}{\frac{1.0}{\sqrt{25}}}\right) = P(Z > 2.0) = 1 - F(2.0) = 0.0228$$

using Table C.3. Only 2.28% of the consignments will be rejected using the quality control procedure above.

It is clear from Examples 4.4 and 4.5 that Theorem 4.1 is very useful in determining probability values for various sample means. In each case, however, the population mean μ and the population standard deviation σ were given. Next we turn to techniques that allow us to use the sample itself, without any knowledge of the population parameters, to assess its own reliability. These techniques apply an understanding of sampling distributions to develop confidence intervals about various parameters.

4.3 Confidence Intervals for the Population Mean

Recall that a sample mean $\overline{X}$ is an unbiased estimate of the population mean μ. $\overline{X}$'s are not all the same due to sampling variability and, in fact, their scatter depends on both the variability in X's, measured by σ, and the sample size n. The standard error of the mean is $\frac{\sigma}{\sqrt{n}}$. From Theorem 4.1 the random variable

$$\frac{\overline{X} - \mu}{\frac{\sigma}{\sqrt{n}}}$$

is distributed as the standard normal or Z distribution. For the sampling distribution of this random variable, consider the following:

$$P(a \leq Z \leq b) = 0.95. \tag{4.1}$$

What two values a and b capture the middle 95% of the Z distribution?

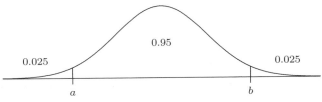

If $F(a) = 0.025$, then looking up 0.025 in the body of Table C.3 we find $a = -1.960$. If $F(b) = 0.975$, then in the body of Table C.3 we find $b = 1.960$. So $P(-1.960 \leq Z \leq$

1.960) = 0.95, or the values ±1.960 capture the middle 95% of the Z distribution. Therefore, we capture the middle 95% of the $\overline{X}$'s if and only if

$$P\left(-1.960 \leq \frac{\overline{X} - \mu}{\frac{\sigma}{\sqrt{n}}} \leq 1.960\right) = 0.95$$

$$\Longleftrightarrow P\left(-1.960\frac{\sigma}{\sqrt{n}} \leq \overline{X} - \mu \leq 1.960\frac{\sigma}{\sqrt{n}}\right) = 0.95$$

$$\Longleftrightarrow P\left(-\overline{X} - 1.960\frac{\sigma}{\sqrt{n}} \leq -\mu \leq -\overline{X} + 1.960\frac{\sigma}{\sqrt{n}}\right) = 0.95$$

$$\Longleftrightarrow P\left(\overline{X} - 1.960\frac{\sigma}{\sqrt{n}} \leq \mu \leq \overline{X} + 1.960\frac{\sigma}{\sqrt{n}}\right) = 0.95,$$

where at the last step we divided through by -1 and rearranged the terms. The probability that the sample mean $\overline{X}$ will differ by no more than 1.960 standard errors $\frac{\sigma}{\sqrt{n}}$ from the population mean μ is 0.95.

$\overline{X}$ is a random variable with a sampling distribution. Because there is an infinite number of values of $\overline{X}$, there is an infinite number of intervals of the form $\overline{X} \pm 1.960\frac{\sigma}{\sqrt{n}}$. The probability statement says that 95% of these intervals will actually include μ between the limits. For any one interval, $\overline{X} \pm 1.960\frac{\sigma}{\sqrt{n}}$, we say we are 95% *confident that μ lies between these limits* or

$$C\left(\overline{X} - 1.960\frac{\sigma}{\sqrt{n}} \leq \mu \leq \overline{X} + 1.960\frac{\sigma}{\sqrt{n}}\right) = 0.95. \qquad (4.2)$$

We use C here for the confidence interval rather than P (denoting probability) because once a confidence interval is calculated, it either contains μ (so $P = 1$) or it doesn't contain μ ($P = 0$).

EXAMPLE 4.6. Suppose a particular species of understory plants is known to have a variance in heights of 16 cm^2 ($\sigma^2 = 16$ cm^2). If this species is sampled with the heights of 25 plants averaging 15 cm, find the 95% confidence interval for the population mean.

SOLUTION. Here $n = 25$, $\overline{X} = 15$ cm, $\sigma^2 = 16$ cm^2, and $\sigma = 4$ cm. So

$$C\left(\overline{X} - 1.960\frac{\sigma}{\sqrt{n}} \leq \mu \leq \overline{X} + 1.960\frac{\sigma}{\sqrt{n}}\right)$$

$$= C\left(15 - 1.960\frac{4}{\sqrt{25}} \leq \mu \leq 15 + 1.960\frac{4}{\sqrt{25}}\right)$$

$$= 0.95,$$

and, therefore, the lower limit $L_1 = 13.432$ cm and the upper limit $L_2 = 16.568$ cm. That is,

$$C(13.432 \text{ cm} \leq \mu \leq 16.568 \text{ cm}) = 0.95.$$

We are 95% confident that the values 13.432 and 16.568 capture the parametric mean, μ.

What if we want to be more confident (say, 99%) that we've included μ in our interval?

SOLUTION. Return to $P(a \leq Z \leq b) = 0.99$. Find a and b so that $F(a) = 0.005$ and $F(b) = 0.995$. From Table C.3 $a = -2.575$ and $b = 2.575$.

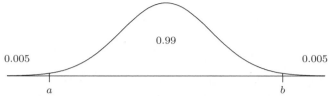

Using equation (4.2) with the substitution of 2.575 for 1.960, the confidence interval is

$$C\left(\overline{X} - 2.575\frac{\sigma}{\sqrt{n}} \leq \mu \leq \overline{X} + 2.575\frac{\sigma}{\sqrt{n}}\right) = 0.99.$$

$\begin{cases} 90\% = 1.645 \\ 95\% = 1.960 \\ 99\% = 2.575 \end{cases}$

z-values

Therefore,

$$L_1 = 15 - 2.575\frac{4}{\sqrt{25}} = 15 - 2.060 = 12.940 \text{ cm}$$

and

$$L_2 = 15 + 2.575\frac{4}{\sqrt{25}} = 15 + 2.060 = 17.060 \text{ cm},$$

so

$$C(12.940 \text{ cm} \leq \mu \leq 17.060 \text{ cm}) = 0.99.$$

Comparing the 99% confidence interval with the 95% confidence interval for the same sample, we see that in order to be more confident about capturing the parametric mean between the limits we must move the limits further apart.

An Overview of Confidence Intervals

We have now developed two types of estimate for μ, the population mean. The first was the sample mean $\overline{X}$, which is called a **point estimate**. A point estimate is a single number which is used to approximate the parameter, in the this case μ.

The second type of estimate is a confidence interval or **interval estimate**. An interval estimate is a range of values between which we expect the parameter (in this case μ) to be located. Now look again at (4.2)

$$C\left(\overline{X} - 1.960\frac{\sigma}{\sqrt{n}} \leq \mu \leq \overline{X} + 1.960\frac{\sigma}{\sqrt{n}}\right) = 0.95.$$

Notice that to determine a confidence interval requires four different numbers:

1. a point estimate, here the sample mean $\overline{X}$;

2. a measure of variability, here the standard error of the mean, $\frac{\sigma}{\sqrt{n}}$;

3. a desired level of confidence $1 - \alpha$, in this case $1 - \alpha = 0.95$, so $\alpha = 0.05$;

4. and the sampling distribution of the point estimate, here the standard normal distribution, which provided the confidence factor with which to adjust the variability for the desired level of confidence, in this case 1.960.

Using this language, the endpoints of the confidence interval have the form

$$\text{point estimate} \pm (\text{confidence factor})(\text{standard error}). \tag{4.3}$$

Each time we develop a confidence interval for a mean, its endpoints will have the form in (4.3). Be sure that you are comfortable with these ideas and refer back to this equation when necessary.

Confidence Intervals for the Mean When the Variance is Unknown

We hope by now that you have questioned the utility of the confidence intervals above. There would be very few times in the real world where you would know σ^2 for the population and not know μ for that population. As developed above, confidence intervals require exactly that situation.

The more likely real-world situation would be that we take a sample from a population with unknown shape, mean, and standard deviation. From this sample we calculate $\overline{X}$ and s. By the Central Limit Theorem, we assume $\overline{X}$'s sampling distribution is approximately normal. We can now use $\frac{s}{\sqrt{n}}$, the *sample standard error*, as our estimate of $\frac{\sigma}{\sqrt{n}}$, the *population standard error*. When $\frac{s}{\sqrt{n}}$ replaces $\frac{\sigma}{\sqrt{n}}$ in the formula

$$\frac{\overline{X} - \mu}{\frac{\sigma}{\sqrt{n}}},$$

we have

$$\frac{\overline{X} - \mu}{\frac{s}{\sqrt{n}}}. \quad \text{— for unknown shape, mean + standard dev.}$$

While the distribution of $\frac{\overline{X}-\mu}{\frac{\sigma}{\sqrt{n}}}$ was known to be the standard normal distribution or Z distribution, replacing σ with s generates a different sampling distribution.

If the value of σ is not known and must be estimated, the distribution of the variable obtained by replacing σ by its estimator is called the **t distribution**. This sampling distribution was developed by W. S. Gossett, while working at the Guinness Brewery, and published under the pseudonym "Student" in 1908. It is, therefore, sometimes called the **Student's t distribution** and is really a family of distributions dependent on the parameter $n - 1$. If we use "$\sim$" as shorthand for "is distributed as," then we've seen that

$$\frac{\overline{X} - \mu}{\frac{\sigma}{\sqrt{n}}} \sim \text{ a normal distribution with } \mu = 0 \text{ and } \sigma = 1,$$

while

$$\frac{\overline{X} - \mu}{\frac{s}{\sqrt{n}}} \sim \text{ a } t \text{ distribution with } \mu = 0 \text{ and } \sigma \text{ depending on the sample size.}$$

As the sample size increases the t distribution approaches the standard normal distribution. See Figure 4.3.

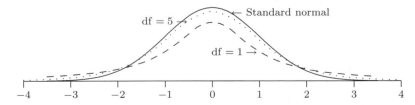

FIGURE 4.3. Two Student's t curves and a standard normal curve.

The exact shape of a Student's t distribution depends on the quantity called **degrees of freedom**. The t distributions are symmetric and bell-shaped like the normal distribution but are a little flatter, i.e., they have a larger standard deviation. The degrees of freedom is just the sample size minus 1: df $= n - 1$ for any t distribution. The Greek letter ν (nu) is also commonly used to indicate the degrees of freedom.

Partial cumulative distribution functions for a variety of t distributions have been collated in Table C.4. The sample size ranges from 2 through 101 with df from 1 to 100. Additional degrees of freedom at various intervals beyond 100 are given. The areas or probabilities given range from 0.0005 to 0.25 for one tail and from 0.001 to 0.50 for two tails.

Since the t distributions are probability density functions, the area under any t distribution is 1. To determine a confidence interval, we have a fixed confidence level or probability (i.e., area under a t distribution) in mind and want to find the endpoints of the interval centered at 0 which has this probability. Let $1 - \alpha$ denote this probability, where α is a small value. Often α is very small, 0.05 or 0.01, so that $1 - \alpha$ is 0.95 or 0.99. Similar to (4.1), we seek to find the number t_0 so that

$$P(-t_0 \leq t \leq t_0) = 1 - \alpha.$$

That is, what two values $-t_0$ and t_0 cut the t distribution such that a middle area of size $1 - \alpha$ will lie between $-t_0$ and t_0? See Figure 4.4.

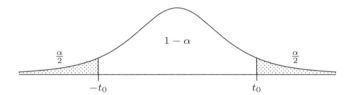

FIGURE 4.4. Locating the middle area of size $1 - \alpha$ in a t distribution.

EXAMPLE 4.7. With $n = 15$, find t_0 such that $P(-t_0 \leq t \leq t_0) = 0.90$.

SOLUTION. From Table C.4, df $= 14$, the two-tailed shaded area equals 0.10, so $-t_0 = -1.761$ and $t_0 = 1.761$.

Find t_0 such that $P(-t_0 \leq t \leq t_0) = 0.95$.

SOLUTION. Again df $= 14$, but the two-tailed shaded area equals 0.05, so $-t_0 = -2.145$ and $t_0 = 2.145$. Since t is a *sampling distribution* for the variable

$$\frac{\overline{X} - \mu}{\frac{s}{\sqrt{n}}}$$

we can say 95% of the t's calculated for samples of size 15 will fall between -2.145 and 2.145.

Now that we have a point estimate $\overline{X}$, its sampling distribution—the t distribution, and an estimate of its standard error $\frac{s}{\sqrt{n}}$, using (4.3) we can generate **confidence intervals for the population mean** for any sample without prior knowledge of any of the population's characteristics.

FORMULA 4.1. A $(1-\alpha)100\%$ confidence interval for the population mean μ is given by

$$C\left(\overline{X} - t_0\frac{s}{\sqrt{n}} \le \mu \le \overline{X} + t_0\frac{s}{\sqrt{n}}\right) = 1 - \alpha,$$

where t_0 has $\nu = n - 1$ degrees of freedom. Thus, the interval endpoints are

$$L_1 = \overline{X} - t_{(1-\frac{\alpha}{2},\nu)}\frac{s}{\sqrt{n}} \quad \text{and} \quad L_2 = \overline{X} + t_{(1-\frac{\alpha}{2},\nu)}\frac{s}{\sqrt{n}}.$$

Notice that t_0 is determined by the degrees of freedom ν and the level of confidence $1 - \alpha$. $\overline{X}$ and $\frac{s}{\sqrt{n}}$ come from the sample itself, so the confidence interval can be determined without any extramundane knowledge.

> **EXAMPLE 4.8.** Return to the measurements of understory plants introduced in Example 4.6. Without any prior knowledge of the plant heights, the ecologist samples 25 plants and measures their heights. He finds that the sample has a mean of 15 cm ($\overline{X} = 15$ cm) and a sample variance of 16 cm^2 ($s^2 = 16$ cm^2). Note here the sample size, sample mean, and sample variance are all determined by the sampling process. What is the 95% confidence interval for the population mean μ?
>
> **SOLUTION.** Simply apply Formula 4.1:
>
> $$C\left(\overline{X} - t_0\frac{s}{\sqrt{n}} \le \mu \le \overline{X} + t_0\frac{s}{\sqrt{n}}\right) = 1 - 0.05.$$
>
> With df $= 25 - 1 = 24$, from Table C.4 we find $\pm t_0 = \pm 2.064$. So
>
> $$L_1 = 15 - 2.064\frac{4}{\sqrt{25}} = 13.349 \text{ cm}$$
>
> and
>
> $$L_2 = 15 + 2.064\frac{4}{\sqrt{25}} = 16.651 \text{ cm}.$$
>
> The plant ecologist is 95% confident that the population mean for heights of these understory plants is between 13.349 and 16.651 cm.

If he chooses a 99% confidence interval,

$$L_1 = 15 - 2.797\frac{4}{\sqrt{25}} = 12.762 \text{ cm}$$

and

$$L_2 = 15 + 2.797\frac{4}{\sqrt{25}} = 17.238 \text{ cm}.$$

His confidence that he has captured the population mean μ is greater, but the price paid is the wider interval required to gain this confidence. Compare the confidence limits above with those calculated in Example 4.6.

	95% C. I.	99% C. I.
Using z and σ	$[13.432, 16.568]$	$[12.940, 17.060]$
Using t and s	$[13.349, 16.651]$	$[12.762, 17.238]$

The t distribution generates confidence limits that are a little further apart reflecting the uncertainty about σ and its estimation from the sample s. However, Example 4.8 is much more realistic than Example 4.6. No parameters are assumed, and the confidence interval for the mean is based solely on the sample collected and its statistics, $\overline{X}$, s, and n. Consequently, **when calculating confidence intervals for the population mean always use Formula 4.1 and Table C.4.**

A final word about the term "confidence interval." A confidence interval for the mean μ of a population is an interval that one can assert with a given probability that it contains μ before the sample, from which the interval was calculated, was drawn. After the sample is drawn and the interval calculated, as noted earlier it either contains μ ($P = 1$) or it doesn't contain μ ($P = 0$). This is the reason confidence (C) rather than probability (P) is used in Formula 4.1. Another way to think about confidence intervals, is that 95% of all such intervals calculated actually capture μ. See Figure 4.5.

FIGURE 4.5. The population mean μ and 95% confidence intervals, not all of which capture μ.

Using 95% confidence levels, only about 1 in 20 intervals will fail to cross μ. Using 99% confidence levels, the lines would be a bit longer and 99 of 100 would cross or capture μ.

When reporting a sample mean in a scientific paper or lab report, one should list it with its standard error and sample size,

$$\overline{X} \pm \frac{s}{\sqrt{n}} \text{ with } n.$$

For example, in Problem 2 of Chapter 1 the mean carapace lengths of the sample of adult female green turtles, *Chelonia mydas*, should be reported as

$$106.3 \pm 3.2 \text{ cm, with } n = 10.$$

Reporting the mean in this way allows readers to generate their own confidence intervals and to interpret the relationship of the statistics to the parameter.

4.4 Confidence Intervals for the Population Variance

Just as we would like to know the population mean μ without measuring every individual in the population, we would also like to have some clear understanding of the population variance σ^2. For a sample we can calculate the point estimate s^2, but each sample will generate a different s^2 and we don't know the sampling distribution that would be generated by the repeated sampling of the population.

To construct a confidence interval for the population variance, we need a random variable that involves this parameter in its expression and whose sampling distribution is well characterized. Fortunately,

$$\frac{(n-1)s^2}{\sigma^2}$$

is such a random variable. If all possible samples of size n are drawn from a normal population with a variance equal to σ^2 and for each of these samples the value $\frac{(n-1)s^2}{\sigma^2}$ is computed, these values will form a *sampling distribution* called a $\boldsymbol{\chi^2}$ **distribution** (chi-square) with $n-1$ degrees of freedom. See Figure 4.6.

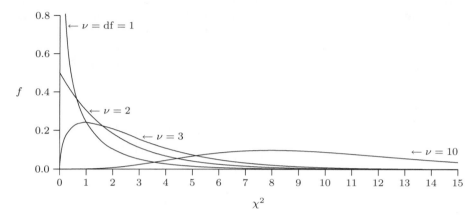

FIGURE 4.6. Chi-square distributions for various degrees of freedom.

As you can see from Figure 4.6, the chi-square distribution is actually a family of distributions (like the t distributions studied earlier) whose shape depends on the degrees of freedom ν. Each chi-square distribution is continuous with values ranging from 0 to positive infinity. These asymmetrical distributions each have

$$\mu = E(\chi^2) = \nu \quad \text{and} \quad \text{Var}(\chi^2) = 2\nu.$$

The fact that the chi-square mean is $n-1$ or ν can be shown using expected value rules. The chi-square population variance equal to $2(n-1)$ or 2ν is somewhat more difficult to derive, and for our purposes will be accepted without proof.

Table C.5 collates some important cumulative distribution function values for chi-square distributions with degrees of freedom from 1 to 60. The CDF probabilities are charted across the top of the table ranging from 0.005 to 0.995 with the degrees of freedom identifying different distributions listed down the left side. For example, the figure below illustrates the case where $\nu = 3$, $\mu_{\chi^2} = 3$, and $\text{Var}(\chi^2) = 6$.

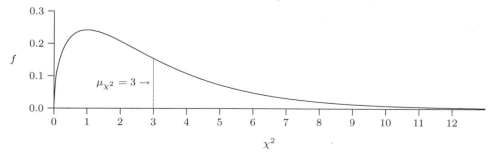

By convention we write χ^2_a, where a is the area to the *left* in the cumulative distribution function. Consider, now, the problem of locating the middle area of size $1-\alpha$ in a χ^2 distribution, much as we did with the t distribution earlier. See Figure 4.7 and compare it to Figure 4.4.

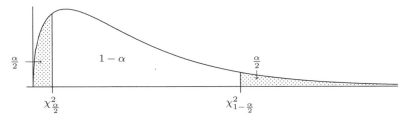

FIGURE 4.7. Locating the middle area of size $1-\alpha$ in a χ^2 distribution.

$$P\left(\chi^2_{\frac{\alpha}{2}} \leq \chi^2 \leq \chi^2_{1-\frac{\alpha}{2}}\right) = P\left[\chi^2_{\frac{\alpha}{2}} \leq \frac{(n-1)s^2}{\sigma^2} \leq \chi^2_{1-\frac{\alpha}{2}}\right] = 1-\alpha.$$

The values of $\chi^2_{\frac{\alpha}{2}}$ and $\chi^2_{1-\frac{\alpha}{2}}$ can be found in Table C.5 and the numerator of the middle term can be calculated from a sample. We now have the materials required to develop a confidence interval for the population variance (σ^2). We need

$$P\left[\chi^2_{\frac{\alpha}{2}} \leq \frac{(n-1)s^2}{\sigma^2} \leq \chi^2_{1-\frac{\alpha}{2}}\right] = 1-\alpha.$$

Taking reciprocals (and remembering that this switches the direction of the inequalities, i.e., $3 < 4 < 5$ but $\frac{1}{3} > \frac{1}{4} > \frac{1}{5}$),

$$P\left[\frac{1}{\chi^2_{\frac{\alpha}{2}}} \geq \frac{\sigma^2}{(n-1)s^2} \geq \frac{1}{\chi^2_{1-\frac{\alpha}{2}}}\right] = 1-\alpha.$$

Multiply through by $(n-1)s^2$ to obtain

$$P\left[\frac{(n-1)s^2}{\chi^2_{\frac{\alpha}{2}}} \geq \sigma^2 \geq \frac{(n-1)s^2}{\chi^2_{1-\frac{\alpha}{2}}}\right] = 1-\alpha.$$

Finally, rearrange the terms to obtain

FORMULA 4.2. A $(1-\alpha)100\%$ confidence interval for the population variance σ^2 is given by

$$C\left(\frac{(n-1)s^2}{\chi^2_{1-\frac{\alpha}{2}}} \leq \sigma^2 \leq \frac{(n-1)s^2}{\chi^2_{\frac{\alpha}{2}}}\right) = 1-\alpha,$$

where the chi-square distribution has $n-1$ degrees of freedom. Explicitly, the interval endpoints are

$$L_1 = \frac{(n-1)s^2}{\chi^2_{1-\frac{\alpha}{2}}} \quad \text{and} \quad L_2 = \frac{(n-1)s^2}{\chi^2_{\frac{\alpha}{2}}}.$$

Because the chi-square distribution is not symmetrical the two limits for any confidence interval for σ^2 are not equidistant from the sample variance (s^2). Note: The confidence interval for σ^2 has a different form from the confidence interval for μ because we started with a sampling distribution that involved the reciprocal of σ^2.

EXAMPLE 4.9. A horticultural scientist is developing a new variety of apple. One of the important traits, in addition to taste, color, and storability, is the uniformity of the fruit size. To estimate the variability in weights she samples 25 mature fruit and calculates a sample variance for their weights, $s^2 = 4.25$ g^2. Develop 95% and 99% confidence intervals for the population variance from her sample.

SOLUTION. For a 95% confidence interval for σ^2, use Formula 4.2 with $\frac{\alpha}{2} = 0.025$. From Table C.5, $\chi^2_{0.975(24)} = 39.4$ and $\chi^2_{0.025(24)} = 12.4$. So

$$L_1 = \frac{(n-1)s^2}{\chi^2_{1-0.025}} = \frac{(25-1)(4.25)}{39.4} = 2.59 \text{ g}^2$$

and

$$L_2 = \frac{(n-1)s^2}{\chi^2_{0.025}} = \frac{(25-1)(4.25)}{12.4} = 8.23 \text{ g}^2.$$

The horticulturalist is 95% confident that σ^2 lies between 2.59 and 8.23 g^2.

For a 99% confidence interval, from Table C.5, $\chi^2_{0.995(24)} = 45.6$ and $\chi^2_{0.005(24)} = 9.89$. So

$$L_1 = \frac{(25-1)(4.25)}{45.6} = 2.24 \text{ g}^2$$

and

$$L_2 = \frac{(25-1)(4.25)}{9.89} = 10.31 \text{ g}^2.$$

The horticulturist increases her confidence of including σ^2 to 99% when she uses the wider limits calculated above. If fruit must have a variance of less than 10.0 g^2, would you say that she has been successful in producing sufficiently uniform fruit in this new variety? We will address this question later with a test of hypothesis.

To obtain a **$1 - \alpha$ confidence interval for the population standard deviation**, simply take the square roots of the confidence limits for σ^2.

$$C\left(\sqrt{\frac{(n-1)s^2}{\chi^2_{1-\frac{\alpha}{2}}}} \leq \sqrt{\sigma^2} \leq \sqrt{\frac{(n-1)s^2}{\chi^2_{\frac{\alpha}{2}}}}\right) = 1 - \alpha,$$

so

$$C\left(\sqrt{\frac{(n-1)s^2}{\chi^2_{1-\frac{\alpha}{2}}}} \leq \sigma \leq \sqrt{\frac{(n-1)s^2}{\chi^2_{\frac{\alpha}{2}}}}\right) = 1 - \alpha.$$

A 95% confidence interval for σ in Example 4.9 is

$$L_1 = \sqrt{2.59 \text{ g}^2} = 1.6 \text{ g}$$

and

$$L_2 = \sqrt{8.23 \text{ g}^2} = 2.9 \text{ g}.$$

4.5   Confidence Intervals for a Population Proportion

Confidence intervals for many other parametric statistics can be developed if the sampling distribution is known or can be reasonably approximated. Let's consider now the confidence interval for p, the population proportion in a binomial distribution.

As we saw in Chapter 1 many statistical studies involve counts (categorical data) rather than measurement data. Among the most common are opinion polls and surveys.

SECTION 4.5: Confidence Intervals for a Population Proportion   119

EXAMPLE 4.10. A local epidemiologist wishes to determine the rate of breast cancer in women under age 35 in a rural county in Ireland. She surveys a random sample of 1200 women in this age group and determines that exactly 6 have had this form of cancer sometime during their lifetime. She wishes to use this information to estimate the population rate p of breast cancer and determine a confidence interval for this estimate.   *Soln on next P.*

If X is the number of "successes" in a sample of size n, then a simple way to estimate the population proportion is to use the **sample proportion**

$$\hat{p} = \frac{X}{n}.$$

The symbol $\hat{p}$ is read as "p hat." In the epidemiologist's case

$$\hat{p} = \frac{X}{n} = \frac{6}{1200} = 0.005.$$

She now has an estimate of the population rate of breast cancer and needs a way of expressing her confidence in this value.

From Section 3.2, a binomial distribution has an approximately normal distribution with mean p and population variance $\sigma^2 = p(1-p)$ if $np > 5$ and $n(1-p) > 5$. The sample proportion, $\hat{p}$, is an unbiased estimator of p, the population parameter.

Since $\hat{p}$ is a sample mean, then by the Central Limit Theorem (Theorem 4.1) the variance of $\hat{p}$ is

$$\sigma_{\hat{p}}^2 = \frac{\sigma^2}{n} = \frac{p(1-p)}{n},$$

and its standard error is

$$SE_{\hat{p}} = \sigma_{\hat{p}} = \sqrt{\frac{p(1-p)}{n}}. \tag{4.4}$$

We have a point estimate $\hat{p}$ for p and its standard error and we know that the sampling distribution is approximately normal (at least for large n). So using (4.3), the end points of a $(1-\alpha)100\%$ confidence interval for p should be

lower limit
$$L_1 = \hat{p} - z_{1-\frac{\alpha}{2}}\sqrt{\frac{p(1-p)}{n}}$$

but includes p not $\hat{p}$

and

upper limit
$$L_2 = \hat{p} + z_{1-\frac{\alpha}{2}}\sqrt{\frac{p(1-p)}{n}},$$

where $z_{1-\frac{\alpha}{2}}$ is the confidence factor (z value) corresponding to $1-\frac{\alpha}{2}$. For example, for a $(1-\alpha)100\% = 95\%$ confidence interval, $\alpha = 0.05$ and the corresponding z value is $z_{1-0.025} = z_{0.975} = 1.960$.

Notice that there is a problem here. These limits include the population parameter p, which we are trying to determine. If we actually knew p, then we wouldn't need any confidence limits! To overcome this paradox, we estimate p by using $\hat{p}$ in the formula for SE. Then

$$L_1 = \hat{p} - z_{1-\frac{\alpha}{2}}\sqrt{\frac{\hat{p}(1-\hat{p})}{n}}$$

this inc.

$\hat{p}$

and

$$L_2 = \hat{p} + z_{1-\frac{\alpha}{2}}\sqrt{\frac{\hat{p}(1-\hat{p})}{n}}.$$

Because the standard error is estimated from the sample, we should also change the z values $(\pm z_{1-\frac{\alpha}{2}})$ to appropriate t values. However, earlier we assumed that sample size n was large. If we assume that $n > 30$, then the t distribution is very nearly equal to the standard normal distribution, so t and z values will be very close. (As df goes to infinity, the t distribution approaches the z distribution.) Thus,

FORMULA 4.3. A good approximation for the $(1-\alpha)100\%$ confidence limits of a population proportion when $n\hat{p} > 5$ and $n(1 - \hat{p}) > 5$ is given by

$$L_1 = \hat{p} - z_{1-\frac{\alpha}{2}} \sqrt{\frac{\hat{p}(1-\hat{p})}{n}} \quad \text{and} \quad L_2 = \hat{p} + z_{1-\frac{\alpha}{2}} \sqrt{\frac{\hat{p}(1-\hat{p})}{n}}.$$

SOLUTION TO EXAMPLE 4.10. Since $n = 1200$ and $\hat{p} = 0.005$, then

$$n\hat{p} = 1200 \times 0.005 = 6 > 5$$

and

$$n(1 - \hat{p}) = 1200 \times 0.995 = 1194 > 5.$$

The approximation in Formula 4.3 will be appropriate here. The 95% confidence limits are

$$L_1 = 0.005 - 1.960\sqrt{\frac{(0.005)(0.995)}{1200}} = 0.005 - 0.00398 = 0.00102$$

and

$$L_2 = 0.005 + 1.960\sqrt{\frac{(0.005)(0.995)}{1200}} = 0.005 + 0.00398 = 0.00898.$$

The 95% confidence interval for p, the proportion of Irish women under age 35 who have experienced breast cancer, is 0.00102 to 0.00898.

We can say that we are 95% confident that the true proportion p is between 0.00102 and 0.00898 or, as these statistics are often stated, between 102 and 898 per 100,000. Remember that we interpret this as meaning that with repeated samples of size 1200 about 95% of the intervals constructed as above would include the true population proportion p.

Choosing Sample Sizes

In some situations a researcher may wish to design a study that will produce a confidence interval for the population proportion p of a certain pre-specified width. This can be accomplished by using a sufficiently large sample size, n.

> **EXAMPLE 4.11.** Using county medical records, the epidemiologist now wishes to determine the five-year survival rate, p, of all women diagnosed with breast cancer in the 1970s. She would like to determine p to within 2%. More precisely, she wishes to construct a 95% confidence interval whose endpoints are within 2% of the five-year survival rate. How large of a sample will she need to do this?

The epidemiologist wants to determine $p \pm 0.02$. Using Formula 4.3 the endpoints of a $(1 - \alpha)100\%$ confidence interval are

$$\hat{p} \pm z_{1-\frac{\alpha}{2}} \sqrt{\frac{\hat{p}(1-\hat{p})}{n}}.$$

We define the **margin of error** for a $(1-\alpha)100\%$ confidence interval for a population proportion to be

$$m = z_{1-\frac{\alpha}{2}}\mathrm{SE}_{\hat{p}} = z_{1-\frac{\alpha}{2}}\sqrt{\frac{\hat{p}(1-\hat{p})}{n}} \qquad (4.5)$$

The researcher specifies m and the confidence level. But having chosen the confidence level, the value of $z_{1-\frac{\alpha}{2}}$ is determined. So to solve for the sample size n we simply need $\hat{p}$. However, since the researcher has not yet done the study, $\hat{p}$ is unknown. But the researcher can't do the study until the sample size is known. We seem to be stuck.

There are two ways around this problem. If resources allow, one could do a pilot study to estimate $\hat{p}$. The other solution is to observe that in (4.5), for any fixed value of n, the margin of error is largest when $\hat{p} = 0.5$.[1] Using $\hat{p} = 0.5$ is the *conservative* approach as it will produce an overestimate of n.

With $\hat{p} = 0.5$ in (4.5), we have

$$m \le z_{1-\frac{\alpha}{2}}\sqrt{\frac{0.5(1-0.5)}{n}}$$

or, squaring both sides of the equation,

$$m^2 \le z_{1-\frac{\alpha}{2}}^2\left(\frac{0.25}{n}\right).$$

Solving for n, we obtain

$$n \le 0.25\left(\frac{z_{1-\frac{\alpha}{2}}}{m}\right)^2 = \left(\frac{z_{1-\frac{\alpha}{2}}}{2m}\right)^2.$$

FORMULA 4.4. A $(1-\alpha)100\%$ confidence interval for a population proportion p will have a margin of error no greater than the pre-specified value m if the sample size is

$$n = \left(\frac{z_{1-\frac{\alpha}{2}}}{2m}\right)^2.$$

As mentioned, this value of n is conservative (larger than necessary). For values of $\hat{p}$ near 0.5 Formula 4.4 will give very good estimates of the sample size required. However, if $\hat{p} < 0.15$ or $\hat{p} > 0.85$, then the sample size given by Formula 4.4 will be at least twice as large as necessary.

SOLUTION TO EXAMPLE 4.11. We may now apply Formula 4.4 to determine the sample size for the breast cancer survival rate study. Since the epidemiologist wants a 95% confidence interval with a 2% margin of error, $\alpha = 0.05$ and $m = 0.02$. Thus,

$$n = \left(\frac{z_{1-\frac{\alpha}{2}}}{2m}\right)^2 = \left(\frac{1.960}{2(0.02)}\right)^2 = 2401.$$

The researcher should use a sample size of roughly 2400.

---

[1]Recall that for a concave down quadratic equation such as $\hat{p}(1-\hat{p}) = \hat{p}-\hat{p}^2$, the maximum value occurs half way between the two roots, which, in this case, is half way between 0 and 1.

4.6   Problems

1. Random samples of size 9 are repeatedly drawn from a normal distribution with a mean of 65 and a standard deviation of 18. Describe the sampling distribution of $\overline{X}$.

2. A dermatologist investigating a certain type of skin cancer induced the cancer in 9 rats and then treated them with a new experimental drug. For each rat she recorded the number of hours until remission of the cancer. The rats had a mean remission time of 400 hours and a standard deviation of 30 hours. Calculate the 95% confidence intervals for μ and σ from these data.

3. In a sample of 48 observations from a normal distribution you are told that the standard deviation has been computed and is 4.0 units. Glancing through the data you notice the lowest observation is 20 and the highest is 80. Does the reported standard deviation appear reasonable? Explain using your knowledge of the characteristics of normal distributions.

4. A wildlife toxicologist studying the effects of pollution on natural ecosystems measured the concentrations of heavy metals in the blood of 25 Galápagos sea lions, *Zalophus californianus.* The sea lions sampled were all females and were 3 to 5 years old. Their mean concentration of heavy metals was 6.2 μg/l and the standard deviation was 1.5 μg/l. Construct 95% confidence intervals for the population mean and variance from these statistics.

5. If X is distributed as normal with $\mu = 20$ and $\sigma^2 = 40$ and a random sample of 16 variates is drawn from the population, what is the probability that their sample mean will be between 17 and 23?

6. The serum cholesterol levels of a certain population of boys follow a normal distribution with a mean of 170 mg/dl and a standard deviation of 30 mg/dl.

   (a) Find the probability that a randomly chosen boy has a serum cholesterol level of 155 mg/dl or less.

   (b) Find the percentage of boys with values between 125 mg/dl and 215 mg/dl.

   (c) Find the probability that the mean serum cholesterol level of a random sample of 25 boys is below 182 mg/dl.

   (d) Determine the probability that the mean serum cholesterol level of a random sample of 100 boys is below 164 mg/dl.

7. An apiculturist investigating nutritional causes of differences in bee morphology weighed 16 worker bee pupae from a commercial hive. She found their average weight to be 530 mg with a standard deviation of 36 mg. Construct the 90% confidence interval for the population mean (μ) and the population variance (σ^2) of worker bee pupal weights.

8. A study of the effects of exercise on the menstrual cycle provides the following ages (in years) of menarche (beginning of menstruation) for 10 female swimmers who began training at least 1 year prior to menarche:

   13.6   13.9   14.0   14.2   14.9   15.0   15.0   15.1   15.4   16.4.

(a) Calculate the mean and standard deviation for this sample. Construct the 95% confidence interval for the mean.

(b) Do you feel that the sample mean is significantly higher than the overall population mean for nonathletes of 12.5 years? Use the confidence interval to provide your rationale here. Compare this to the rationale that you provided in Problem 4 (b) in Chapter 1.

9. An ornithologist studying turkey vultures, *Cathartes aura*, at Hinckley, Ohio, measured the lengths of 16 of these so-called buzzards. She found that the average length was 80.0 cm with a standard deviation of 8.0 cm. Calculate the 99% confidence limits for the mean and variance using her data.

10. Blood coagulation and clot formation in humans is a complicated process with at least 13 different plasma coagulation factors involved. One of the more important protein factors is called prothrombin, which must be converted into the enzyme thrombin for coagulation to occur. Individuals with vitamin K deficiency are thought to have low prothrombin levels as a result.

   Twenty patients with vitamin K deficiency were randomly chosen and their prothrombin levels were measured. The mean prothrombin level for the sample was 20 mg/100 ml and the variance was 121 (mg/100 ml)^2. Calculate 95% confidence intervals for the population mean and variance from these statistics.

11. To determine the frequency of type O blood (the universal donor) in a population, a random sample of 100 people were blood typed for the ABO group. Of this 100, 42 were found to be type O. Calculate the 95% confidence limits for the proportion of the population that has type O blood.

12. In a pilot study, a public health survey of 2000 people aged 40 to 49 in Boston revealed that 29 had AIDS. Estimate the population proportion and find a 98% confidence interval for the estimate. Express your answer as a rate per 100,000.

13. (a) Twenty-five adult black ducks from a large flock were trapped and weighed. Their mean weight was 1500 g and the variance of their weights was 8100 g^2. Find the 95% confidence intervals for the mean, variance, and standard deviation of the flock.

   (b) Dabbling ducks such as black ducks and mallards often have very skewed sex ratios. In the sample above, 10 were found to be females and the rest males. Use this information to generate a 95% confidence interval for the proportion of females in the flock.

14. As part of a review of Irish healthcare web sites, researchers examined 60 different web sites. Of these, 46 contained service information intended for the general public. The Department of Health in Ireland recommends a reading age at or below 12 to 14 years for health information leaflets aimed at the general public. The following table provides the frequency and reading age for the 46 sites with service information. (Based on data reported in: O'Mahony, B. 1999. Irish health care web sites: a review. *Irish Medical Journal*, **92**(4): 334–336.)

Reading age	Web sites
8	1
9	0
10	1
11	2
12	1
13	4
14	2
15	7
16	4
17	24

(a) Find the mean reading age and standard deviation for these data.

(b) Make a bar graph of these data.

(c) From your bar graph, does it appear to you that the reading age for these web sites is at or below 14 years of age?

(d) Find a 95% confidence interval for the reading age. Does the confidence interval support your impression from part (c) above?

15. Hormonal oral contraceptives continue to be one of the most effective reversible means of family planning. Glucose intolerance was one of the first side effects associated with their use. In a study of the effects on blood glucose levels, 100 woman had their levels of fasting blood glucose levels measured before beginning and after completing a 6 month regime of taking an oral contraceptive. Cholesterol and triglyceride levels were also measured. Find 95% confidence intervals for the mean of each baseline and posttrial measurement. (Based on data reported in: Mostafavi, H. et al. 1999. A comparative analysis of three methods of contraception: Effects on blood glucose and serum lipid profiles. *Annals of Saudi Medicine*, **19**(1): 8–11.)

	$\overline{X}$	s
FBG (mmol/l)		
Baseline	3.84	0.50
After 6 months	4.70	1.00
Cholesterol (mmol/l)		
Baseline	3.90	0.80
After 6 months	4.20	0.98
Triglyceride (mmol/l)		
Baseline	1.05	0.39
After 6 months	1.25	0.96

16. The human carcinogen dioxin can be produced by the burning of common trash. Curiously it is hypothesized that trash from homes that recycle will produce more dioxins per kilogram of trash burned than homes that don't recycle. It is thought that a higher proportion of chlorine and metals in the recyclers' trash is the cause of higher levels of dioxin per kilogram. Below are the dioxin levels in picograms/kg of trash from 15 families that are avid recyclers.

Dioxin (picograms/kg)				
251	247	207	281	290
260	265	190	212	150
264	230	219	250	177

(a) Calculate the mean and standard deviation for these data.

(b) Calculate the 95% confidence intervals for the mean and variance of these data.

(c) If a well-run municipal incinerator produces 200 picograms/kg of trash burned, do you feel the recyclers' trash produces significantly more dioxin per kilogram?

17. In a study of the effectiveness of acupuncture to relieve the symptoms of shoulder impingement, 53 of 80 patients claimed to have complete recovery within one year of the end of a series of acupuncture treatments. Is this number significantly higher than the 50% that claim the same result with surgery? Find the 95% confidence interval for the proportion p of those obtaining complete recovery from acupuncture and use this interval to answer the question.

18. In a study of the mating behavior of the tent cobweb weaver spider, *Tidarren sisyphoides*, copulatory duration was one of the features studied. In this species the male dies during insertion and remains coupled to the female epigynum for a period of time. He is not consumed by the female as in many species but is instead eventually removed from the web by the fertilized female.

Data from 15 copulations indicated an average copulation time of 144 min with a standard deviation of 42 min. From the information given construct 99% confidence intervals for the mean and standard deviation of the copulation times.

19. The flightless birds of the genus *Dinornis*, commonly known as moas, were hunted to extinction shortly after the islands of New Zealand were permanently settled around 1310. Because of their recent extinction and the large population sizes prior to human exploitation a great many fossil bones can be found. From a remote site on the North Island, 16 femur bones were recovered and measured. These bones had an average length of 122 cm and a variance of 64 cm^2.

The paleontologist who found the bones believes that they are a new species that is larger than the largest known moa species, *Dinoris giganteus*. *D. giganteus* is described as having a mean femur of 115 cm with a variance of 30 cm^2.

Calculate the 95% confidence intervals for the mean and variance of the newly discovered bones. Use these confidence intervals to argue *for* or *against* the new species hypothesis.

20. In a population where the proportion of people who have the ability to taste umami is unknown, a random sample of 100 people were tested and 74 could clearly distinguish umami from saltiness. Calculate the 95% and 99% population proportion confidence intervals for this finding.

21. (a) A pharmaceutical company wants to obtain a 95% confidence interval for the population proportion p of people who obtain relief from headache pain in under 30 minutes using their product with a margin of error of no more than 3%. What size should their survey sample be?

(b) What sample size would be required for a 99% confidence interval?

22. The unicorn myth may come from a giant rhinoceros, *Elasmotherium sibiricum*, that lived in Southern Russia, Ukraine, and Moldova during the Early Pleistocene. These animals were 6 m long and over 2 m high with single large horns in their foreheads. The horns were claimed to be very long ($\mu = 2.0$ m) and uniform in size ($\sigma^2 = 0.09$ m^2). From fossil remains, 16 horns were recovered and measured and the summary statistics were: $\overline{X} = 1.8$ m, $s^2 = 0.25$ m^2, and $s = 0.5$ m.

(*a*) Find the 95% confidence interval for the horn mean using the fossil data.

(*b*) Find the 95% confidence interval for the horn variance using the fossil data.

(*c*) Do the data and confidence intervals support the claims of size and uniformity?

23. A recent investigation of the long-term physiological effects of child abuse studied the body's stress response in adulthood of individuals who had been abused as children. The abuse included maternal rejection, physical abuse, sexual abuse, and two or more changes in a child's primary caregiver. In formerly abused adults at age 32, researchers studied the blood concentration of the inflammatory substance C-reactive protein (CRP), which fights infections and repairs damaged tissue. The data were collected as mg/l and are summarized below.

$$n = 25, \qquad \sum_{i=1}^{25} X_i = 88, \qquad \sum_{i=1}^{25} X_i^2 = 340$$

(*a*) Find the mean and standard deviation for these data.

(*b*) If normal levels are 2.0 mg/l, would you say these individuals' CRP levels were extraordinarily high? Use your understanding of the Central Limit Theorem and a probability argument here.

24. QCR. True or False. Justify your answer.

(*a*) As long as the sample size is relatively large, $n \geq 30$, the sampling distribution of the sample mean, $\overline{X}$, is normal or approximately normal according to the Central Limit Theorem.

(*b*) Confidence intervals always give limits (L_1, L_2) that are equidistant from the point estimate of the population parameter under investigation.

(*c*) As long as the sampling distribution is known, confidence intervals of any percentage (90, 95, 99, etc.) can be calculated.

(*d*) In general, before it's calculated, a 95% confidence interval has a 0.95 probability of capturing the parameter under investigation. After the calculation is made the interval either contains the parameter (probability $= 1$) or doesn't contain it (probability $= 0$). So we say 95% of such intervals will contain the parameter or we are 95% confident in the interval.

(*e*) Using the same data set, a 99% confidence interval calculation will always generate a wider interval than a 95% confidence interval.

$n = 25$ and assume $\sigma = 1.0$ oz and is widely known. Not a situation that would occur in the real world very often, but then you're not going to be a grocery buyer. This is just a heuristic example!

Suppose the mean of your 25 box sample is 10.36 oz. Is that significantly different from (greater than) 10 oz so that we should reject the claim of 10 oz stated in the H_0? Clearly it is greater than 10 oz, but is this mean *rare enough* under the claim of $\mu \leq 10$ oz for us to reject this claim?

To answer this question we will use the standard normal transformation to find the probability of $\overline{X} \geq 10.36$ oz when the mean of the sampling distribution of $\overline{X}$ is 10 oz. If this probability is less than 0.05 (1 in 20), we consider the result to be too rare for acceptance of H_0.

Remember that the X_i's (individual weights) may or may not be normally distributed but $\overline{X}$'s are approximately normally distributed according to the Central Limit Theorem. So

$$Z = \frac{\overline{X} - \mu}{\frac{\sigma}{\sqrt{n}}}.$$

We can't find the probability of $\overline{X} \geq 10.36$ oz directly, but the probability of its Z transformation is easy to find in Table C.3:

$$Z = \frac{10.36 - 10.00}{\frac{1}{\sqrt{25}}} = \frac{0.36}{0.20} = 1.8.$$

So

$$
\begin{aligned}
P(\overline{X} \geq 10.36 \text{ oz}) &= P(Z \geq 1.8) \\
&= 1 - P(Z < 1.8) \\
&= 1 - F(1.8) \\
&= 1 - 0.9641 \\
&= 0.0359.
\end{aligned}
$$

Illustrated below is the distribution of z values for a true null hypothesis with $\sigma = 1.0$ oz and $n = 25$:

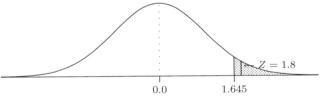

So the probability of a *sample* mean based on 25 observations being as high as 10.36 oz when the *population* mean is 10 oz, is about 3.59%. This $\overline{X}$ has to be considered a *rare* result *if H_0 is true*, so here we should be very suspicious of H_0 and consider the alternative hypothesis more likely. More generally, we say the result is *statistically significant*, so we *reject* H_0 and *accept* the research hypothesis H_a.

Having done the calculation above, you would be armed with an objective criterion to support your view that the salesman is conservative. Your real-life action based on this experiment would probably be to buy cornflakes from this particular salesman because the product is a good value. You would come to this conclusion for any Z value greater than 1.645 because values greater than this number occur only 5% of the time if H_0 is true. (See again, Table C.3.)

deviation from the claim (10 oz) in a particular direction ($\mu > 10$ oz), but in Case III we looked for a deviation from the claim (10 oz) in either direction ($\mu < 10$ oz or $\mu > 10$ oz). Case III is a less focused research question (H_a) that requires us to split the probability of rare events into two tails. Functionally, this requires the result to be more different from the prediction under H_0 in order to be sufficiently rare and reject H_0. Whenever possible you should have a one-tailed research question. In Case III we again accept the salesman's claim of 10 oz because the results of our experiment (sampling of 25 boxes) are not significantly different from his claim.

The famous cornflakes example is meant to show you how to formulate research questions into testable H_0's and H_a's and how to use an objective index (the Z transformation) to decide between the two competing hypotheses. Remember that sometimes extreme values occur even when H_0 is true, and that when this happens we will inappropriately reject H_0. This mistake is called a Type I error and its probability is determined by us before the experiment is run.

In the cornflakes example we were willing to accept this type of error 1 in 20 times or with a 0.05 probability. Remember, also, that in real life you will face problems like the cornflakes example but will generate only one pair of hypotheses (Case I, Case II, or Case III) to investigate a particular situation. It is incorrect and violates the philosophy underlying the scientific method to use the same data to test each possible H_a.

5.2 Typical Steps in a Statistical Test of Hypothesis

Here we list the general steps followed in a test of hypothesis:

1. State the problem. Should I buy cornflakes from this salesman?

2. Formulate the null and alternative hypotheses:

(1)

$$H_0: \mu = 10 \text{ oz}$$
$$H_a: \mu \neq 10 \text{ oz}.$$

3. Choose the level of significance. This means to choose the *probability of rejecting a true null hypothesis*. We chose 1 in 20 in our cornflakes example, that is, 5% or 0.05. When Z was so extreme as to occur less than 1 in 20 times if H_0 were true, we rejected H_0.

(2)

4. Determine the appropriate test statistic. Here we mean the index whose sampling distribution is known, so that objective criteria can be used to decide between H_0 and H_a. In the cornflakes example we used a Z transformation because under the Central Limit Theorem $\overline{X}$ was assumed to be normally or approximately normally distributed and the value of σ was known.

(3)

5. Calculate the appropriate test statistic. Only after the first four steps are completed, can one do the sampling and generate the so-called test statistic. Here

$$Z = \frac{10.36 - 10.00}{\frac{1}{\sqrt{25}}} = \frac{0.36}{0.20} = 1.8.$$

6. (a) Determine the critical values for the sampling distribution and appropriate level of significance.

$\left(4a\right)$

For the two-tailed test and level of significance of 1 in 20 we have critical values of ± 1.960. These values or more extreme ones only occur 1 in 20 times if H_0 is true. The critical values serve as cutoff points in the sampling distribution for regions to reject H_0.

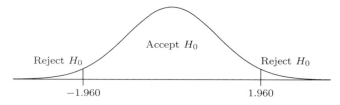

Sampling distribution of Z, if H_0 is true.

$\left(4b\right)$

(b) Determine the **P value** of the test statistic. Alternatively, we can find the probability of our result or one more extreme if H_0 is true and use this so-called P value to choose between the two hypotheses.

From Table C.3, $P(Z \geq 1.8) = 1 - F(1.8) = 1 - 0.9641 = 0.0359$ and $P(Z \leq -1.8) = F(-1.8) = 0.0359$, so the probability of a Z value as extreme as 1.8 or -1.8 is $0.0359 + 0.0359 = 0.0718$. We combine the two tails here because we are looking for extremes in either direction according to the research hypothesis, H_a: $\mu \neq 10$ oz.

Finding critical values as listed in 6(a) is a widely established technique, but determining the P value is a more modern way of looking for extreme results.

$\left(5a\right)$

7. (a) Compare the test statistic to the critical values. In a two-tailed test, the c.v.'s $= \pm 1.960$ and the test statistic is 1.8, so $-1.960 < 1.8 < 1.960$.

(b) Compare the P value to your chosen level of significance. Here the P value is $0.0718 > 0.05$, which is the level of significance.

$\left(5b\right)$

8. Based on the comparison in Step 7, accept or reject H_0.

(a) Since Z falls between the critical values, it is not extreme enough to reject H_0.

(b) Since the P value of Z is greater than 0.05, the Z value is not rare enough to reject H_0. We accept H_0 and, therefore, reject H_a.

$\left(6\right)$

9. State your conclusion and answer the question posed in Step 1. The experiment did not yield results that were so unexpected under H_0 that we could confidently reject H_0. We, therefore, accept H_0 and would buy cornflakes from this particular salesman.

The nine steps represent the general methodology for any statistical hypothesis testing. Steps 1–4 should always be completed before any sampling or experimentation is done. Whether you choose to use the critical value method or the P value method (Steps 6–7), the decision about H_0 will always be the same if you use the same level of significance.

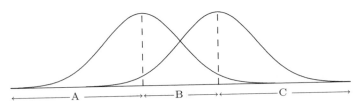

FIGURE 5.2. Means in region A support H_0 while means in region C support H_a.

Suppose now we test H_0 by generating a sample mean based on $n = 25$ and use $\alpha = 0.05$ as the probability of a Type I error. We're saying that only when $\overline{X}$ is so rare that it would occur by chance 1 in 20 times if H_0 were true, will we reject H_0.

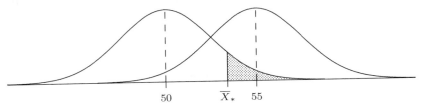

FIGURE 5.3. $\overline{X}_*$ will be the cutoff point for accepting H_0.

Let $\overline{X}_*$ (Figure 5.3) denote the cutoff point for accepting H_0. We can find this value utilizing the standard normal distribution. From Table C.3, $0.05 = P(Z \geq 1.645)$. But

$$Z = \frac{\overline{X}_* - \mu}{\frac{\sigma}{\sqrt{n}}}$$

so

$$1.645 = \frac{\overline{X}_* - 50}{\frac{10}{\sqrt{25}}},$$

where $\mu = 50$ comes from H_0. Therefore

$$\overline{X}_* = 50 + 1.645\frac{10}{\sqrt{25}} = 53.29.$$

Using $\alpha = 0.05$ we would reject H_0: $\mu = 50$ whenever the sample $\overline{X}$ was greater than 53.29 and would make a Type I error. Values of $\overline{X}$ greater than 53.29 will occur 5% of the time when H_0 is true.

What if H_0 were false and H_a were true? Now the sampling distribution is H_a not H_0 in Figure 5.4 and any value below 53.29 would be viewed as supporting H_0. Since H_0 is false, any $\overline{X}$ below 53.29 would lead to a Type II error. For this special case we can now calculate the probability β of a Type II error that corresponds to the lightly stippled area of Figure 5.4.

Using the Z transformation,

$$P(\overline{X} < 53.29) = P\left(Z < \frac{53.29 - 55}{\frac{10}{\sqrt{25}}}\right) = P(Z < -0.855) = 0.1963$$

from Table C.3. So for the special case H_0: $\mu = 50$ and H_a: $\mu = 55$, if H_0 were true and we use $\alpha = 0.05$ we will make a Type I error 5% of the time, but if H_a were

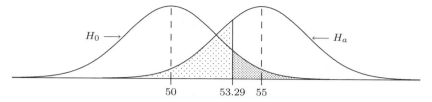

FIGURE 5.4. If H_0 were false, any $\overline{X}$ below 53.29 would lead to a Type II error.

true we will make a Type II error almost 20% of the time (19.63%). The power of the Z test in this case would be $1 - 0.1963 = 0.8037$. Remember that power is the probability of rejecting a false H_0.

Suppose now we have exactly the same situation but reduce α to 0.01. See Figure 5.5. What happens to β and the power of the test?

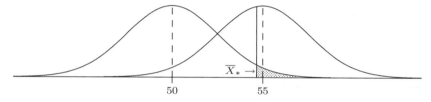

FIGURE 5.5. $\overline{X}_*$ is the cutoff point for accepting H_0 when $\alpha = 0.01$.

With a smaller α, the cutoff point to reject H_0 is shifted to the right because in the standard normal distribution $0.01 = P(Z \geq 2.33)$. So this time

$$\overline{X}_* = 50 + 2.33\frac{10}{\sqrt{25}} = 54.66.$$

So with $\alpha = 0.01$, we would reject H_0: $\mu = 50$ whenever the sample $\overline{X}$ was greater than 54.66 and would make a Type I error 1% of the time if H_0 were true.

What if H_0 were false and H_a were true? The sampling distribution is shown in Figure 5.6, and any value below 54.66 would be viewed as supporting H_0. Again, because H_0 is now false, any $\overline{X}$ below 54.66 would lead to a Type II error β. From Table C.3 this probability is

$$P(\overline{X} < 54.66) = P\left(Z < \frac{54.66 - 55}{\frac{10}{\sqrt{25}}}\right) = P(Z < -0.17) = 0.4325.$$

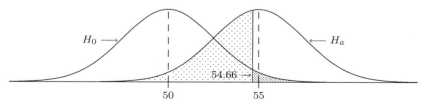

FIGURE 5.6. If H_0 were false, any $\overline{X}$ below 54.66 would lead to a Type II error.

From our special case of H_0: $\mu = 50$ and H_a: $\mu = 55$, if we reduce the probability of Type I error from 0.05 to 0.01, we simultaneously increase the probability of a Type II error from 19.63 to 43.25%. We also reduce the power $(1 - \beta)$ from 80.37 to 56.75%. Take a few minutes to verify that changing α to 0.001 increases β to 71.90% and reduces the power to 28.10%. See the following table.

α	β	$1 - \beta$ (power)
0.05	0.1963	0.8037
0.01	0.4325	0.5675
0.001	0.7190	0.2810

The interrelationship of the magnitude of the two kinds of errors comes from the *overlap* of the two sampling distributions. If H_0: $\mu = 10$ and H_a: $\mu = 1000$ were tested, the sampling distributions would have virtually no overlap, so the probability of a Type II error would become vanishingly small. See Figure 5.7.

FIGURE 5.7. Sampling distributions for $\mu = 10$ and $\mu = 1000$ with $\sigma = 10$ and $n = 25$.

So the power of the test depends on both the level of significance (α) and the value of $\mu_0 - \mu_a$, the true difference between means. Alpha is under our control and is knowable, but in most cases (except artificial ones like above) the difference between means is not known. Remember we usually write hypotheses in the following form: H_0: $\mu \leq 50$ and H_a: $\mu > 50$, implying a *range of possibilities* for the alternative hypothesis.

Looking back at Figure 5.1, the sampling distributions under H_0: $\mu = 50$ and H_a: $\mu = 55$, we might consider other factors influencing the size of the overlap. While the mean positions the distribution on the number line, the spread of the distribution is determined by the standard error $\frac{\sigma}{\sqrt{n}}$. The variability in observations σ and the sample size n both affect this spread. In some cases the standard deviation can be reduced by using more uniform experimental subjects, e.g., rats that are all the same sex or plants that are all the same genotype or clones. Regardless of your ability to reduce σ, you can always increase the power of a test by increasing the sample size (doing more work).

Return to the original test with $\alpha = 0.05$ and $\sigma = 10$:

$$H_0: \mu = 50$$
$$H_a: \mu = 55.$$

Use $n = 100$ instead of $n = 25$. By doing 4 times as much sampling we reduce the standard error from

$$\frac{\sigma}{\sqrt{25}} = 2$$

to

$$\frac{\sigma}{\sqrt{100}} = 1.$$

The sampling distributions for H_0 and H_a are positioned in the same places on the number line as before but exhibit much less spread.

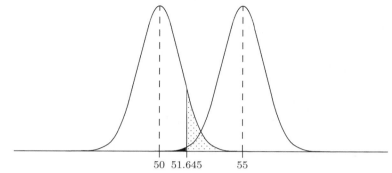

$$50 \quad 51.645 \qquad 55$$

This time with $\alpha = 0.05$,

$$\overline{X}_* = 50 + 1.645 \frac{10}{\sqrt{100}} = 51.645.$$

So from Table C.3,

$$\beta = P(\overline{X} < 51.645) = P\left(Z < \frac{51.645 - 55}{\frac{10}{\sqrt{100}}}\right) = P(Z < -3.355) = 0.0004.$$

With the same H_0 and H_a, the same σ and 4 times as much work ($n = 100$ rather than $n = 25$) the overlap of the sampling distributions is greatly reduced. This results in β being reduced from 19.63 to 0.04% and the power increased from 80.37 to 99.96%. Even in statistical analysis there is no substitute for well-planned work!

To recap, the level of significance, α, is under our control and is usually chosen to be 0.05, 0.01, or 0.001. Rejecting an H_0 at 0.05 we say the result is *significant*, i.e., significantly different from what we expect with H_0 true. Rejecting an H_0 at 0.01 is considered *highly significant,* and rejecting at 0.001 is said to be *very highly significant.* In a test of hypotheses we will either accept or reject the null hypothesis. If we reject H_0 we may have made a Type I error (rejecting a true H_0), and if we accept H_0 we may have made a Type II error (accepting a false H_0). Because we have these two types of error and one is potentially possible in any decision, we *never* say that we've *proved* that H_0 is true or that H_0 is false. Proof implies that there is no possibility for error. Instead we say the data *support* the null hypothesis or the data *support* the alternative hypothesis, when we accept H_0 or reject H_0, respectively.

5.4 Binomial Example of Hypothesis Testing (Optional)

Some readers might like to work through another hypothesis testing situation using a distribution with which many are more comfortable, the binomial. This distribution will allow exact values of α and β to be calculated given various p's and n's.

In a genetics class you make several crosses of fruit flies including Aa × Aa and Aa × aa. Because your lab partner (it's always the lab partner) failed to label the two cultures above, you need to determine which is the Aa × Aa cross and which is the Aa × aa. The aa genotype is recessive and leads to an easily identifiable phenotype called *arc wings.* If progeny are produced by the Aa × Aa mating, according to Mendelian genetics, we expect $\frac{1}{4}$ to have arc wings, and if progeny are produced by the Aa ×

aa mating, again according to Mendelian rules, we expect $\frac{1}{2}$ to have arc wings. Your plan is to sample one of the cultures and determine the phenotypes of 20 randomly chosen flies. See Table 5.2. From this sample you hope to determine which set of parents produced the progeny that you sampled. You have two predictions based on Mendelian genetics:

$$H_0\text{: } p = 0.25$$
$$H_a\text{: } p = 0.50.$$

You've already decided to sample 20 flies, so $n = 20$ and you choose $\alpha = 0.05$. How many arc-winged flies would we need to find in the sample to reject H_0? In other words, when is the number of arc-winged flies high enough that we are sufficiently suspicious of the claim that arc-winged individuals compose 25% of the progeny? From Table C.1, the pdf's for both $p = 0.25$ and $p = 0.50$ with $n = 20$ are determined.

TABLE 5.2. The probability density functions for both $p = 0.25$ and $p = 0.50$ with $n = 20$

x = no. of arc-winged	$p = 0.25$	$p = 0.50$
0	0.0032	0.0000
1	0.0211	0.0000
2	0.0670	0.0002
3	0.1339	0.0011
4	0.1896	0.0046
5	0.2024	0.0148
6	0.1686	0.0370
7	0.1124	0.0739
8	0.0609	0.1201
9	0.0270	0.1602
10	0.0100	0.1762
11	0.0030	0.1602
12	0.0007	0.1201
13	0.0002	0.0739
14	0.0000	0.0370
15	0.0000	0.0148
16	0.0000	0.0046
17	0.0000	0.0011
18	0.0000	0.0002
19	0.0000	0.0000
20	0.0000	0.0000

(The $p = 0.50$ values for $x = 0$ through 8 are bracketed with a sum of 0.2517. The $p = 0.25$ values for $x = 9$ through 20 are bracketed with a sum of 0.0409.)

If we test the hypothesis that $p = 0.25$, we have a one-tailed, right-tailed test. According to the binomial distribution characteristics, we expect the number of arc-winged flies to be $E(X) = np = 20(0.25) = 5$, if H_0 is true. Further, we expect the number of arc-winged flies to be significantly greater than 5, if H_a is true $[E(X) = np = 20(0.5) = 10]$. How far from the expected value of 5 does the number of arc-winged flies have to be to reject H_0? If we use $\alpha = 0.05$, we are saying we will reject H_0 if the result is so extreme as to happen by chance less than 1 in 20 times, if H_0 is true. Looking at the pdfs in Table 5.2 we see that the sum of the probabilities for $X \geq 9$ when $p = 0.25$ is 0.0409. Values of 9 or more arc-winged flies in samples of size 20 occur about 4% of the time when arc wings makes up 25% of the population ($p = 0.25$). Eight or more has a probability of 0.1018, so 9 is the cutoff point.

If we sample the culture and find 9 or more with arc wings in 20 flies, we will assume the culture isn't the Aa × Aa one. The probability of a Type I error (rejecting a true null hypothesis) is then 0.0409.

If we sample the culture and find 8 or fewer with arc wings in the 20 flies, we accept the null hypothesis. When we do this we may be making a Type II error (accepting a false null hypothesis). This probability can be found by summing the probabilities in the last column of Table 5.2 for $X \leq 8$. $P(X \leq 8 \mid p = 0.50) = 0.2517$. So if the number of arc-winged flies is 8 or fewer, there is a 0.2517 probability that the sample came from the population with $p = 0.50$. The probability of Type II error is 0.2517.

Given that $\alpha = 0.0409$ and $\beta = 0.2517$, you should be much more comfortable rejecting H_0: $p = 0.25$ than accepting it in this situation. Looking at Table 5.2 we see the most ambiguous results would be 7 or 8 flies with arc wings in the sample of 20. This is because 7 and 8 fall midway between the expectations for each hypothesis, H_0: $E(X) = 5$ and H_a: $E(X) = 10$. As we move closer to one or the other expectations the ambiguity fades.

5.5   Problems

1. Explain why most researchers are more comfortable rejecting H_0 than accepting it. Use a probability argument in your explanation.

2. Discuss the relationship between the Central Limit Theorem and the sampling distribution for $\overline{X}$.

3. Outline the factors affecting Type II error and the power of a test of the hypothesis that $\mu = 10$.

4. Explain why it is inappropriate to use the term "proof" when performing tests of hypothesis.

5. Why are H_0 and H_a always written as mutually exclusive and all-inclusive predictions?

6. A physical anthropologist studying the effects of nutritional changes on human morphometrics is interested in the impact of Western diet on the heights of adult Japanese males. Studies prior to World War II indicated that the heights of adult Japanese males had a mean of 64.0 inches and a standard deviation of 2.0 inches. The anthropologist measured the heights of a random sample of 25 adult Japanese men and found their height to average 66.5 inches with a standard deviation of 3.0 inches.

   (a) Calculate the 95% confidence interval for μ from the sample.

   (b) Calculate the 95% confidence interval for σ^2 from the sample.

   (c) Has the mean height changed since World War II? Write out the hypotheses associated with this question.

   (d) Has the variance of heights changed since World War II? Again, write out the hypothesis associated with this question.

   (e) Relate the confidence intervals in (a) and (b) to the hypotheses in (c) and (d).

7. Relating the concepts of Type I and Type II error to the American criminal justice system, we always have an H_0 that the person is innocent. In order to convict that person we must be sure beyond a reasonable doubt that H_0 is false. This really means that α is very small. With a small α we automatically accept a larger β. So a great many criminals escape punishment because we view one kind of error to be more egregious than the other kind of error. It is more tragic to convict an innocent person (Type I error: rejecting a true H_0) than to let many guilty people go unpunished (Type II error: accepting a false H_0). Can you think of biological examples where this kind of logic might lead you to choose very small alpha levels?

8. Write appropriate pairs of hypotheses for the following problems.

   (*a*) Chapter 1, Problem 4

   (*b*) Chapter 1, Problem 6

   (*c*) Chapter 4, Problem 14

   (*d*) Chapter 4, Problem 16

9. Return to Table 5.2. Suppose you had selected a smaller value of α for your test of hypotheses, say $\alpha = 0.01$. What is the minimum value of the number of arc-winged flies in a sample of size 20 that would allow you to reject H_0? What is the probability of making a Type II error now? Which value of α seems to be a better compromise between the two types of error in this situation?

10. QCR. True or False. Justify your answer.

   (*a*) The appropriate null and alternative hypotheses must be formulated before an experiment is carried out or any sampling done.

   (*b*) Critical values partition the sampling distribution into region(s) where H_0 will be accepted and region(s) where H_0 will be rejected.

   (*c*) Comparing the P value of a test statistic to the alpha level is equivalent to comparing the test statistic to a critical value at the same alpha level.

   (*d*) In a test of hypothesis it is generally easier to know the probability of a Type I error than the probability of a Type II error.

   (*e*) Since Type I error is under the experimenter's control, it is standard procedure to make it as small as possible.

   (*f*) If the probability of a Type I error is denoted by α, the probability of a Type II error is denoted by $1 - \alpha$.

   (*g*) The power of a statistical test of hypothesis can be increased or the probability of a Type II error decreased by increasing the sample size(s) in the experiment.

One-Sample Tests of Hypothesis

Concepts in Chapter 6:

- Hypothesis Tests Involving a Single Mean
    - Two-tailed t tests
    - Left-tailed t tests
    - Right-tailed t tests
- Hypothesis Tests Involving a Single Variance
    - Two-tailed χ^2 tests
    - Left-tailed χ^2 tests
    - Right-tailed χ^2 tests
- Nonparametric Tests and Hypothesis Testing
    - The One-Sample Sign Test for the Median
    - Confidence Intervals for the Median Based on the Sign Test
    - The One-Sample Wilcoxon Signed-Rank Test for the Median

One-Sample Tests of Hypothesis

We now have the tools needed to develop a variety of tests of hypotheses. Using our understanding of probability, the concept of sampling distributions, and the nine steps outlined in the previous chapter, we present both parametric and nonparametric tests involving a single sample. The tests presented and their inferences are the most common for one-sample situations, but are not meant to be an exhaustive survey of all possible tests.

6.1 Hypotheses Involving the Mean (μ)

We know from Chapter 4 that sample means have a sampling distribution known as the t distribution, and that probabilities for this distribution for various sample sizes are well known (Table C.4). This knowledge makes hypothesis tests on the parametric mean relatively straightforward, as illustrated in the following examples.

> **EXAMPLE 6.1.** A forest ecologist, studying regeneration of rainforest communities in gaps caused by large trees falling during storms, read that stinging tree, *Dendrocnide excelsa*, seedlings will grow 1.5 m/yr in direct sunlight in such gaps. In the gaps in her study plot she identified 9 specimens of this species and measured them in 2005 and again 1 year later. Listed below are the changes in height for the 9 specimens. Do her data support the published contention that seedlings of this species will average 1.5 m of growth per year in direct sunlight?

1.9   2.5   1.6   2.0   1.5   2.7   1.9   1.0   2.0

SOLUTION. With the question clearly stated we now frame H_0 and H_a to match this question. The ecologist is looking for deviations from 1.5 m in either direction, so we have a <u>two-tailed test</u> here:

uses
t value

$$H_0: \mu_d = 1.5 \text{ m/year}$$
$$H_a: \mu_d \neq 1.5 \text{ m/year}$$

If the sample mean for her 9 specimens is *close* to 1.5 m/year, we will accept H_0. If her sample mean is *significantly larger* or *smaller* than 1.5 m/year, we will accept H_a. In this case significantly different would be means that are so rare that they would occur by chance less than 5% of the time, if H_0 is true, i.e., $\alpha = 0.05$. Her test statistic here will be

σ known
↓
$z = \dfrac{\overline{X} - \mu_0}{\sigma/\sqrt{n}}$

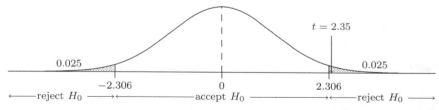

σ
not → $t = \dfrac{\overline{X} - \mu_0}{\frac{s}{\sqrt{n}}}$. *compares claimed value (μ)*
known   *to experimental value ($\overline{X}$)*   (6.1)

This statistic is appropriate because it compares the claimed value (μ) to the experimental value ($\overline{X}$). We expect the statistic (t value) to be zero or close to zero, if H_0 is true because the numerator should be small if H_0 is true. Further, we expect $\overline{X} - \mu$ to be quite different from zero, if H_a is true. Notice that we use t and not z in these tests because σ^2 is unknown and must be estimated from the sample.

If H_0 is true, $E(t) = 0$ and if H_a is true, $E(t)$ is different from zero. We use "different" here because the deviation could be in either direction, greater or less than claimed. The magnitude of t when compared to critical values gives us an *objective index* to use in accepting or rejecting H_0.

The sample statistics are $n = 9$, $\overline{X} = 1.90$ m, $s^2 = 0.260$ m^2, $s = 0.51$ m, and

$$t = \frac{\overline{X} - \mu}{\frac{s}{\sqrt{n}}} = \frac{1.90 - 1.50}{\frac{0.51}{\sqrt{9}}} = \frac{0.40}{\frac{0.51}{3}} = 2.35.$$

Clearly the t value of 2.35 is not zero, but is it far enough away from zero so that we can comfortably reject H_0? With a predetermined α level of 0.05 we must get a t value that is far enough from zero that it would occur less than 5% of the time if H_0 is true.

From Table C.4 we have the following sampling distribution for t with $\nu = n - 1 = 9 - 1 = 8$, and $\alpha = 0.05$ for a two-tailed test.

$t = 2.35$

0.025   0.025

-2.306 0 2.306

←——reject H_0 ——→←————accept H_0 ————→←——reject H_0 ——→

If H_0 *is true* and we sample hundreds or thousands of times with samples of size 9 and each time we calculate the t value for the sample, these t values would form a distribution with the shape indicated above. Two and one-half percent of the samples would generate t values below -2.306 and 2.5% of the samples would generate t values above 2.306. So values as extreme as ± 2.306 are rare, if H_0 is true. The test statistic in this sample is 2.35 and since $2.35 > 2.306$, the result would be considered rare for a true null hypothesis. We reject H_0 based on this comparison and conclude that the average growth of stinging trees in direct sunlight is *different* from the published value and is, in fact, *greater* than 1.5 m/year (we claim greater here because the index (t value) is in the right-hand tail.

5

Introduction to Hypothesis Testing

Concepts in Chapter 5:

- A General Overview of Hypothesis Testing
- The Basics of Hypothesis Testing
- Type I versus Type II errors
- Interrelationship of Type I and Type II errors
- Understanding Significance Levels
- Binomial Test of Hypothesis (Optional)

Introduction to Hypothesis Testing

The principal goal of this chapter is to tie your understanding of probability and sampling distributions to a process that will allow you to make objective decisions about the veracity of a variety of predictions. Recall that early in Chapter 1 we briefly outlined the general steps involved in applying the scientific method to problem solving. *Please take time to review these steps before working through the present chapter.*

5.1 An Overview: The Famous Cornflakes Example

To start the discussion of hypothesis testing we present an example that was used to introduce this topic to one of us many, many years ago by his graduate school mentor.

The Scenario

Upon leaving (hopefully graduating) your undergraduate institution, you begin work in a high paying job as a grocery store buyer (o.k., we know—but it's a pedagogical example). To challenge your education and intellectual ability, the store manager puts you in charge of buying cornflakes. You decide that you will be very objective and scientific in completing this task and prepare to meet the cornflakes salesman. The issue is not the price of cornflakes but the amount of cornflakes in each box. The salesman appears and claims that the cornflakes he is selling are packaged at 10 oz/box. You have *exactly* four possible views of his claim.

1. He is honest and $\mu = 10$ oz.

2. He is conservative and there is more than 10 oz/box; $\mu > 10$ oz.

3. He is trying to cheat you and there is less than 10 oz/box; $\mu < 10$ oz.

4. He is new on the job, just like you, and doesn't really know the amount per box; his claim could be high or low, $\mu \neq 10$ oz.

If you think he is honest you would just go ahead and order your cornflakes from him. You may, however, have one of the other views, he's conservative, he's a liar, or he's clueless. The position you hold regarding the salesman can be any one of these but not more than one. You can't assume he's a liar and conservative, i.e., $\mu < 10$ oz and $\mu > 10$ oz, at the same time.

Proper use of the scientific method will allow you to test one of these alternative positions through a sampling process. Remember you can choose only one to test. How do you decide?

Case I. *Testing that the salesman is conservative*

Suppose the salesman is remarkably shy and seems to lack self-confidence. He is very conservatively dressed and appears to often understate himself. You feel from his general demeanor that he is being conservative in his claim of 10 oz/box. This situation can be summarized with a pair of hypotheses, actually a pair of predictions. The first is the salesman's claim and is the prediction we will directly test. It is usually called H_0 or **null hypothesis**. In this case H_0: $\mu = 10$ oz. The second is called the **alternative** or **research hypothesis**, which is your belief or position. The alternative hypothesis in this case is H_a: $\mu > 10$ oz. By writing the null hypothesis as H_0: $\mu \leq 10$ oz, the predictions take the following form:

$$H_0: \mu \leq 10 \text{ oz}$$

$$H_a: \mu > 10 \text{ oz}$$

and we have generated two *mutually exclusive* and *all-inclusive* possibilities. Therefore, either H_0 or H_a will be true, but not both.

In order to test the salesman's claim (H_0) against your view (H_a), you decide to do a small experiment. You select 25 boxes of cornflakes from a consignment and carefully empty each box, then weigh, and record its contents. This experimental sampling is done after you've formulated the two hypotheses. If the first hypothesis were true you would expect the sample mean of the 25 boxes to be close to or less than 10 oz. If the second hypothesis were true you would *expect* the sample mean to be significantly greater than 10 oz. We have to pause here to think about what *significantly greater* means in this context. In statistics *significantly less* or *more* or *different* means that the result of the experiment would be a rare result if the null hypothesis were true. In other words, the result is far enough from the prediction in the null hypothesis that we feel that we must reject the veracity of this hypothesis.

This idea leads us to the problem of what is a rare result or rare enough result to be sufficiently suspicious of the null hypothesis. For now we'll say if the result could occur by chance less than 1 in 20 times if the null hypothesis were true, we'll reject the null hypothesis and consequently accept the alternative one. Let's now look at how this decision making criterion works in Case I:

$$H_0: \mu \leq 10 \text{ oz}$$

$$H_a: \mu > 10 \text{ oz},$$

Remember that rejecting H_0 may lead to a **Type I error** (rejecting a true H_0) and that we set this probability at 0.05 before analyzing the sample. With the data in hand we realize that we may be incorrect in rejecting H_0, but that this probability is less than 1 in 20.

uses
P value

EXAMPLE 6.2. Documents from the whaling industry of the Windward Islands in the Lesser Antilles indicate that in the 1920s short-finned pilot whales weighed an average of 360 kg. Cetologists believe that overhunting and depletion of prey species has caused the average weight of these whales to drop in recent years. A modern survey of this species in the same waters found that a random sample of 25 individuals had a mean weight of 336 kg and a standard deviation of 30 kg. Are pilot whales significantly lighter today than during the 1920s?

SOLUTION. We are anticipating a deviation from the claimed value *in á particular direction*, that is, a decrease in weight. This is a one-tailed, left-tailed test with hypotheses

$$H_0: \mu \geq 360 \text{ kg}$$
$$H_a: \mu < 360 \text{ kg}.$$

Here $n = 25$, $\overline{X} = 336$ kg, and $s = 30$ kg.

If H_0 is true, $E(t) \geq 0$ and if H_a is true, $E(t) < 0$. Let's use the P value protocol here to decide whether to accept or reject H_0. Recall that the P value is the probability of getting a test statistic as extreme as or more extreme than what we got in the sample *given* that H_0 is true. This technique requires that we calculate the t statistic and determine its P value. If the P value is *sufficiently small*, we will reject H_0 in favor of H_a.

$$t = \frac{\overline{X} - \mu}{\frac{s}{\sqrt{n}}} = \frac{336 - 360}{\frac{30}{\sqrt{25}}} = \frac{-24}{\frac{30}{\sqrt{25}}} = \frac{-24}{6} = -4.0.$$

Is -4.0 sufficiently below the expected value of zero to reject H_0? From Table C.4 with $\nu = n - 1 = 25 - 1 = 24$, $P(t < -3.745) = 0.0005$, so $P(t \leq -4.0) < 0.0005$.

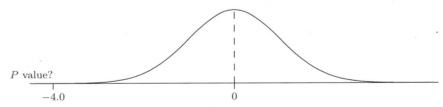

While we can't find the exact probability of $t < -4.0$ from Table C.4, we know that this probability is less than 1 in 1000 or 0.001. If we have a random sample of 25 whales and the mean of the population (μ) really is 360, we would get samples with an average of 336 kg or less only about 1 in 1000 times! We, therefore, reject H_0 here and accept H_a. One would say that the sample indicates that the population mean is significantly lower than 360 kg. The probability of a Type I error is in this case equal to the P value, less than 0.001. We are quite confident here that we are making the correct decision in rejecting H_0.

If we had used a critical value approach here and an alpha level of 0.05, we would have found that the cutoff point for rejecting H_0 from Table C.4 is -1.711 for $\nu = 24$.

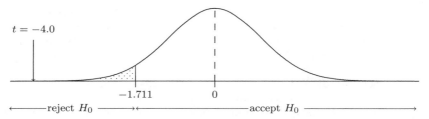

Since $-4.0 < -1.711$, we would reject H_0 at the 0.05 significance level. Both methods (P value and critical value) lead us to the same decision. The P value indicates more precisely the probability of a Type I error and is becoming the preferred approach in tests such as this.

EXAMPLE 6.3. To test the effectiveness of a new spray for controlling rust mites in orchards, a researcher would like to compare the average yield for treated groves with the average displayed in untreated groves in previous years. A random sample of 16 one-acre groves was chosen and sprayed according to a recommended schedule. The average yield for this sample of 16 groves was 814 boxes of marketable fruit with a standard deviation of 40 boxes. Yields from one acre groves in the same area without rust mite control spraying have averaged 800 boxes over the past 10 years. Do the data present evidence to indicate that the mean yield is sufficiently greater in sprayed groves than in unsprayed groves?

SOLUTION. We are anticipating a particular change, an *increase* in yield, so we have a one-tailed, right-tailed test:

$$H_0: \mu \leq 800 \text{ boxes}$$
$$H_a: \mu > 800 \text{ boxes.}$$

Here $n = 16$, $\overline{X} = 814$ boxes, and $s = 40$ boxes:

$$t = \frac{\overline{X} - \mu}{\frac{s}{\sqrt{n}}} = \frac{814 - 800}{\frac{40}{\sqrt{16}}} = \frac{14}{10} = 1.40.$$

If H_0 is true, $E(t) \leq 0$ and if H_a is true, $E(t) > 0$.

Obviously 814 boxes is greater than the average of 800 boxes, but is it significantly greater? Because pesticide sprays are expensive and possibly detrimental to the environment, one must be reasonably sure that they increase the yield enough to offset these added costs. Below is the sample distribution of t if H_0 is true:

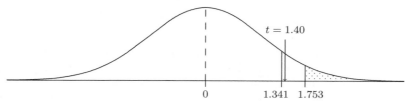

From Table C.4, $P(t > 1.341) = 0.10$ and $P(t > 1.753) = 0.05$, so $0.05 < P(t \geq 1.40) < 0.10$. A mean of 814 boxes in samples of size 16 would be expected between 5 and 10% of the time when the population mean *really is 800 boxes*. So here we don't see 814 as a particularly large deviation from the expected yield and would be uncomfortable rejecting H_0 for this small increase.

Using a critical value approach, with $\alpha = 0.05$ and $\nu = 15$, the cutoff to reject H_0 would be 1.753. Since $1.4 < 1.753$ we would have to accept H_0 here. By accepting H_0

we may be committing a Type II error, but we have no way of knowing the probability of this type of mistake. In this example we simply don't have enough evidence to claim that yield is *significantly* increased using the rust mite control spray. If the t statistic had been significantly different from zero (say, 3.0), then we would have rejected H_0 and would have to decide about *biological* or *economic* significance. Will an increase of 14 boxes pay for the sprays with a little leftover to increase the profits? This question doesn't have to be addressed because the result wasn't statistically significant. Watch for these two types of significance in future examples.

To Summarize

Hypotheses about the population mean can take one of three forms:

a) $H_0: \mu = c$ $E(t) = 0$ b) $H_0: \mu \geq c$ $E(t) \geq 0$ c) $H_0: \mu \leq c$ $E(t) \leq 0$

      $H_a: \mu \neq c$ $E(t) \neq 0$ $H_a: \mu < c$ $E(t) < 0$ $H_a: \mu > c$ $E(t) > 0$,

where c is a real number chosen *before* the data are gathered. Each H_0 above is *tested* with a t *statistic*, and the decision about H_0 is based on how far this statistic deviates from *expectation* under a true H_0. If the t statistic exceeds the critical value(s), H_0 is rejected. Alternatively, if the P value for the t statistic is smaller than the predetermined alpha level, H_0 is rejected.

For any particular experiment *only* one of the sets of hypotheses is appropriate and can be tested. In other words, you can't test (a) then (b) using the same data. H_0 and H_a are predictions that follow naturally from the question posed and the result anticipated by the researcher. Also, *hypotheses contain only parameters (Greek letters) and claimed values, never numbers that came from the sample itself. H_0 always contains the equal sign and is the hypothesis that is examined by the test statistic.*

Before we move on to another type of test of hypothesis, take a moment to review confidence intervals for the population mean (Section 4.3) and see how they relate to the two-tailed t tests presented above. See, in particular, Example 4.8. In a two-tailed t test of the hypothesis, $H_0: \mu = c$, the test statistic

$$\frac{\overline{X} - \mu}{\frac{s}{\sqrt{n}}}$$

is compared to critical values ($\pm t_0$) gleaned from the appropriate t distribution table. A confidence interval with the same α level uses the same table values to generate L_1 and L_2 by calculating $\overline{X} + t_0 \frac{s}{\sqrt{n}}$. So L_1 and L_2 generated in this fashion represent the limits of all the μ values that would not be rejected in a two-tailed t test with the same alpha level.

Sample Size Considerations (Optional)

When designing an experiment one of the most important considerations is how large a sample to collect. Too large a sample results in wasted time and effort. Too small a sample makes it impossible to find real and important differences between the hypotheses. From the discussions of Type I and Type II errors in Chapter 5, it is apparent that sample size greatly influences the power of the test of hypothesis (see Section 5.3).

We return to Example 6.2, the study of weights of short-finned pilot whales; the original hypothesis was

$$H_0: \mu \geq 360 \text{ kg}.$$

How large a sample would have to be collected so that a decrease in weight of 25 kg could be detected 95% of the time? In other words, if the true mean is 335 kg and we want the probability of accepting a false null hypothesis to be $\beta = 0.05$, how many animals have to be weighed?

Since Type I error is interrelated with Type II error, we need to determine an acceptable level of α (rejecting a true null hypothesis) before determining the required sample size. For our present example, let's set $\alpha = 0.01$.

Earlier in this section, hypotheses about a single mean were analyzed using the t statistic. Since critical values for this statistic are determined by sample size, the t distribution cannot be used directly to estimate appropriate sample size for an experiment. The limit of the t distribution is the standard normal distribution or the z distribution, which is not dependent on sample size.

Consider the Z statistic

$$Z = \frac{\overline{X} - \mu}{\frac{\sigma}{\sqrt{n}}}.$$

At $\alpha = 0.01$, because H_a is left-tailed, $z = -2.33$ (see Table C.3).

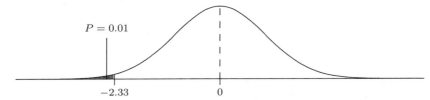

From the null hypothesis $\mu = 360$ kg. While σ is unknown and unknowable without measuring every member of the population, we have a preliminary estimate of σ in $s = 30$ kg. Filling in the various claims and estimates, we have

$$-2.33 = \frac{\overline{X} - 360}{\frac{30}{\sqrt{n}}}.$$

solving for $\overline{X}$,

$$\overline{X} = 360 - 2.33 \left(\frac{30}{\sqrt{n}} \right).$$

We would reject the null hypothesis if $\overline{X}$ is less than this value.

Return now to considerations of the power of the test. If the true mean were actually 335 kg, then the standard normal deviate, Z, could be expressed as

$$\frac{\overline{X} - 335}{\frac{30}{\sqrt{n}}}.$$

We want to reject the null hypothesis with a probability of $1 - \beta$ or 0.95. The Z value that corresponds to $\beta = 0.05$ is 1.645.

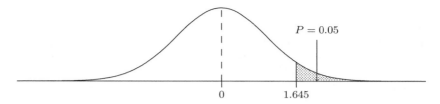

So

$$1.645 = \frac{\overline{X} - 335}{\frac{30}{\sqrt{n}}},$$

and solving for $\overline{X}$ yields

$$\overline{X} = 335 + 1.645\frac{30}{\sqrt{n}}.$$

Next, set the two expressions for the sample mean $\overline{X}$ equal to each other

$$360 - 2.33\left(\frac{30}{\sqrt{n}}\right) = 335 + 1.645\left(\frac{30}{\sqrt{n}}\right).$$

Multiply both sides of the equation by $\sqrt{n}$ and collect the terms to obtain

$$\sqrt{n}(360 - 335) = (2.33 + 1.645)(30).$$

Finally, solve for n,

$$n = \left[\frac{(2.33 + 1.645)30}{360 - 335}\right]^2 = 22.75. \qquad (6.2)$$

Therefore, a sample of 23 would allow detection of a decrease in weight of 25 kg at least 95% of the time.

In the example above, the test was one-tailed and left-tailed. We expected the weight of short-finned pilot whales to decrease. If a more general two-tailed test is carried out (H_0: $\mu = 360$ kg versus H_a: $\mu \neq 360$ kg), we must modify the critical value of z for the appropriate α level. At $\alpha = 0.01$, $\frac{\alpha}{2}$ is 0.005 and the z critical values become ± 2.575 (Table C.3). Substituting this critical value into (6.2),

$$n = \left[\frac{(2.575 + 1.645)30}{360 - 335}\right]^2 = 25.64.$$

A sample of 26 would be required to reject a false null hypothesis 95% of the time. The sample size required for a two-tailed test will always be larger than for a comparable one-tailed test.

To generalize (6.2) for arbitrary α and β, the sample size for a one-tailed test can be estimated by

$$n = \left[\frac{(z_\alpha + z_{1-\beta})\sigma}{\mu_0 - \mu}\right]^2. \qquad (6.3)$$

Clearly, the choices of α and β affect the sample size requirements, but so do the size of the difference in means $\mu_0 - \mu$ you wish to detect and the intrinsic variability among observations, σ.

TABLE 6.1. Sample sizes required for various α and β depending on the size of $\left|\frac{\sigma}{\mu_0-\mu}\right|$

One-sided Two-sided	$\alpha = 0.05$ $\alpha = 0.10$			$\alpha = 0.025$ $\alpha = 0.05$		
$\left|\frac{\sigma}{\mu_0-\mu}\right|$	$1-\beta=0.90$	$1-\beta=0.95$	$1-\beta=0.99$	$1-\beta=0.90$	$1-\beta=0.95$	$1-\beta=0.99$
0.50			4		4	5
0.55		4	5	4	4	6
0.60	4	4	6	4	5	7
0.65	4	5	7	5	6	8
0.70	5	6	8	6	7	10
0.75	5	7	9	6	8	11
0.80	6	7	11	7	9	12
0.85	7	8	12	8	10	14
0.90	7	9	13	9	11	15
0.95	8	10	15	10	12	17
1.00	9	11	16	11	13	19
1.05	10	12	18	12	15	21
1.10	11	14	20	13	16	23
1.15	12	15	21	14	18	25
1.20	13	16	23	16	19	27
1.25	14	17	25	17	21	29
1.30	15	19	27	18	22	32
1.35	16	20	29	20	24	34
1.40	17	22	31	21	26	37
1.45	19	23	34	23	28	39
1.50	20	25	36	24	30	42
1.55	21	27	38	26	32	45
1.60	22	28	41	27	34	48
1.65	24	30	43	29	36	51
1.70	25	32	46	31	38	54
1.75	27	34	49	33	40	57
1.80	28	36	52	35	43	60
1.85	30	38	54	36	45	63
1.90	31	40	57	38	47	67
1.95	33	42	60	40	50	70
2.00	35	44	64	43	52	74
2.10	38	48	70	47	58	82
2.20	42	53	77	51	63	89
2.30	46	58	84	56	69	98
2.40	50	63	91	61	75	106
2.50	54	68	99	66	82	115
2.60	58	74	107	72	88	125
2.70	63	79	115	77	95	134
2.80	68	85	124	83	102	145
2.90	73	92	133	89	110	155
3.00	78	98	142	95	117	166
3.25	91	115	167	111	138	195
3.50	105	133	194	129	160	226
3.75	121	153	222	148	183	259
4.00	138	174	253	169	208	294

Table 6.1 utilizes the z distribution as in (6.3) and summarizes the minimum sample sizes required for selected α levels and powers $(1 - \beta)$ assuming the indicated ratios of population standard deviation to the difference between the means under consideration. The table values are *minimum* sample sizes because the parameters σ, μ, and μ_0 are not really known and are estimated from preliminary experiments or are guesses generated from the experimenter's knowledge of the subject area.

EXAMPLE 6.4. Utilize Table 6.1 to generate the minimum appropriate sample size needed to detect a decrease in the weights of short-finned pilot whales of 22 kg with an alpha level of 0.025 and a power of 0.99.

SOLUTION. Assuming $\sigma = 30$ kg,

$$\left| \frac{\sigma}{\mu_0 - \mu} \right| = \left| \frac{30}{338 - 360} \right| = 1.364.$$

To be conservative, round this value up to 1.40 and find the estimated required sample size of 37 from the row indicated by 1.40 and the last column denoting $\alpha = 0.025$ and $1 - \beta = 0.99$.

6.2   Hypotheses Involving the Variance (σ^2)

Oftentimes the question asked about a sample is not its central tendency but its variability or scatter. This type of question requires one to frame very different H_0's and H_a's and to rely on a different index or test statistic to differentiate between them.

EXAMPLE 6.5. Normally economy-sized boxes of cornflakes average 40 oz with a standard deviation of 4.0 oz. In order to improve quality control a new packaging process is developed that you hope will significantly decrease variability. Thirty boxes packed by the new process are weighed and have a standard deviation of 3.0 oz. Is there evidence that the cornflakes boxes are significantly more uniform when packed by the new process?

SOLUTION. Write H_0 and H_a in terms of the *variance* because we have a sampling distribution for variances but not for standard deviations:

$$H_0: \sigma^2 \geq 16.0 \text{ oz}^2$$
$$H_a: \sigma^2 < 16.0 \text{ oz}^2.$$

Here $n = 30$ and $s^2 = 9.0$ oz^2. The statistic to test H_0 is

$$\chi^2 = \frac{(n-1)s^2}{\sigma_0^2}$$

and it approximates the **chi-square** statistic that was introduced when the confidence interval for σ^2 was developed in Chapter 4. This statistic will be close to $n - 1$ when s^2 is close to the hypothesized value σ^2. Values of χ^2 close to or above $n - 1$ support H_0. If s^2 is significantly smaller than σ^2, the χ^2 value would be much smaller than $n - 1$ and H_a would be supported.

If H_0 is true and we sample hundreds or thousands of times with samples of size 30 and each time we calculate the χ^2 value for the sample, these χ^2 values would form an asymmetrical distribution with the shape indicated below:

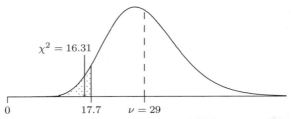

From Table C.5 with $\nu = n - 1 = 30 - 1 = 29$ and choosing $\alpha = 0.05$, the c.v. $= 17.7$. Five percent of these χ^2 values would be below 17.7 when the true population variance is 16.0 oz^2. For the sample in Example 6.5,

$$\chi^2 = \frac{(30 - 1)9}{16} = 16.31.$$

If H_0 is true, then $E(\chi^2) \geq n - 1$ (29 in this example), and if H_a is true, then $E(\chi^2) < n - 1$ (less than 29 in this example). Here the test statistic is less than 29; in fact, it's less than 17.7, so we would reject H_0. We know from the sampling distribution that we would get a value for χ^2 as low as or lower than 17.7 only 1 in 20 times when H_0 is true. By rejecting H_0 at the critical value of 17.7 we are willing to accept the risk of a Type I error at 0.05.

Note that other than the hypotheses, test statistic, and sampling distribution, the methodology here is identical to the tests of hypotheses performed to resolve questions about the mean in Section 6.1.

EXAMPLE 6.6. Dugongs are large, docile marine mammals that normally occur in herds of about 50 animals. Because the populations are being restricted by human activity along the southeastern coast of Queensland the groups are now somewhat smaller and, therefore, thought to be more highly inbred. Inbreeding often leads to an increase in the variability of the size of animals among the various herds. Healthy dugong populations have a mean weight of 350 kg and a variance of 900 kg^2. A sample of 25 Moreton Bay dugongs had a variance of 1600 kg^2. Is the Moreton Bay population significantly more variable than standard populations?

SOLUTION. We are anticipating deviations that are larger than the claimed value making this a right-tailed test:

$$H_0: \sigma^2 \leq 900 \text{ kg}^2$$
$$H_a: \sigma^2 > 900 \text{ kg}^2.$$

Here $n = 25$ and $s^2 = 1600$ kg^2.

Under H_0 we expect the χ^2 statistic to be $n - 1$ (24 here) or less, and under H_a we expect the χ^2 to be greater than $n - 1$:

$$\chi^2 = \frac{(n - 1)s^2}{\sigma_0^2} = \frac{(25 - 1)1600}{900} = 42.67.$$

The test statistic of 42.67 is greater than 24, but is it enough greater that we should abandon H_0 and accept H_a? In other words, is it significantly greater than the expectation under H_0? Let's use the P value approach to make this determination. In Table C.5 with $\nu = 24$ we find the following probabilities:

F	0.900	0.950	0.975	0.990	0.995
χ^2	33.2	36.4	39.4	43.0	45.6

We see that $P(\chi^2 < 39.4) = 0.975$, so $P(\chi^2 > 39.4) = 0.025$ and $P(\chi^2 < 43.0) = 0.990$ so $P(\chi^2 > 43.0) = 0.01$; therefore,

$$0.025 > P(\chi^2 \geq 42.67) > 0.01.$$

In words we would say that the probability of getting a chi-square statistic as large as 42.67 or larger by chance when the population variance is really 900 kg^2 is between 0.025 and 0.01. We would get such an extreme value between 1 in 40 times and 1 in 100 times if H_0 were true. Clearly the test statistic is extreme enough to reject H_0 and accept H_a. The analysis supports the view that the Moreton Bay population is significantly more variable than standard populations. The probability that we have made a mistake in taking this position (Type I error—rejecting a true H_0) is between 1 and 2.5%. This is an acceptable risk if we were to use 0.05 as the significance level. The critical value for $\alpha = 0.05$ and $\nu = 24$ is 36.4, and since $42.67 > 36.4$, we would also reject H_0 using this approach.

EXAMPLE 6.7. A geneticist interested in human populations has been studying growth patterns in American males since 1900. A monograph written in 1902 states that the mean height of adult American males is 67.0 inches with a standard deviation of 3.5 inches. Wishing to see if these values have changed over the twentieth century the geneticist measured a random sample of 28 adult American males and found that $\overline{X} = 69.4$ inches and $s = 4.0$ inches. Are these values significantly different from the values published in 1902?

SOLUTION. There are two questions here—one about the mean and a second about the standard deviation or variance. Two questions require two sets of hypotheses and two test statistics.

For the question about means, the hypotheses are

$$H_0: \mu = 67.0 \text{ inches}$$
$$H_a: \mu \neq 67.0 \text{ inches},$$

with $n = 28$ and $\alpha = 0.01$. This is a two-tailed test with the question and hypotheses (H_0 and H_a) formulated before the data were collected or analyzed:

$$t = \frac{\overline{X} - \mu}{\frac{s}{\sqrt{n}}} = \frac{69.4 - 67.0}{\frac{4.0}{\sqrt{28}}} = \frac{2.4}{0.76} = 3.16.$$

Using an alpha level of 0.01 for $\nu = n - 1 = 27$, we find the critical values to be ± 2.771 (Table C.4). Since $3.16 > 2.771$, we reject H_0 and say that the modern mean is significantly different from that reported in 1902 and, in fact, is greater than the reported value (because the t value falls in the right-hand tail). $P(\text{Type I error}) < 0.01$.

For the question about variances, the hypotheses are

$$H_0: \sigma^2 = 12.25 \text{ inches}^2$$
$$H_a: \sigma^2 \neq 12.25 \text{ inches}^2.$$

The question about variability is answered with a chi-square statistic. Since $n = 28$,

$$\chi^2 = \frac{(n-1)s^2}{\sigma_0^2} = \frac{(28-1)16}{12.25} = 35.3.$$

The χ^2 value is expected to be close to $n - 1 = 27$ if H_0 is true and significantly different from 27 if H_a is true.

From Table C.5 using an alpha level of 0.01 for $\nu = 27$, we find the critical values for χ^2 to be 11.8 and 49.6. Since $11.8 < 35.3 < 49.6$, we don't reject H_0 here. There isn't statistical support for H_a. The P value here for $P(\chi^2 > 35.3)$ is between 0.500 (31.5) and 0.250 (36.7), indicating the calculated value is not a rare event under the null hypothesis. For this example we would conclude that the mean height of adult American males is greater now than reported in 1902, but the variability in heights is not significantly different today than in 1902.

To Summarize

Hypotheses about the population variance can take one of three forms:

a) H_0: $\sigma^2 = c$ $E(\chi^2) = n - 1$ b) H_0: $\sigma^2 \geq c$ $E(\chi^2) \geq n - 1$ c) H_0: $\sigma^2 \leq c$ $E(\chi^2) \leq n - 1$
   H_a: $\sigma^2 \neq c$ $E(\chi^2) \neq n - 1$ H_a: $\sigma^2 < c$ $E(\chi^2) < n - 1$ H_a: $\sigma^2 > c$ $E(\chi^2) > n - 1$,

where c is a real number chosen *before* the data are gathered. Each H_0 above is tested with a chi-square statistic and the decision about H_0 is based on how far this statistic deviates from expectation under a true H_0. If the χ^2 statistic exceeds the critical value(s), H_0 is rejected. Alternatively, if the P value for the χ^2 statistic is smaller than the predetermined alpha level, H_0 is rejected. Again for any particular experiment only one of the sets of hypotheses is appropriate and can be tested.

Tests of hypotheses regarding the standard deviation (σ) are usually done as tests of hypothesis about the variance (σ^2) and utilize the chi-square statistic. Claims supported under this test involving the variance will have the same validity for the standard deviation. In other words, H_0: $\sigma^2 = c$ is equivalent to H_0: $\sigma = \sqrt{c}$.

6.3   Nonparametric Statistics and Hypothesis Testing

By now you should be familiar with statistical tests that are based on certain probability distributions. The normal distribution is the most important of these and it provides the foundation for many of the statistical calculations that you have done. For example, in using a t test to test H_0: $\mu = 50$ or finding a 95% confidence interval for the mean of a population requires an assumption of "normality." Nonparametric statistics are sometimes called "distribution-free" statistics and they don't require the assumption of normality. This makes them ideal for student investigations where the underlying distribution of the population may not be known and the sample size is often too small to accommodate assumptions of the Central Limit Theorem.

Nonparametric tests can be simple to perform; often they reduce to counting or ranking observations. Another appealing characteristic is that the logic of a nonparametric test is often "obvious." The probability distributions of nonparametric test statistics are generally simple, or at least it is easier to see how the distributions are related to the statistics. Knowledge of nonparametric statistics can help you understand the logic of all statistical tests and make you a better user of statistics in general.

While parametric statistics often focus on the mean of a population, nonparametric statistics usually focus on the *median*. Many of the nonparametric tests we describe will follow a basic pattern. Starting with a sample $X_1, \ldots, X_n$ of size n:

- The n observations are used to produce a set of *elementary estimates*.

- The median of these elementary estimates can be used as a *point estimate* of the median of the population.

- The elementary estimates are also used to construct confidence intervals. Recall that a confidence interval for the mean of a population cuts out the most extreme estimates and leaves the central core of most likely values for the mean. The nonparametric case is similar, but uses elementary estimates. The estimates are listed in increasing order. Then, using certain rules, the most extreme estimates at either end are discarded and the remaining ones provide the range of values for the confidence interval.

- Hypothesis tests are carried out with the help of test statistics that simply count or rank the elementary estimates.

The key, then, is to understand how the elementary estimates are constructed and how the discarding and counting is done in each test.

6.4 The One-Sample Sign Test

In some investigations, a population median is known or suspected from earlier work. A question arises as to whether some new population is, in fact, identical to the known population. Consider the following problem.

> **EXAMPLE 6.8.** Southeastern Queensland is home to more than 40 species of frogs. Among them are several species of *Crinia*, which are small, highly variable brownish or grayish froglets usually found in soaks, swamps, or temporary ponds. One way to distinguish these froglets is by their lengths. A population of froglets from a single pond appear to be of the same species. You tentatively identify them as clicking froglets *C. signifera*, a species known to have a median length of 30 mm. To be honest, you are not really sure that's what they are. A random sample of 18 froglets is captured and their lengths (in mm) are recorded in the table below. Is it reasonable to assume that these are clicking froglets?
>
> | 24.1 | 22.9 | 23.0 | 26.1 | 25.0 | 30.8 | 27.1 | 23.2 | 22.8 |
> |------|------|------|------|------|------|------|------|------|
> | 23.7 | 24.6 | 30.3 | 23.9 | 21.8 | 28.1 | 25.4 | 31.2 | 30.9 |

This example can be generalized in the following way. Suppose we have a population (e.g., froglets) and an hypothesized median for some measurement on this population (e.g., the median length). Given a random sample from this population, how can it be used to confirm the hypothesis or support an alternative hypothesis?

The sign test is based on the defining property of the median: half of the population lies above the median and half below. In a random sample, we expect roughly half the observations to be above the suspected median (here, 30 mm) and half below it. In this example, 4 observations are above, the other 14 are below. Using the binomial distribution, one can determine just how likely or unlikely this result is, given the suspected median. We make these ideas explicit below.

Assumptions

The assumptions for the one-sample sign test are:

1. X is a continuous random variable with median M.

2. $X_1, \ldots, X_{n'}$ denotes a random sample of size n' from the distribution of X.

3. M_0 denotes the hypothesized median of the population.

The elementary estimates for the sign test are just the sample observations.

Hypotheses

The underlying assumption or null hypothesis is that the true median for the population M is, in fact, the hypothesized value M_0. The hypothesis test then takes one of three forms:

- *Two-tailed.* H_0: $M = M_0$ and H_a: $M \neq M_0$. The alternative hypothesis states that the values of X_i tend to be different (either smaller or larger) than M_0.

- *Left-tailed.* H_0: $M \geq M_0$ and H_a: $M < M_0$. The alternative hypothesis states that the values of X_i tend to be smaller than M_0.

- *Right-tailed.* H_0: $M \leq M_0$ and H_a: $M > M_0$. The alternative hypothesis in this case states that the X_i tend to be larger than M_0.

The two-tailed test is the most appropriate for Example 6.8. Given the very limited information before the data were collected, there was no reason to expect the lengths to be either larger or smaller than the hypothesized median of 30 mm.

Test Statistic and Theory

If H_0 is true ($M = M_0$), then any observation X_i has probability $\frac{1}{2}$ of lying below M_0 and probability $\frac{1}{2}$ of lying above. So any difference $X_i - M_0$ has probability $\frac{1}{2}$ of being negative, probability $\frac{1}{2}$ of being positive, and probability 0 of being 0.

   Note: Because X is a continuous random variable, X has probability 0 of equaling M_0. However, sometimes observations do equal the hypothesized median M_0 and $X_i - M_0 = 0$. *When calculating the test statistic, the general procedure is to remove such points from the sample, as long as there are not many such observations.*

   Let n be the number of points remaining in the sample after the removal of any values equal to M_0. Let S_- and S_+ denote the number of negative and positive signs, respectively. These are the test statistics.

   If H_0 is true, the probability of a '+' or a '−' is $\frac{1}{2}$, analogous to flipping a fair coin. So S_- and S_+ are random variables that have a binomial distribution with parameters n and $p = \frac{1}{2}$ and an expected value of $np = \frac{n}{2}$.

Decision Rules

Begin by setting the α level at which the hypothesis will be tested. Most often we will use $\alpha = 0.05$.

   In a right-tailed test, if H_a is true (i.e., $M > M_0$), then we should observe more positive signs and fewer negative signs than expected. The test statistic is S_-, the number of negative signs. S_- counts as evidence for H_0, the null hypothesis. Is there so little evidence for H_0, i.e., is S_- so small, that we can safely reject H_0? To decide, enter Table C.1 with parameters n (the reduced sample size) and $p = 0.5$ and find the P value for S_-. If $P < \alpha$, then there is little evidence to support H_0, so H_0 is rejected. If $P > \alpha$, we continue to accept H_0.

In a left-tailed test, if H_a is true (i.e., $M < M_0$), then we should observe more negative signs and fewer positive signs than expected. The test statistic is S_+, the number of positive signs, which acts as evidence for H_0. Test as above.

In a two-tailed test, if H_a is true (i.e., $M \neq M_0$), then we should observe either more negative or more positive signs than expected. Because of the way Table C.1 is constructed, the test statistic here is the minimum of S_- and S_+. Since the test is two-sided, compare the P value from Table C.1 to $\frac{\alpha}{2}$ to determine whether H_0 should be rejected or not.

> **SOLUTION TO EXAMPLE 6.8.** Since we decided on a two-tailed test, we have H_0: $M = 30$ mm versus H_a: $M \neq 30$ mm. Perform the test with $\alpha = 0.05$. Of the 18 observations only 4 exceeded 30 mm so $S_+ = 4$ and 14 were smaller so $S_- = 14$. The test statistic is the smaller of these two, $S_+ = 4$. From Table C.1, the P value is
>
> $$F(4) = P(S_+ \leq 4) = 0.0154.$$
>
> Since 0.0154 is smaller than $\frac{\alpha}{2} = 0.025$, we reject H_0. If H_0 were true, $E(S_+) = E(S_-) = \frac{18}{2} = 9$. But $S_+ = 4$ and $S_- = 14$ and are suffficiently far from this expectation to reject H_0. Based on the evidence, it is reasonable to conclude that these are *not* clicking froglets.

Larger Samples

Notice that Table C.1 only goes up to a sample size of $n = 20$. It may be the case that you have sample sizes larger than that, so how do you proceed? As we have seen, *for large values of n, the binomial distribution with parameters n and p is well-approximated by the normal distribution with mean np and variance $np(1-p)$.* For the sign test, $p = \frac{1}{2}$ so the mean is $\mu = \frac{n}{2}$, the variance is $\frac{n}{4}$, and the standard deviation is $\sigma = \frac{\sqrt{n}}{2}$, where n is the number of observations not equal to M_0. If S is the test statistic (i.e., the number of positive or negative signs, whichever is appropriate), then the P value is computed using the normal approximation to the binomial, with the usual continuity correction factor of 0.5. (To review this approximation method, see Section 3.6.)

$$F(S) = P\left(Z \leq \frac{S + 0.5 - \mu}{\sigma}\right) = P\left(Z \leq \frac{S + 0.5 - \frac{n}{2}}{\frac{\sqrt{n}}{2}}\right),$$

which can be looked up in Table C.3. The P value is then compared to α (or $\frac{\alpha}{2}$ in a two-tailed test) and H_0 is either rejected or not as in the small sample case.

> **EXAMPLE 6.9.** In June of 1996 in the Finger Lakes region of New York, a series of warm days in the middle of the month caused people to comment that it was "unusually" warm. The median normal maximum temperature for Ithaca in June is $75°$F. Do the data (the daily maximum temperatures for all 30 days in June) support the notion that June was unusually warm? (Based on data reported in The Ithaca Climate Page, http://snow.cit.cornell.edu/climate/ithaca/moncrt_06-96.html, July 2000.)
>
> | 72 | 78 | 79 | 75 | 74 | 70 | 77 | 83 | 85 | 85 |
> |----|----|----|----|----|----|----|----|----|----|
> | 84 | 85 | 80 | 78 | 82 | 83 | 83 | 77 | 68 | 69 |
> | 76 | 76 | 80 | 72 | 73 | 70 | 70 | 78 | 78 | 78 |

Assumptions

The assumptions of this test are more restrictive than those of the sign test.

  1. The continuous random variable X is *symmetric* about a median M.

  2. $X_1, \ldots, X_{n'}$ denotes a random sample of size n' from the distribution of X.

  3. M_0 denotes a hypothesized median for X.

The assumption of symmetry forms the basis of the null hypothesis of the test and is crucial in defining the test statistic. (Draw a few continuous probability distributions that are symmetric. What well-known distribution is symmetric?) Also note that when a distribution is symmetric, then the median and the mean are equal. So "mean" can be substituted for "median" throughout this discussion.

Hypotheses

The hypotheses are the same as for the sign test:

  - *Two-tailed.* H_0: $M = M_0$ versus H_a: $M \neq M_0$.

  - *Left-tailed.* H_0: $M \geq M_0$ versus H_a: $M < M_0$.

  - *Right-tailed.* H_0: $M \leq M_0$ versus H_a: $M > M_0$.

Test Statistics

Consider the set of n' differences $X_i - M_0$. Continuity implies that no difference should be exactly 0. But, as when using the sign test, this sometimes happens. *The general procedure is to remove such points from the sample, as long as there are not many such observations.*

Let n be the number of points remaining in the sample after the removal of any values equal to M_0. If the null hypothesis is valid ($M = M_0$), there should be roughly as many positive as negative differences—that's the way the sign test works. However, because of the additional assumption of symmetry, the differences $X_i - M_0$ should be symmetric about 0, since subtracting M_0 from X should shift the center of the distribution curve to 0. Not only should there be roughly the same number of positive and negative differences, but by symmetry the *sizes* of the positive and negative differences should be about the same.

The size of a difference is just its absolute value. So, next evaluate $|X_i - M_0|$ and then rank the resulting values from smallest to largest. Use midranks when there are ties. Finally add a plus or minus sign to each rank depending on whether the difference was positive or negative.

> **EXAMPLE 6.11.** The bridled goby, *Arenigobus frenatus*, occupies burrows beneath rocks and logs on silty or muddy tidal flats in Moreton Bay. Ventral fins join at the base to form a single disc—characteristic of the family Gobidae. The median length is thought to be 80 mm, and the distribution of lengths is thought to be symmetric. Suppose a small sample was collected and the following lengths (in mm) were recorded. Determine the signed ranks for the sample.
>
> 63.0   82.1   81.8   77.9   80.0   72.4   69.5   75.4   80.6   77.9

SOLUTION. Remove the observation that is the same as the suspected median, $M_0 = 80$. The differences of the remaining nine sample values are calculated and their absolute values are given signed ranks. *Note:* Here we use midranks in the case of ties. That is, the 82.1 mm length and the two 77.9 mm lengths all have the same absolute difference, 2.1 mm, from the hypothesized median length of 80 mm. Since they constitute the third, fourth, and fifth smallest absolute ranks, they are all given the value of 4 or the average of the tied ranks (the midrank). The next smallest absolute difference is 4.6 mm (for 75.4 mm) and it has rank 6 after ranks 1, 2, 4, 4, and 4.

X_i	63.0	82.1	81.8	77.9	72.4	69.5	75.4	80.6	77.9
$X_i - 80$	-17.0	2.1	1.8	-2.1	-7.6	-10.5	-4.6	0.6	-2.1
$|X_i - 80|$	17.0	2.1	1.8	2.1	7.6	10.5	4.6	0.6	2.1
Signed rank	-9	$+4$	$+2$	-4	-7	-8	-6	$+1$	-4

The signed ranks are used to create two test statistics.

- W_- = the absolute value of the sum of the negative ranks.
- W_+ = the sum of the positive ranks.

In the example above, check that $W_- = 38$, while $W_+ = 7$. As a general check on your work, note that the sum of all the rankings is

$$W_- + W_+ = \sum_{i=1}^{n} i = \frac{n(n+1)}{2}.$$

In the example above, $W_- + W_+ = 45 = \frac{9(9+1)}{2}$. Under H_0 and the assumption of symmetry, we would expect that $W_- \approx W_+$. That is, the expected value of each test statistic is one-half of the sum of all the ranks, $E(W_-) = E(W_+) = \frac{n(n+1)}{4}$.

Decision Rules

Select the α level at which the hypothesis will be tested. In a right-tailed test, if H_a is true, then X's are larger than expected so there should be more positive ranks than negative ranks. The test statistic is $W = W_-$, the absolute sum of negative ranks, because W_- counts as evidence for H_0 in this case. When W is sufficiently small (meaning there are too few negative ranks), then H_0 should be rejected. Use Table C.6 with the possibly reduced sample size n as the column and find the P value for the test statistic W as the row w. If $P < \alpha$, then there is little evidence to support H_0, so H_0 is rejected. If $P > \alpha$, we continue to accept H_0.

In a left-tailed test, if H_a is true, then X's are smaller than expected so we should observe more negative ranks than positive ones. The test statistic is $W = W_+$, the sum of positive signs, which acts as evidence for H_0. Use Table C.6 to determine when W is sufficiently small to reject H_0.

In a two-tailed test, if H_a is true we should observe either many more negative or positive ranks than expected. Because of the way Table C.6 is constructed, the test statistic W is the minimum of W_- and W_+. Proceed as above to find the critical W value, taking care to compare the P value from Table C.6 to $\frac{\alpha}{2}$ to determine whether H_0 should be rejected or not.

EXAMPLE 6.12. Return to the situation in Example 6.11. Researchers assumed because the sample was collected at a location adjacent to a ferry port where there were higher than normal levels of disturbance and pollution that the bridled gobies would be somewhat stunted, that is, shorter than 80 mm. Do the data collected support this contention?

SOLUTION. The hypotheses are H_0: $M \geq 80$ mm versus H_a: $M < 80$ mm. The test is left-tailed, so the test statistic is W_+, which we previously calculated to be 7. The corresponding P value from Table C.6 is 0.0371. Since $0.0371 < 0.05$, we reject H_0. There is evidence to support that the gobies at this location are stunted.

EXAMPLE 6.13. Adult heights are known to be symmetrically distributed. We measured the heights of 10 male faculty in the science division at Hobart and William Smith Colleges. The median male adult height in the U.S. is supposedly 178 cm. Do the data collected support this assertion? Set up the appropriate two-tailed hypothesis and determine whether H_0 can be rejected at the $\alpha = 0.05$ level.

$$171 \quad 175 \quad 177 \quad 178 \quad 180 \quad 182 \quad 190 \quad 192 \quad 195 \quad 202$$

SOLUTION. The null hypothesis is H_0: $M = 178$ versus the two-sided alternative H_a: $M \neq 178$. Note that one of the observations is the same as the suspected median so it must be removed. The differences of the remaining nine sample values are calculated and their absolute values are ranked.

X_i	171	175	177	180	182	190	192	195	202
$X_i - 178$	-7	-3	-1	2	4	12	14	17	24
$|X_i - 178|$	7	3	1	2	4	12	14	17	24
Signed rank	-5	-3	-1	$+2$	$+4$	$+6$	$+7$	$+8$	$+9$

From the ranks we see that

$$W_+ = 9 + 8 + 7 + 6 + 4 + 2 = 36 \qquad W_- = |-5 - 3 - 1| = 9.$$

As a check, $W_+ + W_- = 45 = \frac{9(9+1)}{2} = \frac{n(n+1)}{2}$.

For a two-sided test, the test statistic is the minimum of W_+ and W_-; so $W = W_- = 9$. From Table C.6, the P value for the test statistic $W = 9$ with $n = 9$ is 0.0645. Since $0.0645 > 0.025 = \frac{\alpha}{2}$, we cannot reject H_0. There is no evidence that the median male height is different from 178 cm.

Theory

Where do the P values in Table C.6 come from? In the case of the sign test, the P values were derived from the binomial distribution. Though the distribution for the Wilcoxon signed-rank test is more complicated, we can get a glimpse of how it is computed by looking at a simple example.

Suppose a sample consists of just five observations. Assume that we have ranked the absolute values of the five differences, $|X_i - M_0|$. The ranks for the Wilcoxon test consist of the numbers 1 through 5, with each number having a + or − sign to indicate whether the associated difference is positive or negative. By assumption, M_0 is the median for the population, so the probability of any difference $X_i - M_0$ being positive or negative is 0.5. Consequently, the (classical) probability of any rank being positive or negative is also 0.5. Since one of the two signs must be chosen for each of the 5 ranks, there are a total of $2^5 = 32$ possible rankings. So the probability of any particular set of ranks, such as $(-1, -2, -3, +4, -5)$, occurring is $\frac{1}{32}$. Using this we can calculate the classical probability of any value of W_+ occurring. Table 6.2 lists all 32 possible rankings and gives the corresponding value for W_+.

TABLE 6.2. The 32 possible rankings for a sample of five observations and the corresponding values of W_+

Ranking	W_+	Ranking	W_+
$-1, -2, -3, -4, -5$	0	$+1, +2, +3, -4, -5$	6
$+1, -2, -3, -4, -5$	1	$+1, +2, -3, +4, -5$	7
$-1, +2, -3, -4, -5$	2	$+1, +2, -3, -4, +5$	8
$-1, -2, +3, -4, -5$	3	$+1, -2, +3, +4, -5$	8
$-1, -2, -3, +4, -5$	4	$+1, -2, +3, -4, +5$	9
$-1, -2, -3, -4, +5$	5	$+1, -2, -3, +4, +5$	10
$+1, +2, -3, -4, -5$	3	$-1, +2, +3, +4, -5$	9
$+1, -2, +3, -4, -5$	4	$-1, +2, +3, -4, +5$	10
$+1, -2, -3, +4, -5$	5	$-1, +2, -3, +4, +5$	11
$+1, -2, -3, -4, +5$	6	$-1, -2, +3, +4, +5$	12
$-1, +2, +3, -4, -5$	5	$+1, +2, +3, +4, -5$	10
$-1, +2, -3, +4, -5$	6	$+1, +2, +3, -4, +5$	11
$-1, +2, -3, -4, +5$	7	$+1, +2, -3, +4, +5$	12
$-1, -2, +3, +4, -5$	7	$+1, -2, +3, +4, +5$	13
$-1, -2, +3, -4, +5$	8	$-1, +2, +3, +4, +5$	14
$-1, -2, -3, +4, +5$	9	$+1, +2, +3, +4, +5$	15

Only one set of the 32 ranks gives a value of $W_+ = 0$, so $P(W_+ = 0) = \frac{1}{32}$. Similarly, $P(W_+ = 1) = \frac{1}{32}$ and $P(W_+ = 2) = \frac{1}{32}$. But $W_+ = 3$ occurs in two ways, so $P(W_+ = 3) = \frac{2}{32}$. In this fashion the entire density function for W_+ can be generated (and by symmetry, W_-). The density function is then used to produce the cumulative distribution. For example,

$$P(W \le 3) = P(W = 0, 1, 2, 3) = \frac{1 + 1 + 1 + 2}{32} = \frac{5}{32} = 0.15625.$$

In exactly this fashion the entire table below is filled in.

w	0	1	2	3	4	5	6	7	8	9	10	11	12	13	14	15
$P(W = w)$	$\frac{1}{32}$	$\frac{1}{32}$	$\frac{1}{32}$	$\frac{2}{32}$	$\frac{2}{32}$	$\frac{3}{32}$	$\frac{3}{32}$	$\frac{3}{32}$	$\frac{3}{32}$	$\frac{3}{32}$	$\frac{3}{32}$	$\frac{2}{32}$	$\frac{2}{32}$	$\frac{1}{32}$	$\frac{1}{32}$	$\frac{1}{32}$
$P(W \le w)$	$\frac{1}{32}$	$\frac{2}{32}$	$\frac{3}{32}$	$\frac{5}{32}$	$\frac{7}{32}$	$\frac{10}{32}$	$\frac{13}{32}$	$\frac{16}{32}$	$\frac{19}{32}$	$\frac{22}{32}$	$\frac{25}{32}$	$\frac{27}{32}$	$\frac{29}{32}$	$\frac{30}{32}$	$\frac{31}{32}$	$\frac{32}{32}$

With a sample size $n = 5$, if the test statistic were 1, then $P(W \le 1) = \frac{2}{32} = 0.0625$. If one were using a significance level of $\alpha = 0.05$, the null hypothesis could not be rejected. With this small sample size only $W = 0$ would allow rejection of the null hypothesis because $P(W \le 0) = \frac{1}{32} = 0.03125 < 0.05 = \alpha$.

Larger Samples

Table C.6 has a maximum sample size of $n = 25$. There is a normal approximation for larger samples. The distribution for the test statistic W when the sample size is $n > 25$ is well-approximated by the normal distribution with mean $\mu = \frac{n(n+1)}{4}$ and variance $\sigma^2 = \frac{n(n+1)(2n+1)}{24}$, where n is the number of observations not equal to M_0. If W is the test statistic and w is its observed value (i.e., W_- or W_+, whichever is appropriate), then the P value is computed as for a normal distribution:

$$P(W \le w) = P\left(Z \le \frac{w - \mu}{\sigma}\right) = P\left(Z \le \frac{w + 0.5 - \frac{n(n+1)}{4}}{\sqrt{\frac{n(n+1)(2n+1)}{24}}}\right),$$

which can be looked up in Table C.3. Note the continuity correction of 0.5. The P value is then compared to α (or $\frac{\alpha}{2}$ in a two-tailed test) and H_0 is either rejected or not as in the small sample case.

EXAMPLE 6.14. Suppose that you were doing a two-sided hypothesis test at the $\alpha = 0.05$ level with a sample size of $n = 100$. You find that $W_- = 2010$, while $W_+ = 3040$. Determine whether H_0 should be rejected.

SOLUTION. The test statistic is $W = W_- = 2010$. Using $n = 100$ in the formula above and then Table C.3,

$$P(W \le 2010) = P\left(Z \le \frac{2010 + 0.5 - \frac{100(101)}{4}}{\sqrt{\frac{(100)(101)(201)}{24}}}\right)$$

$$= P\left(Z \le \frac{-514.5}{290.84}\right)$$

$$= P(Z \le -1.77)$$

$$= 0.0384.$$

Since the test is two-sided we compare the P value of 0.0384 to $\frac{\alpha}{2} = 0.025$. We are unable to reject the null hypothesis in this example.

Comments

The Wilcoxon signed-rank test is a bit more complicated than the sign test. Why use it if the sign test can be applied to all population distributions? The answer is that the Wilcoxon test gains its power from the fact that it makes use of the size of the difference between $X_i - M_0$, while the sign test uses only the sign of this difference. For example, suppose we were doing a two-sided hypothesis test with $n = 20$ with a symmetrically distributed random variable. Suppose that 8 of the differences $X_i - M_0$ were small and negative while 12 were large and positive. The test statistic for the sign test would be $S_- = 8$ and according to Table C.1 we could not reject H_0. However, the test statistic for the Wilcoxon test would be $W_- = |-1-2-3-4-5-6-7-8| = 36$, since the 8 negative differences were the smallest. According to Table C.6, for $n = 20$ with a test statistic of $W_- = 36$, the P value is $0.0042 < 0.025 = \frac{\alpha}{2}$. We are able to reject H_0.

This does not mean that you should keep applying tests until you find one that gives you the result that you want. It simply means that you should use all the information about the distribution that you have. If you have no information, use the sign test. If you know the distribution is symmetric, then use the Wilcoxon signed-rank test. If you know that not only is the distribution symmetric, but it is also normal, then a t test is more appropriate. Each piece of information that you use allows you to make finer discriminations.

Summary of One-Sample Tests

One-sample tests of hypothesis with continuous variables

Hypothesis	Sampling assumptions	Test statistic
$H_0: \sigma^2 = c$ $H_0: \sigma = c$; $H_0: \sigma^2 = c^2$	None	χ^2 test
$H_0: \mu = c$	Normal population or approximately normal under Central Limit Theorem	t test
$H_0: \mu$ or $M = c$	Symmetrical distribution	One-sample Wilcoxon signed-rank test
$H_0: M = c$	None	One-sample sign test

While the vast majority of questions about single samples can be framed using hypotheses involving the mean (μ), the variance (σ^2), or the median (M), occasionally you may need to address questions about proportions or normality, which are treated in later chapters.

6.7 Problems

1. A plant breeder has been selecting Crispin apple trees for many years in an attempt to decrease variability in fruit size. When he began selecting, the variance was 400 g^2 for mature fruit. A random sample of 30 apples from his most recent crop had a variance of 300 g^2. Has the breeder been successful in making Crispin apples less variable? What is the expected value of the test statistic here, if H_0 is true? Find the P value of this test. $300 = $ sample variance $\qquad H_0 \geq 400$
 $400 = $ population variance $\qquad H_a < 400$

2. As part of a benthic community survey of Lady Elliot Island, 16 sea stars, *Linckia laevigata*, were collected and their longest arm was measured to the nearest tenth of a centimeter. Assume normality for this population.

10.3	11.0	10.5	10.0	11.3	14.5	13.0	12.1
12.1	9.4	11.3	12.0	11.5	9.3	10.1	7.6

(a) For these data calculate the sample mean and variance.

(b) Calculate the 95% confidence interval for the population mean. In your own words explain what this confidence interval represents.

(c) Use a t test with $\alpha = 0.05$ to determine whether the mean is significantly different from 12 cm. Discuss the relationship between the confidence interval in (b) above and this two-tailed t test.

(d) Calculate the 95% confidence interval for the population variance. Again, in your own words, explain what this confidence interval means.

(e) Use a χ^2 test with $\alpha = 0.05$ to determine whether or not the variance is significantly different from 5.0 cm^2. Discuss the relationship between the confidence interval in (d) above and this two-tailed χ^2 test.

3. In healthy adult males the sciatic nerve conduction velocity values are normally distributed with a mean of 65 cm/msec and a standard deviation of 5 cm/msec. The conduction velocities of 16 subjects admitted to the poison control center of a metropolitan hospital with a diagnosis of methylmercury poisoning had a mean conduction velocity of 55 cm/msec and a variance of 49 (cm/msec)^2.

   (a) Do these data provide sufficient evidence to indicate that the conduction velocities are significantly slower in the poisoned individuals?

   (b) Do these data provide sufficient evidence to indicate that poisoned individuals are more variable in their sciatic nerve conduction velocities than normal individuals?

4. A television documentary on overeating claimed that adult American males are on average 15 lb overweight. To test this claim, 25 randomly selected adult males were examined, and the average excess weight was found to be 18 lb with a standard deviation of 10 lb. Do these data give us reason to believe the claim of 15 lb is incorrect? What type of error could you have made here? Explain.

5. Before purchasing a large corn farm you wish to determine its profitability by analyzing its productivity. One claim made by the present owner is that the fields are very uniform in their yield per acre. In fact, it is claimed that the standard deviation in bushels per acre of corn yield is 5. A random sample of 25 one-acre plots yields an average of 140 bushels/acre and a standard deviation of 6.5 bushels/acre. Are the fields significantly more variable than is claimed? What type of error could you have made here? Explain.

6. Recently there have been concerns about the effects of phthalates on the development of the male reproductive system. Phthalates are common ingredients in many plastics. In a pilot study a researcher gave pregnant rats daily doses of 750 mg/kg of body weight of DEHP (di-2-ethylhexyl phthalate) throughout the period when their pups' sexual organs were developing. The newly born male rat pups were sacrificed and their seminal vesicles were dissected and weighed. Below are the weights for the eight males (in mg).

1710   1630   1580   1670   1350   1650   1600   1650

   If untreated newborn males have a mean of 1700 mg, can you say that rats exposed to DHEP in utero have a significantly lower weight?

7. The carpet python, *Morelia spilota variegata*, is widely distributed and is common in rainforests, wet or dry eucalypt forests, and pastoral and urban areas. They locate warm-blooded prey by means of heat-sensory pits located along the lower jaw that can detect temperature changes of less than one-thirtieth of a degree ($^\circ$C). Their median length is approximately 250 cm. Because of an extended drought from 1994 through early 1996 at Lamington National Park, it was thought that the median length of these snakes would be smaller than 250 cm. A survey was conducted and the lengths (in cm) below were recorded. Clearly state the null and alternative hypotheses. Assuming that lengths of snakes are symmetrically distributed about the mean, carry out an appropriate test of the hypothesis at the 0.05 level. Interpret the result.

269   221   204   258   211   198   276   223   207   242

8. (a) The ancient Greeks used the term "golden rectangle" to describe those rectangles whose height-to-width ratio is $1 : \frac{1+\sqrt{5}}{2} \approx 0.618034$. They found such rectangles especially pleasing to the eye and often used them in architectural design (e.g., the Parthenon). Other artists and artisans, including musicians, have made explicit use of this ratio in their work. The ratio, unbelievably, arises in biological growth processes. For example, there are claims that the ratio of navel height to total height is this same golden ratio. The data below are breadth-to-length ratios of beaded rectangles used by the Shoshoni Indians to decorate leather goods. Is it reasonable to assume that they are implicitly using golden rectangles? Use the sign test here. *(HSDS, #150; see end-of-book References.)*

0.693	0.662	0.690	0.606	0.570
0.749	0.672	0.628	0.609	0.844
0.654	0.615	0.668	0.601	0.576
0.670	0.606	0.611	0.553	0.933

   (b) Find a 95% confidence interval for the median height-to-width ratio.

9. The wall skink, *Cryptoblepharus virgatus*, is a common insectivore in the Brisbane area. They can be seen sunning themselves on vertical surfaces such as the brick walls of buildings. They can also be heard scurrying into the leaf litter as one walks along a path. Suppose we suspect that their median length (including tail) is 7 cm. Test the hypothesis H_0: $M = 7$ cm versus H_a: $M \neq 7$ cm at the 0.05 level. The data consist of a random sample of 66 skinks that were captured and measured: 41 have length greater than 7 cm, 23 are shorter than 7 cm, and 2 have length equal to 7 cm.

10. (a) The angle that a spider's web makes with a line vertical to the earth's surface varies by species. For example, for *Isoxya cicatricosa* this angle has been shown to have a median of 28.12°. The angles for 10 webs were observed. Based on these data can you reject the hypothesis that *I. cicatricosa* constructed these webs? *(HSDS, #194)*

$$25° \quad 12° \quad 31° \quad 26° \quad 17° \quad 15° \quad 24° \quad 10° \quad 16° \quad 12°$$

   (b) Find a 95% confidence interval for the median angle.

11. Redo Problem 4 in Chapter 1 as a test of hypothesis question.

12. Redo Problem 6 in Chapter 1 as a test of hypothesis question.

13. (a) The fasting blood glucose (FBG) levels of a population of women is thought to have a standard deviation of 0.75 mmol/l. A study is being designed to determine whether FBG levels are increased in women who have taken oral contraceptives for at least 6 months. How large a sample is required to detect a 0.50 mmol/l increase if $\alpha = 0.05$ and the power is 0.95?

   (b) The researchers also wish to determine whether oral contraceptives have any effect on cholesterol levels. If the standard deviation for cholesterol is thought to be 0.80 mmol/l, how large a sample is required to detect a 0.30 mmol/l difference if $\alpha = 0.05$ and the power is 0.95?

14. A Secchi disc (see figure below) is a simple but effective device to measure water transparency. The disc is attached to a rope and slowly lowered through the water until it just disappears from view. This depth is noted. It is lowered some more, then slowly raised until the disc is just observed. This depth is also noted. The Secchi depth is the average of these two depths. The depth is proportional to water transparency and inversely proportional to water turbidity, but it is also influenced by cloudiness, position of the sun, roughness of the water, and observer bias. Plankton biomass and suspended mud particles are the main contributors to water turbidity. In Seneca Lake (New York), the depth is primarily a function of plankton biomass. In 1997, the reported median Secchi depth at a station at the north end of the lake was 8.0 m. The following depths were recorded on 16 dates in 1999. The question is whether there has been a change in Secchi depth since 1997. Use a sign test. (Based on data reported in John Halfman, `http://people.hws.edu/Halfman/Zebras/Zebras.htm`, July 2000.)

9.0	9.0	10.5	8.0	9.2	7.0	8.3	6.0
7.1	4.7	7.8	6.0	7.9	6.5	8.3	7.4

15. The fasting blood glucose levels of nine women who had been using oral contraceptives for at least 6 months prior to sampling are listed below. It is thought that the mean fasting blood glucose level in women not using contraceptives is 3.8 mmol/l. Test whether there is evidence that the use of contraceptives raises glucose levels. Assume symmetry.

4.71	3.61	4.92	4.85	4.66	3.92	5.10	3.55	4.78

16. Create a table giving the density and cumulative distribution function for W_+ for a sample size of $n = 4$. Use the same techniques that were used for a sample size of $n = 5$ in Section 6.6.

17. Bornstein and Bornstein (The pace of life. 1976. *Nature*, **259**: 556–558) proposed a correlation between the population of a city or town and the pace at which people walk in that location. For a particular sized town their data predict a time of 13.9 sec to walk 50 feet. To investigate this prediction the paces of 16 randomly chosen people were measured. Making no special assumptions about the distribution of individual paces, set up the appropriate hypotheses and determine whether the null hypothesis should be rejected at $\alpha = 0.05$.

sign test

8.4	9.8	10.3	10.6	10.8	11.0	11.1	11.5
11.6	11.9	12.4	12.5	13.9	14.0	15.4	15.7

18. (*a*) Revisiting Problem 3 in this chapter regarding sciatic nerve conduction velocities in methylmercury-poisoned individuals, find the minimum sample size required to detect a decrease in conduction velocity of 5 cm/msec if $\alpha = 0.05$ and the power is 0.99.

(*b*) Under the same conditions find the sample size for a 2 cm/msec decrease in conduction. Comment on the sample size needed to detect a 1 cm/msec decrease.

19. Trisomy 21 (Down Syndrome) is a complex phenotype in humans caused by the presence of an extra number 21 chromosome in all the cells of the body. It was first described by John Langdon Down in 1866 and was called mongolism because of the epicanthic fold at the corner of the eye. While diminished mental capacity is a principal characteristic of the syndrome, many organs and tissues are affected. Interestingly, the dermatoglyphic patterns are also disrupted including finger and palm prints. One well-described characteristic of the palm print is the *atd* angle of the triradii illustrated in the figure below.

not told symmetry so either way ok

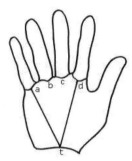

The *atd* angle was measured in eight patients with trisomy 21, see below.

$$70° \quad 46° \quad 75° \quad 83° \quad 69° \quad 80° \quad 79° \quad 65°$$

(*a*) If the *atd* angle among normal individuals has a median of 48°, would you say this small sample provides significant evidence that the *atd* angle is increased in trisomy 21 patients?

(*b*) Find the 95% confidence interval for the median *atd* angle in trisomy 21 individuals.

20. The data below represent the systolic blood pressures (in mmHg) of 14 patients undergoing drug therapy for hypertension. Assuming symmetry of systolic blood pressures, on the basis of these data can you conclude that the median is significantly less than 165 mmHg?

$$
\begin{array}{ccccccc}
183 & 152 & 178 & 157 & 194 & 163 & 144 \\
194 & 163 & 114 & 178 & 152 & 118 & 158
\end{array}
$$

21. A number of studies have related the size of the hippocampus, the brain region involved in memory and stress, to the risk of developing post-traumatic stress disorder (PTSD). Previous studies have indicated that individuals with small hippocampi are more likely to develop chronic forms of stress disorder. The hippocampi of 16 veterans of the Gulf War who were suffering from PTSD were measured using magnetic resonance imaging. The data are presented below as volumes in cubic centimeters.

$$
\begin{array}{cccccccc}
1.4 & 1.5 & 1.4 & 1.7 & 1.4 & 1.3 & 1.8 & 1.3 \\
1.9 & 2.0 & 1.5 & 1.4 & 1.2 & 1.1 & 1.7 & 1.4
\end{array}
$$

If the average size for the hippocampus in the general population is 1.8 cm³, do the above data support the conclusions of the earlier studies? Assume volumes are normally distributed.

22. Newborn mouse pups are rather difficult to sex, but males have a significantly longer distance from the anus to the genitals than do females. The male anogenital mean distance is 1.20 mm with a standard deviation of 0.03 mm, while females average 0.97 mm with a standard deviation of 0.05 mm.

    In many small mammals that give birth to large litters, male and female fetuses are lined up in the uterus like peas in a pod. It is hypothesized that female fetuses with a male on either side (so-called 2m females) will have a significantly longer anogenital distance due to the proximity of male hormones *in utero*. The anogenital distance was measured in 10 female pups who had been bracketed by males *in utero*. The data (in millimeters) are presented below.

    1.05   1.00   1.03   0.99   1.01   1.07   1.03   1.05   1.08   1.06

    Are the 2m female anogenital distances significantly longer than average for all females?

23. A numerial taxonomist studying the branching filamentous alga, *Cladophora glomerata*, believes she has discovered a new species. *C. glomerata* has a median branching angle of 47°. While the new species is quite similar to *C. glomerata*, she believes the branching angle is less than that of *C. glomerata*. Measuring the branching angle on 17 specimens she has collected, she found the following branching angles.

    42°   40°   39°   45°   38°   39°   41°   41°   40°
    45°   49°   43°   42°   40°   35°   40°   47°

    Are the angles of her sample specimens significantly smaller than those of *C. glomerata*?

24. Suppose 14 patients suffering from acute migraine headaches are given a new medication for headache relief. After 20 minutes each patient was asked if their sense of well-being improved, became worse, or did not change.

    10 reported an improvement
    2 reported no change
    2 reported feeling worse

    Can it be concluded that the new medication is effective in the treatment of migraine headaches?

25. It is thought that the clinical onset of autism may be preceded by two phases of abnormal brain growth: A reduced head size at birth followed by an excessively large growth in head size between 2 months and 14 months of life.

    (a) The mean head circumference of newborn males in the U.S. is 35.83 cm. The head circumferences of 15 males who were later diagnosed with autism spectrum disorder are given below. Is there evidence that head circumference at birth for these males was significantly smaller than normal?

    34.8   35.0   36.6   35.9   32.8   33.4   33.3   35.0
    34.9   35.2   35.9   34.9   31.8   34.6   33.1

    (b) The mean head circumference of one-year-old males in the United States is 46.35 cm. The same group of 15 males was measured at age one year with $\overline{X} = 47.64$ cm and $s = 1.19$ cm. Is there evidence that head circumference at one year for these males was significantly larger than normal?

26. A particular drug company claims that its headache remedy stops the average headache in 14 minutes. Being skeptical, you randomly select 25 headache patients from the outpatient clinic, asking them to take one of these pills when they have a headache and record the length of time (in minutes) until it disappears. The results of your study are: $\overline{X} = 19$ minutes and $s = 7$ minutes. Do your results disagree with the drug company's claim?

27. Cardiac output (liters/min) was measured by thermodilution in a simple random sample of 15 postcardiac surgical patients in the left lateral position (see below). May we conclude on the basis of these data that the population median is different from 5.05? Assume symmetry but not normality for the data.

Wilcoxon

4.91 4.10 6.74 7.27 7.42 7.50 6.56 4.64 5.98 3.14 3.23 5.80 6.17 5.39 5.77

28. A healthy life style undoubtedly plays a role in longevity, but so does genetic makeup. Recent studies have linked large cholesterol particles to good health and longevity. A variant of a gene called CETP encoding the cholestryl ester transferase protein apparently causes the formation of large cholesterol particles. In a particular population the life spans for males are normally distributed with a mean of 74.2 yr and a standard deviation of 10.0 yr. A sample of 16 males in this population that had the variant CETP gene lived an average of 81.2 yr with a standard deviation of 8.0 yr.

(*a*) Does this sample indicate people with the CETP variant gene live significantly longer than those without it?

(*b*) Does the sample indicate that the CETP variant carriers are significantly less variable in their life spans than the general population?

29. Inorganic arsenic is a carcinogen, while organic forms of arsenic are less toxic, combat animal diseases, and accelerate growth. Organic arsenic is an approved ingredient in roxarsone, a feed additive used for poultry and swine. About 70 percent of the chickens grown for meat in the U.S. receive roxarsone. They excrete most of the arsenic but retain some, both organic and inorganic forms, in their tissues. Inorganic arsenic in chicken meat at 0.40 ppm is considered a safe level. A large shipment of processed chicken meat was randomly sampled and analyzed with the following results.

0.42   0.30   0.49   0.47   0.35   0.44   0.43   0.58   0.79   0.34

Use the one-sample Wilcoxon signed-rank test to determine if the shipment has significantly high levels of inorganic arsenic. Use H_0: $M \leq 0.40$ ppm.

30. Telomeres are the cap-like DNA structures on the tips of eukaryotic chromosomes. It is thought that these telomeres act as molecular clocks, shortening every cell division until they become so short that the cells enter into a non-dividing state of senescence. Researchers studying the roundworm, *Caenorhabditis elegans*, isolated a gene that promotes lengthening of telomeres. With this gene they increased the telomere length in a line of *C. elegans*. Unmodified *C. elegans* live an average of 20 days. Below are the life spans (in days) of 16 modified *C. elegans*. Do the modified *C. elegans* live significantly longer than the unmodified ones?

25   23   21   24   27   23   20   19
28   26   24   22   26   27   21   28

31. Methyl tert-butyl ether (MTBE) is a common additive used to make unleaded gasoline burn more efficiently. Drinking or breathing MTBE may cause nausea, nose and throat irritation, and nervous system effects. The American Conference of Governmental Industrial Hygienists (ACGIH) has recommended an exposure limit of 40 parts MTBE per million parts of air (40 ppm) for an 8-hour work-day, 40-hour work-week. At a gasoline production facility a series of air samples were found to have the following MTBE values in ppm. Are workers at this facility exposed to more than the recommended MTBE limit?

$$37 \quad 41 \quad 49 \quad 52 \quad 32 \quad 38 \quad 39 \quad 40 \quad 44 \quad 47 \quad 31 \quad 48 \quad 48 \quad 59$$

32. Abuse of substances containing toluene (for example, various glues) can produce neurological symptoms. In an investigation of the mechanism of these toxic effects, researchers measured the concentrations of certain chemicals in the brains of rats who had been exposed to a toluene-laden atmosphere. The concentrations (ng/gm) of the brain chemical norepinephrine (NE) in the medulla region of the brain of 9 toluene-exposed rats was determined and recorded below.

$$543 \quad 523 \quad 431 \quad 635 \quad 564 \quad 580 \quad 600 \quad 610 \quad 550$$

Does the exposure to toluene significantly increase NE levels in rats above the normal average of 530 ng/gm? Assume NE values are symmetrically distributed.

33. The class Polyplacophora includes about 500 species of molluscs called chitons. Chitons live in shallow water along rocky shorelines. The West Indian chiton, *Chiton olivaceus*, is quite variable in color and is often difficult to identify to species. A malacologist recently found an odd colored population of chitons and wanted to determine whether they are *C. olivaceus*. It is well known that the width-to-length ratio in *C. olivaceus* has a median of 0.65. Below are the ratios for 13 of these odd-colored specimens.

Can use either

$$0.75 \quad 0.73 \quad 0.67 \quad 0.69 \quad 0.64 \quad 0.66 \quad 0.71 \quad 0.79 \quad 0.73 \quad 0.68 \quad 0.69 \quad 0.63 \quad 0.70$$

Is their width-to-height ratio significantly different from that of *Chiton olivaceus*? Caution: Ratios may be not distributed normally.

34. QCR. True or False. Justify your answer.

(*a*) Tests of hypothesis about a single mean can take two forms: a left-tailed or a right-tailed test.

(*b*) t tests involve means that are normally distributed or approximately normally distributed according to the Central Limit Theorem.

(*c*) When writing hypotheses, they should contain only claimed values and Greek symbols, never statistics or values derived from sampling or experimentation.

(*d*) The chi-square test is used to test whether a variance or a standard deviation is significantly different from some claimed value.

(*e*) The one-sample sign test only assumes that X is a continuous random variable with a median, M.

(*f*) The Wilcoxon signed-rank test is more restrictive than the one-sample sign test. It requires that the X's are from a normal distribution.

(*g*) One can calculate confidence intervals utilizing parametric statistics such as the t test and chi-square test but the sign test cannot be used to generate confidence intervals.

7

Tests of Hypothesis Involving Two Samples

Concepts in Chapter 7:

- Comparing Two Variances via an F test
- Comparing Two Means for Unpaired Samples

  — Unpaired t Tests with Pooled Variances
  — Welch's Approximate t Tests When Variances Differ

- Confidence Intervals for Differences in Means
- Paired t Tests
- The Wilcoxon Rank-Sum Test for Differences in Medians
- The Sign Test for Paired Data
- The Wilcoxon Signed-Rank Test for Paired Data

Tests of Hypotheses Involving Two Samples

We now move from questions about characteristics of a single population to comparisons of *two* populations. In the next chapter we will address k-sample experiments where $k > 2$. In the current chapter we employ the same basic steps to answer questions about data sets that were used previously. While the hypotheses and test statistics are different, the general approach remains the same as in Chapter 6.

7.1 Comparing Two Variances

We begin the discussion of two-sample hypotheses with questions involving differences between two variances. Sometimes these questions are the focus of the data collection, but often they must be addressed as preliminary analyses to questions about the two means. For this latter reason the variance test is presented prior to tests of two means.

> **EXAMPLE 7.1.** Among the few reptilian lineages to emerge from the Mesozoic extinction are today's reptiles. One of these, the tuatara, *Sphenodon punctatum*, of New Zealand, is the sole survivor of a group that otherwise disappeared 100 million years ago. The mass (in g) of random samples of adult male tuatara from two islets in the Cook Strait are given below. Is the variability in mass of adult males different on the two islets?

Location A		Location B
510	790	650
773	440	600
836	435	600
505	815	575
765	460	452
780	690	320
235		660

The question can be framed as the following pair of hypotheses:

$$H_0: \sigma_A^2 = \sigma_B^2$$
$$H_a: \sigma_A^2 \neq \sigma_B^2.$$

At this point we don't have a test statistic to determine the veracity of H_0, so we need to develop one. Consider the following process: Draw a sample of size n_1 from a normal population with variance σ^2. Then draw a sample of size n_2 from a normal population with the *same* variance σ^2. Calculate s_1^2 and s_2^2 for the samples; most likely these sample variances will not be identical. The ratio of the sample variances $\frac{s_1^2}{s_2^2}$, *if the two populations are normal and have equal variances*, will come from a sampling distribution known as the **F distribution**. Since $E(s_1^2) = \sigma^2$ and $E(s_2^2) = \sigma^2$, the mean or expected value of the F distribution is 1 while the actual shape of this distribution depends on the degrees of freedom, ν_1 and ν_2, for the two sample variances. See Figure 7.1. Because the shape and expected value of the F distribution are known, the F ratio $\frac{s_1^2}{s_2^2}$ is an appropriate test statistic for a null hypothesis of $\sigma_1^2 = \sigma_2^2$.

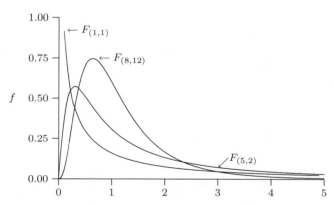

FIGURE 7.1. F distributions for various pairs of degrees of freedom.

The ratio of two sample variances is called the **F test** in honor of the renowned statistician Sir Ronald A. Fisher (1890–1962). The cumulative distribution functions for 0.90, 0.95, 0.975, and 0.99 and various pairs of degrees of freedom are given in Table C.7. More extensive tables are available elsewhere or can be generated by standard statistical programs.

Table C.7 can be used to find the right-hand critical values directly and the left-hand critical value through a conversion process. For Example 7.1 the degrees of freedom would be $\nu_A = n_A - 1 = 13 - 1 = 12$ and $\nu_B = n_B - 1 = 7 - 1 = 6$. If we use the variance for location A as the numerator and the variance for location B as the denominator in the F statistic, $\nu_1 = 12$ and $\nu_2 = 6$. Under H_0: $\sigma_A^2 = \sigma_B^2$, we expect the F value to be close to 1 and when it is extremely large or extremely small we will reject H_0 in favor of the alternative hypothesis $\sigma_A^2 \neq \sigma_B^2$.

Using $\alpha = 0.05$ and $\nu_1 = 12$ and $\nu_2 = 6$, we find the critical values for this F statistic to be:

- Right tail from Table C.7, $F_{0.975(12,6)} = 5.37$.

- Left tail for an F with ν_1 and ν_2 degrees of freedom is the reciprocal of the corresponding right-tail critical value with the degrees of freedom reversed.

$$F_{0.025(12,6)} = \frac{1}{F_{0.975(6,12)}} = \frac{1}{3.73} = 0.27.$$

So if the variance of Location A is less than 0.27 times the variance of Location B *or* if the variance of Location A is more that 5.37 times the variance of Location B, we will assume that H_0 is false because the F value is so far from the expected value of 1. Rejecting H_0 with these critical values will lead to a Type I error 5% of the time.

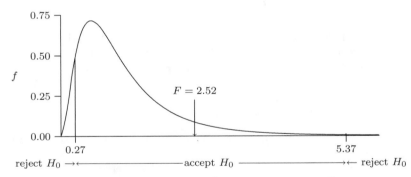

We can now finish the analysis. Since $s_A^2 = 37{,}853.17$ and $s_B^2 = 15{,}037.00$,

$$F = \frac{s_A^2}{s_B^2} = \frac{37{,}853.17}{15{,}037.00} = 2.52.$$

The variance of Location A is 2.52 times the variance of Location B, but because of the small sample sizes here we cannot say that this statistic (2.52) is significantly different from expectation under H_0 $[E(F) = 1]$. Because $0.27 < 2.52 < 5.37$ we accept H_0. In accepting H_0 we may have made a Type II error, but we have no way of quantifying the probability of this type of error here.

Some researchers will calculate the two sample variances *before* determining which is the numerator and which is the denominator. By putting the larger value as the numerator the F statistic will *always* be greater than 1. In two-tailed tests, then, only the right-tail critical value must be found. Remember to use $\frac{\alpha}{2}$ when determining this value.

EXAMPLE 7.2. In a fruit-processing plant the machine that is normally used to package frozen strawberries produces packages with an average of 250 g/box. A new machine is being considered that will process much faster, but may produce more variable results. In order to investigate this concern, the quality control supervisor measured the contents of 50 boxes produced by each machine and found $s_O^2 = 25$ g^2 and $s_N^2 = 64$ g^2. From her results, are her suspicions supported?

SOLUTION. The hypotheses are

$$H_0: \sigma_N^2 \leq \sigma_O^2 \qquad E(F) \leq 1$$
$$H_a: \sigma_N^2 > \sigma_O^2 \qquad E(F) > 1.$$

If we use σ_N^2 as the numerator, under H_0 we expect F to be equal to or smaller than 1. Under H_a we expect F to be significantly larger than one. We need to know the critical value $F_{0.95(49,49)}$, but $\nu_1 = 49$ and $\nu_2 = 49$ are not in Table C.7. We use the next *lower* values found in the table. Note that using lower degrees of freedom makes the critical value larger and the test more conservative:

$$F_{0.95(49,49)} \approx F_{0.95(40,40)} = 1.69.$$

If the new machine's sample variance is at least 1.69 times larger than the old machine's sample variance, we will reject H_0:

$$F = \frac{s_N^2}{s_O^2} = \frac{64}{25} = 2.56.$$

Since $2.56 > 1.69$, we reject H_0. The data support the suspicion that the new machine is producing significantly more variable boxes than the old machine.

To use a P value approach is a bit awkward with the F statistic. If we take each value from Table C.7 for $\nu_1 = 40$ and $\nu_2 = 40$, we find

$$F_{0.90(40,40)} = 1.51, \quad F_{0.95(40,40)} = 1.69, \quad F_{0.975(40,40)} = 1.88, \quad F_{0.99(40,40)} = 2.11.$$

When we compare the statistic 2.56 to these critical values, we find that $2.56 > 2.11$. This indicates that the probability of getting an F value as large as 2.56 is less than 0.01, if H_0 is true. The test statistic here would be a very rare result if H_0 is true.

By rejecting H_0 we are confident that we are making a mistake (Type I error) *less than* 1% of the time: $P(F \geq 2.56) < 0.01$.

To Summarize

Hypotheses about two population variances can take one of three forms:

a) $H_0: \sigma_1^2 = \sigma_2^2$ b) $H_0: \sigma_1^2 \geq \sigma_2^2$ c) $H_0: \sigma_1^2 \leq \sigma_2^2$
     $H_a: \sigma_1^2 \neq \sigma_2^2$ $H_a: \sigma_1^2 < \sigma_2^2$ $H_a: \sigma_1^2 > \sigma_2^2.$

Each H_0 above is *tested* with an F *statistic* and the decision about H_0 is based on how far this statistic deviates from expectation under a *true* H_0. If the F statistic exceeds the critical value(s), H_0 is rejected. Alternatively, if the P value for the F statistic is smaller than the predetermined alpha level, H_0 is rejected.

7.2   Testing the Difference Between Two Means of Independent Samples

There is a somewhat confusing array of tests to resolve questions about two sample means. Careful consideration of the structure of the sampling process will help you choose the proper analysis. In this section we present methods used for two *independent* samples. By independent we mean that each observation in sample 1 has no special relationship to any observation in sample 2. Compare this to measurements on the same individual before and after some treatment that forms two samples that are related or dependent. The first observation in each sample is the same individual! This latter type of design will be considered in Section 7.4.

Equal Population Variances

We first consider two independent random samples of size n_1 and n_2 from two *normally* distributed populations, *with equal population variances assumed.*

> **EXAMPLE 7.3.** Returning to the New Zealand tuataras, we now ask the question: Does the average mass of adult males differ between Location A and Location B?
>
> **SOLUTION.** Clearly the samples are independent, the first observation at Location A has no special relationship to any observation at Location B. We tested the hypothesis of equal variances in Example 7.1 and accepted H_0. We now have new hypotheses about the means
>
> $$H_0\colon \mu_A = \mu_B$$
> $$H_a\colon \mu_A \neq \mu_B.$$
>
> Alternatively, we could write $H_0\colon \mu_A - \mu_B = 0$ and $H_a\colon \mu_A - \mu_B \neq 0$. The descriptive statistics for the two locations are

Location A	Location B
$n_A = 13$	$n_B = 7$
$\overline{X}_A = 618$	$\overline{X}_B = 551$
$s_A^2 = 37{,}853.17$	$s_B^2 = 15{,}037.00$

> We have a null hypothesis involving two population means and we have two estimates of those means ($\overline{X}$'s). As in Section 7.1 we need a test statistic whose sampling distribution is known and that is appropriate for the H_0 presented.
>
> If the two samples come from normal populations with equal variances, then a t statistic analogous to the one presented in Chapter 6 is warranted. $\overline{X}_1 - \overline{X}_2$ will be distributed as normal with a mean equal to $\mu_1 - \mu_2$. It can be shown mathematically that the variance of the difference between two independent variables is equal to the sum of the variances of the two variables. So the variance of $\overline{X}_1 - \overline{X}_2$ is equal to
>
> $$\frac{\sigma_1^2}{n_1} + \frac{\sigma_2^2}{n_2}.$$
>
> The problem is how to best estimate this variance so that we can form a t statistic analogous to (6.1), i.e., a difference over a standard error of the form
>
> $$t = \frac{(\overline{X}_1 - \overline{X}_2) - (\mu_1 - \mu_2)}{\mathrm{SE}_{\overline{X}_1 - \overline{X}_2}}.$$

Since s_1^2 and s_2^2 are estimates for σ_1^2 and σ_2^2 one possibility is to use

$$\frac{s_1^2}{n_1} + \frac{s_2^2}{n_2} \qquad \textit{variances are NOT equal}$$

as an estimate of the variance. In fact, we will do this, but only when we assume that the variances of the two populations are *not equal*. When the variances are assumed *equal*, as in the current situation, we take a weighted average of the two sample variances. This average is called the **pooled sample variance** and is defined as

$$s_p^2 = \frac{(n_1 - 1)s_1^2 + (n_2 - 1)s_2^2}{n_1 + n_2 - 2}. \qquad \textit{equal variances} \qquad (7.1)$$

It is the best estimate of the overall variability in this situation.

FORMULA 7.1. Under H_0, $(\mu_1 - \mu_2) = 0$, making the test statistic for the difference in means of independent samples with equal variances

$$t = \frac{(\overline{X}_1 - \overline{X}_2) - (\mu_1 - \mu_2)}{\sqrt{s_p^2 \left(\frac{1}{n_1} + \frac{1}{n_2} \right)}},$$

with $\nu = n_1 + n_2 - 2$ degrees of freedom.

This statistic is distributed as a t distribution so we can use Table C.4 for critical values or P values. Under the null hypothesis, the *expected value* of t is 0 because $\overline{X}_1$ should be the same as $\overline{X}_2$ making the numerator of the t statistic equal to 0. When the t statistic deviates significantly from 0, we will reject H_0.

SOLUTION TO EXAMPLE 7.3, continued. The pooled sample variance is

$$s_p^2 = \frac{(n_A - 1)s_A^2 + (n_B - 1)s_B^2}{n_A + n_B - 2} = \frac{(13 - 1)37{,}853.17 + (7 - 1)15{,}037.00}{13 + 7 - 2} = 30{,}247.78.$$

Consequently, from Formula 7.1

$$t = \frac{(618 - 551) - 0}{\sqrt{30{,}247.78 \left(\frac{1}{13} + \frac{1}{7} \right)}} = \frac{67}{\sqrt{6{,}647.86}} = \frac{67}{81.5} = 0.82.$$

The test statistic of 0.82 is an index of the difference between two means. Is it significantly different from 0?

At $\alpha = 0.05$ and $\nu = 18$ in Table C.4 we find the critical values to be ± 2.101. Since $-2.101 < 0.82 < 2.101$, we cannot reject H_0. The mean masses for Locations A and B are not significantly different.

EXAMPLE 7.4. An ornithologist studying various populations of the Australian magpie, *Gymnorhina tibicen*, mist-netted 25 adults from an isolated rural area and 36 from an urban picnic area. She measured the total body lengths in centimeters and reported the following summary statistics:

	Rural	Urban
$\overline{X}$	38 cm	35 cm
s	4 cm	3 cm

Because picnickers often feed the magpies, it is thought that the urban population might be somewhat stunted due to the consumption of processed rather than natural foods. Completely analyze the data given to see if it supports this hypothesis.

SOLUTION. Clearly the samples are independent, but do they come from populations with the same variance? We perform a *preliminary* test to answer this question.

$$H_0: \ \sigma_r^2 = \sigma_u^2$$
$$H_a: \ \sigma_r^2 \neq \sigma_u^2$$

Use $\alpha = 0.05$, $\nu_r = 25 - 1 = 24$, and $\nu_u = 36 - 1 = 35$. Then

$$F = \frac{s_r^2}{s_u^2} = \frac{16}{9} = 1.78.$$

The c.v.'s for $\alpha = 0.05$, $\nu_r = 24$, and $\nu_u = 35$ from Table C.7 are

$$F_{0.975(24,35)} \approx F_{0.975(24,30)} = 2.14$$

and

$$F_{0.025(24,35)} \approx \frac{1}{F_{0.975(30,24)}} = \frac{1}{2.21} = 0.45.$$

Since $0.45 < 1.78 < 2.14$, the variances are assumed to be equal.

Now the question about average size can be addressed. Are the magpies smaller in urban areas than in rural areas?

$$H_0: \ \mu_r \leq \mu_u$$
$$H_a: \ \mu_r > \mu_u$$

This is a one-tailed test because we anticipate a *particular* deviation from equality. Since $\sigma_r^2 = \sigma_u^2$ is assumed, we calculate a pooled sample variance:

$$s_p^2 = \frac{(25-1)16 + (36-1)9}{25 + 36 - 2} = \frac{699}{59} = 11.85.$$

So

$$t = \frac{(\overline{X}_r - \overline{X}_u) - (\mu_r - \mu_u)}{\sqrt{s_p^2 \left(\frac{1}{n_r} + \frac{1}{n_u} \right)}} = \frac{(38 - 35) - 0}{\sqrt{11.85 \left(\frac{1}{25} + \frac{1}{36} \right)}} = \frac{3}{\sqrt{0.80}} = 3.35.$$

Under H_0, we expect t to be close to 0 or negative. Under H_a we expect t to be significantly larger than 0.

From Table C.4 with $\nu = 25 + 36 - 2 = 59$, the c.v. is 1.671 for a right-tailed test. Since $3.35 > 1.671$ we consider the t statistic to be *significantly larger* than expected under H_0. We, therefore, accept H_a and say that the analysis indicates the urban magpies are *significantly smaller* than the rural magpies.

The P value approach here would indicate that 3.35 is extreme enough that it occurs with a probability of more than 0.001 and less than 0.025 when H_0 is true. The low probability indicates that when we reject H_0 we are very rarely making an error $(0.001 < P(\text{Type I error}) < 0.025)$.

To Summarize

Hypotheses about two means from independent samples of normal populations *with equal* variances can take the following forms:

a) $H_0: \mu_1 = \mu_2$ b) $H_0: \mu_1 \geq \mu_2$ c) $H_0: \mu_1 \leq \mu_2$
       $H_a: \mu_1 \neq \mu_2$ $H_a: \mu_1 < \mu_2$ $H_a: \mu_1 > \mu_2$

If a preliminary F test of H_0: $\sigma_1^2 = \sigma_2^2$ is *accepted*, then the appropriate H_0 above is tested using a t test of the following form:

$$t = \frac{(\overline{X}_1 - \overline{X}_2) - (\mu_1 - \mu_2)}{\sqrt{s_p^2 \left(\frac{1}{n_1} + \frac{1}{n_2}\right)}}, \qquad \text{with } \nu = n_1 + n_2 - 2.$$

The decision about H_0 is based on how far this statistic deviates from *expectation* under a true H_0. If the t statistic exceeds the critical value(s), H_0 is rejected. Alternatively, if the P value for the t statistic is smaller than the predetermined alpha level, H_0 is rejected.

Note that the F test is *required* here to ascertain the suitability of this t statistic for the hypothesis of equal means.

Unequal Population Variances

If we can assume that the two samples are independent and from normal populations, *but cannot assume that they come from populations with equal variances*, the t test must be modified into a more conservative form.

When the preliminary H_0: $\sigma_1^2 = \sigma_2^2$ is rejected, as mentioned earlier we do not pool the sample variances because they are each estimating a different parameter. Instead, we take their sum.

FORMULA 7.2. The t test for $H_0 : \mu_1 = \mu_2$ when the population variances differ is

$$t = \frac{(\overline{X}_1 - \overline{X}_2) - (\mu_1 - \mu_2)}{\sqrt{\left(\frac{s_1^2}{n_1} + \frac{s_2^2}{n_2}\right)}}$$

with

$$\nu = \frac{\left(\frac{s_1^2}{n_1} + \frac{s_2^2}{n_2}\right)^2}{\frac{\left(\frac{s_1^2}{n_1}\right)^2}{n_1 - 1} + \frac{\left(\frac{s_2^2}{n_2}\right)^2}{n_2 - 1}}.$$

This calculation of the degrees of freedom should be rounded down to the nearest integer and will be somewhat less than $n_1 + n_2 - 2$ in most cases.

The t statistic presented as Formula 7.2 is identical to the t statistic in Formula 7.1 when $n_1 = n_2$ or $s_1^2 = s_2^2$. If both n's and s^2's are equal not only will the t statistics be identical in both procedures but the degrees of freedom will be as well. The modified t given in Formula 7.2 is sometimes called **Welch's approximate t** or the **Smith-Satterthwaite procedure**.

EXAMPLE 7.5. As part of a larger study of the effects of amphetamines, a pharmacologist wished to investigate the claim that amphetamines reduce overall water consumption. She injected 15 standard lab rats with an appropriate dose of amphetamine and 10 with a saline solution as controls. Over the next 24 hours she measured the amount of water consumed by each rat and expressed the results in ml/kg body weight:

	Amphetamine	Saline
n	15	10
$\overline{X}$	115	135
s	40	15

Does the amphetamine significantly suppress water consumption? Use an alpha level of 0.05 for any tests of hypotheses.

SOLUTION. First we do a preliminary F test with $\alpha = 0.05$ for the variances:

$$H_0: \sigma_A^2 = \sigma_S^2$$
$$H_a: \sigma_A^2 \neq \sigma_S^2.$$

$s_A^2 = 1600$ $(\text{ml/kg})^2$ and $s_S^2 = 225$ $(\text{ml/kg})^2$, so

$$F = \frac{s_A^2}{s_S^2} = \frac{1600}{225} = 7.11.$$

Since $\nu_A = 15 - 1 = 14$ and $\nu_S = 10 - 1 - 9$, the critical values from Table C.7 are

$$F_{0.975(14,9)} \approx F_{0.975(12,9)} = 3.87$$

and

$$F_{0.025(14,9)} = \frac{1}{F_{0.975(9,14)}} = \frac{1}{3.21} = 0.31.$$

Since $7.11 \gg 3.87$, we reject H_0 and *cannot* assume the two variances are equal.

The hypotheses for the means are

$$H_0: \mu_A \geq \mu_S$$
$$H_a: \mu_A < \mu_S.$$

We now test the null hypothesis about average water consumption with Welch's approximate t:

$$t = \frac{(\overline{X}_A - \overline{X}_S) - (\mu_A - \mu_S)}{\sqrt{\left(\frac{s_A^2}{n_A} + \frac{s_S^2}{n_S}\right)}} = \frac{(115 - 135) - 0}{\sqrt{\frac{1600}{15} + \frac{225}{10}}} = \frac{-20}{11.37} = -1.759.$$

We expect this t statistic to be close to zero if H_0 is true and significantly less than zero if H_a is true. Is -1.759 far enough below zero that we lose faith in H_0 and accept H_a?

$$\nu = \frac{\left(\frac{s_A^2}{n_A} + \frac{s_S^2}{n_S}\right)^2}{\frac{\left(\frac{s_A^2}{n_A}\right)^2}{n_A - 1} + \frac{\left(\frac{s_S^2}{n_S}\right)^2}{n_S - 1}} = \frac{\left(\frac{1600}{15} + \frac{225}{10}\right)^2}{\frac{\left(\frac{1600}{15}\right)^2}{15-1} + \frac{\left(\frac{225}{10}\right)^2}{10-1}} = \frac{16684.03}{868.95} = 19.2,$$

which is rounded down to 19. Note that $\nu < n_A + n_S - 2 = 23$. The critical value from Table C.4 is $t_{0.05(19)} = -1.729$.

Since $-1.759 < -1.729$, the t statistic is significantly *less than expected* if H_0 is true. We would reject H_0 here and say that the data support the claim that amphetamines significantly suppress water consumption. Probability of a Type I error is close to 5% because the test statistic is close to the 0.05 critical value.

Two-tailed and right-tailed hypotheses using Welch's approximate t are handled in a comparable fashion. Moderate deviations from normality do not seriously affect the validity of the t test. In statistical parlance, the t test is said to be **robust**.

To Summarize

Hypotheses about two means from independent samples *without* the assumption of equal variances can take the following forms:

    a) $H_0: \mu_1 = \mu_2$ b) $H_0: \mu_1 \geq \mu_2$ c) $H_0: \mu_1 \leq \mu_2$

         $H_a: \mu_1 \neq \mu_2$ $H_a: \mu_1 < \mu_2$ $H_a: \mu_1 > \mu_2$.

If a preliminary F test of the $H_0: \sigma_1^2 = \sigma_2^2$ is *rejected*, then the appropriate H_0 above is tested using a t test of the following form:

$$ t = \frac{(\overline{X}_1 - \overline{X}_2) - (\mu_1 - \mu_2)}{\sqrt{\left(\frac{s_1^2}{n_1} + \frac{s_2^2}{n_2}\right)}} \quad \text{with} \quad \nu = \frac{\left(\frac{s_1^2}{n_1} + \frac{s_2^2}{n_2}\right)^2}{\frac{\left(\frac{s_1^2}{n_1}\right)^2}{n_1 - 1} + \frac{\left(\frac{s_2^2}{n_2}\right)^2}{n_2 - 1}}. $$

The decision about H_0 is based on how far this statistic deviates from expectation under a true H_0. If the t statistic exceeds the critical value(s), H_0 is rejected. Alternatively, if the P value for the t statistic is smaller than the predetermined alpha level, H_0 is rejected. Note that the F test is again *required* to ascertain the suitability of this t statistic for the hypothesis of equal means.

7.3   Confidence Intervals for $\mu_1 - \mu_2$

When the null hypothesis of equal means for two independent samples is rejected, confidence intervals for the difference between the two population means can easily be calculated using methods analogous to those presented in Section 4.3 for a single mean.

We have a point estimate $\overline{X}_1 - \overline{X}_2$ for the difference of the means, its sampling distribution is the t distribution, and an estimate of its standard error based on whether the population variances are equal or not. Using (4.3) we can determine a **confidence interval for the difference in the population means**.

FORMULA 7.3. Assuming equal variances, the confidence interval for $\mu_1 - \mu_2$ at the $(1 - \alpha)100\%$ level of confidence is given by

$$ C\left[(\overline{X}_1 - \overline{X}_2) - t_0 s_{\overline{X}_1 - \overline{X}_2} \leq \mu_1 - \mu_2 \leq (\overline{X}_1 - \overline{X}_2) + t_0 s_{\overline{X}_1 - \overline{X}_2}\right] = 1 - \alpha, $$

where

$$ s_{\overline{X}_1 - \overline{X}_2} = \sqrt{s_p^2 \left(\frac{1}{n_1} + \frac{1}{n_2}\right)} $$

equal variances

and $t_0 = t_{1 - \frac{\alpha}{2}, \nu}$ with degrees of freedom $\nu = n_1 + n_2 - 2$.

> **EXAMPLE 7.6.** The magpie body lengths in Example 7.4 for the rural population were deemed significantly longer than the body lengths for the urban population. Calculate a 95% confidence interval for the difference in size.
>
> **SOLUTION.** The difference in the sample means was
>
> $$ \overline{X}_r - \overline{X}_u = 38 - 35 = 3 \text{ cm} $$
>
> with a standard error of
>
> $$ s_{\overline{X}_r - \overline{X}_u} = \sqrt{11.85 \left(\tfrac{1}{25} + \tfrac{1}{36}\right)} = \sqrt{0.80} = 0.89. $$

Here $\nu = 59$, so $t_0 = 2.001$ from Table C.4. Thus, using Formula 7.3, the endpoints of the confidence interval are

$$L_1 = (\overline{X}_r - \overline{X}_u) - 2.001(s_{\overline{X}_r - \overline{X}_u}) = 3 - 2.001(0.89) = 1.219 \text{ cm}$$
$$L_2 = (\overline{X}_r - \overline{X}_u) + 2.001(s_{\overline{X}_r - \overline{X}_u}) = 3 + 2.001(0.89) = 4.781 \text{ cm}.$$

The 95% confidence interval for the difference in body lengths between rural and urban magpies is $[1.219, 4.781]$ cm. Notice that 0 is not included in the interval, and we are 95% confident that rural magpies are on average between 1.219 cm and 4.781 cm longer than urban magpies.

When H_0: $\mu_1 = \mu_2$ is rejected, utiliizing Welch's approximate t, the confidence interval takes a slightly different form. We use the standard error and degrees of freedom from the Welch's approximate t test of hypothesis to configure the appropriate confidence limits in accordance with (4.3).

FORMULA 7.4. When equal variances cannot be assumed, the confidence interval for $\mu_1 - \mu_2$ at the $(1 - \alpha)100\%$ level of confidence is given by

$$C\left[(\overline{X}_1 - \overline{X}_2) - t_0\sqrt{\frac{s_1^2}{n_1} + \frac{s_2^2}{n_2}} \le \mu_1 - \mu_2 \le (\overline{X}_1 - \overline{X}_2) + t_0\sqrt{\frac{s_1^2}{n_1} + \frac{s_2^2}{n_2}}\right] = 1 - \alpha,$$

where $t_0 = t_{1-\frac{\alpha}{2},\nu}$ with degrees of freedom

$$\nu = \frac{\left(\frac{s_1^2}{n_1} + \frac{s_2^2}{n_2}\right)^2}{\frac{\left(\frac{s_1^2}{n_1}\right)^2}{n_1 - 1} + \frac{\left(\frac{s_2^2}{n_2}\right)^2}{n_2 - 1}}.$$

not equal variance

EXAMPLE 7.7. Calculate the 95% confidence interval from Example 7.5 for the reduction in water consumption in lab rats when they are treated with amphetamines.

SOLUTION. Since equal variances couldn't be assumed, the original hypothesis of H_0: $\mu_A \ge \mu_S$ was tested with Welch's approximate t test. The difference in sample means is

$$\overline{X}_A - \overline{X}_S = 115 - 135 = -20 \text{ ml/kg}.$$

Next,

$$\sqrt{\frac{s_A^2}{n_A} + \frac{s_S^2}{n_S}} = \sqrt{\frac{1600}{15} + \frac{225}{10}} = 11.37$$

and

$$\nu = \frac{\left(\frac{1660}{15} + \frac{225}{10}\right)^2}{\frac{\left(\frac{1600}{15}\right)^2}{15-1} + \frac{\left(\frac{225}{10}\right)^2}{10-1}} = 19.2.$$

From Table C.4, $t_0 = 2.093$ for $\alpha = 0.05$ (two-tailed) and $\nu = 19$. So the endpoints of the confidence interval are

$$L_1 = (115 - 135) - 2.093(11.37) = -43.797 \text{ ml/kg}$$
$$L_2 = (115 - 135) + 2.093(11.37) = 3.797 \text{ ml/kg}.$$

We are 95% confident that the reduction in the amount of water consumed lies in the interval $[-43.797, 3.97]$ ml/kg. Notice that 0 (implying no difference) is included in this confidence interval. What does this indicate?

In Example 7.5 the test of hypothesis was one-tailed, investigating a decrease in water consumption with amphetamines. The H_0 claiming no decrease was rejected at $\alpha = 0.05$ with $t = -1.759 < $ c.v. $= -1.729$. The test statistic was just different enough from zero to reject H_0. A confidence interval is really a two-tailed test of hypothesis. Because α is split equally into both tails for the confidence interval, it becomes more difficult to reject H_0.

Sometimes confidence intervals give a different impression of the experimental differences than do one-tailed tests of hypothesis. Confidence intervals for $\mu_1 - \mu_2$ should be calculated only when a significant difference is indicated by the original test of hypothesis (either the t test or Welch's approximate t test). If H_0 is not rejected by these tests, the confidence interval at the same alpha level will always include zero and, therefore, be relatively uninformative.

7.4 The Difference Between Two Means with Paired Data

Sometimes we design experiments with two random samples that *lack independence*. For instance, a pharmacologist might measure the systolic blood pressure in a sample of 25 adults with hypertension *before* administering a drug and again 6 hours *after* administering the drug. Each observation in the first sample is naturally paired with an observation in the second sample that comes from the same person. Studies involving pretesting and posttesting in education also generate naturally paired data, as do twin studies in genetics.

There are several ways in which pairing naturally occurs. **Before and after comparisons** fit into the more general category of **repeated measures**. In such investigations, repeated measurements of the same subject are taken and the differences in the measurements are then analyzed. Sometimes we are interested in the effects of two different treatments upon a single subject, so **simultaneous testing** is appropriate. The two treatments (one may be a placebo) are simultaneously applied to each subject. The individual differences are recorded and analyzed. For example, the efficacy of two different sunblocks might be investigated by having consumers use a different one on each arm. Sometimes it may not be possible to apply both treatments to the same subject. In such cases **matched pairs** may be used. Members of two groups are matched with respect to various extraneous variables and the results for the additional variable of study are compared. For example, in a retrospective study to determine whether bypass surgery or angioplasty is a more effective treatment for arterial blockage, each bypass surgery patient might be matched with respect to age, gender, and blood pressure to a similar patient who had an angioplasty. For each pair a variable of interest, such as survival time, would be compared. Such matched studies are sometimes the only method for doing medical research. In each of these situations two random samples are available but they are *not* independent of each other. Each observation in the first sample is naturally paired with an observation in the second sample. Careful design of experiments, including the use of pairing, can remove a great deal of the variability in data that often obscures meaningful differences.

EXAMPLE 7.8. Watching an infomercial on television you hear the claim that without changing your eating habits, a particular herbal extract when taken daily will allow you to lose 5 lb in 5 days. You decide to test this claim by enlisting 12 of your classmates into an experiment. You weigh each subject, ask them to use the herbal extract for 5 days and then weigh them again. From the results recorded below, test the infomercial's claim of 5 lb lost in 5 days.

Subject	Weight before	Weight after
1	128	120
2	131	123
3	165	163
4	140	141
5	178	170
6	121	118
7	190	188
8	135	136
9	118	121
10	146	140
11	212	207
12	135	126

SOLUTION. Because the data are paired we are not directly interested in the values presented above, but are instead interested in the *differences* or *changes* in the pairs of numbers. Think of the data as in groups.

Group 1	Group 2
X_{11}	X_{21}
X_{12}	X_{22}
X_{13}	X_{23}
$\vdots$	$\vdots$
X_{1n}	X_{2n}

Each observation X is doubly subscripted with the first subscript denoting the group (ith group) and the second subscript denoting position within the group (jth observation). X_{ij} notation will be used extensively in Chapters 8 and 9, so we want to get comfortable with it now.

For the paired data here we wish to investigate the differences, or d_i's, where

$$X_{11} - X_{21} = d_1, \qquad X_{12} - X_{22} = d_2, \qquad \ldots, \qquad X_{1n} - X_{2n} = d_n.$$

Expressing the data set in terms of these differences d_i's, we have the following table. Note the importance of sign of these differences.

Subject	d_i
1	8
2	8
3	2
4	−1
5	8
6	3
7	2
8	−1
9	−3
10	6
11	5
12	9

The infomercial claim of a 5-lb loss in 5 days could be written: H_0: $\mu_B - \mu_A \geq 5$ lb, but H_0: $\mu_d \geq 5$ lb is somewhat more appealing.

$$H_0\text{: } \mu_d \geq 5 \text{ lb}$$
$$H_a\text{: } \mu_d < 5 \text{ lb.}$$

Choose $\alpha = 0.05$. Since the two columns of data collapse into one column of interest, we treat these data now as a one-sample experiment. There is *no preliminary F test* and our only assumption is that the d_i's are approximately normally distributed. Analysis is then analogous to that presented in Section 6.1.

FORMULA 7.5. The test statistic for the paired sample t test is

$$t = \frac{\overline{X}_d - \mu_d}{\frac{s_d}{\sqrt{n_d}}}$$

with $\nu = n - 1$, where n is the number of *pairs* of data points.

SOLUTION TO EXAMPLE 7.8, continued. Here $\overline{X}_d = 3.8$ lb, $s_d = 4.1$ lb, and $n_d = 12$. We expect this statistic to be *close to* 0 if H_0 is true, i.e., the herbal extract allows you to lose 5 lb in 5 days. We expect this statistic to be *significantly less* than 0 if the claim is false:

$$t = \frac{3.8 - 5}{\frac{4.1}{\sqrt{12}}} = -1.01$$

with $\nu = n_d - 1 = 12 - 1 = 11$. The critical value for this left-tailed test from Table C.4 is $t_{0.05(11)} = -1.796$. Since $-1.796 < -1.01$, the test statistic doesn't deviate enough from expectation under a true H_0 that you can reject H_0. The data gathered from your classmates *support* the claim of an average loss of 5 lb in 5 days with the herbal extract. Because you accepted H_0 here, you may be making a Type II error (accepting a false H_0), but we have no way of quantifying the probability of this type of error.

EXAMPLE 7.9. An experiment was conducted to compare the performance of two varieties of wheat, A and B. Seven farms were randomly chosen for the experiment and the yields in metric tons per hectare for each variety on each farm were as follows:

Farm	Yield of variety A	Yield of variety B
1	4.6	4.1
2	4.8	4.0
3	3.2	3.5
4	4.7	4.1
5	4.3	4.5
6	3.7	3.3
7	4.1	3.8

(*a*) Why do you think both varieties were tested on each farm rather than testing variety A on seven farms and variety B on seven different farms?

(*b*) Carry out a hypothesis test to decide whether the mean yields are the same for the two varieties.

SOLUTION. The experiment was designed to test both varieties on each farm because different farms may have significantly different yields due to differences in soil characteristics, microclimate, or cultivation practices. "Pairing" the data points accounts for most of the "between farm" variability and should make any differences in yield due solely to wheat variety.

Farm	Difference (A − B)
1	0.5
2	0.8
3	−0.3
4	0.6
5	−0.2
6	0.4
7	0.3

The hypotheses are

$$H_0: \mu_A = \mu_B \text{ or } \mu_d = 0$$
$$H_a: \mu_d \neq 0.$$

Let $\alpha = 0.05$. Then $n = 7$ and $\overline{X}_d = 0.30$ ton/hectare and $s_d = 0.41$ ton/hectare. So

$$t = \frac{0.30 - 0}{\frac{0.41}{\sqrt{7}}} = 1.94$$

with $\nu = 7 - 1 = 6$. The critical values from Table C.4 are $t_{0.025(6)} = -2.447$ and $t_{0.975(6)} = 2.447$. Since $-2.447 < 1.94 < 2.447$, the test statistic doesn't deviate enough from 0, the expected t value if H_0 is true, to reject H_0. From the data given we cannot say that the yields of varieties A and B are significantly different.

To Summarize

Hypotheses about two samples with *paired* data take the following forms:

a) $H_0: \mu_d = c$ b) $H_0: \mu_d \geq c$ c) $H_0: \mu_d \leq c$
    $H_a: \mu_d \neq c$ $H_a: \mu_d < c$ $H_a: \mu_d > c$,

where c is a real number chosen *before* the data are gathered; c is often 0 under the assumption of *no* change. The data collapse into a single set of differences so there is *no* preliminary F test. The appropriate H_0 above is tested using a t test of the following form:

$$t = \frac{\overline{X}_d - \mu_d}{\frac{s_d}{\sqrt{n_d}}}$$

with $n_d - 1$ degrees of freedom.

The decision about H_0 is based on how far this statistic deviates from expectation under a true H_0. If the t statistic exceeds the critical value(s), H_0 is rejected. Alternatively, if the P value for the t statistic is smaller than the predetermined alpha level, H_0 is rejected.

7.5   The Wilcoxon Rank-Sum (Mann-Whitney U) Test   Table C8

If there are reasons to believe that the distributions involved are normal, then tests on the differences in the means of the two populations are carried out by using one form or another of the t test, depending on whether variances are equal or not. But often we don't know anything about the populations or perhaps we know that the distributions are not normal. In this case, the most common distribution-free hypothesis test used is the **Wilcoxon rank-sum test**, also known as the **Mann-Whitney U test**.

Assumptions

The assumptions for the Wilcoxon rank-sum test are the following:

1. X and Y are continuous random variables.

2. The data consist of two *independent,* random samples. $X_1, \ldots, X_m$ denotes a random sample of size m from the distribution of X, and $Y_1, \ldots, Y_n$ denotes a random sample of size n from the distribution of Y. Note that m and n are often, but not always, different.

3. The null hypothesis is that the X and Y populations are identical.

For convenience, we will assume that the X's represent the smaller sample, that is, $m \leq n$. This makes listing critical values of the test statistic easier. The Wilcoxon rank-sum test examines whether the populations differ in location (median).

Hypotheses

Let M_X and M_Y denote the medians of X and Y, respectively. The hypothesis test can take one of three forms:

- *Two-tailed.* H_0: $M_X = M_Y$ versus H_a: $M_X \neq M_Y$. This tests whether the X's tend to be different from the Y's.

- *Left-tailed.* H_0: $M_X \geq M_Y$ versus H_a: $M_X < M_Y$. This tests whether the X's tend to be smaller than the Y's.

- *Right-tailed.* H_0: $M_X \leq M_Y$ versus H_a: $M_X > M_Y$. This tests whether the X's tend to be larger than the Y's.

Test Statistic and Decision Rules

The approach is to pool the $m+n$ X and Y observations to form a single sample. The observations are ranked from smallest to largest, without regard to which population each value came from. Midranks are used for tied values. The test statistic is the sum of the ranks from the X population (the smaller sample) and is denoted by W_X. If the sum is too small (too large), this is an indication that the values from the X population tend to be smaller (larger) than those from the Y population. The null hypothesis of "no difference" between the populations will be rejected if the sum W_X tends to be extreme in either direction.

We need the distribution of the random variable W_X to know when it is small or large enough to reject H_0. There is one such distribution for each pair of values of m and n. So tables of critical values tend to be complicated and large and, therefore, give only a limited number of critical values. The appropriate table for the Wilcoxon rank-sum test is Table C.8.

> **EXAMPLE 7.10.** Two color morphs (green and red) of the same species of sea star were found at Polka Point, North Stradbroke Island. One might suppose that the red color morph is more liable to predation and, hence, those found might be smaller than the green color morph. The following data are the radial lengths of two random samples of these sea stars (in mm). Are red color morphs significantly smaller than green color morphs?

Red	108	64	80	92	40	
Green	102	116	98	132	104	124

SOLUTION. Let $X = R$ represent the red sea star morphs since this sample has the smaller number of observations and $Y = G$ represent the green morphs. To test whether red color morphs tend to be smaller than the green, the appropriate test is left-tailed with H_0: $M_R \geq M_G$ versus H_a: $M_R < M_G$. The test is done at the $\alpha = 0.05$ level. The first step is to rank the combined data set. We denote the red sea star observations and ranks in **bold** for convenience:

Size	**40**	**64**	**80**	**92**	98	102	104	**108**	116	124	132
Rank	**1**	**2**	**3**	**4**	5	6	7	**8**	9	10	11

The test statistic is the sum of the red sea star ranks,

$$W_X = W_R = 1 + 2 + 3 + 4 + 8 = 18.$$

This does seem small; most red sea stars are smaller than the green. In Table C.8, find the columns for a one-tailed test with $\alpha = 0.05$. Locate the row with $m = 5$ and $n = 6$. The table entry under the W column reads $(20, 40)$. This means that if $W_X \leq 20$, then H_0 can be rejected in favor of the alternative in a left-tailed test at the $\alpha = 0.05$ level. (If a right-tailed test were being performed, H_0 would be rejected if $W_X \geq 40$.) We reject H_0 since $W_X = 18 \leq 20$. There is support for the hypothesis that red color morphs are smaller than the green ones.

Theory

As you can see, the test is fairly easy to carry out; it simply requires care in determining and adding the ranks. As we already noted, the logic of the test is straightforward. Small values of W_X mean that the X's were lower ranked and, hence, smaller than the Y's, indicating a difference in medians. An analogous statement applies for large W_X. But how is the probability of a particular value of W_X determined to create Table C.8? What are the density and cumulative distribution functions for W_X?

The answers to these questions are complicated, but an example with small values of m and n will illustrate the general method. Let's take sample sizes of $m = 2$ for the X's and $n = 3$ for the Y's. Then there are $m + n = 5$ numbers to rank. Two of the ranks will correspond to the X's. So there are

$$\binom{5}{2} = \frac{5!}{2!3!} = \frac{120}{12} = 10$$

different rankings. In general there are $\binom{m + n}{m}$ rankings. The 10 rankings and the corresponding test statistic in our case are

1	2	3	4	5	W_X
X	X	Y	Y	Y	3
X	Y	X	Y	Y	4
X	Y	Y	X	Y	5
X	Y	Y	Y	X	6
Y	X	X	Y	Y	5
Y	X	Y	X	Y	6
Y	X	Y	Y	X	7
Y	Y	X	X	Y	7
Y	Y	X	Y	X	8
Y	Y	Y	X	X	9

Under the assumptions of the test that the samples from X and Y are independent random samples from the same population, all 10 of these rankings are equally likely, so any one of them has a probability of 0.1. Notice that the values of the test statistic range from 3 to 9 and some values arise from more then one ranking. For example, $W_X = 7$ occurs two ways so $P(W_X = 7) = 0.2$. Similarly, eight of the rankings produce values of W_X no greater than 7, so $P(W \leq 7) = 0.8$. In this way, the complete density function and the cumulative distribution for W_X when $m = 2$ and $n = 3$ are produced.

w	3	4	5	6	7	8	9
$P(W_X = w)$	0.1	0.1	0.2	0.2	0.2	0.1	0.1
$P(W_X \leq w)$	0.1	0.2	0.4	0.6	0.8	0.9	1.0

Because the sample sizes are so small in this case, only if $W_X = 1$ (or 9) could we reject H_0 in a left (right) tailed test at the $\alpha = 0.10$ level. Similar calculations to these had to be carried out to produce the entries in Table C.8.

Comments

If the populations of X and Y actually happen to be normal, then the Wilcoxon rank-sum test is testing the same hypothesis as the pooled t test, since we are assuming that both samples come from the same population. This is why the observations are pooled here and a common set of ranks is used.

7.6   The Sign Test and Paired Data Table C.1

Earlier we applied the sign test to a sample drawn from a population with a known or hypothesized median (see Section 6.4). A more common situation is a *paired comparison* in which subjects are matched in pairs and the outcomes are compared within each matched pair. The sign test is easily extended to answer questions about median difference with paired observations as in the question below.

> **EXAMPLE 7.11.** Triglycerides are blood constituents that are thought to play a role in coronary artery disease. An experiment was conducted to see if regular exercise could reduce triglyceride levels. Researchers measured the concentration of triglycerides in the blood serum of 7 male volunteers before and after participation in a 10-week exercise program. The results are shown below. Note the variation in levels from one person to the next, both before and after. But an overall pattern of reduction after exercise is clear. An obvious question to ask is: "How much did exercise reduce triglyceride levels?" This is asking for either a point estimate or a confidence interval for the median difference.
>
> | Before | 0.87 | 1.13 | 3.14 | 2.14 | 2.98 | 1.18 | 1.60 |
> |---|---|---|---|---|---|---|---|
> | After | 0.57 | 1.03 | 1.47 | 1.43 | 1.20 | 1.09 | 1.51 |

Assumptions

The assumptions for the paired sign test are the analogues of the one-sample sign test assumptions:

1. X and Y are continuous random variables.

2. $(X_1, Y_1), \ldots, (X_{n'}, Y_{n'})$ denotes a random sample of size n' from the distribution of (X, Y).

3. Consider the n' continuous differences $X_1 - Y_1, \ldots, X_{n'} - Y_{n'}$. The underlying assumption is that there is no difference between X and Y. More precisely, the assumption is that the median difference $M_{X-Y} = 0$.

The elementary estimates for the paired sign test are just the observed differences $X_i - Y_i$ in pairs.

Hypotheses

Again the hypothesis test can take one of three forms:

- *Two-tailed.* H_0: $M_{X-Y} = 0$ and H_a: $M_{X-Y} \neq 0$. The alternative hypothesis means that X_i tends to be different from Y_i.

- *Left-tailed.* H_0: $M_{X-Y} \geq 0$ and H_a: $M_{X-Y} < 0$. The alternative hypothesis means that X_i tends to be smaller than Y_i.

- *Right-tailed.* H_0: $M_{X-Y} \leq 0$ and H_a: $M_{X-Y} > 0$. The alternative hypothesis means that X_i tends to be larger than Y_i.

Test Statistic

If H_0 is true ($M_{X-Y} = 0$), then any particular difference $X_i - Y_i$ has probability $\frac{1}{2}$ of being negative, probability $\frac{1}{2}$ of being positive, and since the random variables are continuous, probability 0 of being 0. As in the one-sample case, there are times when a difference $X_i - Y_i$ is 0. As long as there are not many such observations, we proceed as before, dropping such pairs from consideration in the hypothesis testing. Let n be the number of points remaining in the sample after the removal of any pairs with 0 difference. Let S_- and S_+ denote the number of negative and positive signs. This is exactly the situation of the one-sample test: The *difference* is a continuous random variable to which the earlier method applies. (For review, see Section 6.4.) If H_0 is true, then the random variables S_- and S_+ have a binomial distribution with parameters n and $p = \frac{1}{2}$ and an expected value of $\frac{n}{2}$.

Decision Rules

Select the alpha level at which the hypothesis will be tested, often $\alpha = 0.05$. Then apply the same decision rules as in the one-sample test.

SOLUTION TO EXAMPLE 7.11. In this case a right-tailed test (H_a: $M_{X-Y} > 0$) is called for because researchers are testing whether regular exercise lowers triglyceride levels, that is, whether the before exercise levels (X) are larger than the after exercise levels (Y). The null hypothesis is H_0: $M_{X-Y} \leq 0$, the triglyceride levels before exercise are no greater than those after. We will use $\alpha = 0.05$. The elementary estimates are the paired differences $X_i - Y_i$:

Before $-$ After $= X_i - Y_i$	0.30	0.10	1.67	0.71	1.78	0.09	0.09

All seven differences are positive. For a right-tailed test, the test statistic is S_-, the number of negative differences, which is 0 here. From Table C.1, the corresponding P value is 0.0078. This is much less than $\alpha = 0.05$. There is a drop in the median triglyceride levels after exercise.

EXAMPLE 7.12. Zonation is a process of adaptation that can occur when various organisms compete for resources in a small area and can be observed in microhabitats such as the intertidal shoreline of a bay. Moving a distance of a meter or two perpendicular to the waterline can drastically change the amounts of moisture, temperature, and salinity. Different organisms tolerate different extremes and zonation occurs.

While at North Stradbroke Island, students carried out exercises to quantify these processes. The intertidal zone (the shore between the lowest low tide and the highest high tide) was divided into five equal-width bands parallel to the waterline. Square-meter quadrats were used to repeatedly sample the various bands for marine organisms. One subset of the data was used to study the vertical zonation of the intertidal zone by various gastropods, such as *Bembecium*, *Austrocochlea*, and *Nerita*. The data below show the distribution of *Bembecium* at 12 locations where counts were made at the mean water mark and at one zone above it. Is there a difference in the median numbers of *Bembecium* in the two zones?

Above mid-shore	81	20	25	62	41	28	43	50	48	31	39	24
Mid-shore	135	15	31	81	59	72	66	101	36	92	59	41

SOLUTION. In this case a two-tailed test is called for since prior to the exercise students knew little about the natural history of the organisms. The null hypothesis is H_0: There is no difference in the median numbers of *Bembecium* in the two zones ($M_{X-Y} = 0$). The alternative is H_a: There is evidence for a difference in the median numbers ($M_{X-Y} \neq 0$). We will use $\alpha = 0.05$. The elementary estimates are the paired differences $X_i - Y_i$ where the X population occurs in the higher zone:

$$-54 \quad 5 \quad -6 \quad -19 \quad -18 \quad -44 \quad -23 \quad -51 \quad 12 \quad -61 \quad -20 \quad -17$$

All 12 differences are nonzero. For a two-tailed test, the test statistic is the minimum of S_-, the number of negative differences, and S_+, the number of positive differences. There are only two positive differences so the test statistic is $S_+ = 2$. From Table C.1, the corresponding P value is $2(0.0193) = 0.0386$ since the test is two-sided. This is less than $\alpha = 0.05$. There is a difference in the median numbers of *Bembecium* per square meter in these two zones.

7.7   The Wilcoxon Signed-Rank Test for Paired Data

The Wilcoxon signed-rank test is easily extended to paired data in a fashion entirely analogous to the way the sign test was extended to paired data.

Assumptions

1. X and Y are continuous random variables.

2. $(X_1, Y_1), \ldots, (X_{n'}, Y_{n'})$ denotes a random sample of size n' from the distribution of (X, Y).

3. The set of n' differences $X_1 - Y_1, \ldots, X_{n'} - Y_{n'}$ is assumed to be drawn from a population that is *symmetric* about 0.

Recall that *symmetry* is crucial to the Wilcoxon test. The differences now satisfy the assumptions of the one-sample signed-rank test. The test is carried out on these differences.

Hypotheses

The hypothesis test takes the same form as for the paired sign test:

- *Two-tailed.* H_0: $M_{X-Y} = 0$ and H_a: $M_{X-Y} \neq 0$.

- *Left-tailed.* H_0: $M_{X-Y} \geq 0$ and H_a: $M_{X-Y} < 0$.

- *Right-tailed.* H_0: $M_{X-Y} \leq 0$ and H_a: $M_{X-Y} > 0$.

Methods

For hypothesis testing, be sure to eliminate any paired differences that are 0. Test statistics and/or elementary estimates are computed exactly as in the one-sample versions of the signed-rank test.

> **EXAMPLE 7.13.** Body height and arm span are related quantities; taller people generally have longer arms. One rule of thumb says that adult height is nearly equal to adult span. The arm spans of the 10 science faculty in Example 6.13 were also measured. Do the data support the contention? (Height and span data are symmetric.)

Height	171	175	177	178	180	182	190	192	195	202
Span	173	178	182	182	188	185	186	198	193	202

> **SOLUTION.** Test the null hypothesis H_0: $M_{H-S} = 0$ versus the two-sided alternative hypothesis, H_a: $M_{H-S} \neq 0$ at the $\alpha = 0.05$ level using the Wilcoxon signed-rank test because of the symmetry assumption. Calculate the 10 differences and rank them using signed ranks. Note the use of midranks and that the observation with a 0 difference has not been ranked:

$H_i - S_i$	-2	-3	-5	-4	-8	-3	4	-6	2	0
$|H_i - S_i|$	2	3	5	4	8	3	4	6	2	0
Signed rank	-1.5	-3.5	-7	-5.5	-9	-3.5	$+5.5$	-8	$+1.5$	$*$

The test statistic is $W_+ = 7$, which is the smaller of W_- and W_+. Using Table C.6 with $n = 9$, the P value is 0.0371. Since $0.0371 > 0.025 = \frac{\alpha}{2}$, we cannot reject H_0 in favor of the alternative hypothesis in this two-sided test. The data support the claim that adult height and arm span are nearly equal.

Summary of Two-Sample Tests

Two-sample tests of hypothesis with continuous variables

Hypothesis	Sampling assumptions	Test statistic
$H_0: \sigma_1^2 = \sigma_2^2$	Independent samples Normal populations	F test
$H_0: \mu_1 = \mu_2$	Independent samples Normal populations Equal variances	Unpaired t test with pooled variance
$H_0: \mu_1 = \mu_2$	Independent samples Normal populations Unequal variances	Welch's approximate t test
$H_0: \mu_d = c$	Paired data Normal populations	Paired t test
$H_0: M_X = M_Y$	Independent samples	Wilcoxon rank-sum test
$H_0: M_{X-Y} = 0$	Paired data	Sign test
$H_0: M_{X-Y} = 0$	Paired data Symmetric populations	Wilcoxon signed-rank test for paired data

7.8   Problems

 Phosphate pollution of lakes and rivers is a considerable problem. Elevated phosphate levels can cause algal blooms and rapid eutrophication of freshwater ecosystems. To protect these ecosystems several strict pollution laws were enacted. The phosphate levels in a river were measured at several randomly chosen sites downstream from a chemical plant before and after these laws were passed. The following measurements were recorded in μg/l units. Was there a significant decrease in phosphate levels after the laws were enacted? Assume the phosphate values are normally distributed.

	Before	After
n	10	12
$\overline{X}$	650	500
$\sum(X_i - \overline{X})^2$	5,760	19,008
s^2	640	1,728

2. A reproductive physiologist studying the effect of photoperiod on reproductive readiness in Japanese quail measured the cloacal gland widths of male quail exposed for 2 weeks to different day lengths. Long days consisted of 16 hours of light and 8 hours of dark. Short days had 8 hours of daylight and 16 hours of dark. Assuming cloacal gland widths are normally distributed, are the glands significantly wider during long days?

Long days	Short days
$\overline{X}_1 = 12.0$ mm	$\overline{X}_2 = 8.5$ mm
$s_1 = 2.5$ mm	$s_2 = 2.0$ mm
$n_1 = 16$	$n_2 = 16$

3. The mitral valve is the heart valve between the left atrium and left ventricle and it has two flaps. In patients with mitral valve prolapse, one or both valve flaps are enlarged. When the heart pumps (contracts), part of one or both flaps collapses backward into the left atrium permitting a small amount of blood to leak backward through the valve and cause a heart murmur. Patients with mitral valve prolapse are at higher risk for infection during dental procedures. Consequently, a single large dose of an antibiotic is usually prescribed for such patients to be taken one hour before any dental work is done. In a pilot study of antibiotics, penicillin was administered to six volunteers and in a later trial the same dose of amoxicillin was given to the same people. The table below gives the concentration of each drug in the bloodstream 1 hour after administration. The measurements are in μg/ml and are assumed to be normally distributed. Is the bloodstream concentration of penicillin significantly different from that of amoxicillin?

Person	Penicillin	Amoxicillin
1	42	36
2	34	44
3	57	61
4	40	35
5	28	35
6	48	50

4. A potential side effect of using oral contraceptives is an increase in blood pressure. Given below are the systolic blood pressures of 10 women measured before beginning and after having taken an oral contraceptive for a 6-month period. Assuming systolic blood pressure to be normally distributed, do these data suggest that using an oral contraceptive significantly increases systolic pressure? What type of error (I or II) could you have made here? Explain.

Woman	Before	While taking
1	113	118
2	117	123
3	111	114
4	107	115
5	115	122
6	134	140
7	121	120
8	108	105
9	106	111
10	125	129

5. Reconsider the data in the previous problem. Suppose that you mistakenly believed that the data sets were independent samples taken from two different populations: Women not taking contraceptives and different women taking contraceptives. Which type of t test would now be appropriate? Carry this test out and determine whether you would have found that women taking contraceptives have a significantly higher systolic pressure. Explain the cause of any differences in your results when compared to the previous problem.

6. Copper sulfate is routinely used to control algal blooms in ponds and lakes. An ichthyologist believes that copper sulfate has an adverse effect on the gill filaments of several species of fish, including largemouth bass, reducing the number of mucus cells in these species. To test her belief, she recorded the number of mucus cells per square micron in the gill filaments of untreated fish and in fish exposed for 24 hours to copper sulfate at 1 mg/l.

(a) Assuming normality for these data, does the ichthyologist have support for her contention?

Untreated	16	17	12	18	11	18	12	15	16	14	16			
Exposed	10	8	10	12	13	14	6	10	9	10	10	11	9	8

(b) Develop a 99% confidence interval for $\mu_U - \mu_E$. Explain in your own words what this calculation represents.

7. Take sample sizes of $m = 2$ for the X's and $n = 4$ for the Y's. Find all possible rankings and corresponding test statistics W_X for these samples. *Hint:* There are $m + n = 6$ numbers to rank. Two of the ranks will correspond to the X's. So there are

$$\binom{m+n}{m} = \binom{6}{2} = \frac{6!}{2!4!} = 15$$

different rankings that are possible. Determine the test statistic W_X for each of these rankings and then determine the density and distribution functions for W_X. *Further hint:* It is sufficient to choose all possible pairs of numbers between 1 and 6 to get the X ranks from which to determine W_X. OK, write them out!

w	3	4	5	6	7	8	9	10	11
$P(W_X = w)$									
$P(W_X \le w)$									

8. Aposematic coloration acts as a signal to warn predators of potentially dangerous or unpleasant features. It is believed that individuals which possess color patterns resembling those of aposematic species gain a defensive advantage because predators have learned to avoid animals with these patterns through encounters with aposematic species. A series of experiments tested this hypothesis using the aposematic marine flatworm, *Phrikoceros baibaiye*, and a potential predator, *Thalassomus lunare* (moon wrasse). Each of seven wrasses was presented separately with two different models of flatworms. One was an accurately colored model of *P. baibaiye* and the other was a colorless control model. Both models were made from brine shrimp and agar and should have been palatable for the wrasses. X_i denotes the number of times the ith wrasse attacked the model of *P. baibaiye* and Y_i the number of times it attacked the colorless model during the experimental period. (Based on data reported in Ang, H. and L. Newman. 1996. Warning colouration in pseudocerotid flatworms Platyhelminthes, polycladida. Personal communication.)

(a) Use the data below to test the aposematic hypothesis H_0: $M_{X-Y} \ge 0$, the model of *P. baibaiye* is attacked at least as frequently as the colorless model, versus H_a: $M_{X-Y} < 0$, the model of *P. baibaiye* is attacked less frequently than the colorless model.

Fish	X: *P. baibaiye* model	Y: Colorless model
1	1	9
2	5	9
3	7	5
4	6	9
5	1	9
6	5	9
7	7	9

(*b*) Comment on the sample-size aspect of the design of this experiment.

9. It has been shown that levels of beta-endorphin in the blood are elevated by emotional stress. For 19 patients scheduled to undergo surgery, blood samples were taken twice: (a) 12–14 hours before surgery and (b) 10 minutes before surgery. Beta-endorphin levels were measured in fmol/ml. Assuming the symmetry of the distributions, determine if there is a significant increase in beta-endorphin levels immediately before surgery using Wilcoxon signed-rank methods. *(HSDS, #232)*

(a)	10.0	6.5	8.0	12.0	5.0	11.5	5.0	3.5	7.5	5.8
	4.7	8.0	7.0	17.0	8.8	17.0	15.0	4.4	2.0	

(b)	6.5	14.0	13.5	18.0	14.5	9.0	18.0	42.0	7.5	6.0
	25.0	12.0	52.0	20.0	16.0	15.0	11.5	2.5	2.0	

10. (*a*) A biochemist studying the activity of the enzyme triose phosphate isomerase (TPI) in a species of *Drosophila* wishes to know if the mean activity of this enzyme is the same at pH $= 5$ and at pH $= 8$. Assuming activities in μM/min are normally distributed, determine if the mean activities of TPI are significantly different at these two pH levels.

$$\text{pH} = 5: \quad 11.1 \quad 10.0 \quad 13.3 \quad 10.5 \quad 11.3$$
$$\text{pH} = 8: \quad 12.0 \quad 15.3 \quad 15.1 \quad 15.0 \quad 13.2$$

(*b*) Develop a 95% confidence interval for $\mu_5 - \mu_8$. What is the appropriate standard error for this confidence interval? What are the appropriate degrees of freedom for this confidence interval?

11. Redo Problem 8 in Chapter 1 as a test of hypothesis question.

12. (*a*) The pit viper, *Porthidium picadoi*, is found in cloud forests in Costa Rica. In a study of its natural history, the following data were reported for adult total body lengths (in cm). Is there a difference between male and female lengths? Analyze the data with the appropriate statistical methods. (Based on data reported in: Solórzano, A. 1990. Reproduction in the pit viper *Porthidium picadoi* Dunn (Serpentes: Viperidae) in Costa Rica. *Copeia*, (4): 1154–1157.)

	n	$\overline{X} \pm s$
Males	34	94.6 ± 11.72
Females	38	96.4 ± 9.92

(b) As part of the same study, neonates (newborn) of the same species were measured and the following data were recorded for length and weight. Analyze these data for differences between genders.

	n	$\overline{X} \pm s$
Length (cm)		
Males	30	21.53 ± 1.76
Females	29	21.47 ± 1.69
Weight (g)		
Males	30	7.55 ± 1.33
Females	29	7.94 ± 1.42

13. Japanese scientists have reported that both fish oil and the pungent compound responsible for chili peppers' bite will turn on fat oxidation—the chemical conversion of flab to heat. A scientist believes that epigallocatechin gallate (EGCG), a flavonoid in green tea, does the same thing. To study the effect of EGCG on energy expenditure, the nutritionist gave volunteers capsules with each meal then measured how much energy they burned over the next 24 hours. One day the capsules consisted of a placebo and one day they consisted of the amount of EGCG in 2–3 cups of green tea. Do you think the EGCG increases energy expenditure? Analyze the data below to support your conclusion. (Think about experimental design before you begin the analysis. Is there a special relationship between the numbers in the data columns?)

	Energy expenditure (calories/day)	
Subject	EGCG	Placebo
1	2310	2200
2	1890	1950
3	2400	2300
4	2100	2110
5	2250	2020
6	1930	1820
7	1770	1710
8	2650	2540
9	2000	1910
10	2010	1950

14. It is common lore in the Adirondack mountain region of New York that Avon's Skin So Soft hand lotion is a very effective repellant for black flies, *Simulium venustum*. To test this claim volunteers applied fixed amounts of a commercial repellant, N,N-diethyl-meta-toluamide (DEET) 20% (Treatment I), or Skin So Soft hand lotion (Treatment II) to their hand and arm, placing it in a chamber with 50 hungry black flies. The time until first bite in minutes is recorded below.

Treatment I	5	14	6	8	8	15	13	2
Treatment II	3	7	10	49	2	38	6	9

Are the repellants significantly different in their efficacy? Analyze using the Wilcoxon rank-sum test.

15. A popular, or at least frequently shown, television commercial claims that eating oatmeal everyday will lower your total cholesterol level significantly in twelve weeks. You and eleven classmates decide to test this claim by eating oatmeal everyday for breakfast for an entire semester. Below are the total cholesterol values (mmol/L) for you and your friends at the beginning and at the end of the semester.

Subject	Start	End
1	190	188
2	210	212
3	185	180
4	175	172
5	193	196
6	187	180
7	280	200
8	170	166
9	165	159
10	232	225
11	194	206
12	183	170

(*a*) Do the data support the claim that oatmeal has cholesterol lowering properties? Analyze both parametrically and nonparametrically.

(*b*) Suppose subject number 7's end value was really 266 instead of 200. Would the analysis on the corrected data indicate the same trend?

(*c*) Suppose subject number 7 had the genetic condition hypercholesterolemia giving him extremely high cholesterol levels and he couldn't be considered a normal or random subject. How is the analysis affected if he is removed from the study?

16. Many bacteria, including the skin dwelling ones responsible for body odor, require iron for growth. A cosmetic company hopes to exploit that reliance by including iron-sequestering additives in their company's personal hygiene products. Butylated hydroxytoluene, which releases iron bound in sweat, and diethylenetriamine pentaacetic acid, which binds free iron thereby keeping it away from microbes, were added to a new deodorant formulation. The new formulation was used on one armpit of 20 volunteers while the original formulation was used on the other. After 24 hours an unfortunate researcher with a keen sense of smell was asked to evaluate the armpits for offensive odor. Sixteen times he chose the armpit with the new formulation as less offensive. Twice he rated the armpits as equal and twice he rated the original formulation armpit as less offensive.

(*a*) From this data set is there evidence that the additives help reduce body odor more effectively than the original deodorant?

(*b*) Suppose the study was redone using 60 subjects and the reformulation was chosen less offensive 41 times. Evaluate the efficacy of the new additives.

17. Discovered in 1995, leptin is a hormone that plays a role in maintenance of weight. Leptin levels in the blood drop when fat loss occurs. Preliminary studies indicate that leptin influences the levels of thyroid hormones. Thyroid hormones also drop during weight loss causing a drop in metabolism making weight loss difficult to maintain.

A group of 32 dieters who had lost 10% of their starting body weight were randomly divided into two equal groups. The control group received saline injections every other day for three months. The experimental group received replacement doses of leptin every other day for three months. The average weight gain in pounds for the two groups is given below. Do the data support the use of leptin injections to maintain weight loss?

Control	Experimental
$\overline{X}_C = 10.1$	$\overline{X}_E = 6.4$
$s_C = 9.3$	$s_E = 8.7$

18. Dioxin, a byproduct of a number of sophisticated manufacturing processes, has been implicated as a cause in a variety of developmental and sexual anomalies in humans. A concern is the tendency of dioxin to biologically magnify, that is, concentrate at higher levels of the food chain in fatty tissues. Remarkably it is thought that Inuit natives of Nunavut, Canada's newest province, might have high levels of dioxin because their diet consists primarily of fatty animals high in the food chain. Samples of breast milk from women in southern Quebec were compared to the breast milk from Inuit women.

Dioxin (ppm)	
Inuit	Quebec
20	11
19	9
25	12
31	10
26	8
24	4
21	7
18	13
14	
10	

Does the breast milk of Inuit women have a significantly higher concentration of dioxin than that of women living in the more industrialized areas of southern Quebec? Use the Wilcoxon rank-sum test.

19. Centipedes are an interesting and diverse group of arthropods. They belong to the subphylum Myriapoda and class Chilopoda. All centipedes have numerous body segments with one pair of legs per segment. The number of leg bearing segments and, hence, pairs of legs ranges from 15 to 177, but is always an odd number. The legs on the first body segment are modified into venom bearing fangs that centipedes use for food acquisition. The North American species, *Orya barbarica*, may reach lengths up to 17 cm (6–7 in)!

An ecologist studying this species sampled two populations. The first was found in an area with great diversity of soil type and particle size while the second was found in an area with very uniform soil characteristics. The ecologist thought that the more diverse environment might lead to more variability in the lengths of the animals. Below are the summary statistics for the lengths of individuals in the two samples. Do the data support his hypothesis?

diverse
uniform

$X_1 = 12.1$ cm	$s_1 = 3.6$ cm	$n_1 = 25$
$X_2 = 14.5$ cm	$s_2 = 2.5$ cm	$n_2 = 30$

20. A study was designed to test whether cigarette smoking is associated with reduced serum-testosterone levels in men. Two random samples of 25–30-year-old non-smokers and heavy smokers were selected. Each man's serum-testosterone level was measured in μg/dl and the data are compiled below. Use the Wilcoxon rank-sum test to analyze these data.

Nonsmokers	0.44	0.44	0.43	0.56	0.85	0.68	0.96	0.72	0.92	0.87
Heavy Smokers	0.45	0.25	0.40	0.27	0.34	0.62	0.47	0.30	0.35	0.54

21. Neuropathic pain has been recognized and studied since the Civil War. This condition is characterized by ongoing, intense pain despite seemingly minor tissue damage. Physicians now know that neuropathic pain derives from nerve damage that never heals properly. An experimental treatment that involves chemical agents that spur nerve growth during fetal development such as glial-derived neurotrophic factor (GDNF) and artemin have shown promise in reducing neuropathic pain in lab animals. Rats were treated with morphine or GDNF administered in random order with several days recovery period between treatments. After each treatment they were evaluated for sensitivity using a subjective scale to measure sensitivity (8 = most sensitive to 1 = least sensitive). The data are presented below.

Morphine	GDNF
4	5
8	7
3	2
2	3
5	2
7	5
7	6
8	5
7	4
3	4
1	1
3	2

Use a sign test for paired data to determine if there are significant differences between the two treatments. Test H_0: $M_{M-GDNF} = 0$.

22. For each of the following null hypotheses, give the appropriate test statistic and its expected value if H_0 is true. For example, for H_0: $\mu = 50$, the appropriate test statistic is t and its expected value is $E(t) = 0$.

(a) H_0: $\sigma^2 = 250$

(b) H_0: $M = 50$ with no assumptions regarding the data

(c) H_0: $M = 50$ with symmetrical data

(d) H_0: $\mu_d = 5$ for paired data

(e) H_0: $\sigma_1^2 = \sigma_2^2$

(f) H_0: $\mu_1 = \mu_2$

23. Most humans are poorly adapted for underwater vision because about two-thirds of the eye's refractive power is lost when air is replaced by water. In Southeast Asia some tribes of sea gypsies whose children collect food from the sea floor without the use of visual aids appear, however, to have evolved superior underwater vision. One method to improve underwater vision is to constrict the pupil because a smaller pupil can significantly improve resolution.

To test this hypothesis 6 Moken children, sea gypsies from the Archipelago of Burma, and 12 European children of the same age had their pupil diameters measured while diving. The results, in millimeters, are recorded below.

Moken	European
1.95	2.50
2.10	2.65
2.10	2.70
1.80	2.65
1.90	2.45
2.00	2.20
	2.40
	2.35
	2.55
	2.05
	2.30
	2.15

Were the pupil diameters significantly smaller in the Moken children? Analyze without the assumption of normality for the data.

24. To help assess the health risks of second-hand smoke, the levels of cotinine (a metabolite of nicotine) were measured in mmol/l in the urine of seven subjects prior to exposure to second-hand smoke and shortly after a two-hour exposure to secondary cigarette smoke. Did the exposure significantly increase the cotinine level? Assume cotinine levels are not normally distributed and analyze with an appropriate sign test.

Cotinine Levels (mmol/l)		
Subject	Before	After
1	13	19
2	17	23
3	10	19
4	21	25
5	15	20
6	9	15
7	32	60

25. A hormone, peptide PYY-36, that suppresses appetite is made in the small intestine in response to food and then carried by the blood stream to the brain where it switches off the urge to eat. In studies to utilize the hormone to develop a weight loss treatment it was learned that PYY breaks down when taken by mouth in pill form. Since shots are difficult to administer, a nasal spray was developed. The preliminary study to test if the spray safely increased PYY levels in the blood stream involved 15 healthy volunteers who were not overweight. They fasted overnight, skipped breakfast, came to the laboratory, had their natural PYY levels measured (in pg/ml) and then squirted the spray up their noses. One half-hour later their PYY levels were measured again (see below).

Subject	Before	After
1	1.9	4.0
2	2.5	3.8
3	3.1	4.2
4	2.0	3.9
5	2.1	3.7
6	1.9	3.5
7	2.0	4.0
8	1.3	2.2
9	2.9	4.1
10	3.4	3.6
11	2.5	3.8
12	3.0	4.2
13	2.6	4.9
14	3.6	4.1
15	2.1	3.7

Did the nasal spray significantly increase the PYY levels in the bloodstream? Assume normality for the data set.

26. Although the study above was not designed to measure PYY's effect on appetite, the researchers asked the subjects to rate their hunger on a scale of 1 to 10 with 10 being the most hungry. Below are the scores prior to administering the nasal spray and one half-hour after administering the spray.

Subject	Before	After
1	10	7
2	8	4
3	7	5
4	9	7
5	8	3
6	10	5
7	6	6
8	7	4
9	4	5
10	10	7
11	10	7
12	9	3
13	5	6
14	7	6
15	10	4

Did the nasal spray significantly decrease the subjects' hunger levels? Analyze nonparametrically.

27. Natural selection can be extremely difficult to document. However, when a log-gerhead shrike, *Lanius ludovicianus*, catches a lizard, the bird often impales it on a thorn or spur of barbed wire to subdue the prey. After pulling off the flesh, the shrike leaves the lizard carcass hanging, so what remains is a record of the successful shrike predation.

Researchers studied a population of flat-tailed horned lizards, *Phyrnosoma mcalli*, in the desert near Yuma, AZ. On its skull the lizard has multiple pairs of horns facing backwards (parietal horns) and to the side (squamosal horns). When a predator such as a shrike tries to grab the lizard's body, the lizard rears its head to stab the predator. One might expect that lizards with longer horns would have a defensive advantage in this situation.

The researchers measured the two types of horns on each of 29 dead lizards and 155 live ones collected from the same area. They found that living lizards had longer horns than the dead ones. Though the difference in lengths was small (1 to 2 mm), this difference was roughly 10% of the horns length. The researchers believe that this difference is enough for natural selection to take place.

Analyze the data below that was collected to determine whether the small differences in horn length are statistically significant. Note: As is typical in such studies, means and standard errors were reported. You will need to determine the variance from the standard error. (Based on data reported in: Young, K. et al. 2004. How the horned lizard got its horns. *Science*, **304**: 65.)

	Parietal horn length			Squamosal horn length	
	Shrike-killed	Live		Shrike-killed	Live
n	29	155	n	29	155
$\overline{X}$	8.77 mm	9.65 mm	$\overline{X}$	21.99 mm	24.28 mm
s.e.	0.21 mm	0.01 mm	s.e.	0.49 mm	0.21 mm

28. Green and black tea are both derived from the same plant, *Camellia sinensis*. Green tea is produced by lightly steaming fresh-cut leaves. Black tea is made by allowing the leaves to oxidize, thereby converting highly therapeutic polyphenols such as epigallocatechin gallate (EGCG) to less active compounds. These polyphe-nols have potent antioxidant and anticarcinogenic properties. Two brands of green tea were analyzed for their concentrations of EGCG (in mg/cup). Do the brands of green tea have significantly different concentrations of EGCG?

Brand A	Brand B
255	245
300	250
275	230
290	235
280	220
245	235

29. An exercise physiology student noticed that people seem to work out longer when high-energy music is played in the weight room. To test the effect of music on exercise she developed an experiment. Using 12 college-aged volunteers she asked each to ride an exercise bike until exhaustion. She covered the dials on the bike and removed the clocks from the room. One day the weight room sound system played elevator music (Kenny G) and the other day it played AC/DC's "Highway to Hell." The music was randomized so that half the volunteers got the AC/DC the first day and half got it the second day. Below are the number of minutes ridden for each test subject. Did the high-energy background music increase the amount of time exercised? Analyze nonparametrically.

Student	AC/DC	Kenny G
1	27	19
2	35	30
3	41	35
4	19	23
5	38	30
6	61	50
7	37	32
8	32	32
9	45	30
10	42	40
11	44	28
12	29	17

30. The highly publicized idea that listening to classical music improves children's ability to reason about spatial relations is termed the "Mozart effect." While researchers have never been able to document such an effect, some studies have indicated that learning to play a musical instrument may boost IQ scores. Fifteen randomly chosen six-year-olds were given free music lessons for one year. Their IQ scores on standardized tests given prior to the music lessons and after one year of lessons are listed below.

Student	Before	After
1	107	114
2	120	126
3	110	112
4	95	103
5	105	115
6	110	120
7	118	119
8	136	133
9	125	132
10	111	109
11	106	117
12	140	143
13	92	96
14	99	105
15	120	127

(a) Did the IQ scores rise significantly? Assume the scores are normally distributed.

(b) Re-analyze assuming the data are from a symmetrical but not normal distribution.

(c) Re-analyze without assumptions of normality or symmetry.

31. It is thought that the gene encoding the angiotensin-converting enzyme (ACE) may strongly influence human physical performance. The endocrine rennin-angiotensin system is important in controlling the circulatory system. ACE degrades vasodilator kinins and converts angiotensin I to the vasoconstrictor angiotensin II. In addition, local rennin-angiotensin systems may influence tissue growth. It is also thought that a particular allele of the ACE gene (allele I) is associated with higher activity by tissue ACE and, consequently, enhanced physical performance.

In a study of UK army recruits, the ACE genotype was determined for 64 individuals. After a 10-week training period, the maximum duration (in seconds) for which they could perform repetitive elbow flexion while holding a 15-kg barbell was assessed for each genotype. Did the II genotype recruits perform significantly better than recruits with other genotypes? Assume normality for these data.

	II Genotype	Not II Genotype
n	26	38
$\overline{X}$	199.2 s	125.2 s
s	25.2 s	20.4 s

32. A plant physiologist studying a small beach-dwelling plant called the Great Lakes sea rocket, *Cakile edentula*, wondered if the plants grew faster in competition with non-kin plants than when in competition with seedlings that are close relatives.

She set up a direct study of kin recognition by growing sets of four seeds in one container. When the seedlings reached a phase of vigorous root growth, about 8 weeks after planting, she carefully removed the soil and weighed the root masses. Plants grouped with siblings had the following root weight statistics:

$$n = 24 \quad \overline{X} = 6.25 \text{ g} \quad s = 0.47 \text{ g}$$

Plants grouped with unrelated seedlings had the following root weight statistics:

$$n = 20 \quad \overline{X} = 7.22 \text{ g} \quad s = 0.55 \text{ g}$$

Are the plants grown in competition with unrelated individuals allocating significantly more resources to root growth than those growing among relatives?

33. Tiny male spiders of the species *Tidarren sisyphoides* routinely amputate one of their two oversized external sex organs called pedipalps. Males in this species grow to about one one-hundredth the size of the females and each pedipalp accounts for about 10% of their body mass. It has been hypothesized that amputation of a pedipalp enables the males to run faster and longer thereby increasing their chances for winning mates. To test the hypothesis that males with one pedipalp have more endurance than those with two, males were chased around a white paper arena until exhausted. Below are summary statistics for their times in seconds.

	One pedipalp	Two pedipalps
n	10	10
$\overline{X}$	245 s	220 s
s	23 s	20 s

Did the spiders with a single pedipalp run significantly longer than those with two pedipalps?

34. In an investigation of possible brain damage due to alcoholism, an x-ray procedure known as a computer assisted tomography (CAT) scan was used to measure brain densities in 8 chronic alcoholics. For each alcoholic, a nonalcoholic control was selected who matched the alcoholic on age, sex, race, height, and other factors. The brain density measurements on the alcoholics and the matched controls are recorded below. Assume normally distributed data here.

Pair	Alcoholic	Control
1	40	42
2	38	40
3	36	37
4	41	46
5	40	43
6	42	42
7	39	43
8	45	44

Use an appropriate test to determine whether there is evidence of diminished brain density due to chronic alcoholism.

35. In zoology, Bergmann's rule claims that for endothermic (warm-blooded) species the body mass of individuals will increase with latitude and colder climate. Basically in mammals and birds, individuals of a particular species living in colder areas will tend to have greater mass than individuals living in warmer areas. The idea is that larger animals have a lower surface to volume ratio than smaller animals, so they radiate less body heat per unit of mass and, therefore, stay warmer in cold climates.

Two students decide to test this rule during spring break. One went home to Old Forge, New York in the Adirondacks and the other went to Fort Lauderdale, Florida. Both took with them mouse traps and scales. While on break each trapped and weighed 20 white-footed mice, *Peromyscus leucopus*. Their data are summarized below.

Old Forge	Fort Lauderdale
$\overline{X} = 24.1$ gm	$\overline{X} = 22.6$ gm
$s^2 = 1.69$ gm^2	$s^2 = 2.25$ gm^2

Do the data support Bergmann's rule? Analyze appropriately.

36. A dental research team wished to know if teaching people how to brush their teeth would be beneficial. Twelve pairs of patients seen in a dental clinic were obtained by carefully matching on such factors as age, sex, intelligence, and initial oral hygiene scores. One member of each pair received instruction on how to brush their teeth and on other oral hygiene matters. Six months later all 24 subjects were examined and assigned an oral hygiene score by a dental hygienist unaware of which subjects had received the instruction. A low score indicates a high level of oral hygiene. The results are shown below. Analyze nonparametrically.

Pair	1	2	3	4	5	6	7	8	9	10	11	12
Instructed	1.5	2	3.5	3	3.5	2.5	2	1.5	1.5	2	3	2
Not Instructed	2	2	4	2.5	4	3	3.5	3	2.5	2.5	2.5	2.5

37. A recent report in *Science* summarizing a large Norwegian study claimed that first-born children have higher IQ's than later born children. To test the hypothesis that birth order influences IQ you decide to do a small study of your own. Thirteen of your friends volunteer to report their scores on a standardized test as well as scores for their siblings. Below is the data which can be assumed to be normally distributed.

First-born	Second-born	d_i	F$-$S
113	109	4	+
127	120	7	+
130	131	-1	$-$
109	105	4	+
115	110	5	+
119	119	0	0
105	109	-4	$-$
117	114	3	+
135	130	5	+
143	139	4	+
111	105	6	+
128	121	7	+
135	130	5	+

(a) Do your data support the claims of the larger study reported in *Science*?

(b) Reanalyze without making the assumption of normality for the data.

38. The leading drug prescribed for male impotence, sildenafil, commonly known as Viagra, was tested for its efficacy to ameliorate the effects of jet lag. To mimic conditions that can lead to jet lag, scientists habituated hamsters to a daily routine of 14 hours of light and 10 hours of darkness. The researchers then abruptly switched on the lights 6 hours early and continued the same light to dark routine from that point onward, simulating the effects of a flight from Chicago to London.

Just before changing the hamsters' routine, the researchers injected each animal with a size-appropriate dose of either sildenafil or a saline solution. Then, they observed how long it took the hamsters to restart their daily habit of running on a wheel.

	Sildenafil	Saline
n	25	25
$\overline{X}$	6 days	12 days
s^2	25 days^2	40 days^2

(a) Was the restart time significantly shorter for the hamsters that received sildenafil? Analyze parametrically. (Note: Cyclic guanine monophosphate (cGMP) plays a role in regulating circadian rhythm. Hamsters receiving sildenafil had double the cGMP of the saline group because sildenafil shuts down enzymes that would limit cGMP production. How this affects the restoration of circadian rhythm is unknown.)

(b) Most travelers experience jet lag more strongly with eastbound flights that shorten the day and push the circadian clock forward. A second experiment was conducted in which the light shift for the hamsters was 6 hours later stimulating a westbound flight. The data are presented below.

	Sildenafil	Saline
n	20	20
$\overline{X}$	10 days	12.5 days
s^2	36 days^2	25 days^2

Does the sildenafil also work on westbound jet lags?

39. The July 2007 issue of *Gourmet* magazine stated that people would eat 28.4% more candy if it's labeled "low fat." To test this claim you design an experiment to allow students to eat as much candy as they would like during statistics lab. One week you supply candy labeled "low fat" and another week you supply identical candies without the label. At the end of lab each week you determine the amount (in ounces) of candy eaten by each student during lab.

Student	Low fat	Unlabelled
1	8.0	6.5
2	5.0	5.0
3	4.5	3.0
4	5.5	4.5
5	6.0	4.0
6	3.5	0.0
7	10.0	5.0
8	9.0	9.0
9	7.0	5.0
10	4.5	3.5
11	11.0	7.0
12	2.5	0.5
13	8.5	9.0
14	9.5	6.5
15	10.0	7.0

Using a parametric test determine if the claimed 28.4% increase is accurate.

40. The stalked crinoid, *Endoxocrinus parrae*, spends most of its time stationary, catching food with the flowerlike wheel of feathery arms that have earned it and its relatives the common name sea lilies. Scientists have known for decades that these animals sometimes move, but they were thought to be very slow, moving no faster than 0.6 m/hr. Recent video studies indicate that their locomotion is much more rapid and the distances covered much greater than originally expected. In a recent study 16 crinoids were crowded together in a 1-meter-square quadrat. The distances they moved in 12 hours were recorded in meters. They averaged 20 meters with a standard deviation of 4.0 meters. Sixteen uncrowded crinoids were spread out uniformly several meters apart on the sea floor. The average distance the uncrowded crinoids traveled in 12 hours was 14 meters with a standard deviation of 5.0 meters. Does crowding cause the crinoids to travel significantly farther?

41. A scientist proposes that obesity can result from the conditions in which fetuses develop. To determine whether fetal changes in gene activity could initiate obesity, he removed a single ovary from female mice. When these mice became pregnant, the eight or so fetuses that typically occupy both tubes of the mothers' uteruses crowded into a single tube. Vessels that feed into both ends of each tube of a mouse uterus supply it with mother's blood. Therefore, in the experiment, fetuses stuffed in the middle of a single uterine tube received less nourishment than fetuses at either end of it. Those at either end of the tube were unusually heavy at birth and those in the center were unusually light. Fetuses located at intermediate positions were born with normal weights.

If the amount of nutrition that the fetuses received permanently affected how their genes functioned, he expected low birth weight individuals to have low adult weights and high birth weight individuals to have high adult weights. Below are the adult weight data for these two groups. Are his expectations supported by the data? Assume normality of data.

Low birth weight	High birth weight
$n = 25$	$n = 25$
$\overline{X} = 12.21$ g	$\overline{X} = 14.32$ g
$s^2 = 2.11$ g^2	$s^2 = 1.92$ g^2

42. QCR. True or False. Justify your answer.

(a) The F test is used to test the equality of two variances and is required as a preliminary test for the unpaired two-sample t tests.

(b) The degrees of freedom for the Welch's approximate t test is $n_1 + n_2 - 2$.

(c) Confidence intervals for the difference between two means ($\mu_1 - \mu_2$) should only be calculated when the null hypothesis of equal means is rejected.

(d) A paired t test is an effective way to remove "noise" or variability among subjects that may obscure differences between treatments.

(e) The Wilcoxon rank-sum test is a nonparametric equivalent to the paired t test.

(f) The Wilcoxon signed-rank test for paired data requires that X and Y are continuous variables and that the $X_i - Y_i$ differences are symmetric about 0.

k-Sample Tests of Hypothesis: The Analysis of Variance

Concepts in Chapter 8:

- Fixed (Model I) versus Random (Model II) One-Way ANOVA
- Partitioning of the Sums of Squares
- Appropriateness of F Test for H_0: $\mu_1 = \mu_2 = \cdots = \mu_k$
- Global F Test and ANOVA Table for Model I
- Mean Separation Techniques

    — Bonferroni t Tests
    — Student-Newman-Keuls Test

- Global F Test and ANOVA Table for Model II
- Kruskal-Wallis k-Sample Test
- Pairwise Comparisons with Kruskal-Wallis Test

k-Sample Tests of Hypothesis: The Analysis of Variance

In Chapter 6 we presented various techniques for making inferences about a *single* sample. Chapter 7 extended the ability to make statistical inferences to *two* sample situations. We now move to designs with *more than two samples*. The first of these experimental designs involves *three or more independent random samples* that are to be compared together. Let's consider three examples.

EXAMPLE 8.1. A gerontologist investigating various aspects of the aging process wanted to see whether staying "lean and mean," that is, being under normal body weight, would lengthen life span. She randomly assigned newborn rats from a highly inbred line to one of three diets: (1) unlimited access to food, (2) 90% of the amount of food that a rat of that size would normally eat, or (3) 80% of the amount of food that a rat of that size would normally eat. She maintained the rats on three diets throughout their lives and recorded their lifespans (in years). Is there evidence that diet affected life span in this study?

Unlimited	90% diet	80% diet
2.5	2.7	3.1
3.1	3.1	2.9
2.3	2.9	3.8
1.9	3.7	3.9
2.4	3.5	4.0

EXAMPLE 8.2. A paleontologist studying Ordovician fossils collected four samples of the trilobite *Paradoxides* (members of an extinct subphylum of the Arthropoda) from the same rock strata, but from different locations. He measured the cephalon length (in mm) of all the specimens he collected and recorded the data below. Are there significant differences in cephalon length among these sites?

	\multicolumn{4}{c}{Site}			
	I	II	III	IV
n	8	5	11	15
$\overline{X}$	7.0	9.0	8.0	5.0
s	2.1	2.1	2.2	2.2
$\sum X_i$	56.0	45.0	88.0	75.0

EXAMPLE 8.3. An endocrinologist studying genetic and environmental effects on insulin production of pancreatic tissue raised five litters of experimental mice. At age 2 months he sacrificed the mice, dissected out pancreatic tissue and treated the tissue specimens with glucose solution. The amount of insulin released from these specimens was recorded in pg/ml. Are there significant differences in insulin release among the litters?

\multicolumn{5}{c}{Litter}				
1	2	3	4	5
9	2	3	4	8
7	6	5	10	10
5	7	9	9	12
5	11	10	8	13
3	5	6	10	11

Each of these examples involves questions that can be framed as looking for significant differences among means. There are subtle, but important, differences in the goals of each of these studies.

Example 8.1 is an experiment with fixed, well-considered treatments. This experiment could be repeated using the same strain of rats and the same diets. The gerontologist is interested in only these treatments and would like to find any individual differences among the treatment means. This type of study is analyzed as a fixed-effects or **Model I ANOVA** (more on this later). Example 8.2 is a field study with four sites that probably represent the *only* sites of interest for the paleontologist. He may consider cephalon lengths so important that any sites with significantly different sizes may be described as having different species. He can't really repeat the study and is interested in discriminating any differences among these four sites. This study would also be considered a Model I ANOVA.

In Example 8.3 the litters were a *random sample* of mouse litters. The goal of the study was to determine if significant between-litter variability exists that could be attributed to genetic differences among litters. The individual litters are not of interest and, in fact, were destroyed in the experimental protocol. The endocrinologist is not interested in determining the significance of the difference between litters 1 and 2 because those litters are gone. He is interested in only a general statement that there is significant variability among litters (a significant genetic component) or not. This kind of study is analyzed as a random-effects or **Model II ANOVA** (again, more on this later).

The preliminary analyses of both Model I and Model II ANOVAs are identical, but there are additional tests required in Model I. All three examples are considered **one-way** analyses because the individual data points are grouped according to a single criterion: diet, site, or litter. Let's consider first the procedures for Model I analysis.

8.1 Model I ANOVA: One-Way Classification, Completely Randomized Design with Fixed Effects

Model Assumptions

1. The k samples represent <u>independent random samples</u> drawn from k specific populations with means $\mu_1, \mu_2, \ldots, \mu_k$, where $\mu_1, \mu_2, \ldots, \mu_k$ are unknown constants.

2. Each of the k populations is normal. *or large enough for CLT* $(n_i > 25)$

3. Each of the k populations has the same variance σ^2. $\sigma_1^2 = \sigma_2^2 = \cdots = \sigma_k^2$

These assumptions look *very much* like those used for the unpaired t tests in Chapter 7. If the null hypothesis is written $H_0: \mu_1 = \mu_2 = \cdots = \mu_k$, why not test this hypothesis with a series of t tests? On the surface this approach seems logical, but doing repeated unpaired t tests here is *invalid*. Consider a three sample experiment like Example 8.1. We would compare $\overline{X}_U$ to $\overline{X}_{90\%}$, $\overline{X}_U$ to $\overline{X}_{80\%}$, and $\overline{X}_{90\%}$ to $\overline{X}_{80\%}$ in three separate t tests. If we set α at 0.05, the probability of wrongly concluding that two of the sample means estimate μ's that are significantly different is 0.14. To see why, note that with an alpha level of 0.05, we will correctly accept a true H_0 95% of the time. For a set of three independent comparisons (H_0's), all of which are true, the probability of correctly accepting all of them is $(0.95)^3 = 0.86$, so the probability of a Type I error (rejecting at least one true H_0) becomes $1 - 0.86 = 0.14$, which is unacceptably high.

For k means there are

$$\frac{k(k-1)}{2} = \binom{k}{2}$$

pairwise comparisons. With four means there would be six comparisons, and the probability of experimentwise Type I error with $\alpha = 0.05$ for each comparison would rise to $1 - (0.95)^6 = 0.26$. With five means the 10 comparisons would have an overall probability of Type I error equal to 0.40 using $\alpha = 0.05$ for each test! Clearly, then, the use of multiple t tests is not appropriate when you wish to discriminate among three or more independent sample means.

Under the model assumptions given above, the technique used to analyze for differences is called **Analysis of Variance (ANOVA)**. Analysis of variance techniques were first developed in the early twentieth century by Sir Ronald Fisher and are used to investigate a wide range of experimental questions and designs. While it may seem counterintuitive to call a technique to investigate differences in *means* analysis of *variance*, the logic of this term will become evident as we develop the statistical protocol.

Model I Hypotheses

The hypotheses are

$$H_0: \mu_1 = \mu_2 = \cdots = \mu_k$$
$$H_a: \text{at least one pair of } \mu\text{'s is not equal.}$$

The null hypothesis assumes all treatments have the same mean, while the alternative hypothesis doesn't claim they're all different but that *at least some* are different. In Example 8.1 we could find the 90% and 80% diets are not different from each other in their effects on life span, but that they are significantly different from the control diet. We would accept H_a in this case.

The Data Layout For Model I

We need to use the double-subscripted notation introduced in Chapter 7. This allows us to keep track of the treatment and the observation position within that treatment. In X_{ij} the i represents the ith treatment and the j indicates that it is the jth observation of the ith treatment. For example, $X_{2,9}$ is the ninth observation in the second treatment.

		Treatment level		
1	**2**	$\cdots$ i $\cdots$		**k**
X_{11}	X_{21}	$\cdots$ X_{i1} $\cdots$		X_{k1}
X_{12}	X_{22}	$\cdots$ X_{i2} $\cdots$		X_{k2}
$\vdots$	$\vdots$	$\ddots$ $\vdots$ $\ddots$		$\vdots$
X_{1n_1}	X_{2n_2}	$\cdots$ X_{in_i} $\cdots$		X_{kn_k}

To simplify writing summations, we now introduce "dot" notation:

$$T_{i.} = \sum_{j=1}^{n_i} X_{ij};$$

that is, $T_{i.}$ is the sum or total of all observations in treatment i. The dot in the notation indicates the index over which the summation has occurred. So the sample mean for the ith treatment becomes

$$\overline{X}_{i.} = \frac{T_{i.}}{n_i}.$$

Two dots are used to indicate a summation over both indices. For example, the sum of all observations is

$$T_{..} = \sum_{i=1}^{k} \sum_{j=1}^{n_i} X_{ij}.$$

An expanded data layout might take the form

Treatment level					
1	**2**	$\cdots$	**i**	$\cdots$	**k**
X_{11}	X_{21}	$\cdots$	X_{i1}	$\cdots$	X_{k1}
X_{12}	X_{22}	$\cdots$	X_{i2}	$\cdots$	X_{k2}
$\vdots$	$\vdots$	$\ddots$	$\vdots$	$\ddots$	$\vdots$
X_{1n_1}	X_{2n_2}	$\cdots$	X_{in_i}	$\cdots$	X_{kn_k}
$T_{1.}$	$T_{2.}$	$\cdots$	$T_{i.}$	$\cdots$	$T_{k.}$
$\overline{X}_{1.}$	$\overline{X}_{2.}$	$\cdots$	$\overline{X}_{i.}$	$\cdots$	$\overline{X}_{k.}$

Because the grand total or sum of all observations is

$$T_{..} = \sum_{i=1}^{k} \sum_{j=1}^{n_i} X_{ij},$$

and the total number of observations is

$$N = \sum_{i=1}^{k} n_i,$$

then the overall or **grand mean** is

$$\overline{X}_{..} = \frac{T_{..}}{N}.$$

Here are the summary statistics for Example 8.1. Check them by hand to make certain that you understand double summation and dot notation. For further help see the last section of Appendix A.1.

Unlimited	90% diet	80% diet	Summary
$T_{1.} = 12.2$	$T_{2.} = 15.9$	$T_{3.} = 17.7$	$T_{..} = 45.8$
$n_1 = 5$	$n_2 = 5$	$n_3 = 5$	$N = 15$
$\overline{X}_{1.} = 2.44$	$\overline{X}_{2.} = 3.18$	$\overline{X}_{3.} = 3.54$	$\overline{X}_{..} = 3.05$
$s_1^2 = 0.19$	$s_2^2 = 0.17$	$s_3^2 = 0.25$	

Sums of Squares

The total variability in the 15 observations for Example 8.1 can be summarized as

$$\sum_{i=1}^{3} \sum_{j=1}^{5} \left(X_{ij} - \overline{X}_{..} \right)^2.$$

Each observation is expressed as the square of its distance from the grand mean. This sum of squared deviations is called the **Total Sum of Squares** or SS_{Total}. If this value were divided by $N - 1$, we would have a sample variance for a single data set ignoring the treatments. In other words, combining all the data into one group we have a measure of the scatter of *that* group about its mean.

The SS_{Total} can be partitioned into two components: the variability *between* the treatments and the variability *within* treatments. Consider the following:

- Each X_{ij} varies about $\overline{X}_{i.}$: observations vary from their treatment means.

- Each $\overline{X}_{i.}$ varies about $\overline{X}_{..}$: treatment means vary from the overall grand mean.

- Each X_{ij} varies about $\overline{X}_{..}$: observations vary from the grand mean.

FORMULA 8.1. The deviation of an observation from the grand mean can be partitioned into the deviation of the treatment mean from the grand mean and the deviation of an observation from its treatment mean,

$$(X_{ij} - \overline{X}_{..}) = (\overline{X}_{i.} - \overline{X}_{..}) + (X_{ij} - \overline{X}_{i.}).$$

Partitioning the Sums of Squares

Squaring both sides of Formula 8.1 and summing over all values of X_{ij}, we have

$$SS_{Total} = \sum_i \sum_j (X_{ij} - \overline{X}_{..})^2$$

$$= \sum_i \sum_j \left[(\overline{X}_{i.} - \overline{X}_{..}) + (X_{ij} - \overline{X}_{i.}) \right]^2$$

$$= \sum_i \sum_j \left[(\overline{X}_{i.} - \overline{X}_{..})^2 - 2(\overline{X}_{i.} - \overline{X}_{..})(X_{ij} - \overline{X}_{i.}) + (X_{ij} - \overline{X}_{i.})^2 \right]$$

$$= \sum_i \sum_j (\overline{X}_{i.} - \overline{X}_{..})^2 - \sum_i \sum_j 2(\overline{X}_{i.} - \overline{X}_{..})(X_{ij} - \overline{X}_{i.}) + \sum_i \sum_j (X_{ij} - \overline{X}_{i.})^2$$

$$(8.1)$$

Consider the middle term above. Since $\overline{X}_{i.}$ and $\overline{X}_{..}$ are constants relative to j,

$$\sum_i \sum_j 2(\overline{X}_{i.} - \overline{X}_{..})(X_{ij} - \overline{X}_{i.}) = 2 \sum_i \left[(\overline{X}_{i.} - \overline{X}_{..}) \sum_j (X_{ij} - \overline{X}_{i.}) \right].$$

But from (1.2), the sum of deviations of observations from the sample mean always equals 0, i.e., $\sum_j (X_{ij} - \overline{X}_{i.}) = 0$. Therefore, the middle term in (8.1) equals 0. Consequently,

FORMULA 8.2. SS_{Total}, the total variability, may be partitioned into SS_{Treat}, the variability due to differences among treatments, and SS_{Error}, the variability within treatments that is unexplainable.

$$\sum_i \sum_j (X_{ij} - \overline{X}_{..})^2 \quad = \quad \sum_i \sum_j (\overline{X}_{i.} - \overline{X}_{..})^2 \quad + \quad \sum_i \sum_j (X_{ij} - \overline{X}_{i.})^2$$

$$\text{SS}_{Total} \qquad = \qquad \text{SS}_{Treat} \qquad + \qquad \text{SS}_{Error}.$$

Some authors call these sources of variability **among treatment sum of squares** and **residual sum of squares**, respectively.

The Computational Formulas for ANOVA Sums of Squares

Analogous to the computational formula for the sample variance first presented in Chapter 1, we have the time-saving computational formulas for each of the sums of squares in ANOVA:

$$SS_{Total} = \sum_i \sum_j X_{ij}^2 - \frac{\left(\sum_i \sum_j X_{ij}\right)^2}{N} = \text{uncorrected SS} - \text{correction term.}$$

This leads to

FORMULA 8.3. The total variability may be calculated as

$$SS_{Total} = \sum_i \sum_j X_{ij}^2 - \frac{T_{..}^2}{N}.$$

Next,

$$SS_{Treat} = \sum_i \left[\frac{\left(\sum_j X_{ij}\right)^2}{n_i} \right] - \frac{\left(\sum_i \sum_j X_{ij}\right)^2}{N}.$$

The algebraic proof of this formula can be found in Appendix A.3. This may be expressed as

FORMULA 8.4. The variability among treatments may be calculated as

$$SS_{Treat} = \sum_i \frac{T_{i.}^2}{n_i} - \frac{T_{..}^2}{N}.$$

Finally

$$SS_{Error} = \sum_i \left[\sum_j X_{ij}^2 - \frac{\left(\sum_j X_{ij}\right)^2}{n_i} \right] = \sum_i CSS_i,$$

so

FORMULA 8.5. The variability within treatments that is unexplainable may also be calculated as

$$SS_{Error} = \sum_i (n_i - 1)s_i^2.$$

Note that once we know any two of the three quantities SS_{Total}, SS_{Treat}, or SS_{Error} we can calculate the final one by using the relationship $SS_{Total} = SS_{Treat} + SS_{Error}$.

Back to Example 8.1, we now calculate the appropriate sums of squares:

$$\sum_i \sum_j X_{ij}^2 = 145.44, \qquad \sum_i \sum_j X_{ij} = 45.8, \qquad \sum_i n_i = 15.$$

From Formula 8.3,

$$SS_{Total} = 145.44 - \frac{(45.8)^2}{15} = 145.44 - 139.84 = 5.60.$$

The value 5.60 represents the total variability in the 15 life spans. From Formula 8.4,

$$\text{SS}_{\text{Treat}} = \text{SS}_{\text{Diets}} = \left(\frac{(12.2)^2}{5} + \frac{(15.9)^2}{5} + \frac{(17.7)^2}{5} \right) - \frac{(45.8)^2}{15} = 3.15.$$

The value 3.15 represents the portion of the total variability that is *explained* by differences in diet. From Formula 8.5 and the summary statistics listed earlier,

$$\text{SS}_{\text{Error}} = (5-1)(0.19) + (5-1)(0.17) + (5-1)(0.25) = 2.44.$$

With rounding error, $\text{SS}_{\text{Total}} = \text{SS}_{\text{Diets}} + \text{SS}_{\text{Error}}$.

Degrees of Freedom for the Sums of Squares

As we saw earlier, dividing the SS_{Total} by $N - 1$ would give us a measure of the total variability existing in the combined data set. Just as the sum of squares can be partitioned into treatment (explainable) variability and error (unexplainable) variability, so the total degrees of freedom can be partitioned. With k treatments, SS_{Treat} has $k - 1$ degrees of freedom, while SS_{Error} has $\sum_i (n_i - 1) = N - k$ degrees of freedom; note that

$$N - 1 = (k - 1) + (N - k).$$

The Concept of Mean Squares

Within each treatment the corrected sum of squares divided by $n_i - 1$ would give us an estimate of σ^2, the overall intrinsic variability. Assuming that the treatments all have the same variance, our best estimate of that variance will be a pooled value similar to what we did in the unpaired t test in the previous chapter. The **pooled CSS** is defined as

$$\text{SS}_{\text{Error}} = \sum_i \text{CSS}_i.$$

The **Error Mean Squares** (MS_{E}) is

$$\frac{\text{SS}_{\text{Error}}}{\sum_i (n_i - 1)} = \frac{\text{SS}_{\text{Error}}}{N - k}$$

and is our best estimate of the variance σ^2 common to all k populations.

Theory

The **expected value of** MS_{E} is σ^2, that is,

$$E(\text{MS}_{\text{E}}) = \sigma^2.$$

SS_{Treat} divided by its degrees of freedom is called the **Treatment Mean Squares:**

$$\text{MS}_{\text{Treat}} = \frac{\text{SS}_{\text{Treat}}}{k - 1}.$$

The expected value of MS_{Treat} is

$$E(\text{MS}_{\text{Treat}}) = \sigma^2 + \sum_i \frac{n_i (\mu_i - \mu)^2}{k - 1}.$$

The derivations of these expected values are given in Appendix A.4. The expected values of MS_E and MS_{Treat} give us an appropriate statistic to test the null hypothesis.

If H_0: $\mu_1 = \mu_2 = \cdots = \mu_k$ is true, then each $\mu_i = \mu$ and so each term $(\mu_i - \mu) = 0$. Consequently, if H_0 is true, then

$$\sum_i \frac{n_i(\mu_i - \mu)^2}{k-1} = 0$$

and so $E(MS_{Treat}) = \sigma^2 + 0 = \sigma^2$.

The F test first introduced early in Chapter 7 can now be used to test the veracity of H_0: $\mu_1 = \mu_2 = \cdots = \mu_k$. The ratio of the two mean squares

$$\frac{MS_{Treat}}{MS_E}$$

is distributed as an F statistic with $\nu_1 = k - 1, \nu_2 = N - k$.

If H_0 is true, then

$$E(F) = \frac{\sigma^2}{\sigma^2} = 1.$$

If H_0 is false, then

$$E(F) = \frac{\sigma^2 + \sum_i \frac{n_i(\mu_i - \mu)^2}{k-1}}{\sigma^2} > 1.$$

Since the additional term in the numerator must be positive, we expect the F statistic to be *bigger* than 1 if H_0 is *false* and some of the means are different. So the F test in ANOVA is *always one-tailed and right-tailed*. This test is usually presented as part of a summary ANOVA table like Table 8.1.

TABLE 8.1. The analysis of variance table

Source of variation	Sum of squares	df	MS	F	c.v.
Treatments	SS_{Treat}	$k-1$	$\frac{SS_{Treat}}{k-1}$	$\frac{MS_{Treat}}{MS_E}$	See Table C.7
Error	SS_{Error}	$N-k$	$\frac{SS_{Error}}{N-k}$		
Total	SS_{Total}	$N-1$			

As we have noted, the sums of squares and degrees of freedom are both additive.

SOLUTION TO EXAMPLE 8.1. The hypotheses are

$$H_0\text{: } \mu_U = \mu_{90\%} = \mu_{80\%}$$
$$H_a\text{: At least one pair of the } \mu_i\text{'s is not equal.}$$

The summary data to test H_0 appear in the table below.

Source of variation	Sum of squares	df	MS	F	c.v.
Diets	3.15	2	1.575	7.76	3.89
Error	2.44	12	0.203		
Total	5.60	14			

Since $7.76 > 3.89$, the **mean squares diets** is significantly bigger than the **error mean squares,** indicating at least one pair of the diets isn't equal. Our conclusion at this point is only that *some* of the diets result in different life spans (reject H_0). To find exactly where the differences lie we must use mean separation techniques presented later in this chapter.

The F test done in conjunction with the ANOVA table is often called the **global F test**. It tells us whether or not there are *any* differences worth examining further. If H_0 is accepted, then the analysis is *complete* and the conclusion is that there are no significant differences among the treatments analyzed. Mean separation techniques are *not* calculated if the global F test indicates that H_0 cannot be rejected.

SOLUTION TO EXAMPLE 8.2. Return to Example 8.2 and calculate the ANOVA table and global F test to see if there are any significant differences among sites with regard to cephalon length. The hypotheses are

$$H_0: \mu_{\mathrm{I}} = \mu_{\mathrm{II}} = \mu_{\mathrm{III}} = \mu_{\mathrm{IV}}$$
$$H_a: \text{At least one pair of the } \mu_i \text{'s is not equal.}$$

Test with $\alpha = 0.05$. The summary statistics are given below:

	I	II	III	IV
n_i	8	5	11	15
$\overline{X}_i$	7.0	9.0	8.0	5.0
$T_{i.}$	56.0	45.0	88.0	75.0
CSS_i	30.0	18.0	50.0	67.0

Further, $\sum_i n_i = 39$ and $\sum_i T_{i.} = T_{..} = 264$. Next we need the three sums of squares: $\mathrm{SS}_{\mathrm{Total}}$, $\mathrm{SS}_{\mathrm{Treat}}$, and $\mathrm{SS}_{\mathrm{Error}}$. From Formula 8.4,

$$\mathrm{SS}_{\mathrm{Treat}} = \mathrm{SS}_{\mathrm{Sites}} = \sum_i \left[\frac{T_{i.}^2}{n_i} \right] - \frac{T_{..}^2}{N}$$

$$= \left[\frac{(56.0)^2}{8} + \frac{(45.0)^2}{5} + \frac{(88.0)^2}{11} + \frac{(75.0)^2}{15} \right] - \frac{(264)^2}{39}$$

$$= 1876 - 1787.08$$

$$= 88.92.$$

From Formula 8.5, we can calculate $\mathrm{SS}_{\mathrm{Error}}$:

$$\mathrm{SS}_{\mathrm{Error}} = \sum_i (n_i - 1)s_i^2 = \sum_i \mathrm{CSS}_i = 30.0 + 18.0 + 50.0 + 67.0 = 165.0.$$

Since the sums of squares are additive, $\mathrm{SS}_{\mathrm{Total}} = 88.92 + 165.0 = 253.92$.

Source of variation	Sum of squares	df	MS	F	c.v.
Sites	88.92	3	29.64	6.29	2.92
Error	165.00	35	4.71		
Total	253.92	38			

Since $6.29 > 2.92$, we reject H_0 and conclude the cephalon length at some of the sites are significantly different. Because we would like to know exactly which ones differ, we would continue this type of analysis with a mean separation technique. Example 8.2, like 8.1, is a fixed effects ANOVA (Model I) in which we are interested in determining *individual* treatment effects. In this analysis the ANOVA table is *only* the first step if the F test is significant.

8.2   Mean Separation Techniques for Model I ANOVAs

When H_0 is rejected as in Examples 8.1 and 8.2, the conclusion at that point is *at least one pair of means is not equal*. The task at hand is to discover exactly which means are different. The techniques to do this are called **mean separation techniques** or multiple comparisons. There are many of these techniques available, each with its own advantages and disadvantages. Considering the scope of this text we present two of the more common approaches: **Bonferroni t tests** and the **Student-Newman-Keuls test**. In Example 8.1 there are three sample means, and so there are

$$\binom{3}{2} = \frac{3!}{2!1!} = 3$$

possible pairwise comparisons:

$$
\begin{array}{lll}
H_0\colon \mu_U = \mu_{90\%} & H_0\colon \mu_U = \mu_{80\%} & H_0\colon \mu_{90\%} = \mu_{80\%} \\
H_a\colon \mu_U \neq \mu_{90\%} & H_a\colon \mu_U \neq \mu_{80\%} & H_a\colon \mu_{90\%} \neq \mu_{80\%}.
\end{array}
$$

The two-sample t test presented in Chapter 7 with the assumption of equal variances had test statistic

$$t = \frac{\overline{X}_1 - \overline{X}_2}{\sqrt{s_p^2\left(\frac{1}{n_1} + \frac{1}{n_2}\right)}} \qquad \text{with } \nu = n_1 + n_2 - 2.$$

We used s_p^2, a pooled variance, because we assumed equal population variances based on the preliminary F test. The assumptions about one-way ANOVAs include equal variances for *all* populations under investigation. The estimate of this variance is MS_E in the ANOVA table. The Bonferroni t test uses this fact to generate the following test statistic.

FORMULA 8.6. The Bonferroni t test statistic for Model I ANOVAs is

$$\text{Bonferroni } t = \frac{\overline{X}_{i.} - \overline{X}_{j.}}{\sqrt{MS_E\left(\frac{1}{n_i} + \frac{1}{n_j}\right)}},$$

where i and j represent any two treatment means to be separated and the degrees of freedom ν is $N - k$.

While we have to calculate only three test statistics for Example 8.1, if $k = 4$, there are six possible comparisons and if $k = 5$, there are 10 possible comparisons. When the number of treatments, k, becomes large the calculation of these Bonferroni t tests becomes quite time consuming. Fortunately, most statistical software programs will have a procedure for calculating these tests quickly.

Bonferroni t tests are usually run at a smaller α level than the global F test of the ANOVA table. This is necessary because we need to manage the probability of Type I error. If all Bonferroni t tests are performed at α, then the overall probability of at least one Type I error, denoted by α', is larger than α and its value is usually unknown. It can be shown that with three tests conducted at the α level, α' is at most $1 - (1 - \alpha)^3$. For example, if $\alpha = 0.05$ and we do three Bonferroni t tests, the probability of making at least one Type I error becomes at most $1 - (1 - .05)^3 = 0.143$. Using $\alpha = 0.05$ with six comparisons, $\alpha' = 0.265$ and with 10 comparisons, $\alpha' = 0.401$. So as the number of means to be separated increases, the overall probability of a Type I error can quickly become unacceptably high.

Most Bonferroni t tests are, therefore, conducted at an α level lower than the global F test. A general rule is to determine an *experimentwise* acceptable upper boundary for the probability of Type I error, say, b, and divide this probability by c, the actual number of comparisons run, to determine the α level for each comparison. For Example 8.1 if we wished the experimentwise α' to be 0.05, we would use $\frac{0.05}{3} = 0.017$ as the α level for each Bonferroni t test.

SOLUTION TO EXAMPLE 8.1: Bonferroni Mean Separation. Let's now complete the analysis for Example 8.1 via Bonferroni t tests. Recall from earlier calculations that

$\overline{X}_U = 2.44$	$\overline{X}_{90\%} = 3.18$	$\overline{X}_{80\%} = 3.54$
$n_U = 5$	$n_{90\%} = 5$	$n_{80\%} = 5$
	$\mathrm{MS_E} = 0.203$	

Since all the sample sizes are equal, the denominator in the expression for the Bonferroni t is the same for all tests,

$$\sqrt{\mathrm{MS_E}\left(\frac{1}{n_i} + \frac{1}{n_j}\right)} = \sqrt{0.203\left(\frac{1}{5} + \frac{1}{5}\right)} = 0.285.$$

For the first comparison:

$$H_0: \mu_U = \mu_{90\%} \qquad \nu = N - k = 15 - 3 = 12$$
$$H_a: \mu_U \neq \mu_{90\%} \qquad \alpha = \frac{0.05}{3} = 0.017.$$

From Formula 8.6, the test statistic is

$$\text{Bonferroni } t = \frac{\overline{X}_u - \overline{X}_{90\%}}{\sqrt{\mathrm{MS_E}\left(\frac{1}{n_U} + \frac{1}{n_{90\%}}\right)}} = \frac{2.44 - 3.18}{0.285} = -2.60.$$

As with earlier t tests, we expect the statistic to be close to zero if H_0 is true and significantly different from 0 if H_a is true. From Table C.4 the critical values for $\nu = 12$ are

$$\alpha = 0.01 \qquad \pm 3.055$$
$$\alpha = 0.02 \qquad \pm 2.681.$$

Since $-2.681 < -2.60 < 2.681$, the P value exceeds the value of $\alpha = 0.017$. So while the Bonferroni t is fairly large, it is not significant at our chosen α level. Therefore, accept H_0.

For the second comparison:

$$H_0: \mu_U = \mu_{80\%} \qquad \nu = 12$$
$$H_a: \mu_U \neq \mu_{80\%} \qquad \alpha = 0.017.$$

This time,

$$\text{Bonferroni } t = \frac{2.44 - 3.54}{0.285} = -3.86.$$

Since $-3.86 < -3.055$, reject H_0. The mean for 80% diet is *significantly different* (higher) than the unlimited diet.

For the third comparison:

$$H_0: \mu_{90\%} = \mu_{80\%} \qquad \nu = 12$$
$$H_a: \mu_{90\%} \neq \mu_{80\%} \qquad \alpha = 0.017.$$

The test statistic is

$$\text{Bonferroni } t = \frac{3.18 - 3.54}{0.285} = -1.26.$$

The test statistic here is not significantly different from zero ($-2.681 < -1.26 < 2.681$), so we again accept H_0.

In summary, the Bonferroni t tests indicate that the mean for unlimited diet is *significantly* different from the 80% diet. The unlimited diet is not significantly different from the 90% diet and the 90% and 80% diets are not significantly different from each other. While the difference between unlimited and 90% is not significant statistically, it is large enough that further study with larger samples might produce data that will allow rejection of H_0.

We now present a second method used to separate means in a fixed treatment or Model I ANOVA called the **Student-Newman-Keuls test (SNK)**. This test uses the rank orders of the sample means in a multiple range test comparing different ranges of means to critical values based on the number of means in the range. This test is rather easy to apply and is among the more popular mean separation techniques available. It involves the following protocol, *which assumes all samples have the same size n.* (We will provide an adjustment later when different sample sizes are involved.)

1. Linearly order the k sample means from smallest to largest.

2. To compare any means in this rank order ($H_0: \mu_i = \mu_j$) calculate the following statistic

$$q = \frac{\overline{X}_i - \overline{X}_j}{\text{SE}},$$

where $\text{SE} = \sqrt{\frac{\text{MS}_\text{E}}{n}}$ with MS_E from the ANOVA table and n is the common sample size.

3. Compare the q statistics to the appropriate critical values from Table C.9. The critical values are based on the experimentwise alpha level, the degrees of freedom for MS_E, and the number of means in the range being investigated.

We now analyze the means in Example 8.1 with the Student-Newman-Keuls test (SNK). In real life we would choose either the Bonferroni's t test or the SNK test, but not both. Each test is attempting the same thing, to separate means that are significantly different.

SOLUTION TO EXAMPLE 8.1: SNK Mean Separation. For Example 8.1 the means are already rank-ordered from smallest to largest.

U	90%	80%
2.44	3.18	3.54

Here $\alpha = 0.05$, $n = 5$ (common sample size), and

$$\sqrt{\frac{\text{MS}_\text{E}}{n}} = \sqrt{\frac{0.203}{5}} = 0.201.$$

For H_0: $\mu_{80\%} = \mu_U$

$$q = \frac{3.54 - 2.44}{0.201} = 5.473.$$

From Table C.9 with $\nu = 12$ and $k = 3$ (because 3.54 and 2.44 are end points of a range of 3 means), the q critical value = 3.773.

Since $5.473 > 3.773$, reject H_0. The means for unlimited diet and 80% diet are significantly different.

For H_0: $\mu_{80\%} = \mu_{90\%}$

$$q = \frac{3.54 - 3.18}{0.201} = 1.791.$$

From Table C.9 with $\nu = 12$ and $k = 2$ (3.54 and 3.18 are adjacent means, a range of two) the q critical value = 3.083.

Since $1.791 < 3.083$, accept H_0. The means for the 80% and 90% diets are not significantly different.

For H_0: $\mu_{90\%} = \mu_U$

$$q = \frac{3.18 - 2.44}{0.201} = 3.682.$$

Again from Table C.9 the q critical value = 3.083 because 3.18 and 2.44 are adjacent means, a range of two. Since $3.682 > 3.083$, reject H_0. The means for the 90% diet and the unlimited diet are significantly different.

We summarize these results with the superscripts a and b:

U	90%	80%
2.44[a]	3.18[b]	3.54[b]

That the unlimited diet is significantly different from both of the restricted diets (90% and 80%) is indicated by the different superscripts used. That the 90% and 80% diets are not significantly different is reflected in the use of the same superscript.

The outcome with the SNK test is slightly different from the Bonferroni t test because we were quite conservative in our choice of α for the Bonferroni statistics. *Both* analyses indicate that the "lean and mean" rats live longer. In general the Bonferroni t tests will be more conservative than the SNK test if both are applied at the same experimentwise alpha level. By more conservative we mean it will be more difficult to reject H_0 and will produce more Type II errors.

To use the SNK test to separate means based on unequal sample sizes requires the following modification: Use

$$\text{SE} = \sqrt{\frac{\text{MS}_\text{E}}{2}\left(\frac{1}{n_i} + \frac{1}{n_j}\right)}$$

for the denominator of the q statistic instead of $\sqrt{\frac{\text{MS}_E}{n}}$. The remainder of the analysis remains the same.

SOLUTION TO EXAMPLE 8.2: SNK Mean Separation. We now complete the analysis of the field study on the cephalon lengths of trilobites, with a **modified SNK test**. The ordered means are

Site	IV	I	III	II
$\overline{X}$	5.0	7.0	8.0	9.0
n	15	8	11	5

As with many field studies the sample sizes are unequal simply because all the fossils found were measured and different numbers were found at different sites. The overall ANOVA indicated that some of the means are significantly different. We now look for specific differences utilizing a modified SNK test.

Let $\alpha = 0.05$. Use Table C.9 with $\nu = 35$ to obtain the q critical values. Since 35 degrees of freedom isn't in the table, use the closest lower value, 30.

k	2	3	4
q c.v.	2.888	3.486	3.845

To compare site IV to site II, H_0: $\mu_{II} = \mu_{IV}$

$$q = \frac{9.0 - 5.0}{\sqrt{\frac{4.71}{2}\left(\frac{1}{5} + \frac{1}{15}\right)}} = \frac{4.0}{0.793} = 5.044 > 3.845.$$

So reject H_0. Mean cephalon lengths at these sites are significantly different.

To compare site IV to site III, H_0: $\mu_{III} = \mu_{IV}$

$$q = \frac{8.0 - 5.0}{\sqrt{\frac{4.71}{2}\left(\frac{1}{11} + \frac{1}{15}\right)}} = \frac{3.0}{0.609} = 4.926 > 3.486.$$

So reject H_0. Mean cephalon lengths at these sites are significantly different. Note the change in critical value here.

To compare site IV to site I, H_0: $\mu_{I} = \mu_{IV}$

$$q = \frac{7.0 - 5.0}{\sqrt{\frac{4.71}{2}\left(\frac{1}{8} + \frac{1}{15}\right)}} = \frac{2.0}{0.672} = 2.976 > 2.888.$$

So reject H_0. Mean cephalon lengths at these sites are significantly different. Again note the new critical value.

To compare site I to site II, H_0: $\mu_{II} = \mu_{I}$

$$q = \frac{9.0 - 7.0}{\sqrt{\frac{4.27}{2}\left(\frac{1}{5} + \frac{1}{8}\right)}} = \frac{2.0}{0.875} = 2.286 < 3.486.$$

So accept H_0. The mean cephalon lengths at sites I and II are not significantly different.

To compare site I to site III, H_0: $\mu_{III} = \mu_{I}$

$$q = \frac{8.0 - 7.0}{\sqrt{\frac{4.27}{2}\left(\frac{1}{11} + \frac{1}{8}\right)}} = \frac{1.0}{0.713} = 1.403 < 2.888.$$

Accept H_0. The mean cephalon lengths at sites I and III are not significantly different. Finally compare site III to site II, H_0: $\mu_{II} = \mu_{III}$

$$q = \frac{9.0 - 8.0}{\sqrt{\frac{4.27}{2}\left(\frac{1}{5} + \frac{1}{11}\right)}} = \frac{1.0}{0.828} = 1.208 < 2.888.$$

Accept H_0. The mean cephalon lengths at these sites are not significantly different.

From these comparisons we find that site IV has a *significantly smaller mean* than the other sites, while sites I, II, and III are *not significantly different*, as is summarized below.

IV	I	III	II
5.0^a	7.0^b	8.0^b	9.0^b

This may mean that site IV represents a different species of trilobite. The biological significance must be added after determination of the statistical significance. Note also the difference of 2mm between means IV and I is significant, but not between means II and I. The larger sample size at site IV is principally the cause for this discrepancy.

In both Examples 8.1 and 8.2, the global F tests indicated some differences in means. The mean separation techniques of Bonferroni t tests or Student-Newman-Keuls tests allowed us to pinpoint the differences. In Model I ANOVAs these types of tests are indicated any time the overall null hypothesis of all means equal is rejected. The Tukey test, Scheffe's test, Duncan Multiple Range test, and others offer alternatives to the mean separation techniques presented here and are discussed in detail elsewhere (see for example, Chapter 11 in Zar, 1999).

8.3 Model II ANOVA

In Example 8.3 the endocrinologist used five litters of mice as his treatments to study insulin release from pancreatic tissue. These particular litters are *not* of interest, only the fact that they might vary significantly indicating a genetic component to insulin release. The individual mice were sacrificed so he couldn't reuse them for further studies. For this experiment the hypotheses are

H_0: There is not significant variability among litters, $\sigma^2_{\text{litters}} = 0$

H_a: There is significant variability among litters, $\sigma^2_{\text{litters}} > 0$

and we will let $\alpha = 0.05$. The data are

	Litter				
	1	**2**	**3**	**4**	**5**
	9	2	3	4	8
	7	6	5	10	10
	5	7	9	9	12
	5	11	10	8	13
	3	5	6	10	11
$T_{i.}$	29	31	33	41	54
$\overline{X}_{i.}$	5.8	6.2	6.6	8.2	10.8

Next, the basic summary statistics are

$$\sum_i \sum_j X_{ij}^2 = 1634, \qquad \sum_i \sum_j X_{ij} = T_{..} = 188, \qquad N = 25.$$

So

$$SS_{Total} = \sum_i \sum_j X_{ij}^2 - \frac{T_{..}^2}{N} = 1634 - \frac{(188)^2}{25} = 220.24$$

and

$$SS_{Treat} = SS_{litters} = \sum_i \frac{T_{i.}^2}{n_i} - \frac{T_{..}^2}{N}$$

$$= \frac{(29)^2 + (31)^2 + (33)^2 + (41)^2 + (54)^2}{5} - \frac{(188)^2}{25}$$

$$= 83.84.$$

Consequently,

$$SS_{Error} = SS_{Total} - SS_{Treat} = 220.24 - 83.84 = 136.40.$$

Theory

The expected value of MS_{Treat} for this test is

$$E(MS_{Treat}) = \sigma^2 + n_0 \sigma_{Tr}^2,$$

where

$$n_0 = \frac{N - \sum_{i=1}^{k} \frac{n_i^2}{N}}{k - 1}.$$

and σ_{Tr}^2 is the variability due to treatments (litters in this case). The expected value of MS_E for this test is

$$E(MS_E) = \sigma^2.$$

So if H_0 were true, then $\sigma_{Tr}^2 = 0$ and the expected value for F would be $E(F) = \frac{\sigma^2}{\sigma^2} = 1$. If H_0 were false, then $E(F) > 1$ because $\sigma_{Tr}^2 > 0$ and the numerator of the F statistic would be expected to be larger than the denominator, $\frac{\sigma^2 + n_0 \sigma_{Tr}^2}{\sigma^2}$. The derivation of the MS_{Treat} expected value is beyond the scope of this text. Knowing its value, however, gives us some confidence that the ANOVA table is an appropriate test of the null hypothesis.

Decisions

The ANOVA table for Example 8.3 is given below:

Source of variation	Sum of squares	df	MS	F	c.v.
Among litters	83.84	4	20.96	3.07	2.87
Error	136.40	20	6.82		
Total	220.24	24			

Since $3.07 > 2.87$, we have a *significant* F test here. We reject H_0 and accept H_a; there is significant variability among the litters.

In Example 8.3 we have evidence that there are significant differences in litters with regard to insulin release from pancreatic tissue. These differences, probably due to different genotypes, should be considered in designing future experiments. The Model II ANOVA, a random effects study, has a single question in mind which is *answered* with the global F test. Mean separation techniques are *not* done in conjunction with this type of ANOVA.

8.4 The Kruskal-Wallis Test: A Nonparametric Analog of the Model I One-Way ANOVA

The basis for the Wilcoxon rank-sum test can be extended to the problem of testing for the location (median) of more than two populations using the Kruskal-Wallis k-sample test. This test was developed in the early 1950s. It is the nonparametric analog of a Model I ANOVA. We will examine data from a student project and use the analogy with the Wilcoxon rank-sum test to guide our efforts.

EXAMPLE 8.4. Preliminary observations on North Stradbroke Island indicated that the gastropod *Austrocochlea obtusa* preferred the zone just below the mean tide line. In an experiment to test this, *A. obtusa* were collected, marked, and placed either 7.5 m above this zone (Upper Shore), 7.5 m below this zone (Lower Shore), or back in the original area (Control). After two tidal cycles, the snails were recaptured. The distance each had moved (in cm) from where it had been placed was recorded. Is there a significant difference among the median distances moved by the three groups?

Upper Shore	Control	Lower Shore
59	19	5
71	32	13
75	55	15
85	62	41
95	82	46
148	114	51
170	144	60
276		106
347		200

SOLUTION. The first step is to rank the data. In a small set like this, one could list all the data in order, but it will prove convenient to leave the data in tabular form.

Upper Shore	Rank	Control	Rank	Lower Shore	Rank
59	10	19	4	5	1
71	13	32	5	13	2
75	14	55	9	15	3
85	16	62	12	41	6
95	17	82	15	46	7
148	21	114	19	51	8
170	22	144	20	60	11
276	24			106	18
347	25			200	23
Sum	162	Sum	84	Sum	79

If the three populations are identical, then the rankings should be randomly allocated to each of the samples. There is no reason to expect any one population to have a large number of high or low ranks. The average rank in each group should be roughly the same.

In this example there is a grand total of $N = 25$ observations. The average or expected rank is the mean (median) value of the numbers from 1 to 25, which is $\frac{N+1}{2} = \frac{25+1}{2} = 13$. Under the assumption of no difference between the groups, we expect the average rank within each group to be roughly the overall average. Since there are 9 observations in the Upper Shore sample, its average rank is $\frac{162}{9} = 18$. Similarly the average for the Control is $\frac{84}{7} = 12$ and for the Lower Shore $\frac{79}{9} = 8.78$. The averages are different from 13, but are they significantly different? The test statistic for this situation is based on these differences, as we describe below.

Assumptions

There are two basic assumptions for the Kruskal-Wallis k-sample test:

1. Independent random samples of sizes n_1, n_2, ..., and n_k are drawn from k populations and the variable of interest is at least ordinal or rankable.

2. The null hypothesis is that all k populations are identical, or equivalently, that all the samples are drawn from the same population.

The point of the test is to determine whether all k populations are the same.

Hypotheses

Only one pair of hypotheses is possible here and it is similar to the global test carried out in a one-way classification, completely randomized design ANOVA.

- H_0: All k population distributions are identical and have the same median versus H_a: At least one of the populations has a median different from the others.

Test Statistic and Decision Rule

We have already derived the basic elements of the test statistic in Example 8.4. We need to set the notation for the general situation. Begin by ranking all observations without regard to the sample they come from. Use midranks for tied values. As with other rank-based tests, a small number of ties is acceptable. (A correction factor, not discussed here, is required for a large number of ties.)

- Let N denote the total number of measurements in the k samples, $N = n_1 + n_2 + \cdots + n_k$.

- Let R_i denote the sum of the ranks associated with the ith sample.

- The *grand mean* or *average rank* is $\frac{N+1}{2}$.

- The *sample mean rank* for the ith sample is $\frac{R_i}{n_i}$.

The test statistic H measures the dispersion of the sample mean ranks from the grand mean rank:

$$H = \frac{12}{N(N+1)} \sum_{i=1}^{k} n_i \left(\frac{R_i}{n_i} - \frac{N+1}{2} \right)^2.$$

This is similar to the calculation done in analysis of variance. The term $\left(\frac{R_i}{n_i} - \frac{N+1}{2} \right)$ measures the deviation from the grand mean rank due to the fact that the ith group received the ith treatment. If the null hypothesis is true, this deviation should be close to 0 since the grand mean rank should be close to each sample mean rank. So if H_0 is true, the sum of these squared deviates and, hence, the test statistic H should be close to 0. If one or more of the groups has a sample mean rank that differs from the grand mean rank, then H will be positive. The larger H is, the less likely that H_0 is true. (The factor of $\frac{12}{N(N+1)}$ standardizes the test statistic in terms of the overall sample size N.)

As usual, we need to know the distribution of the test statistic H to know when it is large enough to reject H_0. As with the other rank test statistics, the exact distribution of H is complicated. However, H is reasonably well-approximated by the chi-square distribution with $k-1$ degrees of freedom, where k is the number of groupings in the study. If the test is done at the α level of significance, then H_0 should be rejected when H exceeds the value in Table C.5 for the $1 - \alpha$ quantile. Note that an easier way to compute the test statistic H is

$$H = \frac{12}{N(N+1)} \left(\sum_{i=1}^{k} \frac{R_i^2}{n_i} \right) - 3(N+1).$$

The equivalence of the two calculations for H just requires a bit of algebra and is given in Appendix A.5.

SOLUTION TO EXAMPLE 8.4, continued. We carry out the test at the $\alpha = 0.05$ level. In this example $k = 3$. The three sample sizes were $n_1 = n_3 = 9$ and $n_2 = 7$, so there were a total of $N = 25$ observations. We computed the sums of the sample ranks to be $R_1 = 162$, $R_2 = 84$, and $R_3 = 79$. Therefore

$$H = \frac{12}{N(N+1)} \left(\sum_{i=1}^{3} \frac{R_i^2}{n_i} \right) - 3(N+1)$$

$$= \frac{12}{25(26)} \left(\frac{(162)^2}{9} + \frac{(84)^2}{7} + \frac{(79)^2}{9} \right) - 3(26)$$

$$= 7.25.$$

There are $k - 1 = 2$ degrees of freedom in this problem, so using Table C.5, the critical value of the test statistic is 5.99. Since $H = 7.25 > 5.99$, there is evidence to reject H_0. That is, there is evidence that at least one of the groups moved a distance different than the others.

EXAMPLE 8.5. Girths (in cm) of *Callitris* trees were sampled by students at three locations at Brown Lake on North Stradbroke Island: on a slope, at a hilltop, and in a valley. Determine whether there was a significant difference in median girths for the various locations.

Slope	Rank	Hilltop	Rank	Valley	Rank
17	1	24	3	29	7
19	2	28	5.5	58	15
26	4	38	11	64	16
28	5.5	41	12.5	78	17
30	8	41	12.5	80	18
31	9	45	14	82	19
33	10			98	20
Sum	39.5	Sum	58.5	Sum	112

SOLUTION. The null hypothesis is H_0: All populations have the same median girth. The alternative is H_a: At least one of the populations has a median different from the others. In this example $k = 3$, $n_1 = n_3 = 7$, and $n_2 = 6$. There were a total of $N = 20$ observations. The sums of the ranks for the three locations are given in the previous table. From the original definition of the test statistic,

$$H = \frac{12}{N(N+1)} \sum_{i=1}^{k} n_i \left(\frac{R_i}{n_i} - \frac{N+1}{2} \right)^2$$

$$= \frac{12}{20(21)} \left[7 \left(\frac{39.5}{7} - 10.5 \right)^2 + 6 \left(\frac{58.5}{6} - 10.5 \right)^2 + 7 \left(\frac{112}{9} - 10.5 \right)^2 \right]$$

$$= 10.86.$$

There are $k - 1 = 2$ degrees of freedom in this problem, so using Table C.5, the critical value of the test statistic is again 5.99. Since $H = 10.86 > 5.99$, there is evidence to reject H_0, i.e., there is evidence to reject the hypothesis that the median girths are the same at all three locations.

Theory

The Kruskal-Wallis test is fairly easy to carry out and the logic of the test is straightforward. Values of H close to 0 indicate that the medians of all k populations are nearly the same while larger values of H mean that at least one median differs from the others. The chi-square distribution provides a good approximation for the distribution of H, but how is the exact probability of a particular value of H determined? What are the density and cumulative distribution functions for H? As usual, the answers are complicated, but an example with small sample sizes will illustrate the general method.

Consider an example with $k = 3$ and $n_1 = 2$, $n_2 = 1$, $n_3 = 1$. One sample has two observations and other two have only one observation. The exact distribution of H is found under the assumption that all observations come from identical populations. This means that the observations from one population are no larger or smaller than those from any other. But this implies that any ranking of the observations is as likely as any other, so a classical probability approach is appropriate.

Let's count the possible number of rankings. There are $N = 4$ total observations or possible ranks. Think of the ranks as slots into which the 4 observations must be placed. There are $\binom{N}{n_1} = \binom{4}{2} = 6$ ways to "place" (rank) the two observations from the first sample. Now there are only 2 slots left in which to put the 1 observation from the second sample. So there are $\binom{N-n_1}{n_2} = \binom{2}{1} = 2$ ways to rank the observation from the second sample. Finally, there is only 1 slot left into which the final observation from the third sample must be placed. So there is $\binom{N-n_1-n_2}{n_3} = \binom{1}{1} = 1$ way to place this last observation. The general pattern is clear now. The total number of rankings is just the product of these intermediate calculations,

$$\binom{N}{n_1}\binom{N-n_1}{n_2}\binom{N-n_1-n_2}{n_3} = \binom{4}{2}\binom{2}{1}\binom{1}{1} = 6 \cdot 2 \cdot 1 = 12.$$

So the probability of any one of these rankings occurring is $\frac{1}{12}$.

Let X, Y, and Z denote observations from the samples 1, 2, and 3, respectively. These 12 rankings and their corresponding test statistics are

1	2	3	4	*H*
X	X	Y	Z	2.7
X	X	Z	Y	2.7
X	Y	X	Z	1.8
X	Z	X	Y	1.8
X	Y	Z	X	0.3
X	Z	Y	X	0.3
Y	X	X	Z	2.7
Z	X	X	Y	2.7
Y	X	Z	X	1.8
Z	X	Y	X	1.8
Y	Z	X	X	2.7
Z	Y	X	X	2.7

Since each ranking is equally likely, the density function for H is now easily calculated. For example, $H = 1.8$ occurs four ways, so $P(H = 1.8) = \frac{4}{12}$. Since 6 of the rankings produce values of H no greater than 1.8, $P(H \le 1.8) = \frac{6}{12}$. In this way, the complete density function and the cumulative distribution for H with the given sample sizes is produced:

h	0.3	1.8	2.7
$P(H = h)$	$\frac{2}{12}$	$\frac{4}{12}$	$\frac{6}{12}$
$P(H \le h)$	$\frac{2}{12}$	$\frac{6}{12}$	$\frac{12}{12}$

Paired Comparisons

The Kruskal-Wallis test is the nonparametric analog of a Model I one-way ANOVA to compare k population means. Once either test has been done, we are in one of the following situations.

1. We are unable to reject H_0; based on the available data we cannot detect any differences among the k population medians (means). In this situation, the analysis of the data is complete.

2. We were able to reject H_0; based on the available data we conclude there are differences among the k population medians (means). The investigation continues as we try to pinpoint where the differences lie.

You are already familiar with the Bonferroni t tests and the SNK test, which are used in one-way ANOVAs. The test outlined below is the nonparametric analog of the Bonferroni t test. If there are k populations (treatments), then there are $\binom{k}{2} = \frac{k(k-1)}{2}$ possible pairs of medians that can be compared, which lead to $\binom{k}{2}$ tests of the form:

- H_0: The medians of the ith and jth populations are the same.

- H_a: The medians of the ith and jth populations are different.

As before, R_i denotes the sum of the ranks from the ith treatment and $\frac{R_i}{n_i}$ is the *average rank* for the ith treatment. As indicated before, our intuition is that treatments i and j do not differ when $\left| \frac{R_i}{n_i} - \frac{R_j}{n_j} \right|$ is close to 0. Let's make this precise.

Choose an overall significance level α' at which to run the test. As with the Bonferroni t tests, the significance level of any *one* of the comparisons that will be made is $\frac{\alpha'}{\binom{k}{2}}$. For example, if $k = 5$, then we might reasonably choose $\alpha' = 0.10$ as the upper bound probability of making at least one incorrect rejection of a null hypothesis. Since $\binom{5}{2} = 10$, then for each of the 10 comparisons we will use the $\alpha = \frac{0.10}{10} = 0.01$ level of significance.

Note that the larger k is, the larger α' must be in order to run the comparisons at a reasonable level of significance, α. But when α' is larger, we run a greater risk of rejecting a null hypothesis inappropriately (Type I error). Thus, in the design of the experiment, the number of treatments k under investigation should be kept to the minimum of *those tests of real interest to the researcher*.

The test statistic for a two-tailed comparison of the ith and jth treatments is based on the difference in mean ranks for the two samples:

$$Z_{ij} = \frac{\left| \frac{R_i}{n_i} - \frac{R_j}{n_j} \right|}{\sqrt{\frac{N(N+1)}{12} \left(\frac{1}{n_i} + \frac{1}{n_j} \right)}},$$

where N is the total number of observations. (Note the similarity with the Bonferroni t test statistic in Formula 8.6.) It can be shown that the test statistic has an approximate standard normal or z distribution, so Table C.3 is used to determine the critical value, $z_{1-\frac{\alpha}{2}}$, of this two-sided test. H_0 is rejected if $Z_{ij} \geq z_{1-\frac{\alpha}{2}}$, that is, if

$$\frac{\left| \frac{R_i}{n_i} - \frac{R_j}{n_j} \right|}{\sqrt{\frac{N(N+1)}{12} \left(\frac{1}{n_i} + \frac{1}{n_j} \right)}} \geq z_{1-\frac{\alpha}{2}}. \quad = \quad Z_{ij} \geq z_{1-\frac{\alpha}{2}}$$

The calculations are simplified if all the sample sizes are the same since the denominator of Z_{ij} will be the same for all comparisons.

EXAMPLE 8.6. Return to Example 8.4 concerning the gastropod *Austrocochlea obtusa*. The Kruskal-Wallis test allowed us to reject H_0 in favor of the alternative hypothesis that the median distances moved by the three groups were not the same. Use the method of paired comparisons to determine which groups are different.

SOLUTION. The number of treatments corresponds to the different groups, so $k = 3$. Thus, there are $\binom{3}{2} = 3$ comparisons to carry out. Thus, it is reasonable to set

$\alpha' = 0.05$ and then $\alpha = \frac{0.05}{3} = 0.0167$. So $1 - \frac{\alpha}{2} = 1 - \frac{0.0167}{2} = 0.9917$. The critical value from Table C.3 is $z_{1-\frac{\alpha}{2}} = z_{0.9917} = 2.40$. The total number of observations is $N = 25$. The three sample sizes are $n_1 = 9$, $n_2 = 7$, and $n_3 = 9$. The means of the sample ranks are $\frac{R_1}{n_1} = \frac{162}{9} = 18$, $\frac{R_2}{n_2} = \frac{84}{7} = 12$, and $\frac{R_3}{9} = \frac{79}{9} = 8.78$. We can now carry out the comparisons.

To compare the Upper Shore and Control groups, we calculate

$$Z_{12} = \frac{\left| \frac{R_1}{n_1} - \frac{R_2}{n_2} \right|}{\sqrt{\frac{N(N+1)}{12} \left(\frac{1}{n_1} + \frac{1}{n_2} \right)}} = \frac{|18 - 12|}{\sqrt{\frac{25(26)}{12} \left(\frac{1}{9} + \frac{1}{7} \right)}} = \frac{6}{3.709} = 1.618.$$

Because $1.618 < 2.40$, we continue to accept H_0. There is not sufficient evidence of a difference in median distances moved by the Upper Shore and Control gastropods.

To compare the Control and the Lower Shore groups calculate

$$Z_{23} = \frac{\left| \frac{R_2}{n_2} - \frac{R_3}{n_3} \right|}{\sqrt{\frac{N(N+1)}{12} \left(\frac{1}{n_2} + \frac{1}{n_3} \right)}} = \frac{|12 - 8.78|}{\sqrt{\frac{25(26)}{12} \left(\frac{1}{7} + \frac{1}{9} \right)}} = \frac{3.22}{3.709} = 0.868.$$

Because $0.868 < 2.40$, again H_0 cannot be rejected.

Finally, to compare the Upper and the Lower Shore groups calculate

$$Z_{13} = \frac{\left| \frac{R_1}{n_1} - \frac{R_3}{n_3} \right|}{\sqrt{\frac{N(N+1)}{12} \left(\frac{1}{n_1} + \frac{1}{n_3} \right)}} = \frac{|18 - 8.78|}{\sqrt{\frac{25(26)}{12} \left(\frac{1}{9} + \frac{1}{9} \right)}} = \frac{9.22}{3.469} = 2.657.$$

Because $2.657 > 2.40$, H_0 is rejected. There is evidence that the median distances traveled by the Upper and Lower Shore gastropods are different.

Observations and Comments

As with the Bonferroni test, always remember to choose an appropriate significance level α' based on the number of treatments. Large values of α' increase the chance of making at least one Type I error in the full set of comparisons, while small values of α make it nearly impossible to reject a false null hypothesis in the set of the comparisons (Type II error). Lastly, do not use the data to "inspire" suddenly just one or two selected comparisons using the earlier two-sample tests. The entire suite of paired comparisons should be carried out.

8.5   Problems

1. In order to understand better the sums of squares of analysis of variance, do the following exercise. Four samples of size 2 are given in the following table:

I	II	III	IV
2	9	3	5
4	5	9	3

(*a*) Express each of the eight observations as the sum of three components, i.e.,

$$X_{ij} = \overline{X}_{..} + (\overline{X}_{i.} - \overline{X}_{..}) + (X_{ij} - \overline{X}_{i.}).$$

(b) Compute the SS_{Total}, SS_{Treat}, and SS_{Error} terms directly from the components and show that SS_{Total} is equal to the sum of SS_{Treat} and SS_{Error}.

(c) Compute the three SS values by the computational formulas and show that the values obtained are the same as those obtained in (b).

2. Sickle cell anemia is a genetically transmitted disease. Normal human red blood cells are "biconcave discs" meaning that they are shaped like microscopic breath mints. People who are homozygous for the sickle cell gene have "sickled" red blood cells. These cells are curved like a sickle or are irregularly shaped and are often pointed. In the smallest blood vessels sickled cells tend to form clots and hamper normal blood flow. Sickled blood cells are also fragile and rupture easily. Those who are heterozygous for sickle cell anemia may exhibit mild signs of the disease, but these are very rarely serious.

As part of a research study on sickle cell anemia and its relation to other diseases, researchers measured hemoglobin levels (g/100 ml) in 30 subjects divided into three groups of 10: normal (control), sickle cell disease (homozygotes), and sickle cell trait (heterozygotes) subjects. Are there significant differences in hemoglobin levels among the three groups?

	Normal subjects	Sickle cell disease	Sickle cell trait
Mean	13.6	9.8	12.7
Std. dev.	1.6	1.6	1.5

$$SS_{Treat} = 78.87 \qquad\qquad SS_{Error} = 66.33$$

3. Zonation can be observed in microhabitats such as the intertidal shoreline of a bay. A study was conducted on the distribution of the common gastropod *Bembecium* at various levels about 3 m apart in the intertidal zone at Polka Point. Several 0.25-m^2 quadrats were placed along zones 1, 2, and 3 (zone 2 is mean water level) and the number of *Bembecium* in each were counted.

Zone 1	Rank	Zone 2	Rank	Zone 3	Rank
0		29		4	
2		33		6	
12		38		16	
16		43		27	
16		47		41	
21		54			
23		67			

(a) State the appropriate null and alternative hypotheses for these data.

(b) Rank all the observations (use midranks when necessary) and compute the sums of the sample ranks R_i.

(c) Compute the Kruskal-Wallis test statistic H for these data and determine whether H_0 should be rejected.

(d) If necessary, carry out paired comparisons.

4. In a project for a botany class, 15 sunflower seeds were randomly assigned to and planted in pots whose soil had been subjected to one of three fertilizer treatments. Twelve of the seeds germinated, and the table below shows the height of each plant (in cm) 2 weeks after germination. Are there significant differences in heights among treatments? Analyze with ANOVA and Bonferroni t tests, if necessary.

Treatment 1	Treatment 2	Treatment 3
23	26	25
27	28	26
32	33	33
34	35	
	38	

5. In a study of lead concentration in breast milk of nursing mothers, subjects were divided into five age classes. Determine if there was a significant difference in the mean lead concentration (in μg/dl) among the classes. Analyze the following summary data using ANOVA and appropriate mean separation techniques if necessary. (Based on data reported in Younes, B. et al. 1995. Lead concentrations in breast milk of nursing mothers living in Riyadh. *Annals of Saudi Medicine*, **15**(3): 249–251.)

Age class	≤ 20	21–25	26–30	31–35	≥ 36	Total
n_i	4	19	13	3	8	47
$\overline{X}_i$	0.515	0.777	0.697	0.832	1.315	0.828
s_i	0.14	0.41	0.31	0.33	0.65	
T_i	2.060	14.763	9.061	2.496	10.520	38.900

6. (*a*) The data below are steady-state hemoglobin levels (g/100 ml) for patients with different types of sickle cell disease, these being HB SS, S/-thalassemia, and HB SC. Is there a significant difference in median steady-state hemoglobin levels for patients with different types of sickle cell disease? *(HSDS, #310)*

(*b*) If appropriate, carry out a set of paired comparisons.

HB SS	Rank	HB S/-thalassemia	Rank	HB SC	Rank
7.2		8.1		10.7	
7.7		9.2		11.3	
8.0		10.0		11.5	
8.1		10.4		11.6	
8.3		10.6		11.7	
8.4		10.9		11.8	
8.4		11.1		12.0	
8.5		11.9		12.1	
8.6		12.0		12.3	
8.7		12.1		12.6	
9.1				13.3	
9.1				13.3	
9.1				13.8	
9.8				13.9	
10.1					
10.3					

7. To test the effectiveness of various denture adhesives, an instrument called the TA.XT2i Texture Analyzer made by Texture Technologies Corp. was used to measure the amount of force required to separate dentures from a mouth and gum cast. The force required for separation was recorded in decigrams. The adhesives were: (A) karaya, (B) karaya with sodium borate, (C) carboxymethylcellulose sodium (32%) and ethylene oxide homopolymer, and (D) carboxylmethylcellulose sodium (49%) and ethylene oxide homopolymer. Are there any significant differences among these denture adhesives holding abilities? Analyze with ANOVA and SNK, if necessary.

	A	B	C	D
	71	76	75	80
	79	70	81	82
	80	90	60	91
	72	80	66	95
	88	76	74	84
	66	82	58	72
$T_{i.}$	456	474	414	504
$\overline{X}_i$	76	79	69	84
s_i^2	62.0	46.0	83.2	66.8
CSS_i	310	230	416	334

8. The primary reason women seek medical attention for menopausal symptoms is hot flashes. Dietary supplements containing isoflavones derived from natural sources such as soy or red clover are marketed as an alternative treatment for such symptoms and are being used increasingly by women in the United States. Isoflavones are polyphenol compounds that are similar in structure to estrogens.

A study was carried out to determine whether two dietary supplements derived from red clover were more effective than placebo in reducing hot flashes in postmenopausal women. The randomized, double-blind trial was conducted using 252 menopausal women, aged 45 to 60 years, who were experiencing at least 35 hot flashes per week. After a 2-week period where all were given placebo, the women were randomly assigned to Promensil (82 mg of total isoflavones per day), Rimostil (57 mg of total isoflavones per day), or an identical placebo, and followed-up for 12 weeks. The table below provides summary data on the number of hot flashes (per day) experienced by the women at the end of the trial.

	Promensil	Rimostil	Placebo
n_i	84	83	85
$\overline{X}_{i.}$	5.1	5.4	5.0
$T_{i.}$	428	448	425
s_i	4.1	4.6	3.2
$\sum_j (X_{ij}^2)$	3580	4155	2985

Analyze these data (assume normality) to determine whether there are any differences in the mean number of hot flashes per day for these treatments. Summarize your conclusions and recommendations concerning these procedures. (Adapted from: Tice, J. et al. 2003. Phytoestrogen supplements for the treatment of hot flashes: The isoflavone clover extract (ICE) study. *JAMA*, **290**(2): 207–214.)

9. Herbal supplements are widely used to treat vasomotor symptoms (hot flashes, night sweats) of menopause. Researchers conducted a randomized study to test the efficacy of three different herbal regimens, hormone therapy, and placebo. The treatment therapies were: (1) 160 mg daily of black cohosh, (2) 200 mg daily of multibotanical with black cohosh, (3) 200 mg daily of multibotanical with black cohosh plus dietary counseling emphasizing soy products, (4) 0.625 mg daily of conjugated equine estrogen, and (5) lactose placebo. All were administered in the form of a pill. The table below provides the results of the study (in numbers of symptoms per day) after six months of compliance. (Adapted from: Newton, K. et al. 2006. Treatment of vasomotor symptoms of menopause with black cohosh: A randomized trial. *Ann. Intern. Med,* **145**: 869–879.)

		Black Cohosh	Multibotanical	Multibotanical +	Estrogen	Placebo
sample size	n	77	72	74	29	75
mean sample	$\overline{X}_i$	5.71	7.00	6.53	2.34	6.20
sd of sample	s_i	3.04	3.56	3.85	4.94	3.72

(*a*) Determine whether there are any differences in outcomes among the treatments by carrying out an ANOVA. In particular, which treatments, if any, differ from placebo? (Reconstruct the required summary statistics for the analysis from the data given.)

(*b*) If you were a physician's assistant, what would your recommendation be to a patient experiencing vasomotor symptoms from menopause in light of the results from the Women's Health Initiative studies which indicate a significant increase in the risk of cancer associated with estrogen therapy?

10. Workers at a tree farm decided to test the efficacy of three fertilizer mixtures on the growth of Norway maple seedlings, *Acer platanoides*. The table below contains the heights of seedlings (in feet) for the three fertilizer treatments. Determine if there are significant differences in the heights among the three treatments.

	Fertilizer mixture		
	A	**B**	**C**
	2.0	2.3	3.1
	2.1	2.9	1.5
	2.4	1.5	2.2
	2.8	1.2	2.9
	2.9	1.9	1.7
	3.1	1.9	2.1
	3.2	3.4	2.8
	3.7	2.1	1.5
	3.8	2.6	2.8
	4.1	2.4	2.2
$T_{i.}$	30.1	22.2	22.8
CSS_i	4.61	3.82	3.20
s_i	0.72	0.61	0.60

$$\sum_i \sum_j X_{ij}^2 = 203.49, \qquad T_{..} = 75.1, \qquad SS_{Total} = 15.49$$

11. The laughing kookaburra, *Dacelo novaguineae*, is a large terrestrial kingfisher native to eastern Australia. Laughing kookaburras live in cooperatively breeding groups of two to eight birds in which a dominant pair is assisted by helpers that are its offspring from previous seasons. These helpers are involved in all aspects of clutch, nestling, and fledgling care. Does increased group size lead to an increase in mean fledging success? Use the data below to determine whether there was a significant difference in the mean number of fledgings among groups of different sizes. Use mean separation techniques if required. (Based on data interpolated from Legge, S. 2000. The effect of helpers on reproductive success in the laughing kookaburra. *Journal of Animal Ecology*, **69**: 714–724.)

Group size	2	3	4	5
$\overline{X}$	1.15	1.43	1.66	1.67
s	1.16	1.09	1.55	0.73
n	46	28	29	15

$$SS_{Treat} = 5.81 \qquad SS_{Total} = 173.17$$

12. A national dairy manufactures products at facilities throughout the U.S. To test the uniformity of its products four local sites were randomly chosen and samples of margarine produced at these sites were analyzed to determine the level of a particular fatty acid as a percentage of the fats. The data are recorded below. Are there significant differences in this characteristic among the manufacturing sites?

	Site		
A	B	C	D
13.5	13.2	16.8	18.1
13.4	12.7	17.2	17.2
14.1	12.6	16.4	18.7
14.2	13.9	17.3	18.4
55.2	52.4	67.7	72.4

$$\sum\sum X_{ij}^2 = 3907.2$$

13. Fifteen tobacco plants of the same age and genetic strain were randomly assigned to 3 groups of 5 plants each. One group was untreated, one was infected with tobacco mosaic virus (TMV), and one was infected with tobacco ringspot virus (TRSV). After one week the activity of odiphenol oxidase was determined in each plant. Does infection by either virus affect the activity of this enzyme?

Enzyme Activity (μg/ml protein/min)		
Control	TMV-Infected	TRSV-Infected
1.47	2.44	2.87
1.62	2.31	3.05
1.06	1.98	2.36
0.89	2.76	3.21
1.67	2.39	3.00

14. Body builders consume large volumes of dietary supplements, especially products that combine creatine with whey proteins, a waste product of cheese making. In a study to assess the impact of creatine and whey on strength and muscle fiber characteristics, 28 highly trained body builders in their mid-20s were recruited for a 13-week dietary study.

The researchers divided the volunteers into four groups, giving each man the same caloric bonus per day: a flavored drink containing a gram of supplement per kilogram of body weight. One supplement contained just carbohydrates, another just whey powder, and the last two a mix of creatine with either carbohydrates or whey. The experiment was conducted as a double blind with neither the athletes nor the scientists knowing who received each supplement until the study was completed.

Throughout, the men performed supervised resistance training three times a week and all experienced strength gains. Below are the data for increases in bench press totals in kilograms for the 28 body builders.

Carbohydrate only	Whey only	Creatine and carbohydrate	Creatine and whey
3	5	7	11
5	3	8	9
2	5	10	15
5	4	7	9
1	7	5	12
4	4	4	10
8	5	9	5

Do the data indicate that the combination of creatine and whey is significantly better than the other supplements?

15. A topic of recent public health interest is whether there is a measurable effect of passive smoking on pulmonary health, that is, exposure to cigarette smoke in the atmosphere among nonsmokers. White and Froeb studied this question by measuring pulmonary function in a number of ways in the following six groups:

- *Nonsmokers* (NS): People who themselves did not smoke and were not exposed to cigarette smoke either at home or on the job.

- *Passive smokers* (PS): People who themselves did not smoke and were not exposed to cigarette smoke in the home, but were employed for 20 or more years in an enclosed working area that routinely contained tobacco smoke.

- *Noninhaling smokers* (NI): People who smoked pipes, cigars, or cigarettes, but who did not inhale.

- *Light smokers* (LS): People who smoked and inhaled 1–10 cigarettes per day for 20 or more years. (*Note:* There are 20 cigarettes in a pack.)

- *Moderate smokers* (MS): People who smoked and inhaled 11–39 cigarettes per day for 20 or more years.

- *Heavy smokers* (HS): People who smoked and inhaled 40 or more cigarettes per day for 20 or more years.

A principal measure used by the authors to assess pulmonary function was forced mid-expiratory flow (FEF) and it is of interest to compare FEF in the six groups. Analyze appropriately. (Based on data reported in: White, J. and H. Froeb. 1980. Small-airways dysfunction in non-smokers chronically exposed to tobacco smoke. *New England Journal of Medicine*, **302**: 720–723.)

FEF data for smoking and nonsmoking males

Group	Mean FEF (liters/sec)	Std. dev. FEF (liters/sec)	n_i
NS	3.78	0.79	200
PS	3.30	0.77	200
NI	3.32	0.86	50
LS	3.23	0.78	200
MS	2.73	0.81	200
HS	2.59	0.82	200

16. To compare the efficacy of three insect repellants, volunteers applied fixed amounts of repellant to their hand and arm then placed them in a chamber with several hundred hungry female *Culex erraticus* mosquitoes. The repellants were citronella (Treatment I), N, N-diethyl-meta-toluamide (DEET) 20% (Treatment II), and Avon Skin So Soft hand lotion (Treatment III). The data recorded below are the times in minutes until first bite.

Time until first bite (min)

Citronella	DEET 20%	Avon Skin So Soft
5	12	6
6	16	7
8	25	9
8	27	10
14	28	11
15	31	13
		17

Use the Kruskal-Wallis Test to determine if there are any significant differences among the times of protection for these three repellants.

17. To obtain a preliminary measure of plant species richness in various habitats, 64-m^2 quadrats were used at three locations on North Stradbroke Island: Brown Lake, 18-Mile Swamp, and Pt. Lookout. The number of different species in each quadrat was recorded.

(*a*) Is there a significant difference in plant species richness among these locations?

Brown Lake	18-Mile Swamp	Pt. Lookout
14	11	16
15	12	20
18	12	22
19	13	24
20	14	29
23	17	

(b) Carry out a set of paired comparisons at the $\alpha' = 0.05$ level or explain why it is inappropriate to do so.

18. While studying the distribution of mosquitofish, *Gambusia holbrooki*, in Florida streams an ecologist realized that the sex ratio in certain streams was dramatically skewed in favor of males. Most of these sites were downstream from pulp and paper mills. It was thought that bacteria in the streams might be chemically altering plant sterols into androstenedione, the anabolic steroid used by weight lifters and notably Mark McGuire in 1999 when he set the major league home run record. The cholesterol-like pine sterols could be biotransformed into an environmental androgen that could affect the sexual development of mosquitofish. To further investigate this possibility, water samples from various streams were assayed for androstenedione. The concentrations reported in nanograms per liter appeared to be normally distributed when tested for a large number of samples at any site.

Spring Creek	Fenholloway River	Elevenmile Creek
5	19	20
8	17	21
7	18	18
9	17	17
0	19	22
3	15	20
10	16	15
1	14	17

(a) Are there significant differences in androstenedione levels among the three streams?

(b) Fenholloway River and Elevenmile Creek have pulp and paper mills on them, while Spring Creek is a neighboring unpolluted creek. Do the data and analysis prove that the mills are causing the sex ratio changes in *Gambusia holbrooki*?

19. Osteoarthritis occurs when the cartilage that cushions joints breaks down, often leading to pain, swelling and loss of mobility. It is estimated 12% of Americans aged 65 and older have osteoarthritis of the knee. When medical therapy such as drugs or exercise therapy fail to work, patients are often referred for arthroscopic lavage or débridement. Lavage involves flushing the knee joint with fluid, while débridement goes further—along with lavage, damaged tissue fragments are trimmed and the area is smoothed. About 650,000 such procedures are performed in the U.S. every year, at a cost of roughly $5,000 each.

While many patients report relief after undergoing such arthroscopic procedures, it is unclear how the procedure actually achieves this result. Recently, a randomized placebo-controlled study was conducted to evaluate arthroscopy for osteoarthritis of the knee. Initially, a total of 180 patients with osteoarthritis of the knee were randomly assigned (with their consent) to receive arthroscopic lavage or arthroscopic débridement, or placebo surgery. Patients in the placebo group underwent a simulated débridement without the insertion of the arthroscope. Outcomes were assessed at multiple times over a two-year period with the use of six scores: three were self-reported scales for pain, two were self-reported scores of function, and one was an objective test of stair climbing and walking. Note: Higher scores indicate more pain and lower function. A total of 164 patients completed the trial and summary data are reported for them below.

	Placebo	Lavage	Débridement
n_i	55	56	53
$\overline{X}_{i.}$	52.5	56.7	54.0
$T_{i.}$	2887.5	3175.2	2862.0
s_i	25.1	24.1	23.3
$\sum_j (X_{ij}^2)$	185,614.29	211,978.39	182,778.28

Analyze the data to determine whether there are any differences in mean scores for the various treatments. Summarize your conclusions and recommendations concerning these procedures. (Based on data reported in: Moseley, J. 2002. A controlled trial of arthroscopic surgery for osteoarthritis of the knee. *New England Journal of Medicine*, **347**(2): 81–88.)

20. While Alzheimer's disease is the most common type of dementia, it is notoriously difficult to diagnose, with the battery of tests often taking months to complete. In an attempt to develop a more precise diagnostic tool researchers looked at the ratios of two spinal-fluid compounds linked to the disease: tau protein and beta-amyloid peptide. The researchers studied the ratios of these compounds in four groups: Alzheimer's patients, people with other forms of dementia, people with brain disorders not associated with dementia, and healthy individuals. All subjects were Caucasian males between 60 and 70 years old and there were 20 in each category.

	Alzheimer's patients	Other dementia	Brain disorders	Healthy individuals
$\overline{X}_i$	147	74	48	39
s_i^2	131	100	118	98
CSS_i	2489	1900	2242	1862

$$\text{SS}_{\text{Error}} = \sum \text{CSS}_i = 8,493$$

Is the tau protein to beta amyloid peptide ratio a good candidate for a new diagnostic tool? Assume the data are normally distributed and analyze via ANOVA with an SNK test if necessary.

21. The white-tailed deer, *Odocoileus virginianus*, is the most ubiquitous cervid species in the Americas, with a range from southern Canada to Bolivia. Wildlife managers studying various populations of this deer gathered the following information from hunter checkpoints in three states. All animals were antlered bucks. Weights were recorded in kilograms and represent 100 randomly chosen records from each state. Are there significant differences in weights of male deer among these states? Assume normality.

	Wisconsin	Ohio	Tennessee
$\overline{X}$	90.9 kg	84.6 kg	82.3 kg
s	7.0 kg	6.9 kg	7.3 kg

22. Upon visiting the Royal Ontario Museum in Toronto, you notice the medieval suits of armor on display seem rather small. With the permission of the curators you decide to measure the heights of all the suits of armor on display and to group the measurements into age categories for the medieval times when full body armor was being utilized. All measurements are in centimeters.

10th & 11th Centuries	12th & 13th Centuries	After 1300 A.D.
153	158	163
155	159	155
154	159	165
149	165	170
163	149	165
151		171
		153
		166

Are there significant differences in the heights of armor suits among the different time periods? Analyze nonparametrically.

23. Before undertaking a large comprehensive study of weight gain utilizing rats, you learn that there are more than twenty companies in North America raising lab rats for experimentation. You randomly choose three of these companies and order 10 rat pups from each. After a set time period the weight gain in each was recorded in grams.

$$T_{1.} = 527, \qquad T_{2.} = 502, \qquad T_{3.} = 480, \qquad T_{..} = 1509, \qquad \sum_{i=1}^{3}\sum_{j=1}^{10} X_{ij}^2 = 76,511$$

Test the hypothesis that lab rats from the different companies are uniform in their growth rates.

24. QCR. True or False. Justify your answer.

(a) In Model I ANOVA the F test is always right-tailed even though the alternative hypothesis is that some means are not equal.

(b) When comparing more than two independent means, multiple t tests can substitute for analysis of variance.

(c) In Model II ANOVA the global F test completes the analysis.

(d) When the global F test is significant in a Model I ANOVA, further analysis via a mean separation technique is indicated.

(e) The Bonferroni t test is a mean separation technique that is very conservative, meaning it is very difficult to reject H_0 using this test and consequently the Type II error is somewhat increased.

(f) The Student-Newman-Keuls test is a mean separation technique that may be used in conjunction with the Bonferroni t tests in Model I ANOVA.

(g) The Kruskal-Wallis Test is considered a nonparametric analog to a Model II, One-Way ANOVA and is often used in place of the more complicated ANOVA.

9

Two-Factor Analysis

Concepts in Chapter 9:

- Randomized Complete Block Design
  - — Data Layout and Model Assumptions
  - — ANOVA Table and Expected Mean Squares
  - — Mean Separation Techniques

- Factorial Design Two-Way ANOVA
  - — Data Layout and Model Assumptions
  - — ANOVA Table and Expected Mean Squares
  - — Hypothesis-Testing Strategies
  - — Mean Separation Techniques

- The Friedman k-Sample Test: Matched Data
  - — Assumptions, Test Statistic, and Theory
  - — Paired Comparisons

Two-Factor Analysis of Variance

In Chapter 8 we considered data sets that were partitioned into k samples. These partitions were either *fixed effects* (treatments) leading to **Model I** analysis or *random effects* leading to **Model II** analysis. The Kruskal-Wallis test was presented as a nonparametric analog of the Model I ANOVA. In each of these designs each datum was considered as part of a single group, that is, the data were factored one way. In Example 8.1 each datum represented a single diet and in Example 8.2 each datum represented a single site. In the present chapter we consider data sets in which each datum can be used to answer two questions. In a sense the data are looked at two ways as two separate partitions of the same data set.

The techniques presented in Chapter 9 are logical extensions of the protocols outlined in Chapter 8. They are presented in a separate chapter only because of space considerations. If you have mastered Chapter 8, you will find the concepts in Chapter 9 accessible although the notation and formulation may initially appear intimidating. Concentrate on the expected values and their relationship to the various tests of hypothesis.

9.1   Randomized Complete Block Design ANOVA

We begin the discussion of two-factor analysis of variance with an experimental design that is an extension of the paired t test (Section 7.4). In a paired t test we had two samples that were related to each other, for instance, measurements of the same individual before and after some treatment. This pairing of data points allowed us to account for and remove considerable extraneous variability. We focused on the *differences* between the pairs of data rather than the actual data values. If we extend the pairing by treating each individual unit three or more ways and then measuring some attribute after each treatment, we generate data sets that can be partitioned two ways: by the treatments and by the individual units. An example will clarify this.

> **EXAMPLE 9.1.** One very effective way to catch salmon commercially is by gill net-ting. Traditionally gill netters hang long monofilament nets vertically over the fishing grounds. As the salmon encounter these nets they become entangled in the mesh by their gill covers and quickly die. Unfortunately many other species also become ensnared in these nets. In particular, diving sea birds such as common murres and rhinoceros auklets fall prey to gill nets. In addition, seals and dolphins can be eas-ily trapped in the almost invisible monofilament nets. The sea birds and marine mammals caught and killed in gill nets are euphemistically called "by-catch" in the fishing industry. Suppose as part of a conservation project on by-catch reduction, three modified techniques are used on boats fishing for salmon with gill nets.
>
> I.   White-topped gill nets (the top 7 feet of the net is white mesh that is considerably easier to see in the water column).
>
> II.   White-topped gill nets with pingers that emit an annoying beep similar to those made by watches with alarms.
>
> III. Unmodified gill nets with pingers.
>
> IV.   Unmodified gill nets (standard method).
>
> The table below lists the quantity of salmon caught per boat per day using each of these techniques.

Day	Catch of salmon (100 kg/boat-day)			
	I	II	III	IV
1	23	21	24	25
2	16	14	19	17
3	19	17	23	21
4	11	12	15	16
5	41	30	40	42
6	32	20	37	35
7	9	11	19	17
8	10	14	19	16

Let's now develop the notation and basic formulas to test this type of hypothesis in a randomized complete block design. **Randomized** means that each treatment is assigned randomly within blocks and **complete** implies that each treatment is used exactly once within each block.

Each of the 32 data points here can be looked at in two ways. $X_{11} = 23$ is part of the column of numbers (23, 16, 19, 11, 41, 32, 9, 10) that represent Treatment I—catch by white-topped gill nets. The X_{11} value is also related to $X_{21} = 21$, $X_{31} = 24$, and $X_{41} = 25$ because they are all catches from the same day. They represent a block of data that is analogous to a pair of data points in a paired t test. The days represent a random variable that we are not directly interested in but have some control over, i.e., we are not interested in differentiating among days. After all, anyone who has ever fished knows that some days are better than others so why demonstrate that again here? We will attempt to remove the day to day variability from the data set in order to investigate fully the differences in fishing methods.

Consequently, the null hypothesis would be H_0: $\mu_I = \mu_{II} = \mu_{III} = \mu_{IV}$ with the research or alternative hypothesis of H_a: At least one pair of the μ_i's is not equal. As with the examples in Chapter 8, this experiment is asking a single question of the data, in this case, about the effectiveness of fishing techniques. By blocking the data to remove day to day variability here we control some of the extraneous variability that may obscure real differences in the treatments of interest (fishing techniques).

The Data Layout

Block	Treatment 1	2	3	...	k	Block Totals	Means
1	X_{11}	X_{21}	X_{31}	...	X_{k1}	$T_{.1}$	$\overline{X}_{.1}$
2	X_{12}	X_{22}	X_{32}	...	X_{k2}	$T_{.2}$	$\overline{X}_{.2}$
3	X_{13}	X_{23}	X_{33}	...	X_{k3}	$T_{.3}$	$\overline{X}_{.3}$
$\vdots$	$\vdots$	$\vdots$	$\vdots$	$\ddots$	$\vdots$	$\vdots$	$\vdots$
b	X_{1b}	X_{2b}	X_{3b}	...	X_{kb}	$T_{.b}$	$\overline{X}_{.b}$
Treatment Totals	$T_{1.}$	$T_{2.}$	$T_{3.}$	...	$T_{k.}$	Grand Totals $T_{..}$	
Means	$\overline{X}_{1.}$	$\overline{X}_{2.}$	$\overline{X}_{3.}$	...	$\overline{X}_{k.}$	$\overline{X}_{..}$	

The Block Totals and Means represent variability among *blocks* (days in Example 9.1) while the Treatment Totals and Means represent variability among *treatments* (fishing methods in Example 9.1). They are calculated as follows:

$$T_{i.} = \sum_{j=1}^{b} X_{ij} = i\text{th Treatment Total} \qquad \overline{X}_{i.} = \frac{T_{i.}}{b} = i\text{th Treatment Mean}$$

$$T_{.j} = \sum_{i=1}^{k} X_{ij} = j\text{th Block Total} \qquad \overline{X}_{.j} = \frac{T_{.j}}{k} = j\text{th Block Mean}$$

$$T_{..} = \sum_{i=1}^{k}\sum_{j=1}^{b} X_{ij} = \text{Grand Total} \qquad \overline{X}_{..} = \frac{T_{..}}{kb} = \text{Grand Mean}$$

$$\sum_{i=1}^{k}\sum_{j=1}^{b} X_{ij}^2 = \text{Uncorrected Sum of Squares}$$

The Randomized Complete Block Model

Each observation in the data matrix can be thought of as the sum of various effects (which themselves can be positive, negative, or zero):

$$X_{ij} = \mu + \tau_i + \beta_j + \epsilon_{ij} \qquad i = 1, 2, \ldots, k; \ j = 1, 2, \ldots, b,$$

where

- μ is the overall mean effect,

- τ_i is the effect due to the fact that the experimental unit received the ith treatment

$$\tau_i = \mu_{i.} - \mu,$$

- β_j is the effect due to the fact that the experimental unit was in the jth block

$$\beta_j = \mu_{.j} - \mu,$$

- ϵ_{ij} is the random error or residual

$$\epsilon_{ij} = X_{ij} - \mu_{ij},$$

where μ_{ij} is the mean for the combination of the ith treatment and the jth block.

So each observation can theoretically be partitioned into the following components:

$$X_{ij} = \mu + (\mu_{i.} - \mu) + (\mu_{.j} - \mu) + (X_{ij} - \mu_{ij}).$$

Equivalently, we may write

$(X_{ij} - \mu)$	$=$	$(\mu_i - \mu)$	$+$	$(\mu_j - \mu)$	$+$	$(X_{ij} - \mu_{ij}).$
Deviation of observation from grand mean	$=$	deviation of treatment mean from grand mean	$+$	deviation of block mean from grand mean	$+$	residual or unexplainable variability

The Model Assumptions

1. Each observation constitutes a random, independent sample from a population with mean μ_{ij}. There are $k \times b$ of these populations sampled.

2. Each of the $k \times b$ populations is normal with the same variance σ^2.

3. The treatment and block effects are *additive*, i.e., there is no interaction between blocks and treatments. Interaction, in the form of both synergy and interference, will be considered in more detail when we outline factorial design experiments in the next section.

The additivity condition above implies that

$$\mu_{ij} = \mu + (\mu_{i.} - \mu) + (\mu_{.j} - \mu).$$

The model components become

$$(X_{ij} - \mu) = (\mu_{i.} - \mu) + (\mu_{.j} - \mu) + \{X_{ij} - [\mu + (\mu_{i.} - \mu) + (\mu_{.j} - \mu)]\}$$
$$= (\mu_{i.} - \mu) + (\mu_{.j} - \mu) + (X_{ij} - \mu_{i.} - \mu_{.j} + \mu).$$

Replacing the various μ's with their unbiased estimators from the data, we have

$$(X_{ij} - \overline{X}_{..}) = (\overline{X}_{i.} - \overline{X}_{..}) + (\overline{X}_{.j} - \overline{X}_{..}) + (X_{ij} - \overline{X}_{i.} - \overline{X}_{.j} + \overline{X}_{..}). \qquad (9.1)$$

Thus, the deviation of each observation from the grand mean can be partitioned into the *explainable* effects from the treatments $(\overline{X}_{i.} - \overline{X}_{..})$ and blocking $(\overline{X}_{.j} - \overline{X}_{..})$ and the remainder or *unexplainable* variable $(X_{ij} - \overline{X}_{i.} - \overline{X}_{.j} + \overline{X}_{..})$.

Sums of Squares Identity

Squaring both sides of equation (9.1) and summing over all values of i and j, we generate a very useful relationship.

$$\sum_i \sum_j (X_{ij} - \overline{X}_{..})^2 = \sum_i \sum_j [(\overline{X}_{i.} - \overline{X}_{..}) + (\overline{X}_{.j} - \overline{X}_{..}) + (X_{ij} - \overline{X}_{i.} - \overline{X}_{.j} + \overline{X}_{..})]^2. \quad (9.2)$$

The three cross products on the left side of the equality vanish on summation because each can be shown to be equal to 0. For example,

$$2 \sum_i \sum_j [(\overline{X}_{i.} - \overline{X}_{..})(\overline{X}_{.j} - \overline{X}_{..})] = 2 \sum_i (\overline{X}_{i.} - \overline{X}_{..}) \sum_j (\overline{X}_{.j} - \overline{X}_{..}) = 2(0)(0) = 0.$$

By similar algebra the other two cross products can likewise be shown to be equal to 0. Consequently, Equation (9.2) becomes:

FORMULA 9.1. The total variability of a data set is denoted by SS_{Total} and can be represented as various sums of squares:

$$SS_{Total} = \sum_i \sum_j (X_{ij} - \overline{X}_{..})^2$$

$$= \sum_i \sum_j (\overline{X}_{i.} - \overline{X}_{..})^2 + \sum_i \sum_j (\overline{X}_{.j} - \overline{X}_{..})^2 + \sum_i \sum_j (X_{ij} - \overline{X}_{i.} - \overline{X}_{.j} + \overline{X}_{..})^2,$$

with df $= \nu = bk - 1$.

As usual, there is a computational formula for this sum of squares.

FORMULA 9.2. The computational formula for SS_{Total} is

$$SS_{Total} = \sum_i \sum_j X_{ij}^2 - \frac{T_{..}^2}{N}.$$

The other sums of squares are denoted similarly with corresponding computational formulas.

FORMULAS 9.3. For the Treatments the Sum of Squares is

$$SS_{Treat} = \sum_i \sum_j (\overline{X}_{i.} - \overline{X}_{..})^2 = \sum_i \frac{T_{i.}^2}{b} - \frac{T_{..}^2}{N},$$

with df $= k - 1$. For the Blocks the Sum of Squares is

$$SS_{Blocks} = \sum_i \sum_j (\overline{X}_{.j} - \overline{X}_{..})^2 = \sum_j \frac{T_{.j}^2}{k} - \frac{T_{..}^2}{N},$$

with df $= b - 1$. For the Errors the Sum of Squares is

$$SS_{Error} = \sum_i \sum_j (X_{ij} - \overline{X}_{i.} - \overline{X}_{.j} + \overline{X}_{..})^2 = SS_{Total} - (SS_{Treat} + SS_{Blocks}),$$

with df $= (k-1)(b-1)$. The total variability is partitioned additively so that Formula 9.1 becomes

$$SS_{Total} = SS_{Treat} + SS_{Blocks} + SS_{Error},$$

with the degrees of freedom partitioned similarly

$$bk - 1 = (k-1) + (b-1) + (k-1)(b-1).$$

The ANOVA Table

The sum of squares divided by the appropriate degrees of freedom generate *mean squares* that can be used to test hypotheses about means:

$$\text{MS}_{\text{Treat}} = \frac{\text{SS}_{\text{Treat}}}{k-1}$$

$$\text{MS}_{\text{Blocks}} = \frac{\text{SS}_{\text{Blocks}}}{b-1}$$

$$\text{MS}_{\text{E}} = \frac{\text{SS}_{\text{Error}}}{(k-1)(b-1)}.$$

These summary statistics usually are presented in an ANOVA Table, see Table 9.1.

TABLE 9.1. The analysis of variance table for a randomized complete block design

Source of variation	Sum of squares	df	MS	F	c.v.
Treatments	SS_{Treat}	$k-1$	$\frac{\text{SS}_{\text{Treat}}}{k-1}$	$\frac{\text{MS}_{\text{Treat}}}{\text{MS}_{\text{E}}}$	See Table C.7
Blocks	$\text{SS}_{\text{Blocks}}$	$b-1$	$\frac{\text{SS}_{\text{Blocks}}}{b-1}$		
Error	SS_{Error}	$(k-1)(b-1)$	$\frac{\text{SS}_{\text{Error}}}{(k-1)(b-1)}$		
Total	SS_{Total}	$kb-1$			

Expected Mean Squares

To test H_0: $\mu_1 = \mu_2 = \cdots = \mu_k$, the ratio

$$\frac{\text{MS}_{\text{Treat}}}{\text{MS}_{\text{E}}} = F_{(k-1),(k-1)(b-1)}$$

is used. The expected value of the test statistic is 1, if H_0 is *true*. To see this, note that if all the μ_i's are equal, then each difference $(\mu_{i.} - \mu) = 0$ so

$$E(\text{MS}_{\text{Treat}}) = \sigma^2 + \frac{b}{k-1}\sum_{i=1}^{k}(\mu_{i.} - \mu)^2 = \sigma^2 + 0 = \sigma^2$$

is the expected value for the numerator of the F test statistic. Since the expected value of the denominator is $E(\text{MS}_{\text{E}}) = \sigma^2$, the expected value of the F test statistic if H_0 is *true* is

$$\frac{\sigma^2}{\sigma^2} = 1.$$

If H_0 is *false*, then

$$\frac{b}{k-1}\sum_{i=1}^{k}(\mu_{i.} - \mu)^2 > 0,$$

making the expected value of the F test

$$\frac{\sigma^2 + \frac{b}{k-1}\sum_{i=1}^{k}(\mu_{i.} - \mu)^2}{\sigma^2} > 1.$$

So the F test is used to test the null hypotheses of equal treatment means in a fashion quite similar to the one-way ANOVAs of Section 8.1. In the randomized complete block design, however, the variability due to blocks is *removed* making treatment effects clearer or easier to find.

SOLUTION TO EXAMPLE 9.1. The hypotheses are

$$H_0: \mu_{\mathrm{I}} = \mu_{\mathrm{II}} = \mu_{\mathrm{III}} = \mu_{\mathrm{IV}}$$

H_a: At least one pair of the μ's is not equal.

To see the utility of the blocking technique, let's first analyze these data as a one-way ANOVA *ignoring the day to day variability*. The summary data are

	I	II	III	IV
$\overline{X}_i$	20.12	17.38	24.50	23.62
T_i	161	139	196	189
s_i	11.42	6.23	9.10	9.86

$$\sum_i \sum_j X_{ij}^2 = 17,367 \qquad T_{..} = 685, \qquad N = 32.$$

So

$$SS_{\text{Total}} = \sum_i \sum_j X_{ij}^2 - \frac{T_{..}^2}{N} = 17,367 - \frac{(685)^2}{32} = 2703.719$$

and

$$SS_{\text{FT}} = \sum_i \frac{T_{i.}^2}{n_i} - \frac{T_{..}^2}{N} = \frac{(161)^2}{8} + \frac{(139)^2}{8} + \frac{(196)^2}{8} + \frac{(189)^2}{8} - \frac{(685)^2}{32} = 259.094.$$

Consequently,

$$SS_{\text{Error}} = SS_{\text{Total}} - SS_{\text{FT}} = 2703.719 - 259.094 = 2444.625.$$

Source of variation	Sum of squares	df	MS	F	c.v.
Fishing Techniques	259.094	3	86.365	0.989	2.95
Error	2444.625	28	87.308		
Total	2703.719	31			

Since $0.989 < 2.95$, accept H_0. There are no significant differences among the means for these four fishing techniques. The very large within treatment sum of squares (SS_{Error}) makes it difficult to demonstrate any meaningful differences among the treatment means. See, also, the individual treatment standard deviations.

Now let's remove the day to day variability by analyzing the same data as a randomized, complete block design.

Day (jth)	Fishing Technique (ith)				$T_{.j}$	$\overline{X}_{.j}$
	I	II	III	IV		
1	23	21	24	25	93	23.25
2	16	14	19	17	66	16.50
3	19	17	23	21	80	20.00
4	11	12	15	16	54	13.50
5	41	30	40	42	153	38.25
6	32	20	37	35	124	31.00
7	9	11	19	17	56	14.00
8	10	14	19	16	59	14.75
$T_{i.}$	161	139	196	189	$T_{..} = 685$	
$\overline{X}_{i.}$	20.12	17.38	24.50	23.62	$\overline{X}_{..} = 21.41$	

Again
$$\sum_i \sum_j X_{ij}^2 = 17,367 \qquad T_{..} = 685, \qquad N = 32.$$

The same data generate the same total sum of squares as in the previous one-way ANOVA.

$$\text{SS}_{\text{Total}} = \sum_i \sum_j X_{ij}^2 - \frac{T_{..}^2}{N} = 17,367 - \frac{(685)^2}{32} = 2703.719$$

Similarly, the fishing technique sum of squares is also the same as previously calculated.

$$\text{SS}_{\text{FT}} = \sum_i \frac{T_{i.}^2}{n_i} - \frac{T_{..}^2}{N} = 259.094.$$

For the blocks, or days in this example, the sum of squares is

$$\text{SS}_{\text{Days}} = \sum_j \frac{T_{.j}^2}{k} - \frac{T_{..}^2}{N} = \left[\frac{(93)^2}{4} + \frac{(66)^2}{4} + \cdots + \frac{(59)^2}{4} \right] - \frac{(685)^2}{32} = 2267.469.$$

SS_{Days} is large when compared to SS_{Total} and accounts for the day to day variability in fish catches regardless of fishing techniques.

The unexplained variability or error now becomes

$$\begin{aligned} \text{SS}_{\text{Error}} &= \text{SS}_{\text{Total}} - (\text{SS}_{\text{Treat}} + \text{SS}_{\text{Blocks}}) \\ &= \text{SS}_{\text{Total}} - (\text{SS}_{\text{FT}} + \text{SS}_{\text{Days}}) \\ &= 2703.719 - (259.094 + 2267.469) = 177.156. \end{aligned}$$

The unexplained variability has gone from 2444.625 in the one-way ANOVA to 177.156 now. The ANOVA table for the blocked analysis becomes

Source of variation	Sum of squares	df	MS	F	c.v.
Fishing Techniques	259.094	3	86.365	10.238	3.07
Days	2267.469	7	323.924		
Error	177.156	21	8.436		
Total	2703.719	31			

Blocking removes the large amount of daily variability allowing us to now demonstrate that the fishing techniques are significantly different—at least some of them.

Compare this ANOVA table to the one-way ANOVA table presented earlier. Note that the degrees of freedom for the sum of squares error have been reduced from 28 to 21 and the critical value for the F statistic has subsequently increased from 2.95 to 3.07. Since $10.238 > 3.07$, the numerator of the F ratio is significantly greater than the denominator, indicating that some treatment means are significantly different. We reject H_0 and accept H_a here and have a P value of less than 0.05. Remember that the P value is the probability of a Type I error, if we reject H_a.

In the experiment above we blocked the data to reduce the SS_{Error} by SS_{Blocks}. If the variability removed due to blocking is substantial, MS_E will be smaller thereby increasing the F statistic and making the ANOVA more powerful. The degrees of freedom for MS_E in a one-way ANOVA is $N - k$, and the degrees of freedom for MS_E in a randomized complete block design ANOVA is $(k - 1)(b - 1)$. Since $kb = N$,

$$(k - 1)(b - 1) = kb - k - b + 1 = (N - k) - (b - 1).$$

This means that blocking reduces the degrees of freedom of MS_E by $b - 1$.

Looking at Table C.7, as the degrees of freedom of the denominator decrease, the critical values of F increase, making it harder to reject H_0. When it is harder to reject H_0, the power of the test is reduced. From a practical standpoint, what this means is that when blocking is done unnecessarily (when it doesn't remove a significant amount of the variability), the power of the ANOVA is adversely affected. Clearly blocking was appropriate and useful in Example 9.1. As with any Model I ANOVA a significant global F statistic indicates some differences among the means. To pinpoint the differences a mean separation technique should be applied.

Mean Separation

The global F test in Example 9.1 indicates at least two of the four means for the fishing techniques are significantly different. A mean separation technique such as the Bonferroni t test, where

$$\text{Bonferroni } t = \frac{\overline{X}_{i.} - \overline{X}_{j.}}{\sqrt{MS_E \left(\frac{1}{n_i} + \frac{1}{n_j} \right)}},$$

with α chosen to keep the experimentwise α' under control or a Student-Newman-Keuls test, with

$$SE = \sqrt{\frac{MS_E}{b}},$$

is now appropriate.

SOLUTION TO EXAMPLE 9.1: Mean Separation. We complete Example 9.1 with a Student-Newman-Keuls test. The ordered means are

II	I	IV	III
17.38	20.12	23.62	24.50

254   CHAPTER 9: Two-Factor Analysis

Using $MS_E = 8.436$ from the ANOVA table and $b = 8$,

$$SE = \sqrt{\frac{MS_E}{b}} = \sqrt{\frac{8.436}{8}} = 1.027.$$

We need the q critical values for $\alpha = 0.05$ and $\nu = 21$ (the degrees of freedom for MS_E). However, since $\nu = 21$ is not in Table C.9, use $\nu = 20$.

k	2	3	4
q c.v.	2.950	3.578	3.958

To test any hypothesis H_0: $\mu_i = \mu_j$, find

$$q = \frac{\overline{X}_i - \overline{X}_j}{SE}$$

and then compare q to the appropriate critical value. This leads to the following conclusions:

Comparison	q statistic	k	c.v.	Conclusion
II to III	6.933	4	3.958	Reject H_0
II to IV	6.076	3	3.578	Reject H_0
II to I	2.668	2	2.950	Accept H_0
I to III	4.265	3	3.578	Reject H_0
I to IV	3.408	2	2.950	Reject H_0
IV to III	0.857	2	2.950	Accept H_0

We can summarize these results using superscript notation. Any means with the same superscript are not significantly different.

II	I	IV	III
17.38^a	20.12^a	23.62^b	24.50^b

The conclusion now is that techniques I and II are not significantly different and techniques III and IV are not significantly different. However, techniques I and II, white-topped gill nets with and without pingers, significantly reduce the salmon catch. Pingers alone, technique III, is not significantly different from the standard method, IV. In conclusion, the data indicate that fishermen could use pingers to reduce "by-catch" and not lower their catch of salmon.

Notice that the conclusions above are very different from the results indicated in the earlier one-way ANOVA. Without blocking we concluded that there were no significant differences among the four fishing techniques ($F = 0.989$). Blocking exposed treatment differences that were previously obscured by the large SS_{Error} in the one-way ANOVA.

9.2   Factorial Design Two-Way ANOVA

While there are many variations of analysis of variance techniques described in the literature (see for example, Zar, 1999), we present one more important design before leaving this topic. This design is called a *factorial experiment with two-way classification, completely random design with fixed effects*. To understand what this description means, let's look at an example.

EXAMPLE 9.2. In an attempt to find the most effective methods for training companion dogs for the physically challenged, an experiment was conducted to compare three different training regimes in combination with three different reward systems. All the animals in the study were Labrador retrievers and were 6 to 8 months old at the start of the experiment. Individual dogs were assigned randomly to a combination of training regime and reward system. At the end of a 10-week training period the dogs were given a standardized test to measure their ability to function as companion dogs for the visually impaired. The results of the test are given below:

Reward	Training regime		
	I	II	III
Praise	45	51	52
	69	50	18
	53	62	25
	51	68	32
Tangible	54	53	51
	72	63	59
	69	67	47
	66	70	42
Praise and tangible	91	69	66
	87	73	68
	89	77	70
	91	74	64

The behaviorial scientist running this study would like to know which training regime, if any, is best and which reward system, if any, is best. Also, she is looking for *interactions* between training regimes and reward systems. These interactions can be either positive (synergy) or negative (interference). Suppose, for example, a training regime increases the test score 10 points and a particular reward system increases the test score 5 points, and when they are both applied the score increases on average by 17 points. Because the change in score is *greater* than the sum of the individual effects, we say the two treatments *synergize* each other. If, instead, using that particular training regime and reward system generates an average increase of 13 points (less than the sum of the individual effects), we say the two treatments *interfere* with each other. Studying both kinds of effects (training regimes and reward systems) simultaneously is the *only* way to investigate interactions such as synergy or interference.

If no significant interactions are found, then one can say which training regime is best regardless of the reward system and also which reward system is best without considering the training regime. If an interaction is found you must qualify your statement about either training regimes or reward systems by saying regime X is best when reward system Y is used, etc.

The experiment above is called **factorial** because we are interested in two different factors' effects on the measured response. It is **completely random** because individuals are assigned to a combination of training regime and reward system randomly. These combinations are called **cells**. The effects are *fixed* here because we are interested in only training regimes I, II, and III and reward systems praise, tangible, and praise with tangible. Neither the training regimes nor the reward systems are chosen at random from a larger group of interest. We have here, then, a Model I design for both factors.

The Data Layout

Each observation must now be triply subscripted as X_{ijk}. The first subscript identifies the factor A level (ith level A). The second subscript identifies the factor B level (jth level B). The final subscript is the position of a particular observation within the ij cell. Look back at the data table given in Example 9.2. $X_{3,1,4}$ is 32 while $X_{2,3,1}$ is 69. The general layout appears in Table 9.2.

TABLE 9.2. The data organization for a factorial design two-way ANOVA: Each observation is triply subscripted: X_{ijk}. The first subscript, i, identifies the level of factor A. Likewise, the second subscript, j, identifies the level of factor B. The final subscript, k, represents the position of this particular observation within the cell of all observations having both factor A level i and factor B level j.

Factor B level	Factor A level				
	1	**2**	**3**	$\cdots$	a
1	X_{111}	X_{211}	X_{311}	$\cdots$	X_{a11}
	X_{112}	X_{212}	X_{312}	$\cdots$	X_{a12}
	$\vdots$	$\vdots$	$\vdots$	$\vdots$	$\vdots$
	X_{11n}	X_{21n}	X_{31n}	$\cdots$	X_{a1n}
2	X_{121}	X_{221}	X_{321}	$\cdots$	X_{a21}
	X_{122}	X_{222}	X_{322}	$\cdots$	X_{a22}
	$\vdots$	$\vdots$	$\vdots$	$\vdots$	$\vdots$
	X_{12n}	X_{22n}	X_{32n}	$\cdots$	X_{a2n}
$\vdots$	$\vdots$	$\vdots$	$\vdots$	$\vdots$	$\vdots$
$\vdots$	$\vdots$	$\vdots$	$\vdots$	$\vdots$	$\vdots$
$\vdots$	$\vdots$	$\vdots$	$\vdots$	$\vdots$	$\vdots$
b	X_{1b1}	X_{2b1}	X_{3b1}	$\cdots$	X_{ab1}
	X_{1b2}	X_{2b2}	X_{3b2}	$\cdots$	X_{ab2}
	$\vdots$	$\vdots$	$\vdots$	$\vdots$	$\vdots$
	X_{1bn}	X_{2bn}	X_{3bn}	$\cdots$	X_{abn}

The data layout for a factorial design experiment generates a series of sample means. We have means for each level of factor A, $\overline{X}_{i..}$ and means for each level of factor B, $\overline{X}_{.j.}$. We also have the grand mean $\overline{X}_{...}$ and, additionally, we now have cell means, $\overline{X}_{ij.}$, which are the averages of all observations receiving the same combination of A and B factors.

Factor B level	Factor A level (i)					Level B
(j)	1	2	3	...	a	Totals and Means
1	$T_{11.}$	$T_{21.}$	$T_{31.}$	...	$T_{a1.}$	$T_{.1.}$
	$\overline{X}_{11.}$	$\overline{X}_{21.}$	$\overline{X}_{31.}$	...	$\overline{X}_{a1.}$	$\overline{X}_{.1.}$
2	$T_{12.}$	$T_{22.}$	$T_{32.}$	...	$T_{a2.}$	$T_{.2.}$
	$\overline{X}_{12.}$	$\overline{X}_{22.}$	$\overline{X}_{32.}$	...	$\overline{X}_{a2.}$	$\overline{X}_{.2.}$
$\vdots$	$\vdots$	$\vdots$	$\vdots$	$\vdots$	$\vdots$	$\vdots$
b	$T_{1b.}$	$T_{2b.}$	$T_{3b.}$	...	$T_{ab.}$	$T_{.b.}$
	$\overline{X}_{1b.}$	$\overline{X}_{2b.}$	$\overline{X}_{3b.}$	...	$\overline{X}_{ab.}$	$\overline{X}_{.b.}$
Level A						Grand Total and Mean
Totals	$T_{1..}$	$T_{2..}$	$T_{3..}$	...	$T_{a..}$	$T_{...}$
Means	$\overline{X}_{1..}$	$\overline{X}_{2..}$	$\overline{X}_{3..}$	...	$\overline{X}_{a..}$	$\overline{X}_{...}$

The $\overline{X}_{i..}$'s represent variability *among* factor levels A and the $\overline{X}_{.j.}$'s represent variability *among* factor levels B. The $\overline{X}_{ij.}$'s represent cell differences, while the variability *within* a cell is unexplainable variability.

The Factorial Design Two-Way Model

Each observation in the data set can be thought of as the sum of various effects (which themselves can be positive, negative, or 0). For $i = 1, 2, \ldots, a, \quad j = 1, 2, \ldots, b, \quad k = 1, 2, \ldots, n$,

$$X_{ijk} = \mu + \alpha_i + \beta_j + (\alpha\beta)_{ij} + \epsilon_{ijk},$$

- μ is the overall mean effect,

- α_i is the effect due to the fact that the experimental unit received the ith level of factor A

$$\alpha_i = \mu_{i..} - \mu,$$

- β_j is the effect due to the fact that the experimental unit received the jth level of factor B

$$\beta_j = \mu_{.j.} - \mu,$$

- $(\alpha\beta)_{ij}$ is the effect of the *interaction* between the ith level of factor A and the jth level of factor B

$$(\alpha\beta)_{ij} = \mu_{ij.} - \mu_{i..} - \mu_{.j.} + \mu,$$

- ϵ_{ijk} is the random error or residual

$$\epsilon_{ijk} = X_{ijk} - \mu_{ij.}.$$

Note: The α's and β's here refer to treatment effects and are not related to Type I or Type II errors that, unfortunately, use the same symbols (but are not subscripted).

So each observation X_{ijk} can be partitioned theoretically into the following components:

$$X_{ijk} = \mu + (\mu_{i..} - \mu) + (\mu_{.j.} - \mu) + (\mu_{ij.} - \mu_{i..} - \mu_{.j.} + \mu) + (X_{ijk} - \mu_{ij.})$$

or

$$(X_{ijk} - \mu) = (\mu_{i..} - \mu) + (\mu_{.j.} - \mu) + (\mu_{ij.} - \mu_{i..} - \mu_{.j.} + \mu) + (X_{ijk} - \mu_{ij.})$$

Deviation of observation from the grand mean	=	deviation of ith A factor from the grand mean	+	deviation of jth B factor from the grand mean	+	interaction of ith A level and jth B level	+	residual or unexplained.

The Model Assumptions

1. The observations in each *cell* constitute an independent random sample of size n from a population with mean $\mu_{ij.}$.

2. Each of the populations represented by the cell samples is normal and has the same variance, σ^2.

Consider now the statistics to estimate the various model components:

$$(X_{ijk} - \overline{X}_{...}) = (\overline{X}_{ij.} - \overline{X}_{...}) + (X_{ijk} - \overline{X}_{ij.}) \tag{9.3}$$

An observation's deviation from the grand mean	=	deviation of the cell mean from the grand mean	+	deviation of the observation from the cell mean.

The deviation $(\overline{X}_{ij.} - \overline{X}_{...})$ represents *explainable* variability because of the combination of A- and B-level treatments. The deviation $(X_{ijk} - \overline{X}_{ij.})$ within a cell is *unexplainable*. After all, the observations within a cell were all treated identically and should respond identically. When they don't, we say that the variability is residual or background noise in the system.

The explainable portion can be further partitioned into what are called *main effects* and their *interaction*:

$$(\overline{X}_{ij.} - \overline{X}_{...}) = (\overline{X}_{i..} - \overline{X}_{...}) + (\overline{X}_{.j.} - \overline{X}_{...}) + (\overline{X}_{ij.} - \overline{X}_{i..} - \overline{X}_{.j.} + \overline{X}_{...}) \tag{9.4}$$

Explainable variability	=	main effect of ith A level	+	main effect of jth B level	+	interaction of the ith A level and jth B level.

These deviations squared and summed give us the statistics to test various experimental effects.

Sums of Squares

FORMULAS 9.4. The various sums of squares utilized in factorial design two-way ANOVA and their equivalent computational formulas are

$$SS_{\text{Total}} = \sum_i \sum_j \sum_k (X_{ijk} - \overline{X}_{...})^2 = \sum_i \sum_j \sum_k X_{ijk}^2 - \frac{T_{...}^2}{abn}$$

$$SS_A = \sum_i \sum_j \sum_k (\overline{X}_{i..} - \overline{X}_{...})^2 = \sum_i \left(\frac{T_{i..}^2}{bn}\right) - \frac{T_{...}^2}{abn}$$

$$SS_B = \sum_i \sum_j \sum_k (\overline{X}_{.j.} - \overline{X}_{...})^2 = \sum_j \left(\frac{T_{.j.}^2}{an}\right) - \frac{T_{...}^2}{abn}$$

$$SS_{\text{Cells}} = \sum_i \sum_j \sum_k (\overline{X}_{ij.} - \overline{X}_{...})^2 = \sum_i \sum_j \left(\frac{T_{ij.}^2}{n}\right) - \frac{T_{...}^2}{abn}.$$

Moreover, from (9.4) it follows that

$$SS_{\text{Cells}} = SS_A + SS_B + SS_{A \times B},$$

so

$$SS_{A \times B} = SS_{\text{Cells}} - (SS_A + SS_B).$$

From (9.3) it follows that

$$SS_{\text{Error}} = SS_{\text{Total}} - SS_{\text{Cells}}.$$

The degrees of freedom associated with the sums of squares are logical extensions of the work in Section 9.1.

$$
\begin{array}{lll}
SS_{\text{Total}} & \nu = abn - 1 \text{ or } N - 1 & \\
\quad SS_{\text{Cells}} & \nu = ab - 1 & \\
\qquad SS_A & \nu = a - 1 & \\
\qquad SS_B & \nu = b - 1 & \left.\right\} \text{ These sum to } ab - 1. \\
\qquad SS_{A \times B} & \nu = (a-1)(b-1) & \\
\quad SS_{\text{Error}} & \nu = ab(n - 1) &
\end{array}
$$

Again, the degrees of freedom are additive in the same way the sums of squares are additive, as can be seen in the partition the $ab - 1$ degrees of freedom for SS_{Cells}. It is also seen in the partition of the degrees of freedom for SS_{Total} into the degrees of freedom for SS_{Cells} and SS_{Error} since

$$abn - 1 = ab - 1 + ab(n - 1).$$

ANOVA Table and Expected Mean Squares

While the expected values for the mean squares given in Table 9.3 are not proved here, they can be used to gain insights into the various tests of hypothesis possible here.

The testing for a factorial design two-way ANOVA is rather more complicated than in the blocked ANOVA presented in Section 9.1.

TABLE 9.3. The analysis of variance table for a factorial design two-way ANOVA

Source of variation	Sum of squares	df	MS	E(MS)
Cells	SS_{Cells}	$ab-1$	MS_{Cells}	$\sigma^2 + n\sum_i\sum_j \frac{(\mu_{ij.}-\mu_{...})^2}{ab-1}$
A factors	SS_A	$a-1$	MS_A	$\sigma^2 + nb\sum_i \frac{(\mu_{i..}-\mu_{...})^2}{a-1}$
B factors	SS_B	$b-1$	MS_B	$\sigma^2 + na\sum_j \frac{(\mu_{.j.}-\mu_{...})^2}{b-1}$
A × B interaction	$SS_{A\times B}$	$(a-1)(b-1)$	$MS_{A\times B}$	$\sigma^2 + n\sum_i\sum_j \frac{(\alpha\beta)_{ij}^2}{(a-1)(b-1)}$
Error	SS_{Error}	$ab(n-1)$	MS_E	σ^2
Total	SS_{Total}	$abn-1$		

Test for interaction between factors

First test for interaction between the two types of factors being investigated. The hypotheses are

$$H_0\colon (\alpha\beta)_{ij} = 0 \text{ for all } i,j$$
$$H_a\colon (\alpha\beta)_{ij} \neq 0 \text{ for some } i,j. \tag{9.5}$$

By perusing the *expected values* in Table 9.3, one can see that the appropriate test for this null hypothesis is

$$F_{A\times B} = \frac{MS_{A\times B}}{MS_E},$$

with degrees of freedom $\nu_1 = (a-1)(b-1)$ and $\nu_2 = ab(n-1)$. As with all ANOVA F tests, if the numerator is significantly larger than the denominator we will reject H_0.

If there are no interactions

If the null hypothesis of no interactions in (9.5) is *accepted*, the analysis is continued by carrying out two tests.

1. Test whether there are differences in means for A-factor treatments,

$$H_0\colon \mu_{i..}\text{'s all equal} \qquad (\alpha_i = 0 \text{ for all } i)$$
$$H_a\colon \text{At least one pair of the } \mu_{i..}\text{'s is not equal,}$$

using the test statistic

$$F_A = \frac{MS_A}{MS_E}$$

with degrees of freedom $\nu_1 = a-1$ and $\nu_2 = ab(n-1)$.

2. Similarly, test whether there are differences in means for B-factor treatments,

$$H_0\colon \mu_{.j.}\text{'s all equal} \qquad (\beta_j = 0 \text{ for all } j)$$
$$H_a\colon \text{At least one pair of the } \mu_{.j.}\text{'s is not equal,}$$

using test statistic

$$F_B = \frac{\text{MS}_\text{B}}{\text{MS}_\text{E}}$$

with degrees of freedom $\nu_1 = b - 1$ and $\nu_2 = ab(n - 1)$.

If there are interactions

If the null hypothesis of no interactions in (9.5) is *rejected*, we omit F_A and F_B tests. Since the levels of Factor A do not behave consistently across the levels of Factor B, and vice versa, we look for the best combination of A factor and B factor. We might use an SNK test of the form

$$q = \frac{\overline{X}_i - \overline{X}_j}{\text{SE}} \quad \text{with} \quad \text{SE} = \sqrt{\frac{\text{MS}_\text{E}}{n}}$$

to separate the various *cell* means.

SOLUTION TO EXAMPLE 9.2. Analyze the data as a factorial design two-way ANOVA. First we must determine the basic summary statistics. These are given in the following table.

Reward system (jth)	Training Regime (ith)			
	I	**II**	**III**	
Praise	45	51	52	
	69	50	18	
	53	62	25	
	51	68	32	
	$T_{11.} = 218$	$T_{21.} = 231$	$T_{31.} = 127$	$T_{.1.} = 576$
	$\overline{X}_{11.} = 54.5$	$\overline{X}_{21.} = 57.8$	$\overline{X}_{31.} = 31.8$	$\overline{X}_{.1.} = 48.0$
Tangible	54	53	51	
	72	63	59	
	69	67	47	
	66	70	42	
	$T_{12.} = 261$	$T_{22.} = 253$	$T_{32.} = 199$	$T_{.2.} = 713$
	$\overline{X}_{12.} = 65.3$	$\overline{X}_{22.} = 63.3$	$\overline{X}_{32.} = 49.8$	$\overline{X}_{.2.} = 59.4$
Praise and tangible	91	69	66	
	87	73	68	
	89	77	70	
	91	74	64	
	$T_{13.} = 358$	$T_{23.} = 293$	$T_{33.} = 268$	$T_{.3.} = 919$
	$\overline{X}_{13.} = 89.5$	$\overline{X}_{23.} = 73.3$	$\overline{X}_{33.} = 67.0$	$\overline{X}_{.3.} = 76.6$
	$T_{1..} = 837$	$T_{2..} = 777$	$T_{3..} = 594$	
	$\overline{X}_{1..} = 69.8$	$\overline{X}_{2..} = 64.8$	$\overline{X}_{3..} = 49.5$	

$\sum_i \sum_j \sum_k X_{ijk}^2 = 145{,}404$ $\qquad\qquad\qquad$ $\sum_i \sum_j \sum_k X_{ijk} = T_{...} = 2208$

Next, the sums of squares can be calculated using Formulas 9.4:

$$SS_{Total} = 145,404 - \frac{(2208)^2}{36} = 9980.00$$

$$SS_{Cells} = \left[\frac{(218)^2}{4} + \frac{(231)^2}{4} + \cdots + \frac{(268)^2}{4}\right] - \frac{(2208)^2}{36} = 8221.50$$

$$SS_{Error} = SS_{Total} - SS_{Cells} = 9980.00 - 8221.50 = 1758.50.$$

Training regimes:

$$SS_A = \left[\frac{(837)^2}{12} + \frac{(777)^2}{12} + \frac{(594)^2}{12}\right] - \frac{(2208)^2}{36} = 2670.50.$$

Reward systems:

$$SS_B = \left[\frac{(576)^2}{12} + \frac{(713)^2}{12} + \frac{(919)^2}{12}\right] - \frac{(2208)^2}{36} = 4968.17.$$

Interactions:

$$SS_{A \times B} = SS_{Cells} - (SS_A + SS_B) = 8221.50 - (2670.50 + 4968.17) = 582.83.$$

Now we organize the information in an ANOVA table:

Source of variation	Sums of Squares	df	MS
Cells	8221.50	8	
Training regimes	2670.50	2	1335.25
Reward systems	4968.17	2	2484.09
Interaction	582.83	4	145.71
Error	1758.50	27	65.13
Total	9980.00	35	

First investigate interactions.   The hypotheses are

$$H_0: (\alpha\beta)_{ij} = 0 \text{ for all } i, j$$
$$H_a: (\alpha\beta)_{ij} \neq 0 \text{ for some } i, j.$$

The test statistic is

$$F_{A \times B} = \frac{145.71}{65.13} = 2.24.$$

From Table C.7 with $\alpha = 0.05$, $\nu_1 = 4$, and $\nu_2 = 27$, the c.v. = 2.73. Since $2.24 < 2.73$, *we don't have evidence for interaction here.* We accept H_0 and reject H_a.

Now test the main effects for training regimes.   The hypotheses are

$$H_0: \mu_I = \mu_{II} = \mu_{III}$$
$$H_a: \text{At least one pair of } \mu\text{'s is not equal.}$$

The test statistic is

$$F_A = \frac{1335.25}{65.13} = 20.50.$$

From Table C.7 with $\alpha = 0.05$, $\nu_1 = 2$, and $\nu_2 = 27$, we find the c.v. = 3.35. Since $20.50 \gg 3.35$, we have evidence that *at least some training regimes are significantly different.*

Finally, test the main effects for reward systems. The hypotheses are

$$H_0: \mu_P = \mu_T = \mu_{P+T}$$

$$H_a: \text{At least one pair of } \mu\text{'s is not equal.}$$

The test statistic is

$$F_B = \frac{2484.09}{65.13} = 38.14.$$

From Table C.7 with $\alpha = 0.05$, $\nu_1 = 2$, and $\nu_2 = 27$, again the c.v. = 3.35. Since $38.14 \gg 3.35$, we have evidence that *at least some reward systems are significantly different*.

In summary, the three F tests generated from the ANOVA table indicate that there is *no significant interaction* between training regimes and reward systems and that *at least some significant differences among training regimes* and *among reward systems* are indicated. This means the best training regime will be the same over all reward systems and, conversely, the best reward system will be the same over all training regimes.

Mean Separation Techniques Now separate the main effects for training regimes with an SNK test. The linearly ordered means are

Training regimes		
$\overline{X}_{III}$	$\overline{X}_{II}$	$\overline{X}_I$
49.5	64.8	69.8

Using the values in the ANOVA table, we have

$$\sqrt{\frac{\text{MS}_E}{bn}} = \sqrt{\frac{65.13}{3 \times 4}} = 2.330.$$

From Table C.9 with $\alpha = 0.05$ with $\nu = 27 \approx 24$ we obtain the following critical values

k	**2**	**3**
q c.v.	2.919	3.532

To test any hypothesis $H_0: \mu_i = \mu_j$, find

$$q = \frac{\overline{X}_i - \overline{X}_j}{\text{SE}}$$

and then compare q to the appropriate critical value. This leads to the following conclusions:

Comparison	q statistic	k	c.v.	Conclusion
III to I	8.712	3	3.532	Reject H_0
III to II	6.567	2	2.919	Reject H_0
II to I	2.146	2	2.919	Accept H_0

Data indicate that training regimes I and II are *not significantly different*, but they are *superior* to regime III. The superscript convention is used to summarize this conclusion. Values with the same superscript are not significantly different.

III	II	I
49.5^a	64.8^b	69.8^b

Similarly, for the rewards systems the linearly ordered means are

Reward systems		
$\overline{X}_P$	$\overline{X}_T$	$\overline{X}_{P+T}$
48.0	59.4	76.6

Since the reward systems are based on the same sample sizes as the training regimes,

$$\sqrt{\frac{\text{MS}_E}{an}} = \sqrt{\frac{65.13}{3 \times 4}} = 2.330.$$

remains the same. The q critical values for the reward systems will be the same as for the training regimes. We proceed with the mean separation as with the training regimes and obtain the following conclusions:

Comparison	q statistic	k	c.v.	Conclusion
P to P+T	12.275	3	3.532	Reject H_0
P to T	4.893	2	2.919	Reject H_0
T to P+T	7.382	2	2.919	Reject H_0

The SNK test indicates that all three reward systems are significantly different from each other. The very large F statistic ($F_B = 38.14$) was an early indication that differences here were dramatic. This is summarized as

P	T	P+T
48.0^a	59.4^b	76.6^c

Finally, we conclude that while praise plus tangible rewards increase the test scores the most, either training regime I or II is better than III and no interaction between training regime and reward system is indicated.

Clearly the calculations to do factorial design ANOVA by hand are tedious. Fortunately most statistical computer programs will find the sums of squares with relative ease. If you understand the expected values of the various mean squares, appropriate F tests are intuitive and relatively easy to interpret.

9.3   The Friedman k-Sample Test: Matched Data

In this section we present the nonparametric analog to the Randomized Complete Block Design ANOVA discussed in Section 9.1.

We thought of the Kruskal-Wallis test as an extension of the Wilcoxon rank-sum test for two independent samples. But earlier we had considered pairs of data sets that were matched and analyzed them with the sign test or the Wilcoxon signed-rank test. The idea of these tests can be extended to the situation of k-matched samples. We can think of tests on matched pairs as being performed on sets of identical twins, one treatment is applied to one of the twins and another to the other. The differences between pairs of twins are then ranked.

Suppose now that we wanted to compare the effects of k treatments. Oftentimes there are variables that will affect the results even though the variable is not the one of interest. The idea is to control these extra variables by **blocking**. The subjects are split into b groups or "blocks" each of size k, where the members of each block are as nearly alike as possible. Then we assign the k treatments randomly to the subjects within each block. For example, if we had three treatments we wished to investigate, it would be ideal if we had b sets of identical triplets. Each block would be a set of triplets and different treatments would be assigned at random to each of the triplets within a block. In this way, any variation measured within a block should be due to the treatment, not the subjects.

> **EXAMPLE 9.3.** A preliminary study at Lady Elliot Island on the Great Barrier Reef was conducted to investigate how different species of encrusting sponge compete for and partition the space available on the underside of small boulders in the intertidal zone. Eight boulders were examined and the percent cover by each of three different species of sponge (determined by color) was estimated by using a grid. Is there a difference in the space colonized by the different species of sponges?

Boulder	Peach	Black	Cream
1	30	35	8
2	17	10	5
3	33	18	12
4	18	34	2
5	25	20	25
6	50	15	11
7	40	20	6
8	19	5	14

On the surface, the data look no different from those for the Kruskal-Wallis test. But it would be incorrect to analyze it using that technique. In the Kruskal-Wallis test, ranks are assigned to all N observations in order to compare every observation with every other. In the current situation, we are trying to measure the effect within each block (on each boulder). So the observations within each block are ranked separately (here, 1 through 3), as follows.

Boulder	Peach	Black	Cream
1	2	3	1
2	3	2	1
3	3	2	1
4	2	3	1
5	2.5	1	2.5
6	3	2	1
7	3	2	1
8	3	1	2
Rank sum	21.5	16	10.5

There are differences in the rankings for the different sponges, but are the differences significant? Based on the ranking information, how might we compare the space occupied by each sponge? The sum of the ranks for each of the treatments gives us a clue as to whether one species is more dominant than another. If a particular sponge occupies a significantly higher amount of space, then the ranks for that sponge (treatment) on each boulder (within each block) will be higher and so the overall rank sum should be larger for it than for the other species (treatments). The question is, how much larger is sufficient for a treatment to be declared significantly different? It's time to make things precise. The test that we now develop was introduced by Milton Friedman in the late 1930s.

Assumptions

There are four assumptions for the Friedman test:

1. The data consist of b mutually independent blocks of samples from k random variables. The data are laid out in b rows and k columns. The ith block consists of $X_{i1}, X_{i2}, \ldots, X_{ik}$. We think of the elements of each block as being "matched" with respect to all extraneous variables.

2. The random variable X_{ij} comes from the ith block and is associated with treatment j.

3. Within each block the k treatments are assigned randomly. We assume that blocks are independent, that is, the results in one block do not affect those in another.

4. Within each block the observations may be ranked according to the variable of interest. (As usual, a small number of ties is permissible.)

Hypotheses

There is only one hypothesis test that is carried out:

- Null hypothesis H_0: All k treatments have identical effects, versus the alternative hypothesis H_a: At least one pair of treatments produces different effects.

Test Statistic and Decision Rule

We compute the test statistic S as follows. Rank the members within a block from 1 to k, using midranks for tied values. Let R_j denote the sum of the ranks associated with the jth treatment. Since the underlying assumption is that treatment effects are identical, the sum of the ranks for any one treatment ought to be approximately the same as for any other. There are k ranks to assign within each block, so the average rank is $\frac{k+1}{2}$. Each treatment is applied to one member of each block, that is, to b elements in total. So the expected or mean rank sum for each treatment is $\frac{b(k+1)}{2}$. The test statistic is based on the difference between the actual sum of the ranks for a treatment and the expected sum. We use the corrected sum of squares,

$$S = \sum_{j=1}^{k} \left[R_j - \frac{b(k+1)}{2} \right]^2.$$

If H_0 is true, then all the treatment rank sums should be close to the expected value of $\frac{b(k+1)}{2}$ and hence S should be close to 0. If, on the other hand, H_0 is false, then at least one sample has larger (smaller) treatment effects and correspondingly higher (lower) ranks. Its rank sum will differ from $\frac{b(k+1)}{2}$ and, hence, increase the test statistic S.

We need to know the probability distribution of S to know when it is sufficiently large so that H_0 may be rejected. When the number of treatments k and the number of blocks b are small, the exact distribution can be calculated. (We do an example of this later.) The test statistic S is usually adjusted for the size of the sample, to create the Friedman test statistic

$$T = \frac{12S}{bk(k+1)} = \frac{12}{bk(k+1)} \sum_{j=1}^{k} \left[R_j - \frac{b(k+1)}{2} \right]^2.$$

It can be shown that T follows a chi-square distribution with $k-1$ degrees of freedom, so Table C.5 may be used to determine critical values of the test. If the test is at the α level of significance, then H_0 should be rejected when T exceeds the value in Table C.5 for the $1 - \alpha$ quantile.

Using a bit of algebra, much as we did for the Kruskal-Wallis test statistic, it is possible to show that

$$T = \frac{12}{bk(k+1)} \left(\sum_{j=1}^{k} R_j^2 \right) - 3b(k+1).$$

This latter form of the test statistic is easier to use when calculating by hand.

SOLUTION TO EXAMPLE 9.3. The null hypothesis is H_0: All species occupy the same amount of space on the underside of boulders versus H_a: At least one species occupies a different amount of space than the others. Let $\alpha = 0.05$. The number of treatments (species) is $k = 3$ and the number of blocks is $b = 8$. The expected rank sum is $\frac{b(k+1)}{2} = \frac{8(3+1)}{2} = 16$. Using the sums of the treatment ranks that were calculated earlier,

$$T = \frac{12}{bk(k+1)} \left(\sum_{j=1}^{k} R_j^2 \right) - 3b(k+1)$$

$$= \frac{12}{8(3)(3+1)} \left[(21.5)^2 + (16)^2 + (10.5)^2 \right] - 3 \cdot 8(3+1)$$

$$= 7.56.$$

From Table C.5, with $k - 1 = 2$ degrees of freedom, the critical value for T is 5.99. Since $7.56 > 5.99$, reject H_0. There is evidence that at least one species of sponge occupies a different amount of underboulder space.

The Friedman test was particularly appropriate for these data because the measurement methods (grid on a boulder) were by no means precise. But even though the percent cover may not have been especially accurate, it is very likely that the rankings were accurate, which is what the test requires. One could tell on each boulder which species of sponge occupied the most space, the next most space, and the least space.

EXAMPLE 9.4. Aposematic coloration acts as a signal to warn predators of potentially dangerous or unpleasant features possessed by a prey species. It is believed that individuals that possess color patterns resembling those of aposematic species gain a defensive advantage, known as Batesian mimicry, because predators have learned to avoid animals with these patterns through encounters with aposematic species. An experiment was conducted to test this hypothesis using three aposematic marine flatworms, *Pseudoceros paralaticlavus*, *Pseudobiceros stellae*, and *Phrikoceros baibaiye*. The predators were moon wrasse, *Thalassomus lunare*. In seven separate trials, three different species of flatworms, colored models of each species, and a colorless control model were placed randomly in a tank with 14 moon wrasse. All models were made from brine shrimp and agar and should have been palatable to the wrasses. The aposematic (i.e., alternative) hypothesis is that the colored flatworm models will gain a defensive advantage that the control model will not have. Attack values corresponding to the number and intensity of the attacks on the flatworms and models were calculated for each 9-hour trial. Was there a significant difference in the attack values for live flatworms and control models? (Based on data reported by Ang, H. and L. Newman. 1996. Warning colouration in pseudocerotid flatworms Platyhelminthes, polycladida. Personal communication.)

Pseudoceros paralaticlavus		*Pseudobiceros stellae*		*Phrikoceros baibaiye*		Control
Real	**Model**	**Real**	**Model**	**Real**	**Model**	**Model**
0	1	1	1	0	1	9
0	1	1	0	1	5	9
0	2	0	1	2	7	9
0	0	1	2	4	6	9
0	0	0	0	0	1	9
0	0	0	2	9	5	9
0	0	7	5	9	7	9

SOLUTION. Use the Friedman test at the $\alpha = 0.05$ level. Rank the data within blocks.

Pseudoceros paralaticlavus		*Pseudobiceros stellae*		*Phrikoceros baibaiye*		Control
Real	**Model**	**Real**	**Model**	**Real**	**Model**	**Model**
1.5	4.5	4.5	4.5	1.5	4.5	7
1.5	4	4	1.5	4	6	7
1.5	4.5	1.5	3	4.5	6	7
1.5	1.5	3	4	5	6	7
3	3	3	3	3	6	7
2	2	2	4	6.5	5	6.5
1.5	1.5	4.5	3	6.5	4.5	6.5
12.5	21	22.5	23	31	38	48

From the rank sums above, the test statistic is

$$T = \frac{12}{bk(k+1)} \left(\sum_{j=1}^{k} R_j^2 \right) - 3b(k+1)$$

$$= \frac{12}{7(7)(8)} \left[(12.5)^2 + (21)^2 + (22.5)^2 + (23)^2 + (31)^2 + (38)^2 + (48)^2 \right] - 3 \cdot 7(8)$$

$$= 26.13.$$

From Table C.5, the critical value with $k - 1 = 6$ degrees of freedom is 12.59. So H_0 is rejected. There is a significant difference in attack values among the flatworms and the models.

Theory

Though we will always use the chi-square approximation for the distribution of the **Friedman test statistic** T, it is instructive to look at an example of how the *exact* distribution of T can be calculated in a simple case. The exact distribution is found under the assumption that each ranking within a block is equally likely, since the effects of the treatments are assumed to be identical. In general, there are $k!$ ways to arrange the k ranks within a block. Since there are b blocks, there are a total of $(k!)^b$ possible arrangements of the ranks in the entire $b \times k$ table. Let's take $k = 2$ treatments and $b = 3$ blocks. This yields a total of $(2!)^3 = 8$ possible arrangements, which are shown in the table below:

Treatment	I	II	I	II	I	II	I	II	I	II	I	II	I	II	I	II
Arrangement	1	2	2	1	1	2	1	2	2	1	2	1	1	2	2	1
	1	2	1	2	2	1	1	2	2	1	1	2	2	1	2	1
	1	2	1	2	1	2	2	1	1	2	2	1	2	1	2	1
R_j	3	6	4	5	4	5	4	5	5	4	5	4	5	4	6	3
T		3		$\frac{1}{3}$		$\frac{1}{3}$		$\frac{1}{3}$		$\frac{1}{3}$		$\frac{1}{3}$		$\frac{1}{3}$		3

The probability of any one of the arrangements occurring is $\frac{1}{8}$. The exact density and distribution functions are given in the following table:

T_0	$\frac{1}{3}$	3
$P(T = T_0)$	$\frac{6}{8}$	$\frac{2}{8}$
$P(T \leq T_0)$	$\frac{6}{8}$	$\frac{8}{8}$

It would take a similar sort of calculation to find the exact distribution in any application of the Friedman test. For example, in a modest size test with 3 treatments and 10 blocks, the exact distribution would involve examining $(3!)^{10} = 60,466,176$ arrangements. You can see why the chi-square approximation is useful!

Paired Comparisons

As with the Kruskal-Wallis test, when the Friedman test indicates that we are able to reject H_0, we can use a method of paired comparisons to pinpoint where the differences lie. As in earlier discussions of paired comparisons, if we have k populations (treatments), then there are $\binom{k}{2} = \frac{k(k-1)}{2}$ possible pairs of treatments that can be compared that lead to $\binom{k}{2}$ tests of the form:

- H_0: The effects of the ith and jth treatments are the same, versus H_a: The effects of the ith and jth treatments are different.

For a randomized block design, the comparisons are carried out using the *rank-sum differences* $|R_i - R_j|$. The calculations are simplified because the block size is a constant, k, and the sample size for each treatment is always b.

Begin by choosing an overall significance level α' at which to run the test. The significance level of any *one* of the comparisons that will be made is $\frac{\alpha'}{\binom{k}{2}}$. The larger k is, the larger the experimentwise α' must be in order to run the comparisons at a reasonable level of significance α. But when α' is larger, there is a greater risk of rejecting a null hypothesis inappropriately (Type I error).

The test statistic for a two-tailed comparison of the ith and jth treatments is simply

$$Z_{ij} = \frac{|R_i - R_j|}{\sqrt{\frac{bk(k+1)}{6}}}.$$

It can be shown that the test statistic has an approximate standard normal distribution, so Table C.3 is used to determine the critical value $z_{1-\frac{\alpha}{2}}$ for this two-sided test. H_0 is rejected if $Z_{ij} \geq z_{1-\frac{\alpha}{2}}$, or more simply if

$$|R_i - R_j| \geq z_{1-\frac{\alpha}{2}} \sqrt{\frac{bk(k+1)}{6}}.$$

EXAMPLE 9.5. Carry out a set of paired comparisons with $\alpha' = 0.10$ for the data in Example 9.4. Interpret the results.

SOLUTION. The hypotheses are H_0: The attack values for the ith and jth treatments are the same, versus H_a: The attack values for the ith and jth treatments are different. There are a total of $\binom{7}{2} = 21$ comparisons to make. This is the reason for selecting $\alpha' = 0.10$ rather than 0.05. Thus, $\alpha = \frac{0.10}{\binom{7}{2}} = \frac{0.10}{21} = 0.0048$. Therefore, $1 - \frac{\alpha}{2} = 0.9976$ and $z_{0.9976} = 2.82$. H_0 is rejected only if

$$|R_i - R_j| \geq z_{1-\frac{\alpha}{2}} \sqrt{\frac{bk(k+1)}{6}} = 2.82\sqrt{\frac{7(7)(8)}{6}} = 22.79.$$

This means that the rank sums for two different treatments must differ by 22.79 for there to be a significant difference. Look back at the rank sums in Example 9.4. It follows that Treatments 1 through 4 differ from the control model and Treatment 1 also differs from Treatment 6. In particular, notice that none of the flatworms differs from its own model, but the first two models do differ from the control model even though all models were made of brine shrimp and agar. This seems to indicate that the models of the first two flatworms have gained an advantage by Batesian mimicry.

9.4   Problems

1. A physician is interested in comparing the effects of three different allergens in persons extremely sensitive to ragweed via a skin allergy test. Five patients received an injection of an antihistamine. Subsequently, the allergens were applied topically on different sections of the left arm. The areas (in mm^2) of redness, measured after two days, are reported in the table below. Is there a significant difference among the patients' responses to the allergens?

Subject (jth)	Allergen (ith)			$T_{.j}$
	A	**B**	**C**	
1	41	25	40	106
2	42	46	37	125
3	37	45	33	115
4	31	34	42	107
5	56	36	65	157
$T_{i.}$	207	186	217	$T_{..} = 610$

2. Three diets for hamsters were tested for differences in weight gain after a specified period of time. Six inbred laboratory lines were used to investigate the responses of different genotypes to the various diets. The lines were blocked and diets assigned randomly. Are there significant differences among the diets in their ability to facilitate weight gain? The units are grams of increase. Was blocking useful here? Explain.

Inbred lines (jth)	Diet (ith)			$T_{.j}$
	I	**II**	**III**	
A	18	14	15	47
B	16	12	14	42
C	17	16	12	45
D	19	15	14	48
E	12	10	12	34
F	13	11	9	33
$T_{i.}$	95	78	76	$T_{..} = 249$

3. Upon returning home from college you are asked to manage the family hog farm. The first day on the job you are visited by four different feed salesmen each claiming to represent the company with the best hog feed. To test their claims you purchase a small amount of feed from each salesmen and set up a small experiment. From six different litters you select four newly-weaned piglets of the same size and sex. Each piglet is randomly assigned one of the commercial feeds. After 60 days the piglets are reweighed and the weight gain in pounds is recorded below.

Litter	Commercial Feed			
	I	**II**	**III**	**IV**
1	60	50	52	61
2	58	48	50	58
3	49	41	38	47
4	47	40	41	49
5	61	52	49	60
6	57	48	47	57

(a) Analyze the data using analysis of variance. Why was a randomized complete block design used here?

(b) From the analysis would a blocked design be necessary in future feed trials?

4. A limnologist documenting the water quality of a large freshwater lake measured water clarity using a Secchi disc. A Secchi disc is a 20-cm-diameter round disc that is divided into quarters that are alternatively painted black and white. The Secchi disc is lowered into the water column until it can no longer be seen. At that point the depth is recorded. Secchi disc readings give an index of turbidity that, in turn, is affected by dissolved particulates and phytoplankton. Below we have Secchi depth readings (in m) taken at random times during January, May, and September at two locations in this lake. Analyze these data as a factorial design ANOVA to determine if the locations are significantly different and if the clarity changes during the year. From this data analysis how would you recommend this limnologist sample in future years? Explain your rationale.

Location (jth)	Month (ith)		
	January	May	September
North	6.5	6.0	4.3
	7.1	6.6	5.0
	8.3	7.1	4.9
	6.6	8.0	4.5
	7.3	7.5	4.1
South	7.3	7.5	4.3
	9.1	8.0	3.9
	8.0	7.7	3.5
	8.1	8.0	7.1
	8.4	8.3	3.6

5. As a project for an exercise physiology class, a student wanted to test various training methods for their effects on upper body strength. He grouped 15 volunteers according to body type and fitness test performance into five groups. One member of each group was assigned to one of three training methods. Method I consisted of cross training with some pushups and pullups, three times a week. Method II consisted of weight training on four Cybex machines that isolate arms and shoulders, three times a week. Method III consisted of weight training all major muscle groups with twelve Cybex machines, twice a week. The volunteers recorded their maximum bench press weights before training and after 10 weeks of training. The numbers recorded below are the changes in maximum bench press weights (After − Before) in pounds. Analyze the data as a blocked design ANOVA. Are there significant differences among the results of the training methods?

Group	I	II	III
	Training Regime		
1	15	15	20
2	20	18	25
3	9	20	26
4	25	20	30
5	10	22	32

6. An agricultural researcher is developing methods to reduce the use of chemical pesticides on the commercial strawberry crop. His approach is to use alternative planting methods, resistant varieties, and beneficial insects to reduce damage and maintain yield rather than using chemical herbicides, fungicides, and insecticides. This applied ecology approach to agriculture is often termed integrated pest management or IPM. The researcher wishes to compare a modified IPM program to traditional IPM and chemical control. In order to control for variables such as microclimate and soil conditions, he plants 3 plots on each of 6 farms. The plots within each farm are matched as closely as possible in terms of soil type, etc., and the agricultural method is randomly assigned to the plots within each farm. Yields are in pints of marketable berries per plot. From your analysis of the data, is IPM or modified IPM as good as or better than traditional chemical control in terms of marketable strawberry yield? Was blocking important to this analysis? Explain.

Farm (*j*th)	Agricultural method (*i*th)		
	Chemical control	IPM	Modified IPM
I	71	73	77
II	90	90	92
III	59	70	80
IV	75	80	82
V	65	60	67
VI	82	86	85

7. An undesirable side effect of some antihistamines is drowsiness, a consequence of the effect of the drug on the central nervous system. These data come from an experiment to compare the effects a placebo and two antihistamines have on the central nervous system. This was done by measuring what is called the flicker frequency in volunteers who have taken the treatment. There were 9 subjects who took each of the 3 treatments. Flicker frequency values were recorded 6 hours after the drug was administered. *(HSDS, #11)*

Subject	Meclastine	Placebo	Promethazine
1	31.25	33.12	31.25
2	26.63	26.00	25.87
3	24.87	26.13	23.75
4	28.75	29.63	29.87
5	28.63	28.37	24.50
6	30.63	31.25	29.37
7	24.00	25.50	23.87
8	30.12	28.50	27.87
9	25.13	27.00	24.63

(*a*) Is there a significant difference in the effects of the treatments? Use a Friedman test.

(*b*) If appropriate, carry out a set of paired comparisons on the data or explain why such an analysis is not needed.

8. The nature of the light reaching the rainforest floor is quite different from that in the canopy and the understory. One difference is wavelength. A study was conducted to examine the adaptations to light wavelength of 10 different fern species found on the rainforest floor. Each block consisted of 4 mature ferns of the same species. Each fern was grown in the dark for two weeks. One plant of each species was assigned to each of the 4 light treatments. Each plant was exposed to a single dose of light (wavelength measured in nm), returned to the dark, and 24 hours later the increase in the cross-sectional area of the fern tip was measured (in μm^2). Was there a significant difference in growth among the light treatments? Analyze with a Friedman test.

Species	400 nm	500 nm	600 nm	700 nm
A	960	800	950	910
B	780	750	700	630
C	1130	1040	1050	1000
D	670	740	630	610
E	790	770	700	720
F	960	720	810	820
G	1110	1000	1040	980
H	930	970	910	860
I	920	850	840	820
J	880	860	920	830

9. Carry out a set of paired comparisons for the data in Example 9.3.

10. In an agricultural experiment using barley, the effects of various commercial fertilizers and planting densities on yield were studied. Six different commercial fertilizers were used and the barley was planted in three different densities. The plots used were quite uniform with respect to soil characteristics, drainage, etc. The yields in kilograms per plot are recorded in the following table. Analyze these data completely as a factorial design two-way ANOVA. Discuss the results in terms of interactions and main effects. What combination of fertilizer and planting density would you recommend to barley growers? Why?

Planting density	Commercial fertilizer					
	I	II	III	IV	V	VI
A	27	26	31	30	25	30
	29	25	30	27	25	32
	26	24	30	33	26	28
	26	29	33	30	24	30
B	30	28	31	32	28	31
	30	29	31	35	29	30
	28	30	30	29	28	32
	32	25	32	32	27	31
C	33	33	35	34	30	32
	33	30	33	34	29	30
	34	34	37	33	31	30
	32	35	35	35	30	34

11. For a project in her physiology class, a student investigates two lifestyle factors affecting blood pressure: diet and smoking. For a number of her classmates (all female) she determines their smoking habits, the amount of sodium in their diet, and records their systolic blood pressure (in mmHg).

	Sodium intake	
	Moderate	High
Nonsmokers	129	132
	125	137
	129	130
	132	148
	125	154
	128	158
Smokers	140	165
	126	152
	120	140
	137	167
	142	142
	147	177

Analyze the data as a factorial design ANOVA. Are mean separation techniques indicated? Explain.

12. A study on cortex development in the Norway rat investigated both diet and physical environment as factors in brain development. Twenty-four two-week old rats were randomly placed in one of two environments, stimulating or control, and they were fed either commercial rat chow or an enriched, fresh food diet. After four months the rats were sacrificed and their individual cortex weights (mg) recorded below.

	Environment	
Food	Stimulating	Control
Enriched	730	710
	710	690
	705	685
	725	680
	690	700
	730	695
Control	705	660
	695	655
	680	690
	680	670
	695	640
	700	675

Analyze as a factorial design two-way ANOVA. Include mean separation techniques if appropriate.

13. The following data are from a famous Cabernet Sauvignon wine tasting held in
1976 in Paris to introduce California wines to experienced French-wine tasters.
This taste test was partly responsible for the ascension of California wines because
one of them was ranked first. Use a Friedman test to determine if there were any
statistically significant differences among the wines. Note that the highest score
should be ranked first. If paired comparisons are required, use $\alpha = 0.001$. (See:
Orley Ashenfelter and Richard Quandt. www.liquidasset.com/tasting.html.)

The wines in the 1976 tasting

Wine	Name	Origin	Rank
A	Stag's Leap 1973	U.S.	1st
B	Ch. Mouton Rothschild 1970	France	3rd
C	Ch. Montrose 1970	France	2nd
D	Ch. Haut Brion 1970	France	4th
E	Ridge Mt. Bello 1971	U.S.	5th
F	Loville-las-Cases 1971	France	7th
G	Heitz Marthas Vineyard 1970	U.S.	6th
H	Clos du Val 1972	U.S.	10th
I	Mayacamas 1971	U.S.	9th
J	Freemark Abbey 1969	U.S.	8th

Judge	A	B	C	D	E	F	G	H	I	J
Pierre Brejoux	14	16	12	17	13	10	12	14	5	7
A.D. Villaine	15	14	16	15	9	10	7	5	12	7
Michel Dovaz	10	15	11	12	12	10	11.5	11	8	15
Pat. Gallagher	14	15	14	12	16	14	17	13	9	15
Odette Kahn	15	12	12	12	7	12	2	2	13	5
Ch. Millau	16	16	17	13.5	7	11	8	9	9.5	9
Raymond Oliver	14	12	14	10	12	12	10	10	14	8
Steven Spurrier	14	14	14	8	14	12	13	11	9	13
Pierre Tari	13	11	14	14	17	12	15	13	12	14
Ch. Vanneque	16.5	16	11	17	15.5	8	10	16.5	3	6
J.C. Vrinat	14	14	15	15	11	12	9	7	13	7
Total	155.5	155	150	145.5	133.5	123	114.5	111.5	107.5	106

14. Four different sunblock formulations (treatments) were applied in randomized
strips to the backs of 5 subjects. At the end of a 4-hour exposure to bright sunlight,
each strip was evaluated for sunburn damage and ranked, with rank 1 indicating
the least damage and rank 4 indicating the most damage on the particular subject.
Carry out an appropriate test to determine whether there were differences among
the treatments, and if so, locate the treatments that are different.

Subject	Sunblock			
	I	II	III	IV
1	3	4	1	2
2	1	4	2	3
3	3	4	1	2
4	2	4	1	2
5	1	3	2	4

15. As part of a much larger study on neonatal health in America, the birth weights of 36 newborn full-term infants were recorded in kilograms. The principal factors of interest were the maternal age and the smoking status of the mother. Analyze the data below appropriately, assuming normality of the birth weights.

Age of Mother	Smoking Status of Mother		
	Non-smoker	Moderate Smoker	Heavy Smoker
Young < 25 years	3.63	3.35	3.23
	3.45	3.30	3.05
	3.40	3.31	3.18
	3.51	3.20	3.40
	3.60	3.25	3.19
	3.55	3.43	3.15
	$\overline{X}_{11.} = 3.523$	$\overline{X}_{21.} = 3.307$	$\overline{X}_{31.} = 3.200$
Older > 25 years	3.61	3.08	3.05
	3.49	3.10	3.00
	3.58	3.37	3.18
	3.67	3.28	3.20
	3.40	3.19	2.99
	3.18	3.31	3.25
	$\overline{X}_{12.} = 3.488$	$\overline{X}_{22.} = 3.222$	$\overline{X}_{32.} = 3.112$
$\sum_i \sum_j \sum_k X_{ijk}^2 = 395.336$		$\sum_i \sum_j \sum_k X_{ijk} = T_{...} = 119.110$	

16. A horticultural scientist evaluated six newly developed varieties of tomatoes for yield using a randomized complete block design with each block representing a different experimental plot. Data are in kilograms of yield per block. Are there significant differences in yield among the six new varieties?

Block	Variety						$T_{.j}$
	A	B	C	D	E	F	
1	32.6	42.0	56.1	62.0	29.0	40.7	262.6
2	38.4	40.7	49.0	61.7	47.0	49.9	286.7
3	46.5	48.2	50.3	60.3	67.0	70.0	342.3
4	41.0	40.0	54.0	49.0	42.2	44.7	270.9
5	39.9	35.6	51.0	58.8	45.5	42.3	273.1
$T_{i.}$	198.4	206.5	260.4	291.8	230.7	247.6	$T_{..} = 1435.4$
	$\sum_i \sum_j X_{ij}^2 = 71,548.36$			$N = 30$			

17. A number of pairs of animals exhibit *cleaning symbiosis* in which so-called cleaners remove and eat ectoparasites from other animals called clients. You may have seen birds eat insects on cattle, for example. In the symbiosis between the cleaner wrasse, *Labroides dimidiatus*, and client fish such as parrotfish, wrasses eat parasites, mostly gnathiid isopods and monogeneans. The monogenea are hermaphroditic, parasitic flatworms that are found most often (but by no means exclusively) on the gills of fish.

In some cleaning symbioses, there is evidence that the cleaners actually prefer to eat client tissue, e.g., oxpecker birds feed on blood from hippopotami wounds. When this is the case, the clients must develop behaviors to control the activity of the cleaners, or else the relationship will cease to be mutualistic.

An experiment was conducted to test the dietary preferences of cleaner wrasses. Each of 14 wrasses was tested three times. In each trial, three samples of four food items were presented to each wrasse: (1) mucus from parrotfish, (2) monogeneans, (3) gnathiids, and (4) a control of boiled flour glue. The items were placed randomly on a plate which had a grid of 12 cells, with only one food item per cell. The plate was then placed inside each cleaner's aquarium and the first three food items eaten in each of the trials were recorded for each wrasse. The data are presented below. (Based on the experimental design of Grutter, A. and R. Bshary. 2003. Cleaner wrasse prefer client mucus: Support for partner control mechanisms in cleaning interactions. *Proc. R. Soc. London* B (Suppl.). S242–S244.)

Wrasse	Mucus	Monogeneans	Gnathiids	Control
1	4	3	1	1
2	6	2	0	1
3	3	4	2	0
4	5	3	1	0
5	3	5	1	0
6	4	3	2	0
7	3	2	2	2
8	3	4	0	2
9	4	3	2	0
10	2	4	1	2
11	3	2	3	1
12	5	1	3	0
13	4	3	0	2
14	3	4	1	1

Analyze nonparametrically to determine if there were differences among the treatments (diet preferences). If so, locate the differences and interpret the results.

18. **QCR**. True or False. Justify your answer.

(a) Blocking in two-way ANOVA removes or controls for variability that is explainable, but not germane to the null hypothesis of interest.

(b) In two-way ANOVA all F tests are right-tailed regardless of the alternative hypotheses.

(c) A randomized complete block design ANOVA allows one to investigate interactions between the two types of main effects.

(d) In a factorial design two-way ANOVA with a significant interaction, the cell means which combine the two types of main effects can be separated with an SNK test.

(e) Of the ANOVA techniques presented in Chapters 8 and 9 only the factorial design ANOVA tests for synergy or interference among treatments.

(f) Blocking will always increase the power of an ANOVA.

(g) The Friedman k-sample test controls an extraneous variable in a fashion similar to the blocking in a randomized complete block design ANOVA.

10

Linear Regression and Correlation

Concepts in Chapter 10:

- Differences Between Regression and Correlation
- Simple Linear Regression

  — Model Assumptions
  — Least Squares Regression Equation
  — ANOVA Test of Significance
  — Confidence Intervals for β and for $\mu_{Y|X}$

- Simple Linear Correlation

  — Model Assumptions
  — Pearson Correlation Coefficient, ρ
  — Coefficient of Determination
  — t tests for H_0: $\rho = 0$
  — Confidence Intervals for ρ and for ζ

- Correlation Analysis Based on Ranks

  — Kendall's Measure of Correlation τ
  — Test for H_0: $\tau = 0$
  — Spearman's Coefficient r_s
  — Test for H_0: $r_s = 0$

Introduction to Linear Regression and Correlation

In this chapter we present analyses to determine the *strength of relationship* between two variables. In the case of linear regression, we will examine the amount of variability in one variable (Y, the dependent variable) that is explained by changes in another variable (X, the independent variable). Specifically, we will look for straight-line or linear changes in Y as X changes. Regression analysis is usually done in situations in which we have control of the variable X and can measure it essentially without error. While there may be curvilinear relationships between variables, e.g., exponential or polynomial, we restrict our consideration to relationships that appear to be linear. Further, we will look at simple linear models rather than multiple variate models. In other words, we explore the relationship between a *single* X and the behavior of Y.

Correlation analysis is used when both variables are experimental and measured with error. It is more preliminary than regression analysis and generally measures the corelationship between two variables of interest. We will present the Pearson product-moment correlation and two nonparametric analogs, the Kendall and Spearman correlation coefficients, as methods of analysis in these situations.

Let's consider two examples to further highlight the differences between regression and correlation analysis.

EXAMPLE 10.1. An introductory biology student wishes to determine the relationship between temperature and heart rate in the common leopard frog, *Rana pipiens*. He manipulates the temperature in 2° increments ranging from 2 to 18°C and records the heart rate (beats per minute) at each interval. His data are presented below.

Recording number	X Temperature (°Celsius)	Y Heart rate (bpm)
1	2	5
2	4	11
3	6	11
4	8	14
5	10	22
6	12	23
7	14	32
8	16	29
9	18	32

How should he proceed to describe the relationship between these variables (temperature and heart rate)? Clearly the two variables have functional dependence—as the temperature increases the heart rate increases. Here the temperature is *controlled* by the student and can take exactly the same values in another experiment with a different frog (see Figure 10.1). Temperature is the independent variable or "predictor"

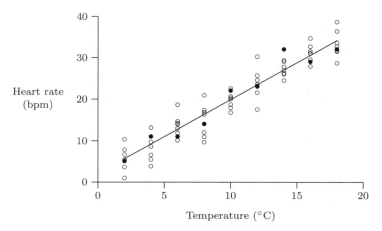

Temperature (°C)

FIGURE 10.1. The temperature and heart-rate data collected for Example 10.1 are marked by solid points ●. These data lie approximately on the line shown. If the experiment were repeated six more times using the same temperatures, the data would be similar, but not precisely the same (see hollow points ○). The line approximating the original data also is a good approximation of all the data. The points line up vertically because the researcher controls the temperatures at which the experiment is run.

variable. Heart rate is determined by temperature and is, therefore, the dependent variable or "response" variable. These data are correctly analyzed by regression analysis with a goal of predicting heart rate at various temperatures.

EXAMPLE 10.2. A malacologist interested in the morphology of West Indian chitons, *Chiton olivaceous*, measured the length (anterior-posterior) and width of the eight overlapping plates composing the shell of 10 of these animals. Her data are presented below:

Animal	Length (cm)	Width (cm)
1	10.7	5.8
2	11.0	6.0
3	9.5	5.0
4	11.1	6.0
5	10.3	5.3
6	10.7	5.8
7	9.9	5.2
8	10.6	5.7
9	10.0	5.3
10	12.0	6.3

This data set is fundamentally different from the data in Example 10.1 because neither variable is under the malacologist's control. To try to predict length from width is as logical (or illogical) as to try to predict width from length. Both variables are free to vary (see Figure 10.2). A correlational study is more appropriate here than

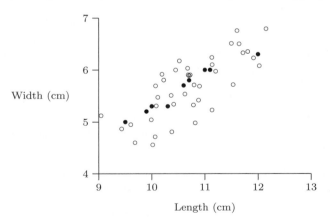

FIGURE 10.2. The chiton length and width data collected for Example 10.2 are marked by solid points •. If 40 more chitons were measured, the pattern of data would be similar, but not precisely the same (see hollow points ○). Because the researcher takes measurements from a random sample of chitons, she is not able to control the length (or width) of the chitons. Therefore, the data points do not line up vertically as in Figure 10.1.

a regression analysis. Because some of the calculations are similar, regression and correlation are often confused. We will emphasize the **Model I** or fixed effect nature of regression to help discriminate between these two protocols.

10.1   Simple Linear Regression

The goal of this analysis is to describe the functional relationship between two variables as a straight line, where X is the independent variable and Y is the dependent variable. We assume X to be measured *without error* and to be *fixed* and *repeatable*. Since Y is the dependent variable, it is free to vary. If data, when graphed, appear to have a linear relationship, we would like to know the true parametric nature of that linear equation.

Linear Model Assumptions

1. X's are fixed and measured without error.

2. The expected or mean value for the variable Y for a given value of X is described by a linear function
$$E(Y) = \mu_{Y|X} = \alpha + \beta X,$$
where α and β are constant real numbers and $\beta \neq 0$.

   The expected value of Y depends on X and the parameters α and β. Note these α and β are not the same as the values used earlier for Type I error and Type II error. In our present context they represent the *intercept* and *slope*, respectively, of the linear relationship between Y and X.

3. For any fixed value of X, there may be several corresponding values of the response variable Y that were measured. (For a fixed temperature, think of making several measurements of the frog's heart rate.) However, we assume that for any such given value X_i, the Y_i's are independent of each other and normally distributed. (See the vertical arrangement of the data in Figure 10.1.) We can represent each Y_i value as
$$Y_i = \alpha + \beta X_i + \epsilon_i \qquad \text{or} \qquad \mu_{Y|X} + \epsilon_i.$$

   Y_i is described as the expected value $(\alpha + \beta X_i)$ plus a deviation (ϵ_i) from that expectation. We assume that ϵ_i's are normally distributed error terms with a mean of zero.

4. The variances of the distributions of Y for different values of X are assumed to be equal. Statisticians would say they are homoscedastic!

To describe the experimental regression relationship between Y and X we need to do the following:

1. Graph the data to ascertain that an *apparent* linear relationship exists.

2. Find the best fitting straight line for the data set.

3. Test whether or not the fitted line explains a significant portion of the variability in Y, i.e., test whether the linear relationship is real or not.

Let's consider each step in more detail.

Graphing the Data

It is always useful to make a preliminary scatterplot of the data to get a sense of whether there is any relation between the two variables and if so, what that relation might be.

Figure 10.3 a) indicates that there is no meaningful relationship between X and Y. Large values of Y are associated with both large *and* small values of X.

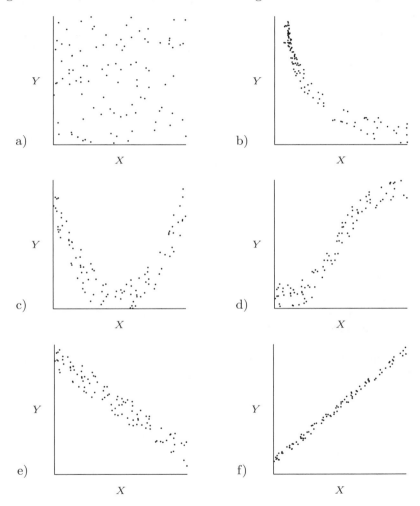

FIGURE 10.3.   Six different data scatterplots.

Figure 10.3 b), c), and d) indicate a relationship between the variables but *not* a straight-line relationship. If they can be transformed by a mathematical function to a linear graph, the regression analysis described below can be done on the transformed data.

Figure 10.3 e) shows a negative linear relationship between Y and X (i.e., as X increases, Y decreases). While the data points are not exactly in a straight line, they give the definite impression of linearity. Figure 10.3 f) demonstrates a very strong positive linear relationship between the variables (i.e., as X increases, Y increases) with little deviation from a straight line. Linear regression analysis is appropriate only for these last two cases.

Fitting the Best Straight Line

Once we determine that regression analysis is appropriate, we have the problem of determining which line best fits the data. For the data plotted in Figure 10.4, three lines have been fitted, a, b, and c. Obviously c "fits" the data better than a or b. The data have a positive trend: as X increases, so does Y. Line b would completely misrepresent the relationship between Y and X and line a implies that there is no relationship.

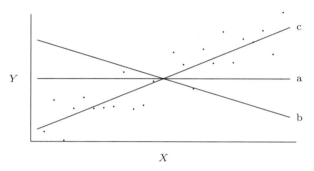

FIGURE 10.4. How are we to determine which line best fits these data?

Now consider the same data and the two lines c and d in Figure 10.5. This time both lines capture the positive trend of the data. To decide which of these lines best fits the data—or whether some other line is an even better fit—requires some standard criterion by which we can judge the fit. Below we will develop this criterion and a method to produce the line which best satisfies it.

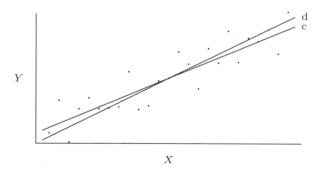

FIGURE 10.5. How are we to determine which of these two similar lines best fits these data?

The goal of regression is to predict values of Y. Begin, for the moment, by ignoring the variable X and think in terms of the earlier one-sample analyses. The predicted value of Y would be $E(Y) = \mu_Y$, which we could estimate by using the sample mean

$\overline{Y}$. So we start our analysis by drawing the horizontal line that passes through $\overline{Y}$. This line has equation $\hat{Y} = \overline{Y}$. See Figure 10.6. We use the notation $\hat{Y}$ (read as "Y hat") rather than Y to indicate that these are *predicted* values of Y rather than actual or observed values. The line $\hat{Y} = \overline{Y}$ has a slope of 0, that is, it is parallel to the X-axis. For $\hat{Y} = \overline{Y}$, the predicted value of Y is the same for any value of X. If this were actually the case, it would mean that there is no relationship between Y and X because the value of Y would not depend on (change with) the value of X.

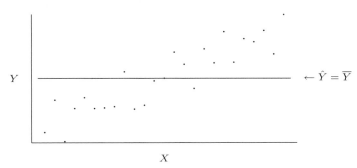

FIGURE 10.6.   Fitting $\hat{Y} = \overline{Y}$ to the data.

However, we believe that Y *does* depend on X and we can measure the difference between our preliminary predictions of Y and the actual values of Y as follows. Draw a vertical segment from each data point to the line $\hat{Y} = \overline{Y}$. The length of any such segment is $(Y_i - \overline{Y})$. See Figure 10.7. If we square and sum these deviations, we have

$$\sum_i (Y_i - \overline{Y})^2 = \text{total sum of squares for } Y.$$

This number divided by $n-1$ would be the variance of Y's *still disregarding* X. Recall that the variability in Y's isn't under our control and we would like to account for as much of it as possible.

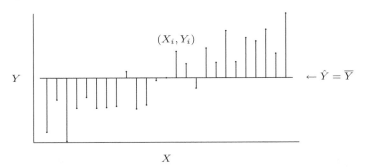

FIGURE 10.7. Draw vertical segments from each data point to the line $Y = \overline{Y}$.

Now draw a sloped line $\hat{Y} = a + bX$ that appears to fit the data. Again, draw a vertical segment from each data point to the sloped line. See Figure 10.8. The length of a segment is $(Y_i - \hat{Y}_i)$. If we square and sum these deviations, we have $\sum_i (Y_i - \hat{Y}_i)^2$. This sum of squares is clearly smaller than $\sum_i (Y_i - \overline{Y})^2$ calculated from Figure 10.7

because the line segments are shorter. The variability that remains after drawing a line that "fits" the data is considered the *residual* or *unexplainable* variability in the system. *The best fit line is the one whose intercept a and slope b together minimize this residual error.*

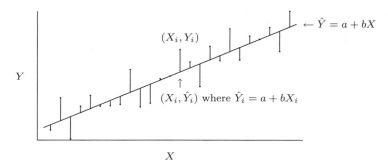

FIGURE 10.8. Draw vertical segments from each data point to the sloped line.

As in the previous two chapters, we can partition sums of squares to quantify the residual error. Each Y_i can be expressed as

$$Y_i = \overline{Y} + (Y_i - \hat{Y}_i) + (\hat{Y}_i - \overline{Y})$$

or, equivalently,

$$(Y_i - \overline{Y}) = (Y_i - \hat{Y}_i) + (\hat{Y}_i - \overline{Y}). \tag{10.1}$$

This is illustrated in Figure 10.9.

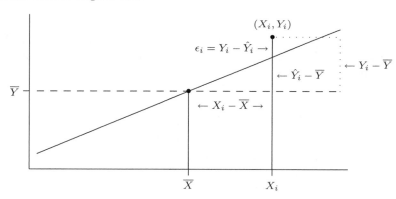

FIGURE 10.9. $Y_i = \overline{Y} + (Y_i - \hat{Y}_i) + (\hat{Y}_i - \overline{Y}).$

Squaring both sides of (10.1) and summing, we obtain

$$\sum (Y_i - \overline{Y})^2 = \sum [(Y_i - \hat{Y}_i) + (\hat{Y}_i - \overline{Y})]^2$$
$$= \sum (Y_i - \hat{Y}_i)^2 + 2\sum (Y_i - \hat{Y}_i)(\hat{Y}_i - \overline{Y}) + \sum (\hat{Y}_i - \overline{Y})^2.$$

As in an ANOVA, using algebra one can show that the cross product term

$$2\sum (Y_i - \hat{Y}_i)(\hat{Y}_i - \overline{Y}) = 0.$$

This leads to the following identity.

FORMULA 10.1. The Sum of Squares (total) is

$$\sum(Y_i - \overline{Y})^2 = \sum(\hat{Y}_i - \overline{Y})^2 + \sum(Y_i - \hat{Y}_i)^2$$

Sum of Squares total	=	Sum of Squares due to regression	+	Sum of Squares residual or error
SS_{Total}		SS_R		SS_E.

We can now state the criterion by which we will determine which line best fits the data: The best regression line is the line that maximizes SS_R and minimizes SS_E, i.e., makes the vertical segments in Figure 10.8 as small as possible. The fitted line

$$\hat{Y} = \hat{\mu}_{Y|X} = a + bX$$

is the sample regression equation that estimates the parametric relationship given earlier, $\mu_{Y|X} = \alpha + \beta X$. Here a is an estimate of the true intercept α, and b is an estimate of the true slope β. Because this line minimizes the sum of squares error term in Formula 10.1, it is known as the **least squares regression line.**

It can be shown (see Appendix A.6) that the best estimate for the intercept is $a = \overline{Y} - b\overline{X}$. Making this substitution for a we obtain

$$\hat{Y} = \hat{\mu}_{Y|X} = (\overline{Y} - b\overline{X}) + bX.$$

This can be further simplified by combining the terms with the factor b,

$$\hat{Y} = \overline{Y} + b(X - \overline{X}).$$

To make use of this equation, we need only know which slope b minimizes the SS_E term of Formula 10.1. This can be accomplished using a technique called the **method of least squares**.

FORMULA 10.2. The **least squares regression equation** is

$$\hat{Y} \text{ or } \hat{\mu}_{Y|X} = \overline{Y} + b(X - \overline{X}).$$

The slope b determined by the **method of least squares** is

$$b = \frac{\sum X_i Y_i - \frac{(\sum X_i)(\sum Y_i)}{n}}{\sum X_i^2 - \frac{(\sum X_i)^2}{n}} = \frac{SS_{XY}}{SS_X}.$$

The numerator SS_{XY} of the slope equation is called the **corrected cross products of X and Y** and is algebraically equivalent to $\sum(Y_i - \overline{Y})(X_i - \overline{X})$. The denominator is the corrected sum of squares of the independent variable X and is denoted by SS_X.

This form of the regression equation is the one found most often in statistical representations. With it each $\hat{Y}$, or predicted value of Y, is given as the mean of Y plus a deviation due to regression. In this form the regression relationship is easier to use in the tests of hypothesis presented later. One derivation of the equation for the slope b requires elementary calculus and is left to Appendix A.6. However, note that the slope b determined by the method of least squares is the corrected cross products SS_{XY} over the corrected sum of squares for X. It is worth repeating that this equation produces the best slope for a linear data set in the sense that the sum of

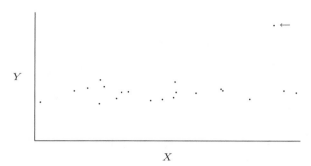

FIGURE 10.10. Why would a positive slope be indicated for these data? Is there a linear relation between X and Y?

squares error term is minimized and the sum of squares regression term in Formula 10.1 is maximized.

If Formula 10.2 gives the best fitting regression equation, we need a method to test the statistical significance of this equation. To see why, consider the data that are graphed in Figure 10.10. They can be fit with a regression line whose slope would be positive because of the data point indicated by the arrow. Would this regression line demonstrate a real linear relationship between X and Y? Probably not. For the equation to be meaningful it must explain a significant portion of the variability in Y as changes in X.

The test of significance derives from analysis of variance and a partitioning of the sums of squares of Y in regression that is analogous to the partition that we did for one-way ANOVA in Chapter 8. We begin by expressing the sum of squares due to regression from Formula 10.1,

$$\sum (Y_i - \overline{Y})^2 = \sum (\hat{Y}_i - \overline{Y})^2 + \sum (Y_i - \hat{Y}_i)^2,$$
$$\text{SS}_{\text{Total}} = \text{SS}_{\text{R}} + \text{SS}_{\text{E}},$$

in a more convenient form. We have

$$\begin{aligned} \text{SS}_{\text{R}} = \sum (\hat{Y}_i - \overline{Y})^2 &= \sum \left[(\overline{Y} + b(X_i - \overline{X})) - \overline{Y} \right]^2 \\ &= \sum b^2 (X_i - \overline{X})^2 \\ &= b^2 \sum (X_i - \overline{X})^2 \\ &= b^2 \text{SS}_X. \end{aligned} \tag{10.2}$$

Since $b = \frac{\text{SS}_{XY}}{\text{SS}_X}$,

$$\text{SS}_{\text{R}} = \left(\frac{\text{SS}_{XY}}{\text{SS}_X} \right)^2 \text{SS}_X = \frac{(\text{SS}_{XY})^2}{\text{SS}_X} = b\text{SS}_{XY}. \tag{10.3}$$

Consequently, we have the following computational expressions.

FORMULAS 10.3. The sum of squares (total) is

$$\text{SS}_{\text{Total}} = \sum (Y_i - \overline{Y})^2 = \sum Y_i^2 - \frac{\left(\sum Y_i \right)^2}{n} = \text{SS}_Y.$$

The sum of squares due to regression is

$$\text{SS}_\text{R} = b^2 \text{SS}_X = b \text{SS}_{XY}.$$

The residual sum of squares is

$$\text{SS}_\text{E} = \sum (Y_i - \hat{Y}_i)^2 = \text{SS}_\text{Total} - \text{SS}_\text{R} = \text{SS}_\text{Total} - b \text{SS}_{XY}.$$

The Global Test of Significance for Regression

The global test of significance for regression is an ANOVA. The hypotheses are

- H_0: The variation in Y is *not* explained by a linear model, i.e., $\beta = 0$.

- H_a: A significant portion of the variation in Y is *explained* by a linear model, i.e., $\beta \neq 0$.

TABLE 10.1. The ANOVA table for a regression analysis

Source of variation	SS	DF	MS	E(MS)	F	c.v.
Regression	SS_R	1	MS_R	$\sigma_Y^2 + \beta^2 \text{SS}_X$	$\frac{\text{MS}_\text{R}}{\text{MS}_\text{E}}$	See Table C.7
Error	SS_E	$n-2$	MS_E	σ_Y^2		
Total	SS_Total	$n-1$				

The expected values listed in Table 10.1 are presented without proof here, but they provide the intuition needed for this test of hypothesis. If H_0 is *true*,

$$E(F) = \frac{\sigma_Y^2}{\sigma_Y^2} = 1,$$

and if H_a is true,

$$E(F) = \frac{\sigma_Y^2 + \beta^2 \text{SS}_X}{\sigma_Y^2} > 1.$$

The term $\beta^2 \text{SS}_X$ is always positive or zero (zero under H_0 and positive under H_a).

Return to Example 10.1, the relationship between heart rate and temperature in Rana pipiens. *Complete the regression analysis.*

SOLUTION. We begin with a scatterplot of the data (see Figure 10.11) that indicates that there is an apparent linear relationship between heart rate and temperature. Next, carry out the preliminary calculations for the regression analysis:

$$n = 9$$

$$\sum X = 90 \qquad \sum Y = 179$$

$$\overline{X} = 10.0 \qquad \overline{Y} = 19.9$$

$$\sum X^2 = 1140 \qquad \sum Y^2 = 4365 \qquad \sum XY = 2216$$

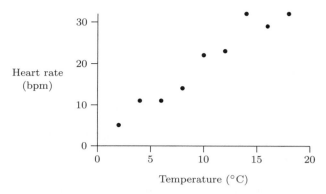

FIGURE 10.11. Scatterplot of temperature and heart rate in *Rana pipiens*.

Now we can calculate the regression coefficient or slope:

$$b = \frac{\text{SS}_{XY}}{\text{SS}_X} = \frac{\sum XY - \frac{(\sum X)(\sum Y)}{n}}{\sum X^2 - \frac{(\sum X)^2}{n}} = \frac{2216 - \frac{(90)(179)}{9}}{1140 - \frac{(90)^2}{9}} = \frac{426}{240} = 1.78.$$

The least squares best fit equation for the data in Figure 10.11 is

$$\hat{Y}_i = \widehat{\mu}_{Y|X} = \overline{Y} + b(X_i - \overline{X}) = 19.9 + 1.78(X_i - 10.0).$$

To plot the line, find two values using two temperatures within the range of values studied. We use $X = 5°$ and $X = 15°$:

$$\widehat{\mu}_{Y|5°} = 19.9 + 1.78(5 - 10.0) = 11.0$$
$$\widehat{\mu}_{Y|15°} = 19.9 + 1.78(15 - 10.0) = 28.8.$$

Place the two points of the form $(X, \widehat{\mu}_{Y|X})$ on the graph and draw the line between them extending to the smallest X and largest X *in the study*. The line drawn should pass through $(\overline{X}, \overline{Y})$. See Figure 10.12.

After the regression equation is calculated, we need to test whether or not it explains a significant portion of the variability in the Y's. The hypotheses are H_0: $\beta = 0$ and H_a: $\beta \neq 0$. Again, some preliminary calculations are required:

$$\text{SS}_{\text{Total}} = \sum Y^2 - \frac{(\sum Y)^2}{n} = 4365 - \frac{(179)^2}{9} = 804.89$$
$$\text{SS}_{\text{R}} = b^2 \text{SS}_X = (1.78)^2(240) = 760.42$$
$$\text{SS}_{\text{E}} = \text{SS}_{\text{Total}} - \text{SS}_{\text{R}} = 804.89 - 760.42 = 44.47.$$

Now we summarize this information in an ANOVA table:

Source of variation	Sum of squares	df	MS	*F*	c.v.
Regression	760.42	1	760.42	119.75	5.59
Error	44.47	7	6.35		
Total	804.89	8			

Since $119.75 \gg 5.59$, we are confident that a significant portion of the variability in heart rate is explained by linear regression on temperature.

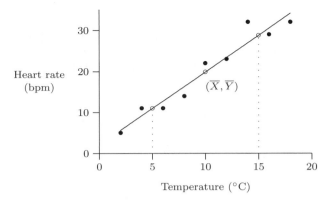

FIGURE 10.12. The least squares regression line is drawn through the points $(5°, \widehat{\mu}_{Y|5°})$ and $(15°, \widehat{\mu}_{Y|15°})$ whose Y coordinates are determined from the regression equation and should pass through $(\overline{X}, \overline{Y})$.

Confidence Intervals for the Linear Model

Since b is a point estimate of the parametric slope β, we would like to have some confidence about this parameter's true value. The variance of b is

$$\sigma_b^2 = \frac{1}{\mathrm{SS}_X}\sigma_Y^2.$$

The estimate of σ_b^2 is

$$s_b^2 = \frac{1}{\mathrm{SS}_X}\left(\frac{\mathrm{SS_E}}{n-2}\right) = \frac{\mathrm{MS_E}}{\mathrm{SS}_X}. \tag{10.4}$$

Consequently,

FORMULA 10.4. The standard error of b is

$$s_b = \sqrt{\frac{\mathrm{MS_E}}{\mathrm{SS}_X}}.$$

Now we can make use of (4.3) to obtain the confidence interval for β.

FORMULA 10.5. A $(1-\alpha)100\%$ confidence interval for the slope β of the least squares regression line is determined by

$$C\left(b - t_{1-\frac{\alpha}{2}}s_b \le \beta \le b + t_{1-\frac{\alpha}{2}}s_b\right) = 1 - \alpha$$

with $\nu = n - 2$. Its endpoints are

$$L_1 = b - t_{1-\frac{\alpha}{2}}s_b \qquad \text{and} \qquad L_2 = b + t_{1-\frac{\alpha}{2}}s_b.$$

Again using (4.3) we determine confidence intervals for the values $\mu_{Y|X}$ (the true mean of the Y's for a given value of X). This time, the standard error of $\widehat{\mu}_{Y|X}$ is

$$s_{\hat{Y}_i} = \sqrt{\frac{\mathrm{SS_E}}{n-2}\left[\frac{1}{n} + \frac{(X_i - \overline{X})^2}{\sum(X_i - \bar{X})^2}\right]}.$$

Notice that $s_{\hat{Y}_i}$ increases as X_i moves away from $\overline{X}$. This is because when X_i is near the mean $\overline{X}$, an erroneous slope in the regression equation produces only a small vertical error in Y. But when X_i is farther from the mean $\overline{X}$, the same erroneous slope produces much larger vertical errors. See Figure 10.13.

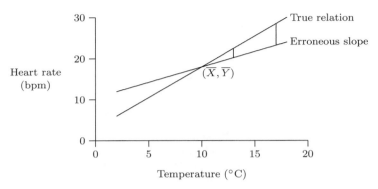

FIGURE 10.13. The size of the vertical error with an erroneous slope in a regression line (vertical segments) increases as X moves away from $\overline{X}$.

FORMULA 10.6. A $(1 - \alpha)100\%$ confidence interval for $\mu_{Y|X}$ is determined by

$$C\left(\widehat{\mu}_{Y|X} - t_{1-\frac{\alpha}{2}} s_{\hat{Y}} \leq \mu_{Y|X} \leq \widehat{\mu}_{Y|X} + t_{1-\frac{\alpha}{2}} s_{\hat{Y}}\right) = 1 - \alpha$$

with $\nu = n - 2$. Its endpoints are

$$L_1 = \widehat{\mu}_{Y|X} - t_{1-\frac{\alpha}{2}} s_{\hat{Y}} \qquad \text{and} \qquad L_2 = \widehat{\mu}_{Y|X} + t_{1-\frac{\alpha}{2}} s_{\hat{Y}}.$$

EXAMPLE 10.3. For Example 10.1, find the following:

- 95% confidence interval for β.
- 95% confidence interval for $\mu_{Y|9°}$, the mean heart rate at $9°$C.
- 95% confidence interval for $\mu_{Y|17°}$, the mean heart rate at $17°$C.

SOLUTION. Begin by finding the standard error of b:

$$s_b^2 = \frac{\text{MSE}}{\text{SS}_X} = \frac{6.35}{240} = 0.026 \implies s_b = \sqrt{0.026} = 0.16.$$

Therefore, the 95% confidence interval for β has endpoints

$$L_1 = 1.78 - 2.365(0.16) = 1.40$$
$$L_2 = 1.78 + 2.365(0.16) = 2.16.$$

While our estimate of the slope is 1.78, we are 95% confident that the true slope β is between 1.40 and 2.16. As expected from the ANOVA table, 0 is not included in this interval.

To find a 95% confidence interval for $\mu_{Y|9°}$ we must find $\hat{Y}_i = \widehat{\mu}_{Y|9°}$ using the regression equation:

$$\hat{Y}_i = \widehat{\mu}_{Y|9°} = 19.9 + 1.78(9 - 10.0) = 18.12.$$

Next, evaluate the standard error

$$s_{\hat{Y}_i} = \sqrt{\text{MS}_\text{E}\left[\frac{1}{n} + \frac{(X_i - \overline{X})^2}{\sum(X_i - \bar{X})^2}\right]} = \sqrt{6.35\left[\frac{1}{9} + \frac{(9 - 10.0)^2}{240}\right]} = 0.86.$$

So from Formula 10.6, the confidence interval limits are

$$L_1 = \hat{Y}_i - t_{1-\frac{\alpha}{2}} s_{\hat{Y}_i} = 18.12 - 2.365(0.86) = 16.09$$
$$L_2 = \hat{Y}_i + t_{1-\frac{\alpha}{2}} s_{\hat{Y}_i} = 18.12 + 2.365(0.86) = 20.15.$$

The confidence interval length is $20.15 - 16.09 = 4.06$ bpm.
    To find a 95% confidence interval for $\mu_{Y|17^\circ}$, first determine $\hat{Y}_i = \hat{\mu}_{Y|17^\circ}$:

$$\hat{Y}_i = \hat{\mu}_{Y|17^\circ} = 19.9 + 1.78(17 - 10.0) = 32.36.$$

The standard error is

$$s_{\hat{Y}_i} = \sqrt{\text{MS}_\text{E}\left[\frac{1}{n} + \frac{(X_i - \overline{X})^2}{\sum(X_i - \bar{X})^2}\right]} = \sqrt{6.35\left[\frac{1}{9} + \frac{(17 - 10.0)^2}{240}\right]} = 1.41.$$

From Formula 10.6, the confidence interval limits are

$$L_1 = \hat{Y}_i - t_{1-\frac{\alpha}{2}} s_{\hat{Y}_i} = 32.36 - 2.365(1.41) = 29.03$$
$$L_2 = \hat{Y}_i + t_{1-\frac{\alpha}{2}} s_{\hat{Y}_i} = 32.36 + 2.365(1.41) = 35.69.$$

The confidence interval length is $35.69 - 29.03 = 6.66$ bpm. Note the increase in confidence interval length as X gets farther from $\overline{X}$.
    Fortunately many statistical packages will make these calculations quickly and graph the confidence bands as indicated in Figure 10.14.

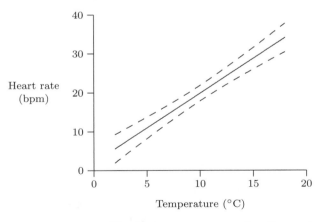

FIGURE 10.14. The least squares regression line and the 95% confidence bands.

Summary

When regression analysis is appropriate, we do the following:

1. Graph data to ascertain that a linear relationship is apparent.

2. Calculate the regression equation using the least squares method.

3. Test the significance of this equation with analysis of variance.

4. If the H_0 is rejected, plot the equation on the graphed data.

5. Finally, calculate any required confidence intervals in support of the linear relationship.

10.2   Simple Linear Correlation Analysis

Correlation analysis is used to measure the intensity of association observed between any pair of variables and to test whether it is greater than could be expected by chance alone. We are largely concerned with whether two variables are interdependent or covary. Here we do not express one variable as a function of the other and do not imply that Y is dependent on X as we did with regression analysis. Both X and Y are measured with error and we wish to estimate the degree to which these variables vary together. See Figure 10.15 and Figure 10.2.

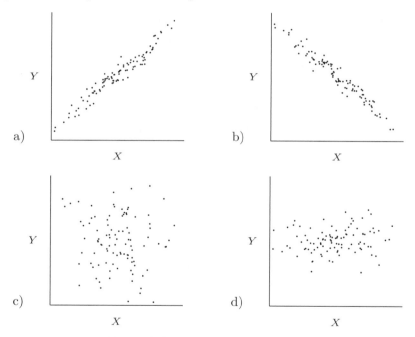

FIGURE 10.15. Panel a) demonstrates a **positive** correlation between X and Y. Panel b) demonstrates a **negative** correlation. Panels c) and d) illustrate **no** correlation between X and Y.

A widely used index of the association of two quantitative variables is the **Pearson product-moment correlation coefficient**, usually just called the **correlation coefficient**, which we now develop.

Consider the standardized normal deviates for X and Y,

$$Z_{X_i} = \frac{X_i - \overline{X}}{s_X} \quad \text{and} \quad Z_{Y_i} = \frac{Y_i - \overline{Y}}{s_Y}. \qquad (10.5)$$

When we multiply corresponding deviates together and sum, we obtain an **index of association**

$$\sum \left(\frac{X_i - \overline{X}}{s_X}\right)\left(\frac{Y_i - \overline{Y}}{s_Y}\right) = \frac{\sum(X_i - \overline{X})(Y_i - \overline{Y})}{s_X s_Y}, \tag{10.6}$$

with the following characteristics:

1. If large X's are associated with large Y's and small X's with small Y's, then both $(X_i - \overline{X})$ and $(Y_i - \overline{Y})$ will have the same sign and so their product in (10.6) will be positive. Consequently, the index will be positive and so we say that there is a **positive correlation** between X and Y in this case.

2. If large X's are associated with small Y's and vice versa, the index will be negative because $(X_i - \overline{X})$ and $(Y_i - \overline{Y})$ will have opposite signs and their product in (10.6) will be negative. We say that there is a **negative correlation** between X and Y in this case.

If we divide the index in (10.6) by $n-1$, we obtain a new index, denoted by r, that satisfies the first two conditions and whose range is from -1 to $+1$ (which we will verify in (10.10)). We have

$$r = \frac{\sum(X_i - \overline{X})(Y_i - \overline{Y})}{(n-1)\cdot s_X \cdot s_Y} = \frac{\sum X_i Y_i - \frac{(\sum X_i)(\sum Y_i)}{n}}{\sqrt{\left[\sum X_i^2 - \frac{(\sum X_i)^2}{n}\right]\left[\sum Y_i^2 - \frac{(\sum Y_i)^2}{n}\right]}} = \frac{SS_{XY}}{\sqrt{SS_X \cdot SS_Y}}.$$
$$\tag{10.7}$$

The index r is singled out for special attention.

FORMULA 10.7. The **Pearson product-moment correlation coefficient** is

$$r = \frac{SS_{XY}}{\sqrt{SS_X \cdot SS_Y}}.$$

That is, r is the corrected cross products over the square root of the product of the corrected sum of squares of X and of Y.

Notice that r is an *estimate* of the parameter ρ defined by

$$\rho = \frac{\sigma_{XY}}{\sigma_X \cdot \sigma_Y}.$$

The Greek letter ρ ("rho") denotes the true population relationship between variables X and Y.

The formula for r in (10.7) is motivated by the regression formula of the previous section. In fact, r is the slope of the least squares regression line for X and Y once they have been converted to the Z scores as in (10.5). To see why, remember the special properties of these Z scores: their means are both 0 and their variances are both 1,

$$\overline{Z}_X = \overline{Z}_Y = 0 \qquad \text{and} \qquad s_{Z_X}^2 = s_{Z_Y}^2 = 1.$$

Since the means are 0, it follows that

$$\sum Z_{X_i} = n\overline{Z}_X = 0 \quad \text{and} \quad \sum Z_{Y_i} = n\overline{Z}_Y = 0 \tag{10.8}$$

and since the variances are both 1

$$\frac{SS_{Z_X}}{n-1} = \frac{\sum (Z_{X_i} - 0)^2}{n-1} = s_{Z_X}^2 = 1$$

so

$$SS_{Z_X} = n - 1 \tag{10.9}$$

and, similarly, $SS_{Z_Y} = n - 1$.

So if b_Z denotes the slope of the least squares regression line with Z_X thought of as the independent variable and Z_Y as the dependent variable, then

$$
\begin{aligned}
b_Z &= \frac{SS_{Z_X Z_Y}}{SS_{Z_X}} && \text{Formula 10.2} \\[2mm]
&= \frac{\sum Z_{X_i} Z_{Y_i} - \frac{(\sum Z_{X_i})(\sum Z_{Y_i})}{n}}{n-1} && \text{From (10.9)} \\[2mm]
&= \frac{\sum Z_{X_i} Z_{Y_i}}{n-1} && \text{From (10.8) } \sum Z_{X_i} = \sum Z_{Y_i} = 0 \\[2mm]
&= \frac{\frac{\sum (X_i - \overline{X})}{s_X} \cdot \frac{\sum (Y_i - \overline{X})}{s_Y}}{n-1} && \text{From (10.5)} \\[2mm]
&= \frac{\sum (X_i - \overline{X})(Y_i - \overline{Y})}{(n-1) \cdot s_X \cdot s_Y} && \text{Combining denominators} \\[2mm]
&= r. && \text{From (10.7)}
\end{aligned}
$$

We can even go one step further. Since $SS_{Z_X} = SS_{Z_Y} = n - 1$, if we switch the roles of Z_X and Z_Y as independent and dependent variables, then

$$r = b_Z = \frac{SS_{Z_X Z_Y}}{SS_{Z_X}} = \frac{SS_{Z_X Z_Y}}{SS_{Z_Y}}.$$

So r is also the slope of the least squares regression line with Z_Y thought of as the independent variable and Z_X as the dependent variable. This symmetry makes sense since a correlation analysis is measuring the association between the pair of variables without assuming that one is dependent on the other.

There is one more connection to regression that we now exploit. Squaring Formula 10.7

$$r^2 = \frac{(SS_{XY})^2}{SS_X \cdot SS_Y} = \frac{\frac{(SS_{XY})^2}{SS_X}}{SS_Y} = \frac{SS_R}{SS_{\text{Total}}},$$

where at the last step we have used equation (10.3) on 288.

FORMULA 10.8. The ratio

$$r^2 = \frac{SS_R}{SS_{\text{Total}}} = \frac{(SS_{XY})^2}{SS_X \cdot SS_Y}$$

is called the **coefficient of determination** and is the amount of variability in one of the variables (either X or Y) that is accounted for by correlating that variable with the second variable.

Notice that since $SS_{Total} = SS_R + SS_E$, it follows that $0 \leq SS_R \leq SS_{Total}$. So in Formula 10.8, we must have

$$0 \leq \frac{SS_R}{SS_{Total}} \leq 1 \qquad \text{or, equivalently,} \qquad 0 \leq r^2 \leq 1.$$

By taking square roots, we find that we must have

$$-1 \leq r \leq 1. \qquad (10.10)$$

In Table 10.2 notice that low correlation coefficients, say less than 0.40, account for very little of the variability (less than 16%). This fact should serve as a warning against assigning much significance to low values of r.

TABLE 10.2. Correlation coefficients and the corresponding coefficients of determination

r	r^2
0.00	0.00
±0.10	0.01
±0.20	0.04
±0.30	0.09
±0.40	0.16
±0.50	0.25
±0.60	0.36
±0.70	0.49
±0.80	0.64
±0.90	0.81
±1.00	1.00

Testing the Significance of ρ

While there are no assumptions required to compute a correlation coefficient, tests of hypothesis and confidence intervals assume sampling from a "bivariate normal distribution." This means that not only are the Y's at each X assumed to be normally distributed but also the X values at each Y are assumed to come from a normal population.

To develop a test of hypothesis for the significance of ρ, we need to determine the standard error for its estimate r. Since r can be thought of as the slope of a regression line, starting from equation (10.4) on page 291 and then using Formula 10.3 we have

$$s_r^2 = \frac{1}{SS_{Z_X}} \left(\frac{SS_E}{n-2} \right) = \frac{1}{SS_{Z_X}} \left(\frac{SS_{Total} - SS_R}{n-2} \right).$$

Utilizing equation (10.2) on page 288 with r rather than b as the slope $SS_R = r^2 SS_{Z_X}$ and noting that $SS_{Total} = SS_{Z_Y}$ we now have

$$s_r^2 = \frac{1}{SS_{Z_X}} \left(\frac{SS_{Total} - r^2 SS_{Z_X}}{n-2} \right) = \frac{1}{SS_{Z_X}} \left(\frac{SS_{Z_Y} - r^2 SS_{Z_X}}{n-2} \right).$$

Now using $\text{SS}_{Z_X} = \text{SS}_{Z_Y} = n - 1$ from (10.9)

$$s_r^2 = \frac{1}{n-1}\left(\frac{(n-1) - r^2(n-1)}{n-2}\right) = \frac{1 - r^2}{n - 2}.$$

Consequently, the standard error of the correlation coefficient is

$$s_r = \sqrt{\frac{1 - r^2}{n - 2}}.$$

Using this standard error we can develop the following

FORMULA 10.9 (Hypothesis Test for ρ). The test for the significance of ρ has hypotheses

$$H_0\colon \rho = 0$$
$$H_a\colon \rho \neq 0$$

with test statistic

$$t = \frac{r - 0}{s_r} = \frac{r}{\sqrt{\frac{1 - r^2}{n - 2}}}, \qquad \text{with } \nu = n - 2.$$

EXAMPLE 10.4. Analyze Example 10.2 as a correlation problem.

SOLUTION. Let X be chiton length (in cm) and Y be chiton width (in cm). The data for the problem and the preliminary calculations appear below:

Chiton	
Length	Width
10.7	5.8
11.0	6.0
9.5	5.0
11.1	6.0
10.3	5.3
10.7	5.8
9.9	5.2
10.6	5.7
10.0	5.3
12.0	6.3

$\sum X = 105.8 \qquad \overline{X} = 10.58$

$\sum X^2 = 1123.9$

$\sum Y = 56.4 \qquad \overline{Y} = 5.64$

$\sum Y^2 = 319.68$

$\sum XY = 599.31$

$n = 10$

From (10.6), the correlation coefficient is

$$r = \frac{\sum XY - \frac{(\sum X)(\sum Y)}{n}}{\sqrt{\left[\sum X^2 - \frac{(\sum X)^2}{n}\right]\left[\sum Y^2 - \frac{(\sum Y)^2}{n}\right]}} = \frac{599.31 - \frac{(105.8)(56.4)}{10}}{\sqrt{\left[1123.9 - \frac{(105.8)^2}{10}\right]\left[319.68 - \frac{(56.4)^2}{10}\right]}}$$

$$= \frac{2.598}{\sqrt{(4.536)(1.584)}}$$

$$= 0.969.$$

Test whether there is a significant correlation with $\alpha = 0.05$ and $\nu = n-2 = 10-2 = 8$. The hypotheses are $H_0\colon \rho = 0$ and $H_a\colon \rho \neq 0$. The standard error of the correlation coefficient is

$$s_r = \sqrt{\frac{1 - r^2}{n - 2}} = \sqrt{\frac{1 - (0.969)^2}{10 - 2}} = 0.087,$$

so the test statistic is

$$t = \frac{r - 0}{s_r} = \frac{0.969 - 0}{0.087} = 11.14.$$

The critical values from Table C.4 for $\nu = 8$ with $\alpha = 0.05$ are ± 2.306. Since $11.14 \gg 2.306$, we find a *strong linear correlation* between length and width of chiton shells.

Confidence Intervals for ρ

To determine a 95% confidence interval for ρ is complicated by the fact that *only* when $\rho = 0$ can r be considered to have come from an approximate normal distribution. For values of ρ other than 0, Fisher's Z transformation, defined below, *must be employed*.

FORMULA 10.10. Fisher's Z is the inverse hyperbolic tangent of r, that is,

$$Z = \tanh^{-1} r = 0.5 \ln \left(\frac{1 + r}{1 - r} \right).$$

If a calculator is not readily available, Table C.10 may be used to make this transformation.

Begin by using Fisher's Z transformation (Table C.10) to transform $r = 0.969$ to the standard normal variable $Z = 2.0756$. The standard error of Z is approximated by

$$\sigma_Z = \sqrt{\frac{1}{n - 3}}.$$

In Example 10.4

$$\sigma_Z = \sqrt{\frac{1}{10 - 3}} = 0.378.$$

We will use the Greek letter ζ ("zeta") to denote the transformation of ρ. First we compute confidence limits for ζ and then transform these limits back to values for ρ. The limits for a 95% confidence interval for ζ are

$$L_1 = Z - 1.960 \sigma_Z = Z - 1.960 \sqrt{\frac{1}{n - 3}}$$

$$L_2 = Z + 1.960 \sigma_Z = Z + 1.960 \sqrt{\frac{1}{n - 3}}.$$

In Example 10.4 the confidence interval for ζ will have limits

$$L_1 = Z - 1.960 \sigma_Z = 2.0756 - 1.960(0.378) = 2.0756 - 0.7409 = 1.3347$$
$$L_2 = Z + 1.960 \sigma_Z = 2.0756 + 1.960(0.378) = 2.0756 + 0.7409 = 2.8159.$$

Use Table C.11 to transform these limits for ζ back into limits for ρ:

$$Z = 1.3347 \longrightarrow r = 0.8625$$
$$Z = 2.8159 \longrightarrow r = 0.9927.$$

So the 95% confidence interval for ρ is $[0.8625, 0.9927]$. This again indicates a very strong co-relationship between the two morphological measurements.

In the future the malacologist could save time and energy by measuring only one of the dimensions, either length or width, because these variables behave in a coordinated and highly predictable fashion.

10.3   Correlation Analysis Based on Ranks

The data for a correlation analysis consist of a bivariate random sample (no assumption of normality) of paired observations of size n, $(X_1, Y_1), \ldots, (X_n, Y_n)$. Examples include the age at entrance to first grade (X_i) and performance on a standardized test (Y_i) or the actual cost of a perscription medication (X_i) and a physician's estimate of its cost (Y_i). In each of these cases we expect some relationship between the two variables. However, we may also be interested in whether two random variables are independent. For example, in a study of student heights (X_i) and GPAs (Y_i) we would expect little or no correlation.

If each (X_i, Y_i) pair is assumed to come from a continuous population, there should be no tied X's or Y's. However, for the nonparametric test described below it is sufficient if the X and Y observations can be ranked from lowest to highest.

By convention, any measure of correlation should satisfy all the characteristics for an index of association listed earlier.

- The measure of correlation should assume values only between -1 and $+1$, inclusive.

- If the larger values of X tend to be paired with the larger values of Y (and, hence, the smaller values of X with the smaller values of Y), then the measure of correlation should be positive, and the stronger this tendency is the closer it should be to $+1$.

- If the larger values of X tend to be paired with the smaller values of Y (and, hence, the smaller values of X with the larger values of Y), then the measure of correlation should be negative, and the stronger this tendency is the closer it should be to -1.

- If the values of X and Y appear to be paired randomly, the correlation measure should be close to 0. In this case we say that X and Y are "uncorrelated" or have "no correlation." For example, this should be the case when X and Y are *independent*.

Kendall's Measure of Correlation τ

The Kendall correlation coefficient depends on a direct comparison of the n observations (X_i, Y_i) with each other. Two observations, for example $(190, 186)$ and $(182, 185)$, are called **concordant** if both members of one pair are larger than the corresponding members of the other pair (here $190 > 182$ and $186 > 185$). A pair of observations, such as $(180, 188)$ and $(182, 185)$, are called **discordant** if a number in the first pair is larger than the corresponding number in the second pair ($188 > 185$), while the other number in the first pair is smaller than the corresponding number in the second pair ($180 < 182$). Pairs with at least one *tie* between respective members are neither concordant nor discordant.

Because there are n paired observations in the sample, there are $\binom{n}{2}$ such comparisons that are made. Let C denote the number of concordant pairs of observations, D the number of discordant pairs, and E the number of ties. Then

$$\binom{n}{2} = \frac{n(n-1)}{2} = C + D + E.$$

The basis of the Kendall correlation coefficient is the following simple notion. In a concordant comparison, the larger X observation occurs in the pair with the larger Y observation and this provides support for a positive correlation between the variables. In a discordant comparison, the larger X observation is associated with the smaller Y observation, so it is evidence for a negative correlation. If the concordant comparisons greatly outnumber the discordant ones ($C \gg D$), there is strong evidence for a positive correlation. Conversely, if the discordant comparisons greatly outnumber the concordant ones ($C \ll D$), there is strong evidence for a negative correlation. If the numbers of concordant and discordant comparisons are roughly equal, there is evidence for no correlation.

FORMULA 10.11. The Kendall correlation coefficient is defined in terms of the difference between C and D divided by the total number of comparisons,

$$\tau = \frac{C - D}{\frac{n(n-1)}{2}} = \frac{2(C - D)}{n(n - 1)}.$$

Although it is a statistic, not a parameter, this calculation is generally denoted by the Greek letter τ ("tau"). There are several points to notice about τ:

- If all $\binom{n}{2} = \frac{n(n-1)}{2}$ comparisons are concordant (a "perfect" positive correlation), then $C = \frac{n(n-1)}{2}$ and $D = 0$, so $\tau = +1$, as it should.

- Similarly, if all $\frac{n(n-1)}{2}$ comparisons are discordant (a "perfect" negative correlation), then $D = \frac{n(n-1)}{2}$ and $C = 0$, so $\tau = -1$.

- In all other cases, $-1 < \tau < +1$.

- Finally, ties are not counted in τ since they do not constitute evidence for either a positive or negative correlation.

EXAMPLE 10.5. There have been several studies of the relative age effect on various types of achievement (e.g., academic, sport, and social). Briefly, relative age effects may occur whenever there are minimum age requirements for participation in an activity. For example, for entrance into kindergarten in Geneva, New York, where the authors reside, a child must be 5 by December 1 of the current school year. Consequently, on December 1, children in the same kindergarten class can range in age from a minimum of 5 years, 0 days (those born on December 1) to 5 years, 364 days (those born on December 2). Similarly, to participate on athletic teams (soccer, little league, etc.) children must attain a minimum age by a certain date, again leading to a 1-year age difference in initial cohorts of participants. Are these differences in ages associated with differences in performance, that is, are there relative age effects?

A study by DeMeis and Stearns examined relative age effects on academic and social performance in Geneva. The data in the following table come from one part of their study, which examined the number of students in grades K through 4 evaluated for the district's Gifted and Talented Student Program.

The first and second columns give the month after the cut-off date in which the student was born. The third column lists the number of students in each month cut-off category in grades K through 4 evaluated for the district's Gifted and Talented Student Program.

Birth month	Month after cut-off	Students evaluated
December	1	53
January	2	47
February	3	32
March	4	42
April	5	35
May	6	32
June	7	37
July	8	38
August	9	27
September	10	24
October	11	29
November	12	27

The table shows that, in general, the older students (those with birth dates in the first few months after the cut-off date) tend to be overrepresented and younger students underrepresented in those evaluated for the district's Gifted and Talented Student Program. For example, the month of December (the first month after the cut-off date) had the most referrals. Is there a correlation between these two measurements? Compute the number of concordant, discordant, and tied pairs of observations. Determine τ. (Based on data reported in: DeMeis, J. and E. Stearns. 1992. Relationship of school entrance age to academic and social performance. *The Journal of Educational Research*, **86:** 20–27.)

SOLUTION. Calculations are simpler if the data are listed in increasing order of the X observations (month number), as was done in the original table. Be sure to carry along the corresponding Y observation if reordering is necessary. Now make the comparisons starting from the top of the new table (see the table below).

To compare the first observation with the 11 observations below it, all one needs to do is compare the Y values, assuming that the X observation is not tied with any X observation below it. Larger Y_i's below Y_1 correspond to concordant comparisons. All 11 comparisons are discordant here. Next compare the second observation in the table with the 10 observations below it. It yields 10 discordant comparisons. The third observation yields 4 concordant, 4 discordant, and 1 tied comparison, and so on down the table. The numbers of concordant, discordant, and tied comparisons at each stage are listed in the last three columns.

X_i	Y_i	Rank X_i	Rank Y_i	Concordant pairs below (X_i, Y_i)	Discordant pairs below (X_i, Y_i)	Tied pairs below (X_i, Y_i)
1	53	1	12	0	11	0
2	47	2	11	0	10	0
3	32	3	5.5	4	4	1
4	42	4	10	0	8	0
5	35	5	7	2	5	0
6	32	6	5.5	2	4	0
7	37	7	8	1	4	0
8	38	8	9	0	4	0
9	27	9	2.5	1	1	1
10	24	10	1	2	0	0
11	29	11	4	0	1	0
12	27	12	2.5	0	0	0

Strictly speaking, the third and fourth columns in the table that rank the X and Y observations are not necessary. But notice that *if we had used the rankings in place of the original data we would have obtained precisely the same set of concordant and discordant comparisons.* This observation will be useful later when we discuss the underlying theory of the test, but it also shows why continuous observations are not required to compute this statistic. One only needs to be able to rank the observations to carry out the test. In fact, this is precisely why the researchers chose this test. In this example, month is a discrete variable that can assume integer values from only 1 to 12, not a continuous variable. Similarly, the number of students referred for evaluation is also discrete, and can assume any integer value from 0 to the number of students enrolled. Thus, it would be inappropriate to analyze these data with Pearson's r.

Using the data in the table, C is just the sum of the entries in the fifth column,

$$C = 0 + 0 + 4 + 0 + 2 + 2 + 1 + 0 + 1 + 2 + 0 + 0 = 12.$$

Similarly, $D = 52$, and $E = 2$. So

$$\tau = \frac{2(C - D)}{n(n - 1)} = \frac{2(12 - 52)}{12(12 - 1)} = -0.606.$$

Kendall's τ is interpreted in the same way that Pearson's r is. It appears that there is a moderately strong negative correlation between month after cut-off date and referrals for gifted evaluation.

A Test for Independence or Zero Correlation

Just as with the Pearson correlation coefficient, it is important to determine whether the coefficient τ is significantly different from 0. Using hypothesis testing, if τ is not significantly different from 0, then it may mean that the two random variables are independent. That is, knowing one component of the (X, Y) pair does not help to predict the other value. Note, however, there are other ways for the correlation to be 0 without the two variables being independent. A more important caution is that even if we show that τ is statistically different from 0, τ may still be so small that it is biologically meaningless.

The usual test performed is two-sided with the hypotheses being:

- H_0: $\tau = 0$ (or "X and Y are independent")

- H_a: $\tau \neq 0$ (or "X and Y are not independent").

If no ties are present, then C and D are natural candidates for a test statistic because they determine τ. In this case, $C + D = \frac{n(n-1)}{2}$, so knowing one of these means that the other is also known. The test statistic used is the smaller of C or D. When there are ties, use the smaller of $C + \frac{1}{2}E$ or $D + \frac{1}{2}E$. The smaller of the pair is used only as a convenience for creating a compact table of P values for the test statistic. Notice that if the test statistic is very small then there are either very few concordances (meaning there is a strong negative correlation) or very few discordances (meaning there is a strong positive correlation). The smaller the test statistic, the stronger the correlation and the more likely we are to reject H_0.

As usual, the exact distribution of the test statistic is difficult to tabulate (but see the next section where we examine one simple case). Table C.12 is used for looking up

P values. Locate the correct sample size and then locate the correct value of the test statistic under the column labelled c. The corresponding P value is for a one-tailed test and so it must be doubled to get our P value for this test.

EXAMPLE 10.6. Determine whether the correlation coefficient $\tau = -0.606$ in Example 10.5 is significantly different from 0 at the $\alpha = 0.05$ level.

SOLUTION. In this problem the sample size is $n = 12$. From earlier calculations, $C = 12$, $D = 52$, and $E = 2$, so the test statistic is $C + \frac{1}{2}E = 13$. In Table C.12, the P value for 13 is 0.0027. So the P value for our test is $2(0.0027) = 0.0054$, which is less than α. Therefore, H_0 is rejected. This is not at all surprising since τ was relatively large. We can make a stronger statement: H_0 can be rejected even at the $\alpha = 0.01$ level.

Large Samples

The largest sample size listed in Table C.12 for Kendall's test of independence is $n = 25$. For larger samples, the test statistic may be modified so that it is well-approximated by the standard normal distribution. If n is large, then τ has an approximately normal distribution with

$$\mu = 0 \qquad \text{and} \qquad \sigma^2 = \frac{2(2n+5)}{9n(n-1)}.$$

Consequently,

$$z_\tau = \frac{\tau - 0}{\sigma} = \frac{\frac{2(C-D)}{n(n-1)}}{\sqrt{\frac{2(2n+5)}{9n(n-1)}}}$$

is closely approximated by the standard normal distribution. Simplifying one finds

$$z_\tau = \frac{C - D}{\sqrt{n(n-1)(2n+5)/18}}.$$

Because $C - D$ is a discrete variable, there is a continuity correction so that more accurate approximations are obtained by using

$$z_\tau = \begin{cases} \frac{C-D-1}{\sqrt{n(n-1)(2n+5)/18}}, & \text{if } C - D > 0 \\ 0 & \text{if } C - D = 0 \\ \frac{C-D+1}{\sqrt{n(n-1)(2n+5)/18}}, & \text{if } C - D < 0. \end{cases} \qquad (10.11)$$

The approximation in (10.11) may be used for sample sizes $n > 10$, though Table C.12 will give more accurate results up to $n = 25$.

EXAMPLE 10.7. Redo Example 10.6 using the approximation in (10.11) to determine whether the correlation coefficient $\tau = -0.606$ in Example 10.5 is significantly different from 0 at the $\alpha = 0.05$ level.

SOLUTION. In Example 10.5, $n = 12$ and $C - D = 12 - 52 = -40 < 0$. From (10.11)

$$z_\tau = \frac{C - D + 1}{\sqrt{n(n-1)(2n+5)/18}} = \frac{-40 + 1}{\sqrt{12(11)(29)/18}} = -2.67.$$

Using Table C.3, the P value for the test is $2(0.0038) = 0.0076$, which is less than α. Therefore, H_0 is rejected. The approximate P value is not all that different than the exact value that was computed in Example 10.6.

Theory

In this section, we show how the cumulative distribution is determined for the test statistics C and D (i.e., the P values in Table C.12). Consider the case when $n = 4$. The sample consists of four pairs (X_1, Y_1), (X_2, Y_2), (X_3, Y_3), and (X_4, Y_4). For convenience, assume that there are no ties and that the X's are ordered so that $X_1 < X_2 < X_3 < X_4$. In other words, the rank of each X is just its subscript.

The test statistic is the minimum of C and D, the number of concordant and discordant comparisons. Now in Example 10.5 we saw that the comparisons could be carried out using the ranks of the X's and Y's instead of their actual observed values. Given that the X's are ranked in the order $(1, 2, 3, 4)$, we must determine the test statistic for each possible ranking of the Y's. There are 4! such rankings because Y_1 can have any one of the four ranks, Y_2 any of the three remaining ranks, Y_3 any of the two remaining ranks, and Y_4 has the one rank that is left. For example, suppose that the order of the Y's were $Y_3 < Y_2 < Y_4 < Y_1$. Then Y_1 would be ranked 4 (largest), Y_2 would be ranked 2, Y_3 would be ranked 1 (smallest), and Y_4 would be ranked 3 or, more concisely, the ranks would be $(4, 2, 1, 3)$. Remember that a pair is concordant when a larger rank appears below a smaller rank, so $C = 0 + 1 + 1 + 0 = 2$ for this ranking. Similarly $D = 3 + 1 + 0 + 0 = 4$. So the test statistic for this ranking of the Y's is 2, the minimum value of C and D. The table below gives similar data for all 24 possible rankings. Only the value of C is listed, since $D = \binom{4}{2} - C = 6 - C$.

Y ranks	C	Y ranks	C	Y ranks	C	Y ranks	C
(1,2,3,4)	6	(2,1,3,4)	5	(3,1,2,4)	4	(4,1,2,3)	3
(1,2,4,3)	5	(2,1,4,3)	4	(3,1,4,2)	3	(4,1,3,2)	2
(1,3,2,4)	5	(2,3,1,4)	4	(3,2,1,4)	3	(4,2,1,3)	2
(1,3,4,2)	4	(2,3,4,1)	3	(3,2,4,1)	2	(4,2,3,1)	1
(1,4,2,3)	4	(2,4,1,3)	3	(3,4,1,2)	2	(4,3,1,2)	1
(1,4,3,2)	3	(2,4,3,1)	2	(3,4,2,1)	1	(4,3,2,1)	0

Because the null hypothesis of the test is that $\tau = 0$ ("no correlation"), this implies that any comparison is just as likely to be concordant as discordant. Thus, all 24 of the rankings above are equally likely, so the probability of any one of them occurring is $\frac{1}{24}$. The distribution can now be read off from the table. For example, there are three rankings that produce a C value of 1, so $P(C = 1) = \frac{3}{24}$. In the same way, there are three rankings that produce a D value of 1. These are the rankings where $C = 5$. So $P(D = 1) = \frac{3}{24}$. There are nine rankings that yield C values less than or equal to 2, so $P(C \leq 2) = \frac{9}{24}$. The entire probability density function and cumulative distribution for C are given below. You might check that the values for the cumulative distribution in the following table are precisely those in Table C.12 for $n = 4$.

c	0	1	2	3	4	5	6
$P(C = c)$	$\frac{1}{24}$	$\frac{3}{24}$	$\frac{5}{24}$	$\frac{6}{24}$	$\frac{5}{24}$	$\frac{3}{24}$	$\frac{1}{24}$
$P(C \leq c)$	$\frac{1}{24}$	$\frac{4}{24}$	$\frac{9}{24}$	$\frac{15}{24}$	$\frac{20}{24}$	$\frac{23}{24}$	$\frac{24}{24}$

Keep in mind that although there were $n!$ rankings, there were only $\binom{n}{2}+1$ different values for C or D ranging from 0 to $\binom{n}{2}$.

Spearman's Coefficient r_s

There are other correlation coefficients that appear in the literature. One of the most common is **Spearman's rank correlation coefficient** r_s. The idea is to rank the X and Y observations (separately) and compute the Pearson correlation coefficient on the *ranks* rather than on the original data. The value r_s is usually different from the value of Pearson's r calculated on the *original data*, but for large sample sizes the two values are usually relatively close. So why bother with the ranks? Well, if there are no ties among the X's and no ties among the Y's, then r_s can be computed much more simply than Pearson's r using the following:

FORMULA 10.12. Spearman's rank correlation coefficient (assuming no ties) is

$$r_s = 1 - \frac{6\sum_{i=1}^{n} d_i^2}{n(n^2-1)},$$

where $d_i = r_{X_i} - r_{Y_i}$ is the difference in the rank of X_i and Y_i.

This is illustrated in the example that follows. The value of r_s is interpreted in the same way as the other correlation coefficients.

> **EXAMPLE 10.8.** A study by Musch and Hay examined relative age effects in soccer in various northern and southern hemisphere countries. For each country, a sample consisting of all players in the highest professional soccer league was investigated. Their data for Germany are given below. A cut-off date of August 1 applies in Germany. Since participation cut-off dates vary by country, foreign players were excluded from their analysis. For each country, the distribution of professional players' birthdays was computed by month. These birthday distributions were then compared with that of the general population of that country.
>
> The first column is the month after the cut-off date in which the player was born. The second column is the number of professional soccer players born in the respective months of the competition year. The third column is the number of soccer players that would be expected on the basis of official birth statistics assuming that the birth distribution of soccer professionals is the same as that of the general population, and that no relative age effect exists. The fourth column is the difference between this expected and the observed number of players.

Month	Actual players	Expected players	Difference
1	37	28.27	8.73
2	33	27.38	5.62
3	40	26.26	13.74
4	25	27.60	−2.60
5	29	29.16	−0.16
6	33	30.05	2.95
7	28	31.38	−3.38
8	25	31.83	−6.83
9	25	31.16	−6.16
10	23	30.71	−7.71
11	30	30.93	−0.93
12	27	30.27	−3.27

The table shows that, in general, the older players in the 1-year cohorts tend to be overrepresented and younger players underrepresented in the total number of professional soccer players in Germany. Compute and interpret r_s for these data where X is the month and Y is the difference between actual and expected numbers of professional players. (Based on data reported by: Musch, J. and R. Hay. 1999. The relative age effect in soccer: Cross-cultural evidence for a systematic discrimination against children born late in the competition year. *Sociology of Sport Journal*, **16**: 54–64.)

SOLUTION. To calculate r_s:

1. Separately rank each group of variables as shown in the following table. The data pairs have been ordered using the months in the first column as X's and the differences in the second column as Y's. Their ranks are then listed in the third and fourth columns.

X: month	Y: difference	r_X	r_Y	d_i
1	8.73	1	11	-10
2	5.62	2	10	-8
3	13.74	3	12	-9
4	-2.60	4	6	-2
5	-0.16	5	8	-3
6	2.95	6	9	-3
7	-3.38	7	4	3
8	-6.83	8	2	6
9	-6.16	9	3	6
10	-7.71	10	1	9
11	-0.93	11	7	4
12	-3.27	12	5	7

2. Compute the differences $d_i = r_X - r_Y$ of the ranks for each pair of variables. (See the final column.)

3. Square these differences and sum them to obtain $\sum_i d_i^2$.

4. Use the Formula 10.12 to obtain r_s,

$$r_s = 1 - \frac{6 \sum_{i=1}^{n} d_i^2}{n(n^2 - 1)},$$

where n is the number of pairs of variables.

From the information in the table,

$$r_s = 1 - \frac{6 \sum_{i=1}^{n} d_i^2}{n(n^2 - 1)}$$

$$= 1 - \frac{6 \left[(-10)^2 + (-8)^2 + (-9)^2 + (-2)^2 + (-3)^2 + (-3)^2 + 3^2 + 6^2 + 6^2 + 9^2 + 4^2 + 7^2\right]}{12(12^2 - 1)}$$

$$= -0.727.$$

This indicates a negative correlation whose strength now can be tested. The hypotheses are

$$H_0: \rho_s = 0$$
$$H_a: \rho_s \neq 0,$$

with $n = 12$. Let $\alpha = 0.05$. From Table C.13, the critical value for the Spearman rank correlation is 0.587. The table lists only positive critical values. We need to compare the magnitude (absolute value) of r_s to the critical value. Since $0.727 > 0.587$, reject H_0 and accept H_a. The negative Spearman rank correlation ($r_s = -0.727$) is significant and indicates that there is an excess in the number of "goliath" players (those born early in the competition year) and a lack of players born late in the competition year among professional soccer players in Germany.

Comment

Spearman's rank correlation coefficient r_s seems somewhat more common in the literature than Kendall's τ. This is probably a result of its relationship to Pearson's correlation coefficient. The main advantage of Kendall's τ is that one can easily test whether it is significantly different from 0. Generally, τ and r_s do not produce radically different correlation values. Both are relatively simple to compute by hand and are nice alternatives to Pearson's r.

10.4   Problems

1. Outline the conditions required for linear regression analysis and contrast them with the conditions for linear correlation analysis.

2. A student undertakes a project to determine the effect of carbon dioxide on respiration rate using the following methodology. A volunteer inhaled air from a bag containing a predetermined amount of carbon dioxide (with partial pressure measured in torr) and the number of breaths per minute was recorded. Her data are given below:

Partial pressure CO$_2$ (torr)	Respiration rate (breaths/min)
30	8.1
32	8.0
34	9.9
36	11.2
38	11.0
40	13.2
42	14.6
44	16.6
46	16.7
48	18.3
50	18.2

(a) Construct a scatterplot of these data. Does there appear to be a linear relationship between the partial pressure of CO_2 and respiration rate?

(b) Compute the linear regression equation so that Y may be predicted from X. Calculate b to 3 decimal places.

(c) Test the significance of this regression equation via ANOVA with H_0: $\beta = 0$. Explain why ANOVA is an appropriate test of hypothesis here. (Hint: Utilize expected values.)

(d) Calculate the 95% confidence interval for β.

(e) Find the predicted respiration rate for a partial pressure of 48 torr, $\widehat{\mu}_{Y|48}$, and the 95% confidence interval for this prediction.

(f) Find the predicted respiration rate for a partial pressure of 38 torr, $\widehat{\mu}_{Y|38}$, and the 95% confidence interval for this prediction.

(g) Why do the two confidence intervals in (e) and (f) have different widths?

3. As noted in Example 10.5, DeMeis and Stearns studied age effect on academic and social performance and found a negative correlation between month of birth after entrance cut-off date and referrals for gifted student evaluation. But what were the consequences of these evaluations? Analyze their data for students actually placed in gifted and talented programs in the Geneva school district. Is there a similar effect?

nonparametric correlation analysis

Spearman

Birth month	Month after cut-off	Gifted placements
December	1	17
January	2	17
February	3	9
March	4	13
April	5	10
May	6	14
June	7	18
July	8	17
August	9	12
September	10	13
October	11	14
November	12	17

4. The following data were collected to investigate the relationship between mean annual temperature and the mortality rate for a type of breast cancer in women. They relate to certain regions of Great Britain, Norway, and Sweden. Determine and interpret Kendall's correlation coefficient for these data. *(HSDS, #102)*

Mean annual temp (°F)	Mortality index
31.8	67.3
34.0	52.5
40.2	68.1
42.1	84.6
42.3	65.1
43.5	72.2
44.2	81.7
45.1	89.2
46.3	78.9
47.3	88.6
47.8	95.0
48.5	87.0
49.2	95.9
49.9	104.5
50.0	100.4
51.3	102.5

5. The proportion of the population made up of elderly persons in the United States is rapidly increasing. The implications for health care expenditures are profound because elderly persons use the health care system more frequently than younger persons. In one study the mean cumulative expenditures per person for acute and long-term care from the age of 65 years until death and the corresponding age at death were recorded. Expenditures are given in 1996 dollars. (Source: Spillman, B. and J. Lubitz. 2000. The effect of longevity on spending for acute and long-term care. *The New England Journal of Medicine*, **342**: 1409–1415.)

**✳ use t-test
not ANOVA**

Age at death	Mean expenditure
65	31,181
70	87,116
75	123,823
80	157,903
85	193,727
90	235,369
95	287,980
100	358,174

(*a*) Construct a scatterplot of these data. Does there appear to be a linear relationship between age at death and mean cumulative health care expenditure?

(*b*) Compute the linear regression equation so that Y may be predicted from X. Calculate b to 1 decimal place. Interpret b.

(*c*) Test the significance of this regression equation via ANOVA with H_0: $\beta = 0$.

(*d*) Calculate the 95% confidence interval for β.

(*e*) Find the predicted mean expenditure for a person age 77 at death, $\hat{\mu}_{Y|77}$, and the 95% confidence interval for this prediction.

(*f*) Find the predicted mean expenditure for a person age 97 at death, $\hat{\mu}_{Y|97}$, and the 95% confidence interval for this prediction.

(*g*) Why are the two confidence intervals in (*e*) and (*f*) different widths?

6. The heights and arm spans of 10 adult males were measured (in cm). Is there a correlation between these two measurements? Carry out an appropriate analysis.

Height (cm)	Span (cm)
171	173
195	193
180	188
182	185
190	186
175	178
177	182
178	182
192	198
202	202

7. The following data were collected at a wetland preserve for a study to investigate the relationship between male dragonfly body size and the system of territories they stake out. Males defend small areas, including small bodies of water, which females visit to oviposit. Do the data below support the hypothesis that high-quality territories, which attract more females, are occupied by larger males? The qualities of the territories are ranked from 1 (highest quality) to 11 (lowest quality). Explain why a nonparametric analysis is most appropriate.

nonparametric correlation analysis

Spearman

Territory rank	Forewing length (mm)
1	15.1
2	14.9
3	14.7
4	15.3
5	14.1
6	14.5
7	14.3
8	15.0
9	13.8
10	14.0
11	13.6

8. Determine the cumulative distribution for C, where C is the number of concordances in Kendall's test, when $n = 3$. This is easy. There are only 3! rankings to consider.

9. The yield of cotton is particularly sensitive to rainfall. Dry weather during the critical growth period in June appears to slow growth considerably judging from these records of an agricultural experiment station about June rainfall and yield of cotton.

normal-distribution based correlation analysis

Pearson

June rainfall (cm)	Yield (lb/acre)
3	1120
6	1750
7	1940
9	2130
11	2380
15	2650
17	2990
19	3130

(*a*) Generate a scatterplot of these data.

(*b*) Calculate r for these data.

(*c*) Are rainfall and cotton yield correlated? Include a test of H_0: $\rho = 0$. What is the P value for this test of hypothesis?

10. A breeder of golden retriever dogs has kept records on litter size for several years. She wishes to predict the litter size from the age of the breeding bitch at conception. From the following data would you say that this is a legitimate goal? Support your explanation with statistical analysis.

Age of bitch	Litter size
2.0	11
2.5	10
4.0	9
3.3	12
6.0	5
5.0	9
4.5	9
4.1	8
8.2	7
2.4	10
2.9	12
3.7	10
4.0	9

11. The same breeder of golden retrievers (see the previous problem) wishes to investigate the possible relationship between the sire's age at conception and litter size. Do you think there is a relationship between these two variables? Again support your answer with the appropriate statistical analysis.

Age of sire	Litter size
9.0	11
7.0	10
6.5	9
4.3	12
5.0	5
4.1	9
3.7	9
8.0	8
7.0	7
4.1	10
4.5	12
7.0	10
6.0	9

12. Suppose that there is a perfect linear relationship between the variables X and Y so that for each observation we have $Y_i = \alpha + \beta X_i$. (You may wish to review Section 1.7 before doing this problem.)

(a) Prove that $\sum \left(X_i - \overline{X} \right) \left(Y_i - \overline{Y} \right) = \beta \sum \left(X_i - \overline{X} \right)^2$.

(b) Prove that $s_Y = |\beta| s_X$. (Why is the absolute value used here?)

(c) Use the fact that $r = \frac{\sum (X_i - \overline{X})(Y_i - \overline{Y})}{(n-1) \cdot s_X \cdot s_Y}$ to prove that r is either -1 or $+1$ in this case.

13. In the table below $Y = X^2$, so the relationship between the variables is quadratic and not linear. However, the larger X's are paired with the larger Y's.

X	Y = X²
0	0
1	1
2	4
3	9
4	16
5	25

do both

(a) Determine the Kendall correlation coefficient τ for these data. Interpret your result. **Do Spearman's, not Kendall's**

(b) Determine the Pearson correlation coefficient r for these data.

(c) Discuss the differences between the two indices of correlation for this type of data.

14. Attention has focused recently on controlling the high costs of perscription medications. General practitioners usually agree that drug costs should be borne in mind when writing scripts, but in reality only one-third of their cost estimates are accurate to within 25% of the true cost of the prescribed medicine. Little information is known with regard to hospital doctors' knowledge of drug costs. Fifty prescribers representing medical and surgical specialties were asked to estimate the cost per day of a standard dose of 13 very commonly prescribed drugs. The extent to which prescribers over- and underestimated the cost of individual medicines was compared to their actual cost by using Kendall's rank correlation test by O'Connell and Feely. They claim that: "We show a strong inverse [negative] relationship between the cost of the medication and the extent to which cost was overestimated." (Based on data reported in O'Connell, T. and J. Feely. 1997. Hospital doctors' knowledge of drug costs and generic names. *Irish Medical Journal*, **90**(8): 298.)

Drug	Actual cost (£ per day)	Estimated cost mean	Percent over- or underestimated
Lanoxin	0.02	0.31	1450
Normison	0.06	0.23	283
Frumil	0.18	0.66	267
Ventolin	0.35	1.07	206
Capoten	0.40	1.65	313
Ponstan	0.48	0.52	8
Voltarol	0.63	1.08	71
Atrovent	0.80	1.24	55
Zantac	1.06	1.49	41
Augmentin	1.10	1.72	56
Losec	1.66	2.48	49
Zinnat	2.02	2.54	26
Claforan	13.50	6.85	−49

(a) Do you agree with O'Connell and Feely's conclusion? Explain.

(b) Would the conclusion change if the difference in price rather than percent difference in price were used? Comment on which test you think is more appropriate.

15. Allometric data often provide instances where we expect a correlation between variables. The data below pertain to a random sample of 14 males of the species *Siganus fuscescens*, a common venomous fish in Moreton Bay. The data give the length (in mm) from the jaw to the tip of the caudal fin and the vertical depth (in mm) of each fish. Determine the Spearman rank correlation coefficient r_s for these data and interpret its meaning. (Based on data from Ian Tibbetts, The University of Queensland Centre for Marine Science, 1996. Personal commumication.)

Length	Depth
104.0	35.6
114.5	38.0
113.2	37.8
115.2	35.2
116.8	38.2
122.3	40.7
120.0	40.0
130.5	49.6
132.4	46.0
133.5	47.2
114.4	38.7
139.0	51.7
144.9	50.0
146.0	47.4

16. To control Eurasian milfoil and other invasive aquatic weeds, the chemical herbicide 2,4-Dichlorophenoxyacetic acid (2,4-D), which is a synthetic auxin, is often applied to lakes in the midwest. A housing development is being built around a recently treated lake. The homes have individual wells tapping into the local aquifer. One week after the 2,4-D treatment, samples of water from 12 wells were analyzed for the presence of 2,4-D. Below are the data in micrograms of 2,4-D per liter of well water and the distances in meters of the wells from the edge of the lake.

2,4-D level (μg/L)	Distance (m)
490	20
455	60
445	90
420	100
400	120
300	160
285	170
260	180
220	220
205	230
90	290
50	340

(*a*) Is there a linear relationship between 2,4-D level and the distance of the well from the lake's edge?

(*b*) What would you predict for the 2,4-D level of a well 250 meters from the lake's edge? Find the 95% confidence interval for this predicted value.

(c) If 85 micrograms/liter is the maximum allowed in well water by the U. S. Environmental Protection Agency, is the well in (b) below the allowable threshold? Explain.

17. As noted in Example 10.8, Musch and Hay studied age effect on soccer in a number of countries. Analyze their data for Australia given below. Is there a similar effect?

Month	Actual players	Expected players	Difference
1	21	17.55	3.45
2	22	16.10	5.90
3	20	18.18	1.82
4	17	17.31	−0.31
5	19	17.86	1.14
6	18	16.56	1.44
7	14	17.50	−3.50
8	23	17.67	5.33
9	15	17.42	−2.42
10	17	17.93	−0.93
11	10	16.49	−6.49
12	11	16.43	−5.43

18. Agarose gel electrophoresis is one of the most widely used techniques in molecular biology. When relatively small pieces of DNA are subjected to an electrical current in an agarose matrix, they migrate through the gel at rates that depend on their lengths. Their positions within the gel can be ascertained by staining the gel with ethidium bromide and viewing it under UV light, as shown below.

The relationship between distance migrated and the size of the DNA fragments can, in fact, be used to form standard curves that can be utilized to estimate the lengths of unknown pieces of DNA. In these standard curves, when the logs of the DNA lengths (thought of as measured without error) are plotted versus the distances migrated, a linear relationship will be evident.

Distance Migrated (cm)	DNA Size (bp)
1.30	23,130
1.65	9,416
1.90	6,557
2.20	4,361
2.90	2,322
3.05	2,027
4.30	564

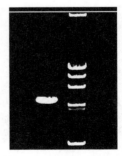

(a) Using the data above graph the log of the DNA lengths versus the distances migrated.

(b) Find the least squares regression equation for these data.

(c) Test the significance of the regression equation with an ANOVA.

(d) If an unknown DNA molecule migrated 2.60 cm in the same gel, estimate its size from the regression equation.

(e) Develop a 95% confidence interval for the DNA molecule size that migrated 2.60 cm.

19. A plant physiologist grew 13 individually potted soybean seedlings in a greenhouse. The table gives measurements of the total leaf area (cm^2) and total plant dry weight (gm) for each plant after 16 days of growth. Calculate the correlation coefficient for these data. Test its significance via a t test. Interpret the result.

Plant	Leaf area, X	Dry weight, Y
1	411	2.00
2	550	2.46
3	471	2.11
4	393	1.89
5	427	2.05
6	431	2.30
7	492	2.46
8	371	2.06
9	470	2.25
10	419	2.07
11	407	2.17
12	489	2.32
13	439	2.12

$$SS_X = 28,465.7, \qquad SS_Y = 0.363708$$

$$\sum(X_i - \overline{X})(Y_i - \overline{Y}) = 82.977$$

20. As part of a study of the environmental effects of large highways on nearby ecosystems, the cadmium concentrations in grasses at different distances from a major highway were measured. The results are presented below.

Distance (m)	Cd Concentration (μg/kg)
1	105
2	48
3	39
4	28
5	18
10	9

$$n = 6 \quad \sum X = 25 \quad \sum X^2 = 155 \quad \sum Y = 247 \quad \sum Y^2 = 16,039 \quad \sum XY = 610$$

For these data compute the least squares regression equation (calculate b to two decimal places). Test the equation's significance with ANOVA. Also calculate the 95% confidence interval for the slope, β. Comment on your findings.

21. It is well known that many species of birds sing in the early morning. Certain species begin very early in the dawn, some sing slightly later, and some wait until the sun crowns the horizon. The range of time from first songs to last is usually more than 100 minutes. The staggering of avian choruses is widely known and was first documented more than 70 years ago. In the 1960s it was proposed that visual acuity determined by eye size governs when birds begin to sing. The hypothesis was the larger the eye the earlier in the twilight the species would begin to sing. To test this proposal 20 species of morning singing birds' eyes were measured with calipers and the time from first light to the beginning of the song was recorded. Do the data support the original hypothesis?

Eye diameter (mm)	Song time from first light (min)
9.2	5
9.0	8
9.0	11
8.9	10
8.8	10
8.7	17
8.6	14
8.4	20
7.5	25
7.2	18
6.8	30
6.7	35
6.7	40
6.5	48
5.9	50
5.8	47
5.5	65
4.3	70
4.2	68
4.0	75

22. The temperature of an object depends on how fast the atoms and molecules of the object vibrate. As an object is cooled, the oscillations of its atoms and molecules slow. In any material, a temperature is eventually reached where oscillations are as slow as they can possibly be. The temperature at which this occurs is called *absolute zero*. Note that the oscillations never come to a complete stop, even at absolute zero. Rather, at absolute zero the molecules and atoms are all in the ground state (i.e., the lowest possible energy state) and the system has the least possible amount of kinetic energy allowed by the laws of physics. On the Kelvin scale, absolute zero is the temperature $0°K$.

The work of the French physicist Jacques Charles (1746–1823) was used to determine absolute zero. Charles discovered that the volume of a gas under constant pressure increases linearly with temperature.

Assume that one mole of hydrogen is held at a constant pressure of one atmosphere. The table below contains the data from an experiment where volume, V, is measured in liters at various temperatures, T, measured in degrees Celsius.

T (°C)	V (l)
−50	18.3
−30	20.0
−10	21.6
0	22.4
10	23.3
30	24.9
50	26.5
70	28.2

(a) Do the data support the contention that V is a linear function of T? Explain.

(b) Clearly, there is a physical lower limit to the volume: It can be no smaller than 0 liters. Use the relation found above with $V = 0$ to find the corresponding temperature of absolute zero (on the Celsius scale). (The actual value is approximately $-273.15°C$. How does your answer compare?)

23. It is well known that blood-testosterone concentration declines after the age of 40. Below are the blood-testosterone levels in nanograms per deciliter (ng/dl) for a small sample of healthy males.

Age (yr)	Testosterone (ng/dl)
40	500
42	490
43	495
44	485
46	460
48	470
49	465
51	465
54	450
60	440
63	415
65	440
69	420
71	415
74	400
77	390

linear regression!

Is there a linear relationship between age and blood-testosterone concentration? If so, describe it with a regression equation and predict the average blood-testosterone concentrations for 50-year-olds and 60-year-olds.

24. QCR. True or False. Justify your answer.

(a) Regression analysis differs from correlation analysis because we assume the independent variable, X, in regression is under our control and is measured without error.

(b) Correlation analysis is used to measure the intensity of association between a pair of variables and to express the dependent variable as a function of the independent variable.

(c) Based on the expected values of MS_R and MS_E, the F test for the hypothesis that $\beta = 0$ is a right-tailed test.

(d) In linear regression the confidence intervals for $\hat{\mu}_{Y|X}$ become narrower as X becomes further from $\overline{X}$.

(e) The calculated regression line when drawn on the scatter plot should extend only to the smallest or largest X values in the study.

(f) The coefficient of determination is actually just a ratio of the variability accounted for as covariability over the total variability.

(g) Spearman's rank correlation coefficient is the Pearson product-moment correlation coefficient calculated on ranks of the data rather than on the original data.

11

Goodness of Fit Tests for Categorical Data

Concepts in Chapter 11:

- The Binomial Test (The Proportions Test)
- Comparing Two Population Proportions
- The Chi-Square Test for Goodness of Fit

    — The Extrinsic Model
    — The Intrinsic Model

- The Chi-Square Test for $r \times k$ Contingency Tables

    — 2×2 Table as a Special Case
    — Partitioning the Overall Chi-Square Test into More Refined Questions

- The Kolmogorov-Smirnov Test
- The Lilliefors Test

Goodness of Fit Tests for Categorical Data

This final chapter introduces some techniques to evaluate *categorical* or *count* data. Data sets in which observations fall into one of a number of mutually exclusive categories must be analyzed with special protocols, some of which are based on familiar distributions, e.g., the chi-square distribution.

There are many situations that require determining whether a sample could have been drawn from a population with a specified distribution. Such tests are based on a comparison of observed frequencies and frequencies that would be expected under the specified conditions. These tests fall under the category of **goodness of fit** tests. Typical of this sort of problem is the following.

> **EXAMPLE 11.1.** Four-o'clocks, *Mirabilis jalapa*, are plants native to tropical America. Their name comes from the fact that their flowers tend to open in the late afternoon. Individual four-o'clock plants can have red, white, or pink flowers. Flower color in this species is thought to be controlled by a single gene locus with two alleles expressing incomplete dominance, so that heterozygotes are pink-flowered, while homozygotes for one allele are white-flowered and homozygotes for the other allele are red-flowered. According to Mendelian genetic principles, self-pollination of pink-flowered plants should produce progeny that have red, pink, and white flowers in a 1:2:1 ratio. A horticulturalist self-pollinates several pink-flowered plants and produces 240 progeny with 55 that are red-flowered, 132 that are pink-flowered, and 53 that are white-flowered. Are these data reasonably consistent with the Mendelian model of a single gene locus with incomplete dominance?

Notice that a theoretical distribution of the data has been specified (i.e., the Mendelian probabilities). The question is whether the observed data fit this specification.

We will work with four different goodness of fit tests for categorical data. The binomial test is used for data that can be grouped into exactly two categories (e.g., male versus female or diseased versus healthy). It is used to determine whether the sample proportions of the two categories are what would be expected with a given binomial distribution. The chi-square test is used when there are several observational categories, as in Example 11.1. This test is used to determine whether the sample proportions are what would be expected if the population were described by some external or *extrinsic* hypothesis.

The chi-square test can also be used with *intrinsic* hypotheses to fit data to binomial, Poisson, or normal distributions whose parameters are determined by the sample itself. Finally, 2×2 and $r \times k$ *contingency* tables can be used to test the independence of various categories.

11.1   The Binomial Test

As mentioned, the binomial test applies to data that can be classified into two mutually exclusive categories. The test presumes that there are estimates for the proportions of the population falling into each of the categories prior to conducting the test. The purpose of the test is to obtain evidence that either supports or refutes the supposition that these hypothesized proportions are the actual proportions. When the proportions for the two categories are hypothesized to both equal 0.5, the binomial test is essentially the sign test in disguise.

Assumptions and Hypotheses

The assumptions for the binomial test are that:

1. An independent random sample of size n is drawn from the population.

2. The population is composed of two mutually exclusive categories.

3. p is the actual proportion of the population in the first of the two categories.

4. p_0 denotes the hypothesized value of p. The value p_0 must be proposed *prior* to conducting the study.

The hypothesis test can take one of three forms:

- Two-tailed. H_0: $p = p_0$ and H_a: $p \neq p_0$.

- Left-tailed. H_0: $p \geq p_0$ and H_a: $p < p_0$.

- Right-tailed. H_0: $p \leq p_0$ and H_a: $p > p_0$.

Test Statistic and Theory

The test statistic X is the number of "successes," i.e., observations falling into the first category. Under the assumptions of the test, X is a binomial random variable with parameters n and $p = p_0$.

Let x be the observed value of the test statistic. For a right-tailed test, H_0 is rejected if and only if $P(X \geq x) < \alpha$, where P is calculated using the binomial distribution with parameters n and $p = p_0$ (see Table C.1). For a left-tailed test, H_0 is rejected at the α level if and only if $P(X \leq x) < \alpha$. For a two-tailed test, H_0 is rejected if and only if either $P(X \leq x) < \frac{\alpha}{2}$ or $P(X \geq x) < \frac{\alpha}{2}$.

> **EXAMPLE 11.2.** The severe drought of 1987 in the U.S. affected the growth rate of established trees. It is thought that the majority of the trees in the affected areas each have a 1987 growth ring that is less than half the size of the tree's other growth rings. A sample of 20 trees is collected and 15 have this characteristic. Do these data support the claim?

> **SOLUTION.** Use the binomial test with $\alpha = 0.05$. Let p denote the proportion of trees with a 1987 growth ring that is less than half their usual size. The alternative hypothesis is that "the majority of the trees" have this property, that is, $H_a\colon p > 0.5$. The null hypothesis is $H_0\colon p \leq 0.5$. The test statistic is $X = 15$. This is a right-tailed test, so using Table C.1, we find

$$P(X \geq 15) = 1 - P(X \leq 14) = 1 - 0.9793 = 0.0207.$$

Since this value is less than $\alpha = 0.05$, we reject H_0. There is evidence that the majority of trees have growth rings for 1987 less than half their usual size.

> **EXAMPLE 11.3.** Pardalotes are small (8–12 cm) birds that feed mostly in the outer canopy, high in eucalyptus trees. However, they nest in holes dug in earth banks. ("Pardalote" comes from the Greek word for "spotted.") There are two different races of striated pardalote, *Pardalotus striatus*, in southeastern Queensland. Suppose that historical records indicate that race A comprised 70% of the population. A small census at Mt. Coot-tha locates 18 pardalotes: 11 of race A and 7 of race B. Do these figures indicate any difference from the historical pattern?

> **SOLUTION.** A two-sided binomial test with $\alpha = 0.05$ is appropriate. The hypothesized proportion of race A is $p = 0.70$. The null hypothesis is $H_0\colon p = 0.7$ while the alternative is $H_a\colon p \neq 0.7$. The test statistic is $X = 11$. Using Table C.1 with $n = 18$ and $p = 0.7$, the probability of lying in the right tail is

$$P(X \geq 11) = 1 - P(X \leq 10) = 1 - 0.1407 = 0.8593,$$

and the probability of lying in the left tail is

$$P(X \leq 11) = 0.2783.$$

Neither of these values is less than $\frac{\alpha}{2} = 0.025$, so there is not sufficient evidence to support a claim of a change in the population proportions of the two races. Note that the two probabilities that were computed do not sum to 1 because both tails overlap at $x = 11$.

Large Samples and the Proportions Test

The binomial test may be carried out with large samples by using the normal approximation to the binomial. When the normal approximation is used, the test is essentially identical to what is sometimes called the **proportions test**. The assumptions and hypotheses for the proportions test are the same as for the binomial test, with the addition of the single assumption that:

- The sample size n is sufficiently large so that $np_0 > 5$ and $n(1 - p_0) > 5$.

Let X continue to denote the number of successes or observations falling into the first category. Since $np_0 > 5$ and $n(1 - p_0) > 5$, then, as we saw at the end of Section 3.6, X is approximately normal with mean $\mu = np_0$, variance $\sigma^2 = np_0(1-p_0)$, and standard deviation $\sigma = \sqrt{np_0(1 - p_0)}$. Consequently,

$$z = \frac{X - \mu}{\sigma} = \frac{X - np_0}{\sqrt{np_0(1 - p_0)}} \tag{11.1}$$

has an approximately standard normal distribution. Thus, we can use z as the test statistic for large sample binomials. For increased accuracy one should also use a continuity correction, as described in Section 3.6.

> **EXAMPLE 11.4.** According to census data 24.4% of people residing in the United States have at least a bachelor's degree. A high-tech firm is considering relocating its headquarters to a city where it is believed that there is a high proportion of college graduates. In a random sample of $n = 480$ individuals in this city, 144 claim to have a bachelor's degree. Is there evidence that citizens of this city are more highly-educated than the population as a whole?
>
> **SOLUTION.** A right-tailed binomial test with $\alpha = 0.05$ is appropriate with null and alternative hypotheses H_0: $p \leq 0.244$ and H_a: $p > 0.244$, respectively. The large sample size requires the use of the normal approximation with
>
> $$\mu = np_0 = 480(0.244) = 117.12$$
>
> and
>
> $$\sigma = \sqrt{np_0(1 - p_0)} = \sqrt{480(0.244)(0.756)} = 9.41.$$
>
> Using the continuity correction for a right-tailed test and Table C.3, then
>
> $$P(X \geq 144) = 1 - P\left(Z \leq \frac{143.5 - 117.12}{9.41}\right)$$
> $$= 1 - P(Z \leq -2.80)$$
> $$= 1 - 0.9974$$
> $$= 0.0026.$$

Since this value is much less than $\frac{\alpha}{2} = 0.025$, we reject H_0. There is evidence that citizens of this city are more highly-educated than the U.S. population as a whole.

Let's see why the binomial test is sometimes referred to as the proportions test. Recall from Section 4.5 that the sample proportion is $\hat{p} = \frac{X}{n}$. If we take the test statistic from the large sample binomial test

$$z = \frac{X - np_0}{\sqrt{np_0(1 - p_0)}}$$

and divide the numerator and denominator by n we obtain

$$z = \frac{\frac{X}{n} - p_0}{\frac{\sqrt{np_0(1-p_0)}}{n}}.$$

Simplifying the square root and using $\frac{X}{n} = \hat{p}$, we conclude that

$$z = \frac{\frac{X}{n} - p_0}{\sqrt{\frac{p_0(1-p_0)}{n}}} = \frac{\hat{p} - p_0}{\sqrt{\frac{p_0(1-p_0)}{n}}}. \tag{11.2}$$

The formulas for z in (11.2) and in (11.1) for the binomial test with large samples are identical, but written in different algebraic forms. Notice in (11.2), z is expressed entirely in terms of proportions: the numerator is the difference between the sample and hypothesized proportions while the denominator is the standard error of the hypothesized proportion. Hence, the test is called the **proportions test**.

We may carry out the proportions test by using (11.2) as the test statistic as long as the sample satisfies the additional size assumption that np_0 and $n(1-p_0)$ are both greater than 5. We proceed as in the binomial test for large samples using the normal approximation.

> **EXAMPLE 11.5.** Assume that the hypotheses are the same as in Example 11.3, but that the sample size is $n = 180$ birds: 110 of race A and 70 of race B. Would these figures indicate any difference from the historical pattern?
>
> **SOLUTION.** A two-sided proportions test with $\alpha = 0.05$ is appropriate and the null and alternative hypotheses remain the same as in Example 11.3. The sample proportion is $\hat{p} = \frac{110}{180} = 0.611$ and the test statistic is
>
> $$z = \frac{\hat{p} - p_0}{\sqrt{\frac{p_0(1-p_0)}{n}}} = \frac{0.611 - 0.7}{\sqrt{\frac{0.7(0.3)}{180}}} = -2.60.$$
>
> Using Table C.3, for the right tail,
>
> $$P(Z \geq -2.60) = 1 - 0.0047 = 0.9953$$
>
> while for the left tail,
> $$P(Z \leq -2.60) = 0.0047.$$
>
> Since this latter value is less than $\frac{\alpha}{2} = 0.025$, we reject H_0 this time. There is evidence that there has been a change in the historical pattern.

There are two comments to make about this last example. First, it points out the importance of the sample size. Notice that the proportions of the two races were the same in Example 11.3 and Example 11.5. However, the larger sample size (10 times more work) in the second case allowed us to discern a significant difference from historical patterns, whereas with the smaller sample we could not rule out that the difference might have been due to chance.

Second, when using the proportions version of the test, there is no simple way to make a continuity correction so your answer will differ slightly from what the binomial test for large samples would produce.

11.2 Comparing Two Population Proportions

There are many situations where we may want to compare the proportions of two groups that have some specified characteristic. To do this, we can modify the proportions test using the ideas encountered in Chapter 7.

> **EXAMPLE 11.6.** A stent is a small wire mesh tube, often inserted in an artery, that acts as a scaffold to provide support to keep the artery open. Stents are a common therapy used to keep open totally occluded arteries in coronary patients. They may be implanted even several days after a heart attack under the assumption that any procedure that increases blood flow will lead to an improvement in survival rate.
>
> A study was carried out to determine whether the time when stents are implanted in patients' arteries changes the effectiveness of the treatment. In the study 2166 stable patients who had had heart attacks in the previous 3 to 28 days and had a total occlusion of an artery related to this attack were randomly assigned to two groups. Group 1 consisted of 1082 patients who received stents and optimal medical therapy, while Group 2 consisted of 1084 patients who received medical therapy alone with no stents. The results were that 171 of the patients in Group 1 and 179 patients in Group 2 had died by the end of four years.
>
> Researchers had expected to find a reduction in death rate for patients receiving stents under the conditions described. Was there any evidence for this?

Hypotheses, Test Statistic, and Theory

To carry out the two-sample proportions test an independent random sample of size n_1 is drawn from the first population and of size n_2 from the second. Note that these are not matched or paired samples as one might have in a before-after study. Let p_1 and p_2 denote the actual proportion of each population in the category of interest and $\hat{p}_1$ and $\hat{p}_2$ the sample proportions. The hypothesis test can take the usual three forms:

- Two-tailed. H_0: $p_1 = p_2$ and H_a: $p_1 \neq p_2$.

- Left-tailed. H_0: $p_1 \geq p_2$ and H_a: $p_1 < p_2$.

- Right-tailed. H_0: $p_1 \leq p_2$ and H_a: $p_1 > p_2$.

To estimate the true difference $D = p_1 - p_2$ between the two population proportions, we use the difference between the respective sample proportions,

$$\hat{D} = \hat{p}_1 - \hat{p}_2.$$

When the sample sizes are large, the sampling distribution of this difference is approximately normal.

Recall that when testing the difference between two populations means, the standard error of $\overline{X}_1 - \overline{X}_2$ depended on whether the populations had equal variances and could be pooled or not. While there is no preliminary F test to carry out in the current situation, there are two reasonable ways to define the standard error of $\hat{D}$. The first method corresponds to pooling the sample variance. A pooled or *combined proportion* is created. It may be calculated in either of two equivalent ways:

$$p_c = \frac{X_1 + X_2}{n_1 + n_2} = \frac{n_1\hat{p}_1 + n_2\hat{p}_2}{n_1 + n_2}. \tag{11.3}$$

Then the standard error of the difference is

$$\text{SE}_{\hat{D}_c} = \sqrt{p_c(1 - p_c)\left(\frac{1}{n_1} + \frac{1}{n_2}\right)}. \tag{11.4}$$

The second *unpooled* method does not create a common sample proportion but simply weights the individual sample variances,

$$\text{SE}_{\hat{D}_u} = \sqrt{\frac{\hat{p}_1(1 - \hat{p}_1)}{n_1} + \frac{\hat{p}_2(1 - \hat{p}_2)}{n_2}}. \tag{11.5}$$

The most common practice is to use the pooled version in hypothesis testing and the unpooled version when developing confidence intervals for the difference in population proportions. While this is inconsistent, there are technical reasons to take this approach and generally the difference between the two is small.

Using (11.4) the test statistic is

$$z = \frac{\hat{D}}{\text{SE}_{\hat{D}_c}} = \frac{\hat{p}_1 - \hat{p}_2}{\sqrt{p_c(1 - p_c)\left(\frac{1}{n_1} + \frac{1}{n_2}\right)}}, \tag{11.6}$$

which has an approximately standard normal distribution.

SOLUTION TO EXAMPLE 11.6. A left-tailed two-sample proportions test at the $\alpha = 0.05$ level is appropriate. The hypotheses are H_0: $p_1 \geq p_2$ and H_a: $p_1 < p_2$. Next

$$\hat{p}_1 = \frac{X_1}{n_1} = \frac{171}{1082} = 0.1580 \quad \text{and} \quad \hat{p}_2 = \frac{X_2}{n_2} = \frac{179}{1084} = 0.1651$$

and

$$p_c = \frac{171 + 179}{2166} = 0.1616.$$

The test statistic is

$$z = \frac{\hat{p}_1 - \hat{p}_2}{\sqrt{p_c(1 - p_c)\left(\frac{1}{n_1} + \frac{1}{n_2}\right)}} = \frac{0.1580 - 0.1651}{\sqrt{0.1616(0.8384)\left(\frac{1}{1082} + \frac{1}{1084}\right)}} = -0.45.$$

Using Table C.3 the probability of lying in the left tail is

$$P(Z \leq -0.45) = 0.3264.$$

Since $0.3264 \gg 0.05$ we cannot reject H_0; there is no evidence for a significant reduction in death rate with the implantation of stents in patients 3 to 28 days after heart attack.

Confidence Intervals for the Difference in Proportions

As we saw in (4.3), to develop a confidence interval we need a point estimate and its sampling distribution, a measure of the variability, and a desired level of confidence. The endpoints of the interval have the form

point estimate ± (confidence factor)(standard error).

The point estimate of the difference in population proportions is $\hat{D} = \hat{p}_1 - \hat{p}_2$. As mentioned above, when developing confidence intervals it is common practice to use

$$\text{SE}_{\hat{D}_u} = \sqrt{\frac{\hat{p}_1(1 - \hat{p}_1)}{n_1} + \frac{\hat{p}_2(1 - \hat{p}_2)}{n_2}}.$$

as the standard error of $\hat{D}$. The sampling distribution is normal. Consequently we have

FORMULA 11.1. The endpoints for a $(1 - \alpha)100\%$ confidence interval for the difference in proportions in two populations, $D = p_1 - p_2$, are given by

$$L_1 = \hat{D} - z_{1-\frac{\alpha}{2}} \sqrt{\frac{\hat{p}_1(1 - \hat{p}_1)}{n_1} + \frac{\hat{p}_2(1 - \hat{p}_2)}{n_2}}$$

and

$$L_2 = \hat{D} + z_{1-\frac{\alpha}{2}} \sqrt{\frac{\hat{p}_1(1 - \hat{p}_1)}{n_1} + \frac{\hat{p}_2(1 - \hat{p}_2)}{n_2}}.$$

EXAMPLE 11.7. Find a 95% confidence interval for the difference in death rates in Example 11.6.

SOLUTION. From earlier work, $\hat{D} = 0.1580 - 0.1651 = -0.0071$. From Table C.3, $z_{0.975} = 1.960$.

$$\text{SE}_{D_u} = \sqrt{\frac{0.1580(0.8420)}{1082} + \frac{0.1651(0.8349)}{1084}} = 0.0158.$$

So

$$L_1 = -0.0071 - 1.960(0.0158) = -0.0381$$

and

$$L_2 = -0.0071 + 1.960(0.0158) = 0.0239.$$

Since the confidence interval contains 0, there is no significant difference between these proportions, as we saw earlier.

11.3   The Chi-Square Test for Goodness of Fit

The first chi-square test that we consider is a test of differences between distributions. Its purpose is to compare the observed frequencies of a discrete, ordinal, or categorical data set with those of some theoretically expected discrete distribution such as the binomial, uniform, or Poisson. The situation in Example 11.1 is typical of problems where the chi-square test is useful.

Assumptions and Hypotheses

The assumptions for the χ^2 **test for goodness of fit** are that:

1. An independent random sample of size n is drawn from the population.

2. The population can be divided into a set of k mutually exclusive categories.

3. The expected frequencies for each category must be specified. Let E_i denote the expected frequency for the ith category. The sample size must be sufficiently large so that each E_i is at least 5. (Categories may be combined to achieve this.)

The hypothesis test takes only one form:

- H_0: The observed frequency distribution is the same as the hypothesized frequency distribution.

- H_a: The observed and hypothesized frequency distributions are different.

Generally speaking, this is an example of a statistical test where one wishes to *confirm* the null hypothesis.

Test Statistic and Theory

Let O_i denote the observed frequency of the ith category. The test statistic is based on the difference between the observed and expected frequencies, $O_i - E_i$. It is defined as

$$\chi^2 = \sum_{i=1}^{k} \frac{(O_i - E_i)^2}{E_i}.$$

The intuition for the test is that if the observed and expected frequencies are nearly equal for each category, then each $O_i - E_i$ will be small and, hence, χ^2 will be small. Small values of χ^2 should lead to acceptance of H_0 while large values lead to rejection. This test is always right-tailed. H_0 is rejected only when the test statistic exceeds a specified value. This statistic has an *approximate chi-square distribution* when H_0 is true; the approximation improves as sample size increases. The values of the chi-square family of distributions are tabulated in Table C.5.

There are two variations of this test. In the **extrinsic model** no population parameters need to be estimated from the data, while in the **intrinsic** form at least one population parameter is unknown and must be estimated. The difference in the two is reflected in the *degrees of freedom* of the test statistic. It is easier to describe this process in the context of specific examples, so let's look at a few.

The Extrinsic Model

EXAMPLE 11.8. Return to the situation in Example 11.1. The progeny of self-fertilized four-o'clocks were expected to flower red, pink, and white in the ratio 1:2:1. There were 240 progeny produced with 55 red plants, 132 pink plants, and 53 white plants. Are these data reasonably consistent with the Mendelian model?

SOLUTION. The hypotheses are

- H_0: The data are consistent with a Mendelian model (red, pink, and white flowers occcur in the ratio 1:2:1).

- H_a: The data are inconsistent with a Mendelian model.

The three colors are the categories. In order to calculate expected frequencies, no parameters need to be estimated, so this is an example of an extrinsic model. The Mendelian ratios are given: 25% red, 50% pink, and 25% white. Using the fact that there are 240 observations, the number of expected red four-o'clocks is $0.25(240) = 60$, that is $E_1 = 60$. Similar calculations for pink and white yield the following table.

Category	O_i	E_i	$\frac{(O_i - E_i)^2}{E_i}$
Red	55	60	0.42
Pink	132	120	1.20
White	53	60	0.82
Total	240	240	2.44

From the final column in the table,

$$\chi^2 = \sum_{i=1}^{3} \frac{(O_i - E_i)^2}{E_i} = 0.42 + 1.20 + 0.82 = 2.44.$$

For an extrinsic chi-square goodness of fit test, the test statistic has approximately a chi-square distribution with

$$\nu = \mathrm{df} = (\text{number of categories}) - 1.$$

In this example df $= 3 - 1 = 2$.

Assume the test is carried out at the $\alpha = 0.05$ level. Because the test is right-tailed, the critical value occurs when $P(\chi^2 > X_2^2) = \alpha$. Thus, in Table C.5 for df $= 2$ and $P = 1 - \alpha = 0.95$ the critical value is found to be 5.99. Since $2.44 < 5.99$, H_0 is accepted. There is support for the Mendelian genetic model.

The flower-color problem was an example of an *extrinsic* model because no parameters for the model distribution had to be calculated from the observed data. The Mendelian probabilities were "given," i.e., specified in advance.

The Intrinsic Model

The next example we consider is an *intrinsic* model. The intrinsic model requires an estimation of some population parameter(s) from the data collected. The degrees of freedom for the distribution is further reduced by 1 for each such parameter estimated. Thus, if there are k categories, and j parameters are estimated from the data,

$$\nu = \mathrm{df} = (\text{number of categories}) - 1 - j = k - 1 - j.$$

This formula is also valid for the extrinsic model. In this case because no population parameters are estimated from the data, $j = 0$ and df $= k - 1 - 0 = k - 1$ as required.

EXAMPLE 11.9. The Poisson distribution is useful for describing rare, random events such as severe storms. To review the characteristics of the Poisson distribution see Chapter 3. In the 98-year period from 1900 to 1997, there were 159 U.S. landfalling hurricanes. Does the number of landfalling hurricanes/year (see the table below) follow a Poisson distribution? (Based on data reported in: Bove, M. et al. 1998. Effect of El Niño on U.S. landfalling hurricanes, revisited. *Bulletin of the American Meteorological Society*, **79**: 2477–2482.)

Hurricanes/year	0	1	2	3	4	5	6
Frequency	18	34	24	16	3	1	2

SOLUTION. The hypotheses are

- H_0: The annual number of U.S. landfalling hurricanes follows a Poisson distribution.

- H_a: The annual number of U.S. landfalling hurricanes is inconsistent with a Poisson distribution.

To use the Poisson density formula we need to specify the mean number of hurricanes/year. Here this parameter must be estimated from the data, and this is what makes this an "intrinsic" model. From the information given

$$\overline{X} \text{ or } \hat{\mu} = \frac{159}{98} = 1.622 \text{ hurricanes/year.}$$

(As a check that a Poisson distribution might be reasonable here, recall that in a Poisson process, the mean and variance will be equal. Now check that the sample variance for the data above is $s^2 = 1.660$, which is very close to the mean.)

Using $\hat{\mu} = 1.622$ hurricanes/year, the Poisson density function or probability of observing x hurricanes/year is

$$f(x) = \frac{e^{-\mu}(\mu)^x}{x!} = \frac{e^{-1.622}(1.622)^x}{x!}.$$

In particular,

$$f(0) = P(0 \text{ hurricanes/year}) = \frac{e^{-1.622}(1.622)^0}{0!} = 0.198$$

$$f(1) = P(1 \text{ hurricane/year}) = \frac{e^{-1.622}(1.622)^1}{1!} = 0.320$$

$$f(2) = P(2 \text{ hurricanes/year}) = \frac{e^{-1.622}(1.622)^2}{2!} = 0.260$$

$$f(3) = P(3 \text{ hurricanes/year}) = \frac{e^{-1.622}(1.622)^3}{3!} = 0.140$$

$$f(4) = P(4 \text{ hurricanes/year}) = \frac{e^{-1.622}(1.622)^4}{4!} = 0.057$$

$$f(5) = P(5 \text{ hurricanes/year}) = \frac{e^{-1.622}(1.622)^5}{5!} = 0.018.$$

The calculation for the final category requires some explanation. The last "observed" category is 6 hurricanes/year. However, theoretically there is a small probability that more than 6 hurricanes might be observed in a year, so we use "6 or more hurricanes/year" as the last category.

$$P(\geq 6 \text{ hurricanes/year}) = 1 - P(< 6 \text{ hurricanes/year}) = 1 - 0.993 = 0.007,$$

where $P(< 6 \text{ hurricanes/year})$ is the sum of the previously calculated probabilities.

To determine the *expected frequency* for observing x hurricanes/year, multiply the probability of x hurricanes/year by the number of years, 98. For example, the expected frequency of observing no hurricanes/year is

$$E_0 = 98 \cdot f(0) = 98(0.198) = 19.40$$

and the expected frequency of observing 1 hurricane/year is

$$E_1 = 98 \cdot f(1) = 98(0.320) = 31.36.$$

In the same way the rest of the fourth column in the table below is filled in:

Count	Probability	O_i	E_i
0	0.198	18	19.40
1	0.320	34	31.36
2	0.260	24	25.48
3	0.140	16	13.72
4	0.057	3	5.59
5	0.018	1	1.76
≥ 6	0.007	2	0.69

(handwritten annotation: not >5 so have to collapse)

The expected frequencies for 5 and for 6 or more hurricanes/year are both less than 5. But the χ^2 test requires that all E_i be at least 5. To get around this difficulty, adjacent categories should be combined until the expected number is at least 5. In this case, it requires collapsing the last three categories into a single category "≥ 4." (See the table below.)

Count	O_i	E_i	$\frac{(O_i - E_i)^2}{E_i}$
0	18	19.40	0.101
1	34	31.36	0.222
2	24	25.48	0.086
3	16	13.72	0.379
≥ 4	6	8.04	0.518

Sum the entries in the last column in the table above to obtain the test statistic

$$\chi^2 = \sum_{i=1}^{5} \frac{(O_i - E_i)^2}{E_i} = 1.306.$$

Now determine the degrees of freedom. There are five categories in the final analysis and one parameter, μ, was estimated from the data. So df $= 5 - 1 - 1 = 3$. From Table C.5, the critical value for a test performed at the $\alpha = 0.05$ level with 3 degrees of freedom is 7.81. Since 1.306 is well under that value, there is evidence to support the claim that the annual number of U.S. landfalling hurricanes is described by a Poisson process with $\mu = 1.622$. Accept H_0. See Figure 11.1.

The Extrinsic and Intrinsic Models Compared

Even though the calculation of the chi-square statistic is the same for both the extrinsic and intrinsic models, the degrees of freedom for the intrinsic model are reduced by one for each parameter estimated from the data. The effect of reducing the degrees of freedom is to decrease the critical value of the chi-square test statistic for any confidence level α. This reduction occurs because by using parameters estimated from the data we have moved the hypothesized distribution closer to the empirical (sample) distribution. Consequently, this decreases the differences between the observed and expected values which in turn decreases the test statistic. As a result, the critical value must be decreased as well.

11.4   The Chi-Square Test for $r \times k$ Contingency Tables

There are several similar statistical tests that fall under the category of "chi-square tests." We will now examine a second important type which is used to test whether the distribution of a categorical variable is the same in two or more populations.

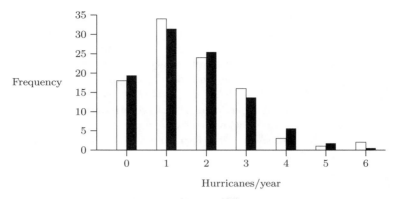

FIGURE 11.1. The observed numbers of U.S. landfalling hurri-
canes/year from 1900 to 1997 (unshaded bars) is well-approximated
by a Poisson distribution with $\mu = 1.622$ (shaded bars). (Adapted
from data reported in: Bove, M. et al. 1998. Effect of El Niño on U.S.
landfalling hurricanes, revisited. *Bulletin of the American Meteorological Society*, **79**: 2477–2482.)

Contingency Tables

Oftentimes we are interested in how categorical variables are distributed among two or
more populations. Let r be the number of categories and k the number of populations
or treatments. We can form a table or matrix with r rows, one for each category, and
k columns, one for each population. The entry O_{ij} in row i and column j represents
the number of observations of category i in population j. Such an arrangement of
data is called an $r \times k$ **contingency table**. The question at hand is whether there
is a relationship or dependency between the row and column variables. (Contingent
means dependent.)

EXAMPLE 11.10. In a study of hermit crab behavior at Point Lookout, North
Stradbroke Island, a random sample of three types of gastropod shells was collected.
Each shell was then scored as being either occupied by a hermit crab or empty. Shells
with living gastropods were not sampled. Do hermit crabs prefer a certain shell type?
Or do hermit crabs occupy shells in the same proportions as they occur as empty
shells? In other words, is the shell species independent of whether it is occupied?

Species	Occupied	Empty	Total
Austrocochlea	47	42	89
Bembicium	10	41	51
Cirithid	125	49	174
Total	182	132	314

The table is a 3×2 contingency table. It is common to include row and column
totals, but they are not counted in the dimensions of the table. Here we view the
occupied shells as one population and the empty ones as another. At a glance, it is
difficult to discern differences in the distributions in part because the samples drawn
from each population are of different sizes.

One method to determine whether such differences exist is to calculate the expected number of observations for each category based on the assumption that the row and column variables are *independent*. For example, to calculate the expected number of occupied *Austrocochlea*, take the fraction of occupied shells in the entire sample, $\frac{182}{314} = 0.5796$, and multiply it by the fraction of *Austrocochlea* shells in the entire sample, $\frac{89}{314} = 0.2834$. This yields $0.5796 \times 0.2834 = 0.1643$. In other words, if shell species and whether a shell were occupied or not were independent events, then we could expect 16.43% of all observations to be occupied *Austrocochlea* shells. Since there were a total of 314 observations, the expected number of occupied *Austrocochlea* shells is $0.1643 \times 314 = 51.59$.

More generally, let E_{ij} denote the expected number of observations falling into row i and column j. Then

$$E_{ij} = \frac{r_i}{N} \times \frac{c_j}{N} \times N = \frac{r_i c_j}{N},$$

where r_i is the sum of the observations in row i, c_j is the sum of the observations in column j, and N is the size of the entire sample. Verify that the table of expected observations for Example 11.10 is

Species	Occupied	Empty	Total
Austrocochlea	51.59	37.41	89
Bembicium	29.56	21.44	51
Cirithid	100.85	73.15	174
Total	182	132	314

An important consequence of the independence assumption is that in the table of expected values, the distribution of the categorical variables is identical in all populations. In fact, *independence of the row and column variables is equivalent to the distributions being identical for each population.* It is easy to see why in Example 11.10. Notice that *Austrocochlea* shells are 28.3% of the total sample ($\frac{r_1}{N} = \frac{89}{314}$), *Bembicium* are 16.2% of the total ($\frac{r_2}{N} = \frac{51}{314}$), and *Cirithids* are 55.4% ($\frac{r_3}{N} = \frac{174}{314}$). Because E_{ij} is just $(\frac{r_i}{N})(c_j)$, the expected entries in the "occupied column" ($j = 1$) are in a ratio of 28.3:16.2:55.4, just as they are in the "empty" column ($j = 2$). The expected ratios within the columns are the same, but the numerical values are different because they are scaled differently by the respective sample sizes (c_j). The fact that independence of the row and column variables is equivalent to the distributions being identical for each population means that the null hypothesis for the chi-square test for $r \times k$ contingency tables can take either of two forms.

Assumptions and Hypotheses

An $r \times k$ contingency table can arise from two different sampling methods.

1. A single random sample is drawn and two categorical variables are observed, one with k categories and the other with r categories.

2. Independent random samples are drawn from k different populations and a categorical variable is observed.

The data in Example 11.10 were collected using the first method. A single random sample was collected and categorized in two ways. By looking at the data table alone, however, we would not be able to determine this. Two independent samples might have been collected, shells occupied by hermit crabs and the other unoccupied shells.

The statements of the null and alternative hypotheses are slightly different in the two cases, reflecting the equivalence discussed earlier:

1)

- H_0: The row and column variables are independent.

- H_a: The row and column variables are not independent.

Alternatively, the hypotheses may take the form:

2)

- H_0: The distribution of the row categories is the same in all k populations.

- H_a: The distribution of the row categories is not the same in all k populations.

Test Statistic and Theory

In either case (and the difference in emphasis can be subtle) the test statistic is calculated in the same way. The χ^2 statistic is based on the difference between the observed and expected numbers of observations.

$$\chi^2 = \sum_{ij} \frac{(O_{ij} - E_{ij})^2}{E_{ij}}, \tag{11.7}$$

where the sum is taken over all entries in the $r \times k$ contingency table.

The intuition for the test is that if the observed and expected frequencies are similar for each category, then each $O_{ij} - E_{ij}$ will be small and, hence, χ^2 will be small. So small values of χ^2 lead to acceptance of H_0 (i.e., the distributions are the same, or row and column variables are independent), while large values lead to rejection of H_0. This test is always right-tailed. H_0 is rejected only when the test statistic *exceeds* a specified value. This statistic has an approximate chi-square distribution when H_0 is true; the approximation improves as sample size increases. For an $r \times k$ contingency table, the test statistic has approximately a chi-square distribution with

$$\nu = \mathrm{df} = (r - 1) \times (k - 1).$$

SOLUTION TO EXAMPLE 11.10. The hypotheses are

- H_0: The status (occupied or not) is independent of the shell species.

- H_a: The status is not independent of the shell species.

There are 2 degrees of freedom since $\mathrm{df} = (3 - 1) \times (2 - 1) = 2$. Using the earlier tables of observed and expected values we calculate

$$\chi^2 = \sum_{ij} \frac{(O_{ij} - E_{ij})^2}{E_{ij}}$$

$$= \frac{(47 - 51.59)^2}{51.59} + \frac{(10 - 29.56)^2}{29.56} + \frac{(125 - 100.85)^2}{100.85}$$

$$+ \frac{(42 - 37.41)^2}{37.41} + \frac{(41 - 21.44)^2}{21.44} + \frac{(49 - 73.15)^2}{73.15}$$

$$= 45.52.$$

Because the test is right-tailed, the critical value occurs when $P(\chi^2 > X_2^2) = \alpha$. In Table C.5 for df $= 2$ and $P = 1 - \alpha = 0.95$ the critical value is found to be 5.99. Since $45.52 \gg 5.99$, H_0 is rejected. There is reason to believe that shell species and occupancy are not independent. That is, hermit crabs are "selective" in the species of shell that they occupy.

EXAMPLE 11.11. A study was conducted in the reef flat of Lady Elliot Island to examine the distribution of the animals associated with different species of coral. Three species of coral were selected and appropriate coral heads were sampled. The number of snails of species A and B associated with each were recorded. Is the distribution of snail species the same for all three coral species? Use the data below to test the hypothesis at the $\alpha = 0.05$ level. Clearly state the hypotheses and interpret the results.

	Coral			
	Pocillopora eydouxi	*Acropora* sp.	*Acropora aspera*	Total
Species A	6	2	14	22
Species B	7	21	1	29
Total	13	23	15	51

SOLUTION. The hypotheses are

- H_0: The distribution of snail species A and B is the same in all three types of coral heads.

- H_a: The distribution of snail species A and B is not the same in all three types of coral heads.

Note that this is an example of the second form of a contingency table analysis because $k = 3$ independent samples were taken corresponding to the different species of coral. The null hypothesis is that the distributions of the animals associated with different species of coral are the same for all three species; the alternative hypothesis is that the distributions vary with species of coral. We will perform the test at the $\alpha = 0.05$ level. Verify that the expected values are as given in the following table.

	Coral			
	Pocillopora eydouxi	*Acropora* sp.	*Acropora aspera*	Total
Species A	5.61	9.92	6.47	22
Species B	7.39	13.08	8.53	29
Total	13	23	15	51

It follows that

$$\chi^2 = \sum_{ij} \frac{(O_{ij} - E_{ij})^2}{E_{ij}} = 26.58.$$

Since df $= (3 - 1) \times (2 - 1) = 2$, the critical value for the test is the same as in the previous example, 5.99. Again $\chi^2 = 26.58 > 5.99$, so the null hypothesis is rejected. Based on these data, the distribution of snail species A and B varies with coral species.

Comment and Caution

Because the critical values in Table C.5 are only approximately correct for determining the P value associated with χ^2, the sample size must be sufficiently large to ensure the approximation is valid. The rule of thumb is that each *expected* frequency E_{ij} should be at least equal to 5. Note that Example 11.11 satisfies this criterion even though some of the *observed* frequencies fall below 5.

2 × 2 Contingency Tables

for class, dont need correction

A special case of the $r \times k$ contingency table is the 2×2 table. Because of the small number of cells (4), a correction factor for discontinuity is usually employed and the computational formula is somewhat more direct than for the larger $r \times k$ tables.

EXAMPLE 11.12. A study of kidney damage during organ retrieval for transplantation was conducted in the United Kingdom using data from the UK National Transplant Database for the 1992 to 1996 period. In many cases of organ donation, when the kidneys are retrieved the liver is retrieved as well in a single surgical procedure. When both types of organs were retrieved, the researchers categorized the surgical team, based on the operating surgeon's specialty, as either a renal retrieval team or a liver retrieval team. Their data are given below. Was the rate of reported damage to kidneys independent of the type of surgical team? Analyze the data with a 2×2 contingency table. *Note:* 94% of damaged organs were still transplanted. (Based on data from: Wigmore, S. et al. 1999. Kidney damage during organ retrieval: Data from the UK National Transplant Database. *The Lancet,* **354**: 1143–1146.)

Team	Damaged kidneys	Undamaged kidneys	Total
Renal retrieval	454	1692	2146
Liver retrieval	415	2054	2469
Total	869	3746	4615

SOLUTION. The hypotheses are

- H_0: Surgical team and condition of kidneys are independent.

- H_a: An association between type of surgical team and condition of kidneys exists (one team experiences more success than the other in these circumstances).

The expected values for the individual cells can be calculated as in Examples 11.10 and 11.11.

$$\text{df} = (r-1)(k-1) = (2-1)(2-1) = 1 \times 1 = 1.$$

The test statistic, however, should have a correction for discontinuity. Remember the χ^2 distribution is continuous while these data are discrete. The correction factor of 0.5 improves the approximation in situations with 1 degree of freedom, much like the 0.5 correction in the normal approximation to the binomial.

The test statistic becomes

$$\chi^2 = \sum_{ij} \frac{(|O_{ij} - E_{ij}| - 0.5)^2}{E_{ij}}. \tag{11.8}$$

This formula can be modified with some careful algebra into a form that avoids computation of expected values. Consider the following table:

n_{11}	n_{12}	$n_{1.}$
n_{21}	n_{22}	$n_{2.}$
$n_{.1}$	$n_{.2}$	$n_{..}$

The marginal sums are expressed in the now familiar dot notation. The formula for the 2×2 contingency χ^2 with the correction factor can be shown to be equivalent to

$$\chi^2 = \frac{n_{..}\left(|n_{11}n_{22} - n_{12}n_{21}| - \frac{n_{..}}{2}\right)^2}{n_{1.}n_{2.}n_{.1}n_{.2}}.$$

While initially this formula appears complicated, it avoids expected values and summation. It is the preferred method if you are doing this analysis by hand.

Returning to Example 11.12,

$$\chi^2 = \frac{4615\left(|454 \times 2054 - 1692 \times 415| - \frac{4615}{2}\right)^2}{2146 \times 2469 \times 869 \times 3746} = 13.91.$$

From Table C.5 with df $= 1$, the P value associated with a test statistic of 13.91 is less than 0.005. We reject H_0. The kidney damage rate for renal surgical teams (21.2%) is significantly higher than the kidney damage rate for liver surgical teams (16.8%), a rather unexpected result.

STOP

Partitioning the Chi-Square Test

The final example in this section demonstrates the use of various chi-square analyses to pinpoint discrepancies from expectation.

EXAMPLE 11.13. A genetics student wished to repeat one of Gregor Mendel's classic experiments with garden peas, *Pisum sativum*. She decided to study two characteristics: stem length and seed pod color. From her research she knows that a single gene locus with two alleles controls stem length. AA or Aa produces tall plants (about 1 m) while aa produces short plants (approximately 0.5 m). Also a single locus with two alleles controls seed pod color with BB and Bb producing green seed pods and bb producing yellow seed pods. In other words, both loci exhibit complete dominance. From Mendel's published work these two gene loci are assumed to be independently assorting. The student crosses together plants that are tall with green seed pods and are known to be heterozygous at both loci:

tall, green pods (AaBb)   ×   tall, green pods (AaBb)

↓

Experimentally produced offspring:

178 tall, green pods (A_B_)
30 tall, yellow pods (A_bb)
62 short, green pods (aaB_)
10 short, yellow pods (aabb)

If these genes behave according to Mendel's laws, she expects the offspring to be in a 9:3:3:1 ratio. Test this hypothesis.

SOLUTION. The hypotheses are

- H_0: The results are in a 9:3:3:1 phenotypic ratio.

- H_a: The results deviate significantly from a 9:3:3:1 ratio.

Category	O_i	E_i	$\frac{(O_i - E_i)^2}{E_i}$
A_B_	178	157.5	2.668
A_bb	30	52.5	9.643
aaB_	62	52.5	1.719
aabb	10	17.5	3.214
Total	280	280	17.244

From Table C.5 with $\nu = 4 - 1 = 3$ and $\alpha = 0.05$, the c.v. $= 7.81$. Since $17.244 > 7.81$, we reject H_0, the data deviate significantly from a 9:3:3:1 ratio.

The student would like to know exactly why the data set failed to meet the expectation that she had so carefully researched. Thinking about this situation, she realized that the 9:3:3:1 ratio was predicated on three assumptions.

1. The gene locus for plant height produced offspring in a 3 A_:1 aa ratio.
2. The gene locus for seed pod color produced offspring in a 3 B_:1 bb ratio.
3. The 2 gene loci, A and B, are independent of each other.

We can test each of these assumptions separately and independently by the following methods:

1. First we test

- H_0: Offspring have a 3 tall:1 short ratio.
- H_a: Offspring deviate significantly from the expected ratio of 3 tall:1 short.

Category	O_i	E_i	$\frac{(O_i - E_i)^2}{E_i}$
Tall (A_B_ and A_bb)	208	210	0.18
Short (aaB_ and aabb)	72	70	0.57
Total	280	280	0.76

From Table C.5 with $\nu = 1$ and $\alpha = 0.05$, c.v. $= 3.84$. Since observed and expected values very similar, $\chi^2 = 0.076$ and H_0 is accepted. Plant heights in offspring are in a 3:1 ratio.

2. Next we test

- H_0: Offspring have a 3 green seed pod:1 yellow seed pod ratio
- H_a: Offspring deviate significantly from the expected ratio of 3 green:1 yellow.

Category	O_i	E_i	$\frac{(O_i - E_i)^2}{E_i}$
Green (A_B_ and aaB_)	240	210	4.286
Yellow (A_bb and aabb)	40	70	12.857
Total		280	280	17.143

So $\chi^2 = 17.143$. From Table C.5 with $\nu = 1$ and $\alpha = 0.05$, c.v. $= 3.84$. Since $17.143 > 3.84$, we reject H_0. The phenotypic ratio for the seed pod color is significantly different from 3:1.

3. Finally we test

- H_0: Gene locus A is independent of gene locus B.
- H_a: The loci are not independent.

We test this H_0 with a 2×2 contingency table:

	B_	bb	Total
A_	178	30	208
aa	62	10	72
Total	240	40	280

$$\chi^2 = \frac{n_{..}\left(|n_{11}n_{22} - n_{12}n_{21}| - \frac{n_{..}}{2}\right)^2}{n_{1.}n_{2.}n_{.1}n_{.2}} = \frac{280\left[|(178 \times 10) - (30 \times 62)| - \frac{280}{2}\right]^2}{(208)(72)(240)(40)} = 0.007.$$

Actually, because the sample size is large (> 200) the correction $\frac{n_{..}}{2}$ is unnecessary. Without this correction the χ^2 value becomes 0.012.

Again from Table C.5 with $\nu = 1$ and $\alpha = 0.05$, c.v. $= 3.84$. The calculated value of χ^2 is obviously less than the critical value, so we accept H_0 that the two loci are behaving independently.

Looking at the three single degree of freedom chi-square analyses, one can see that the discrepancy in the overall χ^2 testing 9:3:3:1 is due to a distortion in the green to yellow seed pod ratio. There may be many reasons for this within-locus distortion, including differential survival of the two phenotypes. Mendelian ratios assume equal fitness among the various phenotypes. Here we may be seeing a Darwinian concept obscuring the Mendelian principle! Notice the distorted ratio for the B locus has no effect on the 2×2 contingency test. Also the single degree of freedom chi-squares are a decomposition of the overall chi-square and sum to that chi-square value with some small rounding differences. The analysis done here is analogous to the partitioning of the sums of squares in analysis of variance. This type of decomposition can be used on chi-square analyses as long as each test is independent of the other tests involved in the decomposition.

11.5   The Kolmogorov-Smirnov Test

As with the previous goodness of fit tests, the purpose of this test is to determine whether a random sample from some unknown distribution is, in fact, actually from some specified distribution. While the χ^2 test is specifically designed for use with discrete or categorical data, the Kolmogorov-Smirnov test is used to test whether a random sample comes from a population with a particular *continuous* distribution (e.g., normal). Why might such a question arise in a data analysis?

To help understand the analysis used in the Kolmogorov-Smirnov test, we begin with a simple problem.

EXAMPLE 11.14. The heights of 18- to 22-year-old females in the US are normally distributed with $\mu = 163$ cm and $\sigma = 8$ cm. As part of a data analysis exercise on the Kolmogorov-Smirnov test, 10 female students were selected at random in one of our statistics classes and their heights measured. Do the data below follow the same distribution as the national data or are the heights of women at our college distributed differently?

149   157   159   160   160   163   168   170   178   185

The sample mean and standard deviation are $\overline{X} = 164.9$ cm and $s = 10.6$ cm. In Chapter 6 we developed tests to determine whether the mean or variance of a sample differed significantly from hypothesized values. You can check that neither does in this case. But these statistics do not tell us whether the heights are normally distributed. We are interested in the *shape* of the entire distribution, not just point estimates.

One possible way to get at this question is to use the fact that for a normal distribution about 68% of the observations should fall within a single standard deviation of the mean μ and about 95% of the observations should fall within two standard deviations of the mean. The table below shows that this is nearly the case with these data.

Std. dev.	Range	Observations	Observed pct.	Expected pct.
±1	[155, 171]	7	70	68.3
±2	[147, 179]	9	90	95.4
±3	[139, 187]	10	100	99.7

The observed percentages for the various standard deviation ranges do not differ all that much from the expected percentages. But are they "sufficiently close" to a normal distribution in the sense that any differences could be solely due to chance? The Kolmogorov-Smirnov test employs an expanded version of the table above to answer this question.

Assumptions and Hypotheses

The χ^2 test for goodness of fit uses expected values, i.e., the probability density function of the hypothesized distribution. In contrast, the Kolmogorov-Smirnov test uses the cumulative distribution function. The assumption of the test is that the data consist of a random sample of size n from a specified continuous cumulative distribution $F(x)$. The sample itself is used to create the **empirical cumulative distribution**, which is denoted by $S(x)$. The values of $S(x)$ are the tabulations of the relative cumulative frequencies of Chapter 1 stated as proportions rather than as percentages. Using this notation the hypotheses of the test can be stated as:

- H_0: $S(x) = F(x)$ for all x. The actual and hypothesized cumulative distributions are identical.

- H_a: $S(x) \neq F(x)$ for at least one value of x. The actual and hypothesized cumulative distributions differ significantly for at least one point.

Test Statistic and Intuition

The test statistic is computed in the following way. Begin by determining the empirical cumulative distribution function $S(x)$ by tabulating the relative cumulative frequencies for the sample, as described in Section 1.6.

Next, $S(x)$ is compared to the hypothesized distribution $F(x)$ by listing the differences $S(x) - F(x)$ for all observed values x. (See Table 11.1.) The largest absolute value of these differences is used as the test statistic K. Geometrically K measures the maximum distance between the graphs of the observed distribution $S(x)$ and the hypothesized distribution $F(x)$ (see Figure 11.2). The closer K is to 0, the closer the distribution of the sample is in *shape* to the specified distribution. Thus, when K is sufficiently small, we accept H_0. On the other hand, if K is large, then $S(x)$ and $F(x)$ differ significantly for at least one point, which supports the alternative hypothesis H_a that the two distributions are different.

SOLUTION TO EXAMPLE 11.14. Select a confidence level for the test, say, $\alpha = 0.05$. Let $S(x)$ denote the empirical distribution and $F(x)$ denote the normal distribution with $\mu = 163$ cm and $\sigma = 8$ cm. The hypotheses are H_0: $S(x) = F(x)$ for all x versus H_a: $S(x) \neq F(x)$ for at least one value of x. The actual and hypothesized cumulative distributions differ significantly at some point.

To carry out the Kolmogorov-Smirnov test, we construct a table (see Table 11.1) of preliminary calculations, as described below.

Column 1. Begin by listing every observation, from smallest to largest, in the first column. Notice that the two observations of 160 cm each appear in the column as the fourth and fifth entries, respectively.

TABLE 11.1. A table used to determine the Kolmogorov-Smirnov test statistic K for Example 11.14. The column calculations are explained in the text. K is the largest value appearing in the final two columns.

x_i	$S(x_i)$	$F(x_i)$	$|S(x_i) - F(x_i)|$	$|S(x_{i-1}) - F(x_i)|$
149	0.10	0.0401	0.0599	0.0401
157	0.20	0.2266	0.0266	0.1266
159	0.30	0.3085	0.0085	0.1085
160	0.40	0.3520	0.0480	0.0520
160	0.50	0.3520	0.1480	0.0480
163	0.60	0.5000	0.1000	0.0000
168	0.70	0.7357	0.0357	0.1357
170	0.80	0.8106	0.0106	0.1106
178	0.90	0.9699	0.0699	**0.1699**
185	1.00	0.9970	0.0030	0.0970

Column 2. This column contains the values of the empirical cumulative distribution $S(x)$ (or in the language of Chapter 1, the relative cumulative frequencies) for the lengths specified in the first column. For example, 159 cm is the third smallest of the 10 observations, so the relative cumulative frequency is $S(159) = \frac{3}{10} = 0.30$. The last entry in the column is necessarily 1.00 since all observations are less than or equal to the largest observation. The first two columns form a relative cumulative frequency table for the sample.

Column 3. The third column contains the values of the hypothesized cumulative distribution $F(x)$ and requires the most calculation. In this case, $F(x)$ is the normal distribution with $\mu = 163$ cm and $\sigma = 8$ cm. Convert each x_i to a standard normal value in the usual fashion. For example,

$$F(x_2) = P(X \leq 157) = P\left(Z \leq \frac{157 - 163}{8}\right) = P(Z \leq -0.75) = 0.2266.$$

A spreadsheet or calculator with a table function is very useful at this stage. Depending on when you roundoff values, your answers may differ slightly from those in the table.

Column 4. Now compare the two distributions. In particular, column 4 contains the absolute values of the differences between the empirical distribution (column 2) and the corresponding hypothesized distribution (column 3). This difference is $|S(x_i) - F(x_i)|$.

Column 5. There's one more column in the table. It compares the previous empirical distribution value $S(x_{i-1})$ (column 2, previous row) to the current expected distribution value $F(x_i)$ (column 3, current row). This difference is $|S(x_{i-1}) - F(x_i)|$. Note: Because there is no observation x_0, we treat $S(x_0)$ as if it were 0 so the first entry in this column is just $F(x_1)$.

Figure 11.2 indicates why this final comparison is necessary. When plotted, the values of $S(x_i)$ form a step function. Remember, we are trying to determine the maximum difference between this step function $S(x)$ and the hypothesized distribution $F(x)$. This maximum difference can occur at either end of a step. The differences between $F(x)$ and the front end of a step are found in column 4, while the differences between $F(x)$ and the back end of a step are found in column 5.

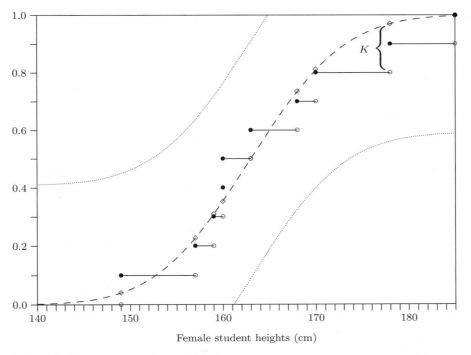

Female student heights (cm)

FIGURE 11.2. The test statistic K is the maximum vertical distance between the graph of $S(X)$ and $F(X)$ [the solid —— and dashed - - - curves, respectively]. This can be visualized as the largest of the vertical distances between the solid • points $(x_i, S(x_i))$ and the diamond ◇ points $(x_i, F(x_i))$, which correspond to the fourth column of Table 11.1, or between the hollow ○ points $(x_i, S(x_{i-1}))$ and the diamond ◇ points $(x_i, F(x_i))$, which correspond to the fifth column of Table 11.1. The vertical distance from $F(X)$, the dashed line, to the corridor boundary is the critical value of the test. We accept H_0 only if $S(X)$ remains within this critical value corridor.

The test statistic K is the maximum value that occurs in the last two columns of Table 11.1. That is,

$$K = \max_{x_i} \left\{ |S(x_i) - F(x_i)|, \; |S(x_{i-1}) - F(x_i)| \right\}. \tag{11.9}$$

Looking back at Table 11.1 we find that

$$K = |S(x_8) - F(x_9)| = |S(170) - F(178)| = 0.1699.$$

This is illustrated in Figure 11.2.

Use Table C.14 to locate the critical value for the test. The test is two-sided. At the $\alpha = 0.05$ level with a sample size of $n = 10$, the c.v. $= 0.4092$. H_0 is rejected only if K is greater than the critical value. Since $0.1699 \leq 0.4092$, H_0 is accepted. We may continue to assume that female student heights in our class are from a normally distributed population with a mean of 163 cm and a standard deviation of 8 cm.

EXAMPLE 11.15. We have suggested that a binomial distribution in which $np > 5$ and $n(1 - p) > 5$ is well approximated by a normal distribution with $\mu = np$ and $\sigma^2 = np(1 - p)$. We used a Microsoft Excel spreadsheet to generate a random sample of the number of successes in 20 trials from a binomial distribution with $n = 20$ and $p = 0.5$. The results are listed below.

$$
\begin{array}{cccccccccc}
10 & 9 & 9 & 13 & 10 & 11 & 12 & 8 & 6 & 12 \\
10 & 10 & 7 & 14 & 10 & 11 & 12 & 10 & 9 & 13 \\
\end{array}
$$

Does a Kolmogorov-Smirnov test indicate that these data could reasonably have come from the normal distribution $F(x)$ with $\mu = np = 20(0.5) = 10$ and $\sigma^2 = np(1 - p) = 20(0.5)(0.5) = 5$?

SOLUTION. Notice that even though the sample comes from a *discrete* binomial distribution, we are asking whether we may act as if the sample were drawn from a *continuous* normal distribution. Thus, a Kolmogorov-Smirnov test is appropriate.

The hypotheses are H_0: $S(x) = F(x)$ for all x and H_a: $S(x) \neq F(x)$ for at least one value of x. We will carry out the test with $\alpha = 0.05$. We construct a table exactly as in Example 11.14. With a sample size of $n = 20$, using a spreadsheet (or statistical software package) is very helpful.

The test statistic is the largest value in either of the last two columns of Table 11.2 so $K = 0.2000$. From Table C.14 with $n = 20$ and $\alpha = 0.05$, the c.v. $= 0.2941$. Since $K < 0.2941$, we accept H_0. We conclude that we can reasonably act as if the data came from a normal distribution with $\mu = 10$ and $\sigma^2 = 5$, as was suggested.

The next example shows how the Kolmogorov-Smirnov test may be applied to test whether a sample comes from some distribution other than a normal one, as long as that distribution is continuous and entirely specified in advance.

EXAMPLE 11.16. It is a tradition for the students in a certain statistics class to organize a charity raffle at the end of the term. Students may "buy" any number between 0 and 1, inclusive. Numbers may be chosen to any level of precision. For example, 0.1, 0.125, $\frac{2}{9} = .\overline{2}$, and $\frac{\pi}{10} = .31415\ldots$ are all valid entries. Entries are written on a piece of paper and submitted in a sealed envelope to the instructor so that no one other than the entrant knows which number he or she has chosen. The winner is determined by using a statistical package to generate a random number from the uniform distribution on $[0, 1]$. The winner is the person having "purchased" the number closest to the random number. The prize is split if two or more people win. (Think about what your strategy would be in selecting a number!)

The winning numbers for the last decade are listed below. Do they appear to have come from the uniform distribution with range limits 0 to 1?

$$
\begin{array}{ccccc}
0.422742393 & 0.264503922 & 0.775292215 & 0.665913877 & 0.570848720 \\
0.536332286 & 0.326822718 & 0.546311838 & 0.748924223 & 0.834620197 \\
\end{array}
$$

TABLE 11.2. The table used to determine the Kolmogorov-Smirnov test statistic K for Example 11.15.

x_i	$S(x_i)$	$F(x_i)$	$|S(x_i) - F(x_i)|$	$|S(x_{i-1}) - F(x_i)|$
6	0.05	0.0368	0.0132	0.0368
7	0.10	0.0899	0.0101	0.0399
8	0.15	0.1855	0.0355	0.0855
9	0.20	0.3274	0.1274	0.1774
9	0.25	0.3274	0.0774	0.1274
9	0.30	0.3274	0.0274	0.0774
10	0.35	0.5000	0.1500	**0.2000**
10	0.40	0.5000	0.1000	0.1500
10	0.45	0.5000	0.0500	0.1000
10	0.50	0.5000	0.0000	0.0500
10	0.55	0.5000	0.0500	0.0000
10	0.60	0.5000	0.1000	0.0500
11	0.65	0.6726	0.0226	0.0726
11	0.70	0.6726	0.0274	0.0226
12	0.75	0.8145	0.0645	0.1145
12	0.80	0.8145	0.0145	0.0645
12	0.85	0.8145	0.0355	0.0145
13	0.90	0.9101	0.0101	0.0601
13	0.95	0.9101	0.0399	0.0101
14	1.00	0.9632	0.0368	0.0132

SOLUTION. In the uniform distribution on $[0,1]$, every number is equally likely to occur. That makes its cumulative distribution easy to determine: $F(x) = x$. For example, $F(0.27) = P(0 \le X \le 0.27) = 0.27$ because 27% of the interval $[0,1]$ lies at or below 0.27. The distribution is completely specified (and is continuous), so a Kolmogorov-Smirnov test is appropriate. The analysis appears in Table 11.3.

TABLE 11.3. The table used to determine the Kolmogorov-Smirnov test statistic K for Example 11.16.

x_i	$S(x_i)$	$F(x_i)$	$|S(x_i) - F(x_i)|$	$|S(x_{i-1}) - F(x_i)|$
0.26450392	0.1	0.26450392	0.16450392	**0.26450392**
0.32682272	0.2	0.32682272	0.12682272	0.22682272
0.42274239	0.3	0.42274239	0.12274239	0.22274239
0.53633229	0.4	0.53633229	0.13633229	0.23633229
0.54631184	0.5	0.54631184	0.04631184	0.14631184
0.57084872	0.6	0.57084872	0.02915128	0.07084872
0.66591388	0.7	0.66591388	0.03408612	0.06591388
0.74892422	0.8	0.74892422	0.05107578	0.04892422
0.77529221	0.9	0.77529221	0.12470779	0.02470779
0.83462020	1.0	0.83462020	0.16537980	0.06537980

The test statistic is $K = 0.26450392$. From Table C.14 with $n = 10$ and $\alpha = 0.05$, the c.v. $= 0.4092$. Since $K < 0.4092$, accept H_0. The students may continue to act as if the random numbers are from a uniform distribution on $[0.1]$.

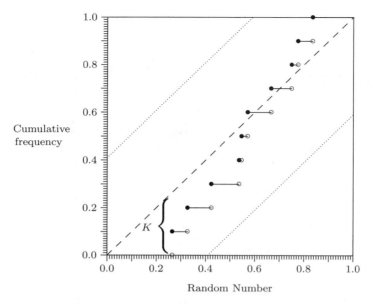

FIGURE 11.3. A graphical illustration of the data in Table 11.3. The step function is the graph of the empirical distribution $S(x)$. The graph of the uniform distribution on $[0, 1]$ is the diagonal line. For $n = 10$ and $\alpha = 0.05$, the c.v. $= 0.4092$. H_0 is accepted since the graph of the empirical distribution lies entirely within the c.v.-corridor.

11.6   The Lilliefors Test

We finish our discussion of goodness of fit tests with a description of the Lilliefors test, which can be viewed as a variant of the Kolmogorov-Smirnov test. Its purpose is to determine whether a sample could have come from a population with a normal distribution *without having specified the mean and variance in advance*. This distinguishes it from the Kolmogorov-Smirnov test.

The intuition behind the Lilliefors test is the same as for the Kolmogorov-Smirnov test. The maximum distance between the empirical distribution S and a normal distribution F with parameters determined from the data is used as the test statistic.

Assumptions, Hypotheses, and Test Statistic

The assumption of the test is that a random sample has been drawn from a population X whose distribution is normal. The hypotheses of the test are

- H_0: The random sample is drawn from a normal distribution, with unspecified mean and variance.

- H_a: The distribution of X is not normal.

Like the Kolmogorov-Smirnov test, the Lilliefors test uses the empirical cumulative distribution S. *The key difference is that the Lilliefors test uses the sample mean $\overline{X}$ and variance s^2 as estimates of μ and σ^2 for the hypothesized normal distribution F.*

The test compares the empirical cumulative distribution S to the normal cumulative distribution F. H_0 is rejected only if the difference between the two distributions is sufficiently large. This is best illustrated in the context of an example.

EXAMPLE 11.17. In problem 2 of Chapter 6 we listed data that were part of a benthic community survey of Lady Elliot Island. Sixteen sea stars, *Linckia laevigata*, were collected and their longest arms were measured to the nearest tenth of a centimeter.

$$
\begin{array}{cccccccc}
10.3 & 11.0 & 10.5 & 10.0 & 11.3 & 14.5 & 13.0 & 12.1 \\
12.1 & 9.4 & 11.3 & 12.0 & 11.5 & 9.3 & 10.1 & 7.6
\end{array}
$$

The problem said to assume normality for the arm lengths. Is this assumption justified?

SOLUTION. Let $\alpha = 0.05$ be the significance level of the test. Verify that $\overline{X} = 11.00$ cm and $s^2 = 2.644$ cm. So the Lilliefors test will test whether the sample is drawn from a normally distributed population with $\mu = 11.00$ cm and variance $\sigma^2 = 2.644$ cm. Proceed as in the Kolmogorov-Smirnov test, creating the same data table (see Table 11.4) with x_i and $S(x_i)$, in the first two columns.

TABLE 11.4. The table used to determine the Lilliefors test test statistic L for Example 11.17.

x_i	$S(x_i)$	$F(x_i)$	$|S(x_i) - F(x_i)|$	$|S(x_{i-1}) - F(x_i)|$
7.6	0.063	0.018	0.045	0.018
9.3	0.125	0.149	0.024	0.087
9.4	0.188	0.164	0.024	0.039
10.0	0.250	0.271	0.021	0.084
10.1	0.313	0.291	0.022	0.041
10.3	0.375	0.334	0.041	0.022
10.5	0.438	0.378	0.060	0.003
11.0	0.500	0.500	0.000	0.063
11.3	0.563	0.571	0.008	0.071
11.3	0.625	0.571	0.054	0.008
11.5	0.688	0.622	0.066	0.003
12.0	0.750	0.729	0.021	0.042
12.1	0.813	0.749	0.064	0.001
12.1	0.875	0.749	**0.126**	0.064
13.0	0.938	0.891	0.047	0.016
14.5	1.000	0.984	0.016	0.047

In the third column calculate $F(x_i)$, where F is the normal distribution with mean $\mu = \overline{X}$ and variance $\sigma^2 = s^2$;

$$F(x_i) = \frac{x_i - \overline{X}}{s}.$$

For example,

$$F(7.6) = \frac{7.6 - 11.0}{\sqrt{2.644}} = -2.09.$$

In the fourth column calculate the absolute differences $|S(x_i) - F(x_i)|$. The last column contains the offset differences $|S(x_{i-1}) - F(x_i)|$ as in the Kolmogorov-Smirnov test.

The Lilliefors test statistic L is the maximum value that appears in these last two columns,

$$L = \max_{x_i} \{|S(x_i) - F(x_i)|, |S(x_{i-1}) - F(x_i)|\}. \tag{11.10}$$

If L is less than or equal to the critical value for the test from Table C.15, then H_0 is accepted. Otherwise, H_0 is rejected. Geometrically, L is the maximum distance between the sample distribution and the standard normal curve (as plotted in Figure 11.4).

In this example $L = 0.126$. Obtain the critical value for the Lilliefors test in Table C.15. With $\alpha = 0.05$ and $n = 16$, the critical value is 0.213. Since $0.126 < 0.213$, we continue to accept H_0.

Figure 11.4 illustrates this result geometrically. The empirical distribution S, which consists of a series of steps, lies within the c.v.-corridor (dotted) of the hypothesized normal distribution (solid curve). Only if S were to cross the dotted corridor boundaries would H_0 be rejected.

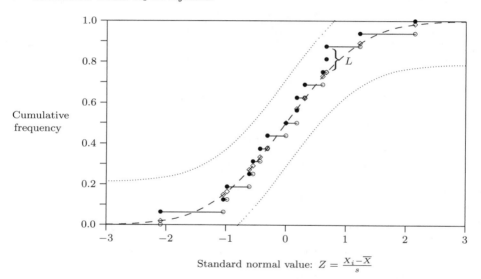

Standard normal value: $Z = \frac{X_i - \overline{X}}{s}$

FIGURE 11.4. The test statistic L is the maximum distance between $S(Z)$ and $F(Z)$[the solid ——— and dashed - - - curves, respectively]. This can be visualized as the largest of the vertical distances between the solid • points $(z_i, S(z_i))$ and the diamond ◇ points $(z_i, F(z_i))$, which correspond to the fourth column of Table 11.4, or between the hollow ○ points $(z_i, S(z_{i-1}))$ and the diamond ◇ points $(z_i, F(z_i))$, which correspond to the fifth column of Table 11.4. The vertical distance from $F(x)$ to the corridor boundary is the critical value of the test. We accept H_0 only if $S(Z)$ remains within this critical value corridor.

Fortunately, most statistical packages will carry out the Lilliefors test without requiring the calculations above. Another option is to use a spreadsheet to easily calculate the required values.

The Kolmogorov-Smirnov and Lilliefors Tests Compared

It may seem that the Kolmogorov-Smirnov test is identical to the Lilliefors test, however, this is not the case. While the test statistic is computed in a similar way, the critical values for the Lilliefors test are smaller. Why? Because in the Lilliefors test the hypothesized distribution $F(x)$ has been moved closer to the empirical distribution function $S(x)$ by estimating μ and σ^2 based on the data. The maximum discrepancy between $F(x)$ and $S(x)$ has been made smaller than it would have been if we had started the analysis with a specified normal distribution. Return to Example 11.14 where the *specified* normal distribution had mean $\mu = 163$ cm and standard deviation $\sigma = 8$ cm, while the sample mean and standard deviation were $\overline{X} = 164.9$ cm and $s = 10.6$ cm. The test statistic in Example 11.14 was calculated using the specified values not the sample values.

We can think about the difference in the two tests in another way. Because the discrepancy between $F(x)$ and $S(x)$ has been made smaller in the Lilliefors test by using the sample mean and variance, the critical values for the Lilliefors test should be smaller than those for the Kolmogorov-Smirnov test for a given sample size n and fixed α level. In Example 11.17, $n = 16$ and $\alpha = 0.05$ and the critical value for the Lilliefors test was 0.213. If we had mistakenly carried out a Kolmogorov-Smirnov test, the critical value would have been 0.3273, which is much larger and could lead to a Type II error of accepting a false null hypothesis.

This difference between the two tests is similar to the difference between the extrinsic and intrinsic model chi-square tests. Analogous to the Kolmogorov-Smirnov test is the extrinsic model in which no parameters are estimated from the data. The Lilliefors test resembles the intrinsic model chi-square test in which at least some parameters are estimated from the sample data. In the later case, the degrees of freedom were reduced for each parameter estimated, which then decreased the critical value of the test.

The other major difference between the Kolmogorov-Smirnov test and Lilliefors test is that the Lilliefors test assumes that the hypothesized distribution is normal. The Kolmogorov-Smirnov test may be used with continuous distributions other than normal ones, but all parameters of the distribution must be specified in advance and not estimated from the data.

11.7   Problems

1. Heron Island is a coral cay in the Great Barrier Reef. Around Heron Island (and elsewhere in Australia) there are 2 color morphs of the reef heron, *Egretta sacra*, a white morph and a dark or blue morph. It is generally accepted that farther north in the southern hemisphere there are many more white than dark morphs, while farther south just the opposite is true. A preliminary study was carried out to test the hypothesis that the ratio of white to dark herons on the island was 3:1.

   (*a*) A small census found 16 white morphs and 4 dark. Can the assumption of a 3:1 ratio be rejected?

   (*b*) What if the census were larger with 160 white morphs and 40 dark?

2. A random sample of 67 *Protoreaster linckii* sea stars had two color morphs, red and green. It was assumed that 25% of the population were red color morphs. Of the sample, 12 were red. Do these data contradict the stated assumption?

3. At a large urban birthing center there were a total of 671 births over a 6-month period. Of these, 230 occurred on the weekends (Saturday and Sunday). Do these data indicate that weekend births are occurring more often than mere chance would indicate?

(*a*) If the births are random, what proportion p would be expected to occur on weekends?

(*b*) Carry out an appropriate test of this model.

4. An investigator set up an experiment involving the effect of a light gradient on the gastropod *Littorina*. The gradient was established with three zones of light intensity, and a sample of 33 gastropods. Fifteen were noted in zone 1, 12 in zone 2, and 6 in zone 3. Test the hypothesis that the light gradient had no effect on the distribution of this organism.

5. Here's a very famous data set (believe it or not). Data were collected that summarize the numbers of Prussian military personnel killed by horse-kicks for each of 14 corps in each of 20 consecutive years, 1875–1894. There were a total of 196 deaths in $14(20) = 280$ corps-years. Are horse-kick deaths a random phenomenon that can be estimated by a Poisson distribution? Carry out a test at the $\alpha = 0.05$ level. Begin by checking that the mean and variance of this sample are similar. (*HSDS, #283*)

Deaths/corps-yr	0	1	2	3	4	≥ 5
Frequency	144	91	32	11	2	0

6. In a study of fruit shape in summer squash, *Cucurbita pepo*, two disc-shaped fruited plants heterozygous for two genes (AaBb) were crossed together. The geneticist believes that a capital allele at both loci results in disc-shaped fruit (A_B_), a capital at only one locus (A_bb or aaB_) results in sphere-shaped fruit, while no capital alleles (aabb) produces a plant with long fruit. If the 2 loci A and B are independently assorting, what is the phenotypic ratio expected in the progeny of this cross? The resulting offspring are listed below. Are the data consistent with the Mendelian predictions of independent assortment and gene interaction (9:6:1 ratio)? Use an appropriate goodness of fit test with $\alpha = 0.05$.

Shape	Number
Disc	117
Sphere	90
Long	17
Total	224

7. Refer to the previous problem. Suppose the number of progeny had been 10 times as large, i.e., 1170 disc-shaped, 900 sphere-shaped, and 170 long. Are the data consistent with the Mendelian model? Discuss the relationship between sample size and goodness of fit chi-square analysis in light of your results.

8. (*a*) The nests of the wood ant, *Formica rufa,* are constructed from small twigs and wood chips. As part of a study of the distribution of these nests, the direction of greatest slope was recorded for 42 such nests in Pound Wood (Essex), England. Nests were recorded facing all quadrants, as shown in the following figure. Using an appropriate statistical test, determine whether the distribution of directions is random or whether wood ants prefer a certain direction of exposure. (Adapted from: Cook, R. Wood Ants of Pound Wood. `ibs.uel.ac.uk/ibs/envmath/resources/multi/PoundWood/WoodAnts/CookAnts.htm`.   July 2000.)

(*b*) Why might there be significantly more nests with southerly exposures?

9. Twenty-five leaves were selected at random from each of six similar apple trees for a total of 150 leaves. The number of adult female European red mites, *Panonychus ulmi,* on each was counted. Do the data come from a Poisson distribution? *(HSDS, #227)*

Mites/leaf	0	1	2	3	4	5	6
Frequency	70	38	17	10	9	3	3

10. Coal mining records from the period 1851 to 1962 revealed 186 explosions that killed 10 or more workers. If the distribution of accidents by day of the week were uniform, then approximately one-sixth of the accidents could be expected to have occurred on any work day. Do the data support the uniform hypothesis? *(HSDS, #204)*

Day	Mon	Tue	Wed	Thu	Fri	Sat
Frequency	19	34	33	36	35	29

11. The following table contains data collected on the Hobart and William Smith campus in Geneva, NY at Houghton House for a project on bird activity. The data are numbers of sightings of common bird species in the early morning (6:00–8:00 AM) and in the afternoon (2:00–4:00 PM). Is the distribution of bird species the same at both times of day?

Species	AM	PM
Starling	51	108
Grackle	22	40
Crow	36	12
Robin	7	15
Goldfinch	31	3
Cardinal	14	3

12. In another study of hermit crab behavior carried out at Lady Elliot Island, a random sample of two types of gastropod shells was collected. Each shell collected was occupied by one of two species of hermit crab. Do both species of hermit crabs occupy each shell type in the same proportion, i.e., are the distributions of the two species of crabs the same in both types of shell? Carry out the test at the $\alpha = 0.05$ level.

Species	Clusterwink	Cirithid	Total
Crab A	158	37	195
Crab B	3	38	41
Total	161	75	236

13. The dinoflagellate zooxanthellae has a symbiotic relation with both clams and coral. For example, the mantle of *Tridacna crocea,* the crocus giant clam, is colored by enclosed zooxanthellae. Green and brown corals are also colored by their symbiotic zooxanthellae. The corals and clams depend on the photosynthetic activity of the zooxanthellae as a source of carbohydrates. The zooxanthellae obtain protection from their host. A study was conducted at Lady Elliot Island to determine whether there was a relationship between the mantle color of *T. crocea* and the depth at which the clam is found. Is the distribution of colors the same for all depths? Use the data below to test the hypothesis at the $\alpha = 0.05$ level. Clearly state the hypotheses and interpret the results.

Color	< 5 m	> 5 m
Browns	26	21
Greens or blues	7	5
Brown w/ green or blue	11	13

14. A study was conducted in the tidal flat at Polka Point, North Stradbroke Island to examine the distribution of the animals associated with the seagrass *Sargassum* at different distances from the shoreline. Samples of *Sargassum* were taken at 5, 10, and 15 m from the shore and these were examined for amphipods and isopods. The observations are recorded below. Is the distribution of the organisms associated with *Sargassum* the same at all three distances? Use the data below to test the hypothesis at the $\alpha = 0.05$ level. Clearly state the hypotheses and interpret the results.

Distance (m)	Amphipods	Isopods
5	2	12
10	27	16
15	47	22

15. (*a*) Another part of the organ retrieval study cited in Example 11.12 compared kidney damage frequency by renal surgical teams removing only the donor's kidneys to the damage frequency when both the kidneys and liver were retrieved. Analyze the data with a 2 × 2 contingency table. (Based on data in: Wigmore, S. et al. 1999. Kidney damage during organ retrieval: Data from the UK National Transplant Database. *The Lancet*, **354**: 1143–1146.)

Organs retrieved	Damaged kidneys	Undamaged kidneys	Total
Liver and kidneys	454	1692	2146
Kidneys only	503	1432	1935
Total	957	3124	4081

(*b*) In the same study, it was noted that some hospitals performed many more kidney retrievals than others. Researchers categorized the site of the retrieval by the number of retrievals done annually, with centers doing: fewer than 20 organ retrievals per year; between 20 and 49 organ retrievals, inclusive, per year; and 50 or more organ retrievals per year. Was the frequency of kidney damage independent of the number of retrievals at a center? Analyze the data below and explain your results.

Annual retrievals	Damaged kidneys	Undamaged kidneys	Total
Fewer than 20	86	350	436
Between 20 and 49	634	1984	2618
50 or more	1006	4954	5960
Total	1726	7288	9014

16. As part of a larger study of population genetics, the importance of cryptic coloration in peppered moths was studied. Two color morphs (one light and one dark) were placed on an intermediate background. Hungry starlings were then given 10 minutes to eat their fill. The results of the preliminary experiment are given below.

Fate	Light morph	Dark morph	Total
Not eaten	32	12	44
Eaten	14	22	36
Total	46	34	80

(*a*) Analyze the data with a 2 × 2 contingency table.

(*b*) Analyze the data with a two-sample proportions test.

In this classic set of data, Ernest Rutherford and Hans Geiger counted the number of scintillations in 72-second intervals caused by the radioactive decay of the element polonium. Altogether there were 10,097 scintillations during 2608 intervals. Do these reflect a Poisson distribution? Experiments such as this one helped physicists understand the nature of radioactive decay. *(HSDS, # 279)*

No. per 72 s	0	1	2	3	4	5	6	7	8	9	10	≥ 11
Frequency	57	203	383	525	532	408	273	139	45	27	10	6

18. Data were collected on Durroc Jersey pigs on the number of males and females in litters that contained exactly 8 piglets. Let X denote the number of males in each litter. Determine whether X is a binomial random variable with $p = 0.5$. *(HSDS, #176)*

No. of males	0	1	2	3	4	5	6	7	8
Frequency	1	8	37	81	162	77	30	5	1

19. Snakes are often preyed upon by other species of snakes. Consequently, the ability of prey snakes to detect chemical cues of predator snakes in order to avoid them and, thereby, minimize encounters with such predators should confer a selective advantage. A series of trials were carried out to examine the ability of pine snake hatchlings, *Pituophis melanoleucus,* to discriminate among the odors of: (1) other pine snakes, (2) corn snakes, *Elaphe guttata,* which cohibernate with and sometimes lay their eggs in the nests of pine snakes, and (3) predators including king snakes, *Lampropeltis getulus.* A Y-shaped maze with clean wood shavings in the base section was used for each trial. In the trials one arm of the maze was designated as the control arm and contained wood shavings from a cage with no snake in it (clean shavings). The other arm of the maze contained shavings from the cage of one of the following: an adult pine snake, an adult king snake, or an adult corn snake. Pine snake hatchlings were placed at the base of the Y-shaped maze and the number of snakes selecting each arm of the maze was then recorded. (Based on data in: Burger, J. Response of hatchling pine snakes *(Pituophis melanoleucus)* to chemical cues of sympatric snakes. *Copeia,* 1990(4), 1160–1165.)

Experiment	Control	Pine	Corn	King
Control and pine	2	24		
Control and corn	14		12	
Control and king	26			0

If pine snakes are unable to detect any chemical odors of other snakes, then there should be a 50% chance that the pine snake hatchlings would select the control arm. Use a binomial test with $p = 0.5$ for each of the three experiments to test whether there is any difference in the number of times pine snake hatchlings select the control arm of the maze versus the experimental arm. How would you summarize the results?

20. One hundred random mud samples were taken from a lake bottom in order to determine whether two species of the genus *Stylaria* are associated, that is, tend to occur together. *Stylaria* are oligochaete worms related to earthworms. Analyze the results below to determine if a significant association between these two species exists.

		Species B		Total
		Present	Absent	
Species A	Present	50	10	60
	Absent	25	15	40
	Total	75	25	100

21. William Bateson and Reginald Punnett working during the first decade of the twentieth century discovered an interesting inheritance pattern for flower color in sweet peas, *Lathryus odoratus*. They crossed two white-flowered plants together and unexpectedly the F_1 was all purple-flowered. The subsequent F_2 had both purple- and white-flowered plants. Bateson and Punnett hypothesized that 2 independently assorting gene loci were complementing each other to produce the purple pigment

$$P: \quad \text{white, AAbb} \quad \times \quad \text{white, aaBB}$$
$$\downarrow$$
$$F_1: \quad \text{purple, AaBb} \quad \times \quad \text{purple, AaBb}$$
$$\downarrow$$
$$F_2: \quad \text{A\_B\_} \quad\quad\quad \text{purple}$$

$$\text{A\_bb} \quad\quad\quad \text{white}$$
$$\text{aaB\_} \quad\quad\quad \text{white}$$
$$\text{aabb} \quad\quad\quad \text{white}$$

(*a*) What ratio of purple-flowered to white-flowered F_2 progeny is expected if Bateson and Punnett were correct in their understanding of the genetic basis of flower color in this plant?

(*b*) If the F_2 had 205 purple-flowered plants and 131 white-flowered plants, do the data support their hypothesis? Analyze the data with both the binomial test and a single degree of freedom χ^2 goodness of fit test. Omit the correction for discontinuity in the χ^2 test because the sample size is greater than 200. Do both analyses lead you to the same conclusion?

22. (*a*) Suppose that out of 20 consumers 12 prefer Coke to Pepsi. Discuss whether Coca-Cola can justifiably advertise, "People prefer Coke!"

(*b*) Suppose that out of 100 consumers 60 prefer Coke to Pepsi. Discuss whether Coca-Cola can justifiably advertise, "People prefer Coke!"

23. It is thought that angioplasty patients may survive longer, with fewer recurring heart problems, if they take cholesterol-lowering drugs called statins. Millions of Americans take statins every day to lower cholesterol. These statins have familiar brand names such as Lipitor, Zocor, and Pravachol. A large study was designed to assess the effectiveness of statin in reducing risk of fatal and non-fatal heart

problems in angioplasty patients. Balloon angioplasty is a very common medical procedure with more than one million performed in the U.S. each year. Most patients gain immediate relief from their chest pain but a significant portion develop serious heart problems within ten years. In the study 900 patients started taking 80 mg/day of statin within two days of their angioplasty. Nine hundred similar patients were given a placebo. The patients were followed for five years with the following outcomes.

Treatment	No Heart Problems	Serious Heart Problems	Total
Statin	731	169	900
Placebo	667	233	900
	1398	402	1800

Do the data indicate that taking statin significantly reduces the risk of serious heart problems following angioplasty?

24. Beer drinkers and brewmiesters have known for a long time that light can initiate a skunky taste and smell in beer. Most beer is sold in green or brown bottles to prevent light-induced skunking. Sophisticated chemical studies have shown how the light sensitive compounds in hops called isohumulones degrade forming free radicals that bond to sulfur to form the skunky molecules. Miller Genuine Draft (MGD) is claimed to be made from chemically altered hops that don't break down into free radicals in light and, therefore, the beer can be sold in less expensive clear bottles. In a taste test of MGD stored for six months in light-tight containers or exposed to light, a panel of 20 tasters chose the light-tight contained beer to be tastier 16 times. Is this an indication that MGD should switch to colored bottles? (Aside: Why do you suppose the manufacturers of Corona encourage drinkers to add lime?)

25. Because there are so many species of mosquitoes (63 species have been described in New Jersey alone), control depends on an accurate understanding of the ecology of the various species. A survey of 160 small ponds with willow trees on the banks, yielded the following data.

		Anopheles quadrimaculatus		
		Present	Absent	Total
Culex erraticus	Present	58	27	85
	Absent	45	20	65
	Total	103	47	150

Is there an association between the distributions of the two species? Discuss the results of the analysis in terms of potential control for these species.

26. Many human diseases have a complex etiology that includes a genetic predisposition. To test the relationship between the APOE4 allele and Alzheimer's disease, a large number of nursing home patients over the age of 70 were screened for this allele and the presence of Alzheimer's. The data are presented below.

Subjects	APOE4/APOE4	APOE4/+	+/+
Normal	47	190	1385
Alzheimer's	151	270	281

Is there an association between the presence of the APOE4 allele and the presence of Alzheimer's?

27. While the death cap mushroom, *Amanita phalloides*, is the world's most dangerous mushroom, it is generally rare in most parts of the United States. It is, however, locally common in the San Francisco Bay area under imported cork trees and in Rochester, NY, under imported Norway maple trees in Durand-Eastman Park. In an effort to understand more about the ecology of this species a distribution study was conducted in Durand-Eastman Park by dividing a section of the mushroom's range into 10×10 meter quadrats and noting the number of death caps per quadrat. Data are presented below.

# Death Caps	0	1	2	3	4	5
# quadrats	360	300	220	80	30	10

Are death caps Poissonally distributed? What are the criteria for a Poisson distribution?

28. In Kerry cattle a single gene with complete dominance controls the presence of horns (pp) or the absence of horns (PP or Pp), a condition called polled. A second gene controlling coat color exhibits incomplete dominance with homozygous (R_1R_1) being white, homozygous (R_2R_2) being red, and (R_1R_2) being roan (exhibiting patches of red and white). In a large number of crosses between cattle heterozygous for both of these genes, the following progeny were produced.

55	polled, roan
27	polled, red
31	polled, white
23	horned, roan
11	horned, red
13	horned, white
160	

(a) What are the expected ratios for the offspring phenotypes assuming equal fitness and independent assortment?

(b) Utilize a chi-square goodness of fit test to determine whether the data are significantly different from the independent assortment expectations.

29. While working as a volunteer at a Red Cross blood drive, you learn that the ABO blood groups are distributed in the following way: 40% A, 35% O, 20% B, and 5% AB. At the end of the drive 200 units of blood were collected and typed with the following results:

Type	Donors
A	90
O	75
B	25
AB	10
	200

Do the units collected in the drive fit the hypothesized frequencies for the various blood types?

30. In a preliminary study of methods to ameliorate the rashes and ulcers that sometimes occur as graft versus host disease (GVHD) in patients receiving bone marrow transplants, use of UV light treatment was investigated. Residual Langerhaus cells are thought to be the cells that trigger inflammation and skin lesions. It is believed that UV light treatment will inactivate Langerhaus cells and, therefore, reduce the symptoms of GVHD in 75% of transplant patients. Twenty bone marrow transplant patients with mild GVHD skin lesions were treated with UV light during their recovery. Twelve had notable improvement in their GVHD skin lesions. Does this result support the claim of improvement in 75% of the patients?

31. An ophthalmologist performing laser-assisted *in situ* keratomileusis (LASIK) surgeries wondered if humidity levels affected the probability of success. In LASIK surgery the outer layer of the cornea is peeled back and a laser pulses to vaporize portions of the inner cornea or stroma to reshape it before the flap is replaced. High humidity may make the cornea swell temporarily and induce a surgeon to remove less stroma than is needed to fully correct an individual's eyesight. When this occurs a follow-up LASIK or touch-up surgery is required. Below are data for surgeries performed by this ophthalmologist. Do you think humidity levels are a factor in the success of his LASIK procedures?

Touch-up surgery	Humidity		Total
	High	Normal/Low	
Required	25	16	41
Not required	55	109	164
Total	80	125	205

32. In 2006 *The New York Times* reported that a National Institutes of Health study indicated that circumcising adult men may cut in half their risk of getting the AIDS virus through heterosexual intercourse. The studies were carried out in Kenya and Uganda. The link between male circumcision and HIV prevention was noted as early as in the late 1980s. Male circumcision plays a role in HIV transmission because cells in the foreskin of the penis are particularly susceptible to the HIV virus. Also, the foreskin is more fragile than the tougher skin surrounding it, providing a surface that the virus may penetrate more easily.

Researchers enrolled 2,784 HIV-negative men in Kenya, and 4,996 HIV-negative men in Uganda, into the studies. Half were circumcised; others were just monitored. Over two years, 22 of the circumcised Kenyans became infected with HIV compared with 47 uncircumcised men. In Uganda, 22 circumcised men became infected versus 43 of the uncircumcised. For each of the two studies, carry out an appropriate test of proportions and determine the corresponding P values. Were the researchers right to stop the study early and offer the men who had not been circumcised the opportunity to do so? (Based on data reported in: *The New York Times.* `www.nytimes.com/aponline/us/AP-HIV-Circumcision.html`. 13 December 2006.)

33. The mygalomorph family Theraphosidae includes the world's largest spiders such as bird-eating spiders and tarantulas. As one part of a preliminary study of the tarantula, *Avicularia avicularia*, in Trinidad the heights at which they build their silken retreats were recorded. A total of $n = 12$ spiders' retreats were censused with $\overline{X} = 2.803$ m and $s = 1.500$ m. The researcher claims that these heights were normally distributed. Use the data below (heights are in meters) to test this claim.

0.67   1.30   1.83   2.12   2.40   2.47   2.47   2.77   3.10   3.37   5.11   6.03

34. Recall from Example 6.8 that southeastern Queensland is home to several species of *Crinia*, which are small, highly variable brownish or grayish froglets usually found in soaks, swamps, or temporary ponds. A population of froglets from a single pond was sampled and tentatively identified as clicking froglets *C. signifera*. This species is known to have normally distributed lengths with a mean of 30 mm and variance of 10 mm^2. Based on the data (in mm) below is it reasonable to assume that the sample comes from this population? Does your analysis support the hypothesis that these are clicking froglets?

24.1	22.9	23.0	26.1	25.0	30.8	27.1	23.2	22.8
23.7	24.6	30.3	23.9	21.8	28.1	25.4	31.2	30.9

35. Are the froglets in the previous problem from some population with normally distributed lengths?

36. The brush box tree, *Lophostemon confertus*, occurs in a variety of locations. In a field exercise the lengths (in mm) of 15 leaves were measured at Simpson's Falls, Brisbane Forest Park. The sample mean is $\overline{X} = 136.9$ mm and the variance is $s^2 = 394.55$ mm^2. Carry out a test to determine whether the sample comes from a normally distributed population. Interpret the results.

95	110	116	125	128
135	140	140	141	145
146	150	150	157	175

37. In Example 6.12 we examined a sample of 10 bridled gobies, *Arenigobus frenatus*, from a population adjacent to a ferry port where there were higher than normal levels of disturbance and pollution. It is known that for general populations the lengths of bridled gobies are normally distributed with $\mu = 80$ mm and $\sigma = 100$ mm^2. Does the sample appear to come from the general population?

63.0   82.1   81.8   77.9   80.0   72.4   69.5   75.4   80.6   77.9

38. Are the bridled gobies in the previous problem from *some* population with normally distributed lengths?

39. Female sea turtles lay between 50 and 170 eggs in nests dug in the sand on oceanic beaches. Most loggerhead or green sea turtles within a clutch pip (i.e., cut through their eggshell) within a few hours of each other and typically spend another 1 to 7 days buried before emerging from their nest. Emergence seems to be a synchronized activity among the hatchlings and may be triggered by the lower temperatures after sunset, since most hatchlings emerge at night.

The following data were collected on $n = 25$ loggerhead emergence events (i.e., the movement of hatchlings from a nest). The mean time of emergence was $\overline{X} = 0.4$ hours (about 24 minutes) after midnight (time 0). The sample variance was $s^2 = 3.72$ hours^2. Carry out an appropriate test to determine whether the emergence events are normally distributed. Times were recorded to the nearest half hour, negative times represent hours before midnight. (Adapted from: Witherington, B. et al. 1990. Temporal pattern of nocturnal emergence of loggerhead turtles hatchlings for natural nests. *Copeia*, **4**: 1165–1168.)

Time	Emergences
−3.0	1
−2.5	1
−2.0	1
−1.5	2
−1.0	2
−0.5	3
0.0	3
0.5	1
1.0	3
1.5	2
2.0	2
2.5	1
3.0	1
3.5	0
4.0	1
4.5	1

40. Limited literacy is common and its prevalence increases with age. Researchers have found that poor health literacy, which is especially prevalent among the elderly, results in poor adherence to prescription instructions, infrequent use of preventive medical services, increased hospitalizations and visits to the emergency room, and poor control of chronic diseases.

A study was conducted to assess whether limited literacy is associated with increased mortality in older adults. The 2512 participants, who ranged in age from 70 to 79 and lived independently, were followed for a five-year period from 1999 to 2004. Among the 603 participants with limited literacy (a reading level of less than ninth grade), the death rate was 19.73%. Among all others, the death rate was 10.58%. Is there an increased mortality rate among this age group for those with limited literacy? (Based on: Sudore, R. et al. 2006. Limited literacy and mortality in the elderly: The health, aging, and body composition study. *J. Gen. Internal Med.*, **21**(8): 806–812.)

41. (*a*) Drs. Lisa Bodnar and Hyagriv Simhan analyzed the records of women who delivered babies at Magee-Women's Hospital over the 10-year period 1995 to 2005. They grouped the women by season of last menstrual period and used this to estimate time of conception. They found that women who conceived in the springtime had a preterm birth rate (delivery before 37 weeks gestation) of 9.2% while those who conceived in the fall had a rate of 8.8%. If there were 18,500 women in each of the two groups, was there a significant difference in preterm birth associated with season of conception? (Rauscher, M. Spring conception may up risk of preterm birth. `www.nlm.nih.gov/medlineplus/news/fullstory-45021.html`)

(*b*) The rate of very preterm birth, or birth before 32 weeks gestation, was 2.7% for spring conception and 2.3% with fall conception. Was this difference significant?

(*c*) Notice that the difference in birthrates is the same in both parts, 0.4%. Were your answers to both parts the same? Explain.

42. The Tuamotus are a chain of atolls in French Polynesia and are the largest chain of atolls in the world, spanning an area of the Pacific Ocean roughly the size of Western Europe. Two of the more important atolls are Makemo and Rangiroa. The beaches at both atolls are light in color leading to a prevalence of the white color morph of the reef heron, *Egretta sacra*, at both locations. A survey at both atolls in 1972 produced the following data:

Location	White morph	Dark morph	Total
Makemo	41	23	64
Rangiroa	81	62	143

Is there a significant difference in proportions of the white color morph at the two atolls? (Based on data reported in: Holyaok, D. 1973. Significance of colour dimorphism in Polynesian populations of *Egretta Sacra*. *Ibis*, **115**: 419–420.)

43. Typically service industries operate on a daily basis. However, in many hospitals staffing is reduced on weekends; many hospitals provide routine care on weekdays but only emergency care on weekends. This means that management of heart attacks, which requires rapid diagnostic and therapeutic procedures, may be somewhat compromised on weekends, possibly leading to less successful outcomes for such patients.

A study was conducted to test this hypothesis. The study included 59,786 patients admitted to New Jersey hospitals between 1999 and 2002 with acute myocardial infarction for the first time as the primary diagnosis. (Source: Kostis, W. et al. 2007. Weekend versus weekday admission and mortality from myocardial infarction. *The New England Journal of Medicine*. **356**(11): 1099–1109.)

(*a*) Of the 15,542 patients admitted on weekends, 2,005 were deceased by 30 days. By comparison, of the 44,244 patients admitted on weekdays, 5,309 were deceased by 30 days after admission. Did patients admitted on the weekend have a significantly higher mortality rate?

(*b*) For this same group, the mortality 365 days after admission was 3,715 and 10,132, respectively. Find a 95% confidence interval for the difference in mortality rates. Based on the confidence interval calculation were the mortality rates different for the two groups of patients? Explain.

44. Some studies have suggested that exposure to allergens during infancy may predispose a child to later allergic disease. Other studies involving children growing up on farms, who were routinely exposed to animal allergens, indicate the opposite effect.

   A study was conducted to investigate the relationship between early dog exposure and later allergic sensitization. In the study 184 children who had some dog exposure during their first year of life were compared to 220 children who had not been exposed to cats or dogs before their first birthday. All children were given a skin test between their sixth and seventh birthdays. The results were that 6 of the children exposed to dogs had a positive skin test while 19 of those with no exposure had a positive test.

   Calculate a 95% confidence interval for the difference in proportions. Would you say that there was a significant difference? (Based on: Ownby, D. et al. 2002. Exposure to dogs and cats in the first year of life and risk of allergic sensitization at 6 to 7 years of age. *JAMA*, **288**(8): 963–972.)

45. A study was conducted to compare the outcomes associated with primary cesarean births with no labor (planned cesareans) to vaginal and cesarean births with labor (planned vaginal). Analysis was based on records of 244,088 mothers with no prior cesarean and no documented prenatal risk in Massachusetts between 1998 and 2003. The mothers were divided into two groups: those with no labor and a planned cesarean delivery (3,334 women) and those with labor and either a vaginal birth or a cesarean delivery (240,754 women). In the first group 640 women were rehospitalized within 30 days due to complications compared to 18,307 in the second group. Find a 99% confidence interval for the difference in the rehospitalization rates for those undergoing planned cesarean delivery when compared to those with planned vaginal births. Interpret your answer. (Based on: Declercq, E. et al. 2007. Maternal outcomes associated with planned primary cesarean births compared with planned vaginal births. *Obstetrics & Gynecology*, **109**: 669–677.)

46. We are all familiar with the placebo response, the tendency of an inert "medication" or ineffective "treatment" to have positive results simply because the recipient believes that it will work. The nocebo effect is less familiar and is the opposite of the placebo response. It is defined as the harmful effects seen in a person that occur in response to a placebo, i.e., effects that cannot be considered due to the specific treatment used. (Nocebo is Latin for "I will harm" while placebo means "I will please.") Some researchers argue that the nocebo effect is as important as the placebo effect when conducting clinical trials and research into new therapies.

   As part of a test to determine a safe drug for 600 patients who had previously had adverse drug reactions (ADRs), the subjects were given capsules of an inert substance in capsule form but were told that it was a low dose of a different medication than they had previously used. Of particular interest was whether the nocebo effect is related to gender. Of the 418 women in the study, 124 had ADRs while of the 182 males 34 had ADRs. Determine a confidence interval for the difference in proportions between males and females who experience the nocebo effect. Is there a significant difference in these proportions?

47. Preliminary research has suggested that women who reach puberty at an unusually early age are more likely to have children who are overweight. To test this relationship researchers recorded the age of 240 girls at menarche and later noted whether their first child was classified as obese by age nine using a standard body mass index (BMI).

BMI	Early menarche by age 11	Normal/late menarche after age 15	Total
Normal	41	131	172
Obese	28	40	68
Total	69	171	240

(a) Is there a relationship between early menarche and obesity of offspring?

(b) Are the percentages of obesity the same for children of early puberty moms as normal/late puberty moms?

Results raise the question of whether or not children are programmed to take a specific growth path. Mothers who reach puberty early might pass on genetic traits to their children that affect appetite and satiety. Or they may imprint their offspring *in utero*.

48. QCR. True or False. Justify your answer.

(a) The binomial goodness of fit is structurally just a sign test.

(b) Chi-square goodness of fit tests are always right-tailed and have an expected value of zero under a true null hypothesis.

(c) Degrees of freedom for all goodness of fit chi-square tests is the number of categories minus 1.

(d) Contingency tables utilize empirical probabilities to test the independence of two variables.

(e) Use of chi-square analysis on discrete count data is an approximation because the chi-square distribution is actually continuous.

(f) The chi-square goodness of fit test can be done on percentages, relative frequencies, or actual counts.

(g) In general the expected values for each category in a chi-square goodness of fit test should be at least 5.

(h) The principle difference between the Kolmogorov-Smirov test and the Lilliefors test is that the Lilliefors test can be used only if a sample came from a particular normal distribution with a known mean and variance.

Proofs of Selected Results

A.1   Summation Notation and Properties

Introduction

The Greek letter sigma, Σ, is used as a shorthand notation for the operation of addition of variables in a sample, treatment, or population.

Consider a sample consisting of the five observations

$$X_1 = 3, \qquad X_2 = 5, \qquad X_3 = 4, \qquad X_4 = 6, \qquad X_5 = 7. \qquad \text{(A.1)}$$

The subscript on the variable X indicates its *position* in the sample or population. X_3 is the third observation, which in this case has a *value* of 4. The sum of the X values beginning with the first observation and including the nth or final observation is denoted

$$\sum_{i=1}^{n} X_i.$$

The subscript of the $\sum$ indicates the start of the addition and the superscript indicates the stopping point for the process of addition. If the $\sum$ is not scripted, it is implicitly assumed that the summation starts at $i = 1$ and goes to n, inclusive, where n is the last observation. For example,

$$\sum_{i=1}^{n} X_i = \sum X_i = (X_1 + X_2 + X_3 + X_4 + X_5) = (3 + 5 + 4 + 6 + 7) = 25.$$

Other possible sums are

$$\sum_{i=2}^{4} X_i = X_2 + X_3 + X_4 = 5 + 4 + 6 = 15$$

or, since $n = 5$ here,

$$\sum_{i=3}^{n} X_i = X_3 + X_4 + X_5 = 4 + 6 + 7 = 17.$$

Sums of complex expressions may be denoted compactly using summation notation. For example, the sum of squares is

$$\sum_{i=1}^{n} X_i^2 = 3^2 + 5^2 + 4^2 + 6^2 + 7^2 = 135.$$

Another complicated sum is

$$\sum_{i=2}^{4}(X_i^2 - 5) = (X_2^2 - 5) + (X_3^2 - 5) + (X_4^2 - 5)$$

$$= (5^2 - 5) + (4^2 - 5) + (6^2 - 5)$$
$$= 20 + 11 + 31 = 62.$$

The Associative Law

The associative law of addition means that the terms in a sum can be rearranged. Because the order of addition doesn't matter, the terms of the sum can be reclustered into covenient groupings. For example,

$$\sum_{i=1}^{n}(X_i + Y_i) = (X_1 + Y_1) + (X_2 + Y_2) + \cdots + (X_n + Y_n)$$

$$= (X_1 + X_2 + \cdots + X_n) + (Y_1 + Y_2 + \cdots + Y_n)$$
$$= \sum_{i=1}^{n} X_i + \sum_{i=1}^{n} Y_i.$$

In other words,

$$\sum_{i=1}^{n}(X_i + Y_i) = \sum_{i=1}^{n} X_i + \sum_{i=1}^{n} Y_i. \qquad (A.2)$$

Sums of Constants

Summation of constants is particularly easy. For any constant c,

$$\sum_{i=1}^{n} c = c + c + \cdots + c = nc. \qquad (A.3)$$

Note that the constant is *not* subscripted. For example, still assuming that $n = 5$,

$$\sum_{i=1}^{n} 8 = 5 \times 8 = 40$$

or

$$\sum_{i=2}^{4} 17 = 3 \times 17 = 51.$$

between 2+4 = 3 numbers

Sums and Sample Means

The formula for a sample mean $\overline{X}$ can be written in summation notation as

$$\overline{X} = \frac{1}{n} \sum_{i=1}^{n} X_i.$$

Notice that this means that $n\overline{X} = \sum_{i=1}^{n} X_i$, or using the fact that $\overline{X}$ is constant for any sample,

$$n\overline{X} = \sum_{i=1}^{n} \overline{X} = \sum_{i=1}^{n} X_i. \tag{A.4}$$

Because $\overline{X} = 5$ in for the sample in (A.1),

$$\sum_{i=1}^{n} \overline{X} = 5 \times 5 = 25.$$

The Distributive Law

The distributive law for real numbers, i.e., the fact that $a(x + y) = ax + ay$, means that certain sums can be expressed in more than one way. For instance, constants can be factored out of sums:

$$\sum_{i=1}^{n} cX_i = c\sum_{i=1}^{n} X_i. \tag{A.5}$$

With the particular sample in (A.1), we could calculate

$$\sum_{i=1}^{n} 5X_i = 5X_1 + 5X_2 + 5X_3 + 5X_4 + 5X_5$$
$$= (5 \times 3) + (5 \times 5) + (5 \times 4) + (5 \times 6) + (5 \times 7)$$
$$= 125$$

or, more simply,

$$\sum_{i=1}^{n} 5X_i = 5\sum_{i=1}^{n} X_i = 5(3 + 5 + 4 + 6 + 7) = 5(25) = 125.$$

Note that summation *cannot* be distributed under an exponent,

$$\sum_{i=1}^{n} X_i^2 \neq \left(\sum_{i=1}^{n} X_i\right)^2.$$

The first expression is a sum of squares, while the second expression is a squared sum:

$$\sum_{i=1}^{n} X_i^2 = 135 \qquad \text{versus} \qquad \left(\sum_{i=1}^{n} X_i\right)^2 = (25)^2 = 625.$$

The Corrected Sum of Squares

The **corrected sum of squares** is defined as

$$\text{CSS} = \sum_{i=1}^{n} (X_i - \overline{X})^2. \tag{A.6}$$

Each observation is "corrected" for its distance from the mean and then squared. Using the associative and distributive properties, as well as (A.4), we can rewrite the corrected sum of squares as

$$\sum_{i=1}^{n}(X_i - \overline{X})^2 = \sum \left(X_i^2 - 2\overline{X}X_i + \overline{X}^2 \right)$$

$$= \sum_{i=1}^{n} X_i^2 - 2\overline{X}\sum_{i=1}^{n} X_i + n\overline{X}^2$$

$$= \sum_{i=1}^{n} X_i^2 - 2\overline{X}(n\overline{X}) + n\overline{X}^2$$

$$= \sum_{i=1}^{n} X_i^2 - 2n\overline{X}^2 + n\overline{X}^2$$

$$= \sum_{i=1}^{n} X_i^2 - n\overline{X}^2$$

$$= \sum_{i=1}^{n} X_i^2 - n\left(\frac{\sum_{i=1}^{n} X_i}{n} \right)^2$$

$$= \sum_{i=1}^{n} X_i^2 - n\frac{\left(\sum_{i=1}^{n} X_i\right)^2}{n^2}.$$

Simplifying the final expression we obtain the formula

$$\sum_{i=1}^{n}\left(X_i - \overline{X}\right)^2 = \sum_{i=1}^{n} X_i^2 - \frac{\left(\sum_{i=1}^{n} X_i\right)^2}{n}. \tag{A.7}$$

The term $\sum X_i^2$ is the "uncorrected sum of squares" while $\frac{(\sum X_i)^2}{n}$ is the "correction term."

Multiple Summations and Subscripts

Suppose that we have sample observations in tabular form with c columns and r rows, as below:

	$i = 1$	2	3	$c = 4$
$j = 1$	6	5	4	5
2	4	3	8	7
3	7	5	9	8
4	9	7	6	6
$r = 5$	8	7	3	4

(A.8)

Each X will now be doubly subscripted because it is a member of two groupings, the ith column and the jth-row. The general form[1] is X_{ij}. For example, $X_{14} = 9$ while $X_{41} = 5$.

---

[1]This is reversed from the usual mathematical matrix convention because in practice the columns represent different treatments in an experiment and are the focus of the analyis.

A double summation of the form

$$\sum_{i=1}^{c}\sum_{j=1}^{r} X_{ij}$$

works from the inside out. So in the expression above, i would be held constant at 1 while all values of X_{1j} are added, then i would change to 2 with all X_{2j} added, then j would change to 3 with all X_{3j} added, and so on. For example, the sum of columns 2 and 3 in (A.8) is given by

to column 3   *to row 5*   *start column 2*   *row 1*

$$\sum_{i=2}^{3}\sum_{j=1}^{5} X_{ij} = (X_{21} + X_{22} + X_{23} + X_{24} + X_{25}) + (X_{31} + X_{32} + X_{33} + X_{34} + X_{35})$$
$$= (5 + 3 + 5 + 7 + 7) + (4 + 8 + 9 + 6 + 3)$$
$$= 57.$$

To add all the observations in the data set, use

$$\sum_{i=1}^{c}\sum_{j=1}^{r} X_{ij} = \sum_{i=1}^{4}\sum_{j=1}^{5} X_{ij} = 121.$$

Dot Notation

To simplify notation, sometimes we replace a summation sign with T (for total) and an index with a dot. The notation indicates that the total represented by T is a summation over the missing index. For example, in (A.8) $T_{1.}$ is the total of the elements in category-column 1:

$$T_{1.} = \sum_{j=1}^{r} X_{1j} = 6 + 4 + 7 + 9 + 8 = 34.$$

The sum of the elements in the second row is

$$T_{.2} = \sum_{i=1}^{c} X_{i2} = 4 + 3 + 8 + 7 = 22.$$

Using the dot convention, the sum of all of the elements in the table is

$$T_{..} = \sum_{i=1}^{c}\sum_{j=1}^{r} X_{ij}.$$

Dot notation is useful in the more complicated calculations of analysis of variance.

A.2   Expected Values

Recall that in Chapter 3 we made the following

DEFINITION. The **expected value** or **mean** for a discrete random variable Y with density function f is given by

$$\mu = E(Y) = \sum_{\text{all } y} y f(y).$$

In other words, in the sum each value y is weighted by its density or probability.

Assumption. In the material that shortly follows, we assume that each object (whose measurement of interest is) y_i is equally likely to be drawn from a finite population Y of size N. In this situation $f(y_i) = \frac{1}{N}$, and the expected value is the ordinary average or population mean

$$E(Y) = \sum_{i=1}^{N} y_i \cdot \frac{1}{N} = \frac{1}{N} \sum_{i=1}^{N} y_i.$$

Notation

In subsequent sections we develop several expected value relationships. We will use the following notation here and throughout the text:

- Let X denote a population (random variable) X of size N. We denote its mean by μ and its variance by σ^2 (or by $\text{Var}(X)$ or σ_X^2). X_i denotes a random element of X (so X_i is also a random variable). We may then write $X_i = \mu + x_i$, where $x_i = X_i - \mu$ is the deviate of X_i from the mean.

- We denote an arbitrary element from a sample of size n from X by X_j, where $j = 1, 2, \ldots, n$. We denote the sample mean by $\overline{X}$ and sample variance by s_X^2.

- $E(X)$ denotes the expected value of X.

- c denotes an arbitrary constant.

Expected Value Relationships

FACT 1. [Another way to regard μ.] If X_i represents a random element from a population X with mean μ, then

$$E(X_i) = \frac{1}{N} \sum_{k=1}^{N} X_k = \mu.$$

Similarly, if X_j is an arbitrary element from a sample drawn from the same population, then

$$E(X_j) = \frac{1}{N} \sum_{i=1}^{N} X_i = \mu.$$

FACT 2. If c is a constant, then $E(c) = c$ because

$$E(c) = \frac{1}{N} \sum_{i=1}^{N} c = \frac{1}{N} \cdot Nc = c.$$

FACT 3. $E(cX_i) = c\mu$ and $E(X_i + c) = \mu + c$. See Theorem 3.1, part 1.

FACT 4. If $x_i = X_i - \mu$, since μ is constant, then using Facts 2 and 3

$$E(x_i) = E(X_i - \mu) = E(X_i) - E(\mu) = \mu - \mu = 0.$$

This is equivalent to the statement that $\sum_{i=1}^{N} x_i = 0$, which we saw in a slightly different form in (1.2) in Chapter 1.

FACT 5. [Another way to regard σ^2.] Recall that the population variance is defined as

$$\sigma^2 = \frac{\sum_{k=1}^{N}(X_k - \mu)^2}{N}.$$

The variance can be written in the form of an expectation as

$$E[(X_i - \mu)^2] = \frac{\sum_{k=1}^{N}(X_k - \mu)^2}{N} = \sigma^2$$

or, since $x_i = X_i - \mu$,

$$E\left(x_i^2\right) = \frac{1}{N}\sum_{k=1}^{N}x_k^2 = \sigma^2.$$

Note: Fact 4 says $E(x_i) = 0$, but $E\left(x_i^2\right) \neq [E(x_i)]^2 = 0$.

FACT 6. Because of the associative law for addition,

$$\sum_{i=1}^{N}(X_i + Y_i) = \sum_{i=1}^{N}X_i + \sum_{i=1}^{N}Y_i.$$

It now follows that

$$E(X + Y) = E(X) + E(Y).$$

This is just a particular case of Theorem 3.1, part 2.

FACT 7. $E\left(x_i x_j\right) = 0$, if $i \neq j$. To see this recall that $x_i x_j$ is the product of two random deviates. In a population of N individuals or variates, there are N^2 such products. Therefore, using Fact 1

$$E(x_i x_j) = \frac{1}{N^2}\sum_{m=1}^{N}\sum_{k=1}^{N}x_m x_k$$

$$= \frac{1}{N^2}[x_1(x_1 + x_2 + \cdots + x_N) + x_2(x_1 + x_2 + \cdots + x_N)$$

$$+ \cdots + x_N(x_1 + x_2 + \cdots + x_N)].$$

However, from Fact 4, $x_1 + x_2 + \cdots + x_N = 0$. So

$$E(x_i x_j) = \frac{1}{N^2}\left[x_1(0) + x_2(0) + \cdots + x_N(0)\right] = 0.$$

FACT 8. $E(X_i^2) = \mu^2 + \sigma^2$. This is just Theorem 3.1, part 3. We give a different proof using the facts above. Since $X_i = \mu + x_i$, then

$$
\begin{aligned}
E\left(X_i^2\right) &= E\left[(\mu + x_i)^2\right] \\
&= E\left(\mu^2 + 2\mu x_i + x_i^2\right) \\
&= E(\mu^2) + E(2\mu x_i) + E(x_i^2) && \text{Fact 6} \\
&= \mu^2 + 2\mu E(x_i) + \sigma^2 && \text{Facts 2, 3, and 5} \\
&= \mu^2 + \sigma^2 && \text{Fact 4.}
\end{aligned}
$$

FACT 9. $E(X_i X_j) = \mu^2$ whenever $i \neq j$. To see this use $X_i = \mu + x_i$ and $X_j = \mu + x_j$. Then

$$
\begin{aligned}
E(X_i X_j) &= E[(\mu + x_i)(\mu + x_j)] \\
&= E(\mu^2 + \mu x_j + \mu x_i + x_i x_j) \\
&= E(\mu^2) + E(\mu x_j) + E(\mu x_i) + E(x_i x_j) \qquad \text{Fact 6} \\
&= \mu^2 + \mu E(x_j) + \mu E(x_i) + 0 \qquad\qquad \text{Facts 2, 3, and 7} \\
&= \mu^2. \qquad\qquad\qquad\qquad\qquad\qquad\quad \text{Fact 4}
\end{aligned}
$$

FACT 10. $E(\overline{X}) = \mu$, that is, the expectation of a sample mean is the population mean. Using Facts 3 and 6 and then Fact 1,

$$
E\left(\overline{X}\right) = E\left(\frac{1}{n}\sum_{j=1}^{n} X_j\right) = \frac{1}{n}\sum_{j=1}^{n} E\left(X_j\right) = \frac{1}{n}\sum_{j=1}^{n} \mu = \frac{1}{n}(n\mu) = \mu.
$$

Because $E(\overline{X}) = \mu$, we say that $\overline{X}$ is an **unbiased estimator** of μ. To determine an unbiased estimator of the population variance, we need one more result.

FACT 11. $E(\overline{X}^2) = \mu^2 + \frac{\sigma^2}{n}$. The proof depends on rewriting the sample mean.

$$
\overline{X}^2 = \left(\frac{1}{n}\sum_{j=1}^{n} X_j\right)^2 = \frac{1}{n^2}\left(\sum_{j=1}^{n} X_j^2 + \sum_{i\neq j} X_i X_j\right), \qquad (A.9)
$$

that is, the square of a sum is the sum of the n squares plus each of the $n^2 - n$ cross terms. Then using this fact

$$
\begin{aligned}
E(\overline{X}^2) &= E\left[\frac{1}{n^2}\left(\sum_{j=1}^{n} X_j^2 + \sum_{i\neq j} X_i X_j\right)\right] \qquad\qquad (A.9) \\
&= \frac{1}{n^2}\left[E\left(\sum_{j=1}^{n} X_j^2\right) + E\left(\sum_{i\neq j} X_i X_j\right)\right] \qquad \text{Facts 3 and 6} \\
&= \frac{1}{n^2}\left[\sum_{j=1}^{n} E(X_j^2) + \sum_{i\neq j} E(X_i X_j)\right] \qquad\quad \text{Fact 6} \\
&= \frac{1}{n^2}[n(\mu^2 + \sigma^2) + (n^2 - n)\mu^2] \qquad\qquad \text{Facts 8 and 9} \\
&= \frac{n^2\mu^2 + n\sigma^2}{n^2} \\
&= \mu^2 + \frac{\sigma^2}{n}
\end{aligned}
$$

Fact 11 is an instance of a more general result.

THEOREM A.1. $E\left(Q^2\right) = \mu_Q^2 + \text{Var}(Q) = \mu_Q^2 + \sigma_Q.$

Though we will not give the proof of this result here, note that we have already seen several instances of this result:

$$\text{Fact 5:}\quad E(x_i^2) = \mu_x^2 + \text{Var}(x) = 0 + \sigma^2$$

$$\text{Fact 8:}\quad E(X_i^2) = \mu^2 + \text{Var}(X) = \mu^2 + \sigma^2$$

$$\text{Fact 11:}\quad E(\overline{X}^2) = \mu^2 + \text{Var}\left(\overline{X}\right) = \mu^2 + \frac{\sigma^2}{n}$$

FACT 12. $E(s^2) = \sigma^2$, that is s^2 is an unbiased estimator of the population variance.

To prove this, start by rewriting $E(s^2)$ using Facts 3 and 6.

$$E(s^2) = E\left(\frac{\sum_{j=1}^n (X_j - \overline{X})^2}{n-1}\right) = \frac{1}{n-1}\sum_{j=1}^n E\left[(X_j - \overline{X})^2\right].$$

Clearing the denominator by multiplying by $n-1$ and then employing the facts above,

$$(n-1)E(s^2) = \sum_{j=1}^n E\left[(X_j - \overline{X})^2\right]$$

$$= \sum_{j=1}^n E\left(X_j^2 - 2\overline{X}X_j + \overline{X}^2\right)$$

$$= \sum_{j=1}^n E(X_j^2) - 2\sum_{j=1}^n E(\overline{X}X_j) + \sum_{j=1}^n E(\overline{X}^2) \qquad \text{Facts 3, 6}$$

$$= \sum_{j=1}^n (\mu^2 + \sigma^2) - 2E\left(\overline{X}\sum_{j=1}^n X_j\right) + nE(\overline{X}^2) \quad \text{Facts 8, 6}$$

$$= n(\mu^2 + \sigma^2) - 2E\left[\overline{X}(n\overline{X})\right] + nE(\overline{X}^2) \qquad \text{(A.4)}$$

$$= n(\mu^2 + \sigma^2) - 2nE(\overline{X}^2) + nE(\overline{X}^2) \qquad \text{Fact 3}$$

$$= n(\mu^2 + \sigma^2) - nE(\overline{X}^2)$$

$$= n(\mu^2 + \sigma^2) - n\left(\mu^2 + \frac{\sigma^2}{n}\right) \qquad \text{Fact 11}$$

$$= (n-1)\sigma^2.$$

Since $(n-1)E(s^2) = (n-1)\sigma^2$, then $E(s^2) = \sigma^2$ and so the sample variance $E(s^2)$ is an **unbiased estimator** of the population variance.

Fact 12 might seem a bit surprising at first. In analogy with Fact 10 (also see the discussion after Formula 1.5), we might have "expected" the mean of the deviates squared

$$\frac{1}{n}\sum_j (X_j - \overline{X})^2$$

to be an unbiased estimator of σ^2. However, from above

$$E\left[\frac{1}{n}\sum_j (X_j - \overline{X})^2\right] = \frac{n-1}{n}\left\{E\left[\frac{\sum_j (X_j - \overline{X})^2}{n-1}\right]\right\}$$

$$= \frac{n-1}{n}E(s^2)$$

$$= \frac{n-1}{n}\sigma^2$$

and so it is a *biased* estimator of σ^2. This is why the factor of $\frac{1}{n-1}$ is used in the sample variance formula. See Formula 1.4.

FACT 13. Just as the population mean μ may be defined as an expectation, so can the population variance, $\text{Var}(X)$.

$$\text{Var}(X) \equiv E(x_i^2) = E\left\{[X_i - E(X_i)]^2\right\}$$

$$= E\left\{X_i^2 - 2X_iE(X_i) + [E(X_i)]^2\right\}$$

$$= E\left(X_i^2 - 2X_i\mu + \mu^2\right) \qquad\qquad \text{Fact 1}$$

$$= E(X_i^2) - 2\mu E(X_i) + E(\mu^2) \qquad\quad \text{Facts 6 and 3}$$

$$= \mu^2 + \sigma^2 - 2\mu^2 + \mu^2 \qquad\qquad \text{Facts 8, 1, and 2}$$

$$= \sigma^2.$$

Similarly, the variance of $\overline{X}$ is

$$\text{Var}(\overline{X}) = E\left\{\left[\overline{X} - E(\overline{X})\right]^2\right\}$$

$$= E\left\{\overline{X}^2 - 2\overline{X}E(\overline{X}) + [E(\overline{X})]^2\right\}$$

$$= E\left(\overline{X}^2 - 2\overline{X}\mu + \mu^2\right) \qquad\qquad \text{Fact 10}$$

$$= E(\overline{X}^2) - 2\mu E(\overline{X}) + E(\mu^2) \qquad\quad \text{Fact 6}$$

$$= \mu^2 + \frac{\sigma^2}{n} - 2\mu^2 + \mu^2 \qquad\qquad \text{Facts 10 and 11}$$

$$= \frac{\sigma^2}{n}.$$

There is a decrease in variability in $\overline{X}$ as compared to X; $\text{Var}(\overline{X}) = \frac{\sigma^2}{n}$ while $\text{Var}(X) = \sigma^2$.

Note: Fact 13 is part of the Central Limit Theorem (Theorem 4.1)

$$\text{Var}(\overline{X}) = \sigma_{\overline{X}}^2 = \frac{\sigma_{\overline{X}}^2}{n}.$$

Retain the concept of expected values for use in analysis of variance.

A.3 The Formula for SS$_{\text{Treat}}$ in a One-Way ANOVA

We now give the derivation of the computational formula for SS$_{\text{Treat}}$ in a one-way ANOVA:

$$\begin{aligned}
\text{SS}_{\text{Treat}} &= \sum_i \sum_j (\overline{X}_{i.} - \overline{X}_{..})^2 \\
&= \sum_i \sum_j (\overline{X}_{i.}^2 - 2\overline{X}_{i.}\overline{X}_{..} + \overline{X}_{..}^2) \\
&= \sum_i n_i (\overline{X}_{i.}^2 - 2\overline{X}_{i.}\overline{X}_{..} + \overline{X}_{..}^2) \\
&= \sum_i (n_i \overline{X}_{i.}^2 - 2n_i \overline{X}_{i.}\overline{X}_{..} + n_i \overline{X}_{..}^2).
\end{aligned}$$

Substituting the summation formulas for the means,

$$\begin{aligned}
\text{SS}_{\text{Treat}} &= \sum_i \left[n_i \frac{\left(\sum_j X_{ij}\right)^2}{n_i^2} - \frac{2n_i \left(\sum_j X_{ij}\right)}{n_i} \frac{\left(\sum_i \sum_j X_{ij}\right)}{\sum_i n_i} + \frac{n_i \left(\sum_i \sum_j X_{ij}\right)^2}{\left(\sum_i n_i\right)^2} \right] \\
&= \sum_i \left[\frac{\left(\sum_j X_{ij}\right)^2}{n_i} - \frac{2\left(\sum_j X_{ij}\right)\left(\sum_i \sum_j X_{ij}\right)}{\sum_i n_i} + \frac{n_i \left(\sum_i \sum_j X_{ij}\right)^2}{\left(\sum_i n_i\right)^2} \right].
\end{aligned}$$

Distributing the summation over i,

$$\begin{aligned}
\text{SS}_{\text{Treat}} &= \sum_i \left[\frac{\left(\sum_j X_{ij}\right)^2}{n_i} \right] - \frac{2\left(\sum_i \sum_j X_{ij}\right)\left(\sum_i \sum_j X_{ij}\right)}{\sum_i n_i} + \frac{\sum_i n_i \left(\sum_i \sum_j X_{ij}\right)^2}{\left(\sum_i n_i\right)^2} \\
&= \sum_i \left[\frac{\left(\sum_j X_{ij}\right)^2}{n_i} \right] - \frac{2\left(\sum_i \sum_j X_{ij}\right)^2}{\sum_i n_i} + \frac{\left(\sum_i \sum_j X_{ij}\right)^2}{\sum_i n_i} \\
&= \sum_i \left[\frac{\left(\sum_j X_{ij}\right)^2}{n_i} \right] - \frac{\left(\sum_i \sum_j X_{ij}\right)^2}{\sum_i n_i}.
\end{aligned}$$

Changing to dot notation, we obtain

$$\text{SS}_{\text{Treat}} = \sum_i \left(\frac{T_{i.}^2}{n_i} \right) - \frac{T_{..}^2}{N}.$$

A.4 ANOVA Expected Values

If the assumptions of analysis of variance have been satisfied, then

$$\frac{\text{MS}_{\text{Treat}}}{\text{MS}_{\text{E}}} \quad \text{is distributed as} \quad F_{[(k-1), \sum_i (n_i - 1)]}.$$

F is always *one-tailed, right-tailed* in ANOVA. To see why the test is right-tailed and to determine what F actually is testing, we return to expected values. For simplicity let n_i be equal for all i, i.e., an equal number n of observations in each of the k treatments, and use Model I, the fixed treatment effects model. The hypotheses are

- H_0: $\mu_1 = \mu_2 = \cdots = \mu_k$

- H_a: At least one pair of μ_i's are different.

Let $\alpha_i = \mu_i - \mu$, the so-called treatment effect. So $\mu_i = \mu + \alpha_i$. Under H_0 all $\alpha_i = 0$.

$$E(SS_{\text{Treat}}) = E \left\{ \sum_{i=1}^{k} \left[\frac{\left(\sum_j X_{ij} \right)^2}{n} \right] - \frac{\left(\sum_{i=1}^{k} \sum_j X_{ij} \right)^2}{kn} \right\}.$$

Remember, $\sum_{j=1}^{n} X_{ij} = n\overline{X}_{i.}$ and $\sum_{i=1}^{k} \sum_{j=1}^{n} X_{ij} = kn\overline{X}_{..}$. Substituting these into the previous equation we obtain

$$E\left(SS_{\text{Treat}}\right) = E \left(\sum_{i=1}^{k} \frac{n^2 \overline{X}_{i.}^2}{n} - \frac{(kn)^2 \, \overline{X}_{..}^2}{kn} \right)$$

$$= E \left(\sum_{i=1}^{k} n\overline{X}_{i.}^2 - kn\overline{X}_{..}^2 \right)$$

$$= E \left(\sum_{i=1}^{k} n\overline{X}_{i.}^2 \right) - knE(\overline{X}_{..}^2). \tag{A.10}$$

Consider the first term of the difference in (A.10). From Fact 11 in Section A.2,

$$E \left(\sum_{i=1}^{k} n\overline{X}_{i.}^2 \right) = \sum_{i=1}^{k} nE \left(\overline{X}_{i.}^2 \right) = \sum_{i=1}^{k} n \left[(\mu + \alpha_i)^2 + \frac{\sigma^2}{n} \right]$$

$$= \sum_{i=1}^{k} n \left(\mu^2 + 2\mu\alpha_i + \alpha_i^2 + \frac{\sigma^2}{n} \right)$$

$$= \sum_{i=1}^{k} n\mu^2 + 2n\mu \sum_{i=1}^{k} \alpha_i + n \sum_{i=1}^{k} \alpha_i^2 + \sum_{i=1}^{k} \sigma^2$$

$$= kn\mu^2 + 2n\mu \sum_{i=1}^{k} \alpha_i + n \sum_{i=1}^{k} \alpha_i^2 + k\sigma^2. \tag{A.11}$$

Because the α_i's are deviations from a mean, $\sum_{i=1}^{k} \alpha_i = 0$ under the original assumptions of ANOVA. Substituting this into (A.11) gives

$$E \left(\sum_{i=1}^{k} n\overline{X}_{i.}^2 \right) = kn\mu^2 + n \sum_{i=1}^{k} \alpha_i^2 + k\sigma^2. \tag{A.12}$$

Applying Fact 11 to the second term in (A.10) we obtain,

$$knE(\overline{X}_{..}^2) = kn \left(\mu^2 + \frac{\sigma^2}{kn} \right) = kn\mu^2 + \sigma^2. \tag{A.13}$$

Substitute (A.12) and (A.13) into (A.10) to obtain

$$E(\text{SS}_{\text{Treat}}) = E\left(\sum_{i=1}^{k} n\overline{X}_{i.}^2\right) - knE(\overline{X}_{..}^2)$$

$$= kn\mu^2 + n\sum_{i=1}^{k}\alpha_i^2 + k\sigma^2 - \left(kn\mu^2 + \sigma^2\right)$$

$$= n\sum_{i=1}^{k}\alpha_i^2 + k\sigma^2 - \sigma^2$$

$$= n\sum_{i=1}^{k}\alpha_i^2 + (k-1)\sigma^2. \tag{A.14}$$

If we substitute (A.14) into the expression for $E(\text{MS}_{\text{Treat}})$, we obtain

$$E(\text{MS}_{\text{Treat}}) = E\left(\frac{\text{SS}_{\text{Treat}}}{k-1}\right)$$

$$= \frac{1}{k-1}E(\text{SS}_{\text{Treat}})$$

$$= \frac{n\sum_{i=1}^{k}\alpha_i^2 + (k-1)\sigma^2}{(k-1)}$$

$$= \frac{n\sum_{i=1}^{k}\alpha_i^2}{k-1} + \sigma^2.$$

Now we develop the expression for $E(\text{SS}_{\text{Error}})$. Using Facts 3 and 6 from Section A.2,

$$E(\text{SS}_{\text{Error}}) = E\left[\sum_{i=1}^{k}\left[(n_i-1)s_i^2\right]\right] = \sum_{i=1}^{k}E\left[(n_i-1)s_i^2\right].$$

In ANOVA the σ_i^2 are all assumed to be equal to σ^2 and for simplicity we chose equal treatment sizes (i.e., $n_i = n$), so using Fact 12

$$E(\text{SS}_{\text{Error}}) = \sum_{i=1}^{k}(n-1)E(s_i^2) = \sum_{i=1}^{k}(n-1)\sigma^2 = k(n-1)\sigma^2.$$

Therefore,

$$E(\text{MS}_{\text{E}}) = E\left[\frac{\text{SS}_{\text{Error}}}{k(n-1)}\right] = \frac{1}{k(n-1)}E(\text{SS}_{\text{Error}}) = \frac{1}{k(n-1)}k(n-1)\sigma^2 = \sigma^2.$$

So in the fixed model, one-way ANOVA,

$$E(\text{MS}_{\text{Treat}}) = \sigma^2 + \frac{n\sum_{i=1}^{k}\alpha_i^2}{k-1} \qquad \text{and} \qquad E(\text{MS}_{\text{E}}) = \sigma^2.$$

The term $\sum_{i=1}^{k}\alpha_i^2 = 0$ if and only if every $\alpha_i = 0$, that is, if and only if $\mu_i = \mu$ for all treatments. In other words, all treatments must produce the same mean effect. So under a *true* H_0,

$$E(F) = E\left(\frac{\text{MS}_{\text{Treat}}}{\text{MS}_{\text{E}}}\right) = \frac{\sigma^2}{\sigma^2} = 1.$$

Under a *true* H_a, substituting $\mu_i - \mu$ for α_i,

$$E(F) = E\left(\frac{\text{MS}_{\text{Treat}}}{\text{MS}_{\text{E}}}\right) = \frac{\sigma^2 + \frac{\sum_{i=1}^{k} n(\mu_i - \mu)^2}{k-1}}{\sigma^2}$$

and the numerator is larger than the denominator. Only when the numerator is significantly larger do we reject H_0. So the F expected value will be 1 or greater than 1 depending on which hypothesis is true, making the test always right-tailed. (See page 219.)

A.5   Calculating H in the Kruskal-Wallis Test

In Chapter 8 (see page 230) we introduced a shorthand method for calculating H, the test statistic for the Kruskal-Wallis test. We now show that the original definition of H and the shorthand method are algebraically equivalent. The only advantage of the first and longer method of calculating H is as an aid in understanding how the test statistic "measures" the similarity or the differences among the various samples. A bit of algebra will show that the two methods of calculating H give the same answer. First we simplify the sum in the first expression for H by squaring out the terms:

$$\sum_{i=1}^{k} n_i \left(\frac{R_i}{n_i} - \frac{N+1}{2}\right)^2 = n_1 \left(\frac{R_1}{n_1} - \frac{N+1}{2}\right)^2 + \cdots + n_k \left(\frac{R_k}{n_k} - \frac{N+1}{2}\right)^2$$

$$= n_1 \cdot \frac{R_1^2}{n_1^2} + \cdots + n_k \cdot \frac{R_k^2}{n_k^2}$$

$$- 2\left(\frac{N+1}{2}\right)\left(n_1 \cdot \frac{R_1}{n_1} + \cdots + n_k \cdot \frac{R_k}{n_k}\right)$$

$$+ (n_1 + \cdots + n_k)\left(\frac{N+1}{2}\right)^2$$

$$= \frac{R_1^2}{n_1} + \cdots + \frac{R_k^2}{n_k} - (N+1)(R_1 + \cdots + R_k) + \frac{N(N+1)^2}{4}.$$

In the last line we have used the fact that $n_1 + \cdots + n_k = N$. Recalling that the sum of all the ranks $R_1 + \cdots + R_k$ is really just the sum of the first N integers, which is $\frac{N(N+1)}{2}$, we can rewrite the last equation as

$$\sum_{i=1}^{k} n_i \left(\frac{R_i}{n_i} - \frac{N+1}{2}\right)^2 = \frac{R_1^2}{n_1} + \cdots + \frac{R_k^2}{n_k} - (N+1)\frac{N(N+1)}{2} + \frac{N(N+1)^2}{4}$$

$$= \left(\sum_{i=1}^{k} \frac{R_i^2}{n_i}\right) - \frac{N(N+1)^2}{2} + \frac{N(N+1)^2}{4}$$

$$= \left(\sum_{i=1}^{k} \frac{R_i^2}{n_i}\right) - \frac{N(N+1)^2}{4}.$$

Multiplying both sides by $\frac{12}{N(N+1)}$ gives us the two expressions for H,

$$\frac{12}{N(N+1)} \sum_{i=1}^{k} n_i \left(\frac{R_i}{n_i} - \frac{N+1}{2} \right)^2 = \frac{12}{N(N+1)} \left[\left(\sum_{i=1}^{k} \frac{R_i^2}{n_i} \right) - \frac{N(N+1)^2}{4} \right]$$

$$= \frac{12}{N(N+1)} \left(\sum_{i=1}^{k} \frac{R_i^2}{n_i} \right) - 3(N+1).$$

A.6   The Method of Least Squares for Linear Regression

This section presents the derivation of the least squares linear regression equation using differential calculus. To predict Y for various values of X when the data indicate a straight-line relationship between the two variables, we need to find a and b in the standard equation for a straight line, $Y = a + bX$, with a being the constant equal to the line's intercept on the Y-axis and b being equal to the line's slope.

Each Y can be expressed as the regression equation value for X_i plus a deviation from that value, ϵ_i:

$$Y_i = a + b \left(X_i - \overline{X} \right) + \epsilon_i.$$

The best-fitting equation will minimize the deviations ϵ_i's. More accurately, it will minimize the sum of the squares of the deviations. If $\hat{Y}_i$ is the predicted value for Y_i, that is, if $\hat{Y}_i = a + b \left(X_i - \overline{X} \right)$, then

$$Y_i = \hat{Y}_i + \epsilon_i.$$

Equivalently,

$$\epsilon_i = Y_i - \hat{Y}_i = Y_i - \left[a + b \left(X_i - \overline{X} \right) \right]$$

is the difference between the actual and predicted Y values for a particular X_i. Therefore,

$$\sum_{i=1}^{n} \epsilon_i^2 = \sum_{i=1}^{n} \left(Y_i - \hat{Y}_i \right)^2$$

$$= \sum_{i=1}^{n} \left\{ Y_i - \left[a + b \left(X_i - \overline{X} \right) \right] \right\}^2$$

$$= SS_{\text{Error}}.$$

Let's call $\sum_{i=1}^{n} \epsilon_i^2 = \Phi$. First we find the intercept a. We can minimize Φ by setting the partial derivatives $\frac{\partial \Phi}{\partial a} = 0$ and $\frac{\partial \Phi}{\partial b} = 0$. Since

$$\Phi = \sum_{i=1}^{n} \left\{ Y_i - \left[a + b \left(X_i - \overline{X} \right) \right] \right\}^2,$$

then using basic rules of differentiation,

$$\frac{\partial \Phi}{\partial a} = 0 \iff -2 \sum_{i=1}^{n} \left[Y_i - a - b \left(X_i - \overline{X} \right) \right] = 0$$

$$\iff \sum_{i=1}^{n} \left[Y_i - a - b \left(X_i - \overline{X} \right) \right] = 0$$

$$\iff \sum_{i=1}^{n} Y_i - \sum_{i=1}^{n} a - b \left(\sum_{i=1}^{n} X_i - \sum_{i=1}^{n} \overline{X} \right) = 0$$

$$\iff \sum_{i=1}^{n} Y_i - na = 0 \qquad \text{By (A.3) and (A.4)}$$

$$\iff \sum_{i=1}^{n} Y_i = na$$

$$\iff a = \overline{Y}.$$

Now we determine the slope b. Again using basic derivative rules,

$$\frac{\partial \Phi}{\partial b} = 0 \iff 2 \sum_{i=1}^{n} \left[Y_i - a - b \left(X_i - \overline{X} \right) \right] \left[- \left(X_i - \overline{X} \right) \right] = 0$$

$$\iff -\sum_{i=1}^{n} Y_i \left(X_i - \overline{X} \right) + \sum_{i=1}^{n} a \left(X_i - \overline{X} \right) + \sum_{i=1}^{n} b \left(X_i - \overline{X} \right)^2 = 0$$

$$\iff b \sum_{i=1}^{n} \left(X_i - \overline{X} \right)^2 = \sum_{i=1}^{n} Y_i \left(X_i - \overline{X} \right) \qquad \text{By (A.4) and (A.5)}$$

$$\iff b = \frac{\sum_{i=1}^{n} Y_i \left(X_i - \overline{X} \right)}{\sum_{i=1}^{n} \left(X_i - \overline{X} \right)^2}. \tag{A.15}$$

Notice that the denominator is the corrected sum of squares for X. We use (A.4) again to simplify the numerator for the expression for b in (A.15).

$$\sum_{i=1}^{n} Y_i \left(X_i - \overline{X} \right) = \sum_{i=1}^{n} X_i Y_i - \overline{X} \sum_{i=1}^{n} Y_i = \sum_{i=1}^{n} X_i Y_i - \frac{\left(\sum_{i=1}^{n} X_i \right) \left(\sum_{i=1}^{n} Y_i \right)}{n}.$$

We can now rewrite (A.15) using this last expression as the numerator and the sum of squares formula (A.7) as the denominator to obtain the expression for the slope,

$$b = \frac{\sum X_i Y_i - \frac{(\sum X_i)(\sum Y_i)}{n}}{\sum X_i^2 - \frac{(\sum X_i)^2}{n}}. \tag{A.16}$$

Using (A.16) for the slope makes

$$\hat{Y} \text{ or } \widehat{\mu}_{Y|X} = \overline{Y} + b \left(X - \overline{X} \right)$$

the best-fitting straight line according to the least squares criterion (see page 287).

Answers to Even-Numbered Problems

Answers for Chapter 1

2. Sample mean: 106.3 cm, corrected sum of squares: 898.1, sample variance: 99.79 cm^2, standard deviation: 10.0 cm, and the range is 27 cm: 93–120 cm.

4. (*a*) Sample mean: 14.75 yr, variance: 0.703 yr^2, standard deviation: 0.84 yr, median: 14.95 yr.

(*b*) The difference between the claimed mean 12.5 yr and the sample mean (14.75 yr) is 2.25 yr or approximately 2.7 standard deviations. Is this a large distance? We will consider methods to evaluate this distance in Chapter 3. Also, no value in the sample is as low as the claimed mean. The mean appears significantly higher for the female endurance athletes.

6. $\overline{X} = 25.87$ hr; $s^2 = 0.849$ hr^2. All values in the sample are greater than 24 hr. The sample mean is over 2 standard deviations from 24 hr. It appears the average day length in isolation chambers is longer.

8. (*a*) Blackburnian warblers forage significantly higher in the canopy than Bay-Breasted warblers. Variability in foraging height is similar between the two species.

Bay-Breasted	**Blackburnian**
$n = 10$	$n = 12$
$\overline{X} = 14.1$ ft	$\overline{X} = 17.8$ ft
$s = 3.0$ ft	$s = 3.2$ ft

(*b*) Standard deviation is a better reflection of variability because it is more efficient, using all the observations in the sample.

Bay-Breasted	**Blackburnian**
$\tilde{X} = 13$ ft	$\tilde{X} = 17$ ft
Range limits = 10–19 ft	Range limits = 15–24 ft
Range = 9 ft	Range = 9 ft

10. This transformation from inches to centimeters is a form of (multiplicative) coding. $\overline{X} = 68$ in $\times$ 2.54 cm/in = 172.72 cm and $s^2 = 100$ in^2 $\times$ $(2.54$ cm/in$)^2 = 645.16$ cm^2.

12. The answers are

n	Median depth	Quartile depth
22	11.5	6
23	12	6
24	12.5	6.5
25	13	6.5

14. $\overline{X} = 13.2$ cm, $\tilde{X} = 13.0$ cm, $s^2 = 2.44$ cm^2, and $s = 1.6$ cm.

X_i	f_i	$f_i X_i$	$f_i X_i^2$
10	2	20	200
11	8	88	968
12	17	204	2448
13	22	286	3718
14	14	196	2744
15	10	150	2250
16	7	112	1792
17	1	17	289
	81	1073	14,409

16. Problems 5 and 7 involve discrete variables and are measured exactly so the 30–300 rule doesn't apply.

18. The frequency table and histogram are given below:

X	Freq	Rel freq	Cum. freq	Rel cum freq
10	2	2.47	2	2.47
11	8	9.88	10	12.35
12	17	20.99	27	33.33
13	22	27.16	49	60.49
14	14	17.28	63	77.78
15	10	12.35	73	90.12
16	7	8.64	80	98.77
17	1	1.23	81	100.00

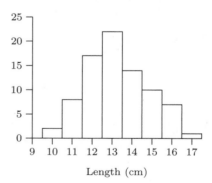

Length (cm)

20. The data describe populations, not samples.

	Females	Males
Mean	49.17 kg	86.73 kg
Pop. std. dev.	3.09 kg	4.94 kg
Pop. var.	9.525 kg^2	24.367 kg^2
Max	55.0 kg	92.3 kg
Min	44.9 kg	77.3 kg
Range	10.1 kg	15.0 kg

22. The lengths do not satisfy the 30–300 rule (since $29 - 19 = 10$ cm) and should have been measured to the nearest millimeter. The diameters do satisfy the 30–300 rule (since $11.3 - 7.5 = 3.8$ cm or 38 mm). The data are a sample; one would assume that more such adzes exist but have not been found. The basic summary information is given below.

	Length	Diameter
$\overline{X}$	23.2 cm	9.97 cm
s^2	11.48 cm^2	1.433 cm^2
s	3.4 cm	1.20 cm

24. Make sure to use equal areas to represent equal frequencies in both histograms.

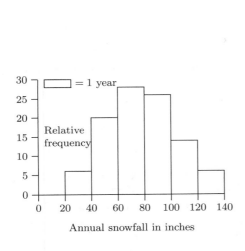
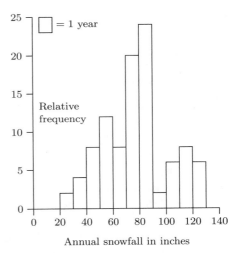

26. The summary statistics are:

	O. virginianus	*O. hemionus*
n	20	20
$\overline{X}$	82.4 kg	97.0 kg
s	7.4 kg	11.4 kg

The five-number summaries are:

	O. virginianus			*O. hemionus*	
Median:	83.5			99.5	
Quartiles:	75.5	89.5		91.0	104.0
Extremes:	70.0		92.0	73.0	120

In general, the white-tailed deer have lower weights and are less variable than the mule deer. There are no outliers. The box plot also indicates more asymmetry in the mule deer data set than in the white-tailed deer data set.

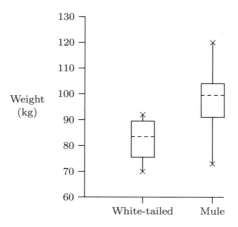

28. (a) $\overline{X} = 91.9$ cm, and $s = 7.6$ cm.

(b) $s^2 = 8.88$ kg^2, $s = 3.0$ kg, and standard error 0.8 kg.

30. (a) True, see page 5. (b) True, see page 6. (c) False, see page 10. (d) False, see page 12. (e) True, see page 23. (f) False, see pages 16–18.

Answers for Chapter 2

2. (a) $\binom{5}{3} = \frac{5!}{3!(5-3)!} = \frac{5!}{3!2!} = 10$, while $\binom{5}{2} = \frac{5!}{2!(5-2)!} = \frac{5!}{2!3!} = 10$.

(b) This makes sense from the viewpoint of the selection process *because every time you choose 3 items from 5, you implicitly **choose** 2 items from 5 to exclude.*

(c) You can do the math, but again, every time you choose k items from n, you implicitly choose $n - k$ items to exclude. The number of ways to choose (to exclude these) $n - k$ items must be the same as the way to choose k items.

4. $\binom{15}{8} = 6,435$.

6. $\binom{30}{5} = 142,506$.

8. $P(D_1 \cap D_2) = P(D_1) \times P(D_2) = 0.6 \times 0.6 = 0.36$ since D_1 and D_2 are independent. $P(D_1 \cup D_2) = P(D_1) + P(D_2) - P(D_1 \cap D_2) = P(D_1) = 0.6 + 0.6 - 0.36 = 0.84$.

10. See pages 48 and 41.

12. The student assumed he was defining a classical probability space, i.e., all the outcomes were equally likely but $P(H) = \frac{1}{2}$ and $P(TH) = \frac{1}{4}$. So the classical method is not applicable. $\{HH, HT, TH, TT\}$ is the classical space required. Then $P(\text{at least one } H) = \frac{n(H)}{n(S)} = \frac{3}{4}$.

14. (a) $P(D \cup St) = P(D) + P(St) - P(D \cap St) = 0.4 + 0.7 - 0.2 = 0.9$. So $P(D' \cap St') = 1 - P(D \cup St) = 1 - 0.9 = 0.1$.

(b) $P(St|D) = \dfrac{P(St \cap D)}{P(D)} = \dfrac{0.2}{0.4} = 0.5$.

(c) Symptoms are independent if and only if $P(D \cap St) = P(D) \cdot P(St)$. Here $0.2 \neq (0.4)(0.7) = 0.28$, so they are not independent.

16. (*a*)

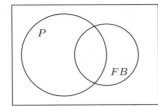

$$P(P) = 0.60 \qquad P(P') = P(W) = 0.40$$

$$P(FB) = 0.30 \qquad P(FB') = P(NFB) = 0.70$$

$$P(P \cap FB) = 0.10$$

(*b*) $P(W \cap FB) = P(FB) - P(P \cap FB) = 0.30 - 0.10 = 0.20.$

(*c*) $P(FB|W) = \frac{P(FB \cap W)}{P(W)} = \frac{0.20}{0.40} = 0.50.$

(*d*) $P(FB|P) = \frac{P(FB \cap P)}{P(P)} = \frac{0.10}{0.60} = 0.17.$

(*e*) Not independent; $P(FB|P) \neq P(FB)$ since $0.17 \neq 0.30$.

(*f*) Only 17% of the pink-flowered trees are infected, while 50% of the white-flowered trees are. Pink-flowered trees appear to be more resistant to fire blight. Recommend buying them.

18. (*a*) $P(M) = 0.80$, $P(T) = 0.30$, and $P(M \cap T) = 0.20$. So $P(M \cup T) = P(M) + P(T) - P(M \cap T) = 0.80 + 0.30 - 0.20 = 0.90.$

(*b*) $P(T'|M) = \frac{P(T' \cap M)}{P(M)} = \frac{0.60}{0.80} = 0.75.$ (Note: $P(T' \cap M) = P(M) - P(M \cap T) = 0.80 - 0.20 = 0.60.$)

(*c*) Not independent because $P(M \cap T) \neq P(M) \times P(T)$.

20. (*a*) $P(\text{Tay-Sachs}) = \frac{1}{2} \times \frac{1}{2} = 0.25.$

(*b*) $P(\text{Tay-Sachs and brachydactyly}) = P(\text{aaB}_-) = \frac{1}{4} \times \frac{1}{2} = \frac{1}{8}.$

(*c*) $P(\text{Aa}|\text{A}_-) = \frac{P(\text{Aa} \cap \text{A}_-)}{P(\text{A}_-)} = \frac{2/4}{3/4} = \frac{2}{3}.$

(*d*) $P(\text{normal}) = P(\text{A}_-\text{bb}) = \frac{3}{4} \times \frac{1}{2} = \frac{3}{8}.$ So $P(2 \text{ normal}) = \left(\frac{3}{8}\right)^2.$

22. $\dfrac{\binom{1}{1}\binom{269}{3}}{\binom{270}{4}} = 0.0148.$

24. Use classical probabilities.

(*a*) $\dfrac{76}{198}$ (*b*) $\dfrac{122}{198}$ (*c*) $\dfrac{84}{198}$

26. (*a*) 0.24

(*b*) 0.105

(*c*) $55/500 = 0.11.$

(*d*) $145/500 = 0.29.$

(*e*) $\frac{300}{1000} \times \frac{299}{999} = 0.0\overline{897}.$

(*f*) $P(\text{Hi Nitrate} \cap \text{Lo DDT}) = 0.055$ while $P(\text{Hi Nitrate}) \times P(\text{Lo DDT}) = 0.24 \times 0.5 = 0.12.$ No, these two contaminant levels are not independent.

28. (*a*)

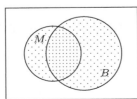

$$P(B) = 0.65 \qquad P(B') = 0.35$$

$$P(M) = 0.40 \qquad P(M') = 0.60$$

$$P(M \cap B) = 0.30$$

(b) $P(M) \times P(B) = 0.40 \times 0.65 = 0.26$. Not independent; $P(M) \times P(B) \neq P(M \cap B) = 0.30$.

(c) $P(B \cap M') = P(B) - P(B \cap M) = 0.65 - 0.30 = 0.35$.

(d) $P(B \cup M) = P(B) + P(M) - P(B \cap M) = 0.65 + 0.40 - 0.30 = 0.75$.

(e) $P(M|B) = \frac{P(M \cap B)}{P(B)} = \frac{0.30}{0.65} = 0.46$.

30. $\frac{\binom{10}{3}\binom{15}{3}}{\binom{25}{6}} = \frac{120 \times 455}{177,100} = 0.308$.

32. (a)

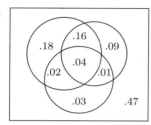

Smokers (upper left cricle) $P(S) = 0.40$

Obese (upper right circle) $P(O) = 0.30$

Heart Disease (lower circle) $P(HD) = 0.10$

(b) 0.03

(c) $P(S \cup O) = P(S) + P(O) - P(S \cap O) = 0.40 + 0.30 - 0.20 = 0.50$.

(d) $P(HD|O) = \frac{0.05}{0.30} = 0.1\overline{6}$.

(e) $P(HD|S) = \frac{0.06}{0.40} = 0.15$.

(f) $P(HD \cap S) \neq P(HD) \times P(S)$ while $P(HD \cap O) \neq P(HD) \times P(0)$. They are not independent.

34. We wish to determine P(Is a tax cheat). The probability of giving a forced "Yes" answer is

$$P(\text{Replied Yes} \,|\, \text{Rolled 1 or 2}) = 1.$$

Analogous to (2.7)

$$P(\text{Is a tax cheat}) = \frac{P(\text{Replied Yes}) - P(\text{Replied Yes} \,|\, \text{Rolled 1 or 2}) \cdot P(\text{Rolled 1 or 2})}{P(\text{Rolled 3, 4, or 5})}$$

$$= \frac{0.40 - (1)(0.333)}{0.5}$$

$$= 0.1333.$$

36. (a) False, see page 36. (b) True, see page 36. (c) True, see page 43. (d) False, see page 44. (e) True, see page 48. (f) True, see page 48. (g) False, see pages 49–51.

Answers for Chapter 3

2. The probabilities are

(a) 0.22   (b) 0.25   (c) 0   (d) 0.47   (e) 0.53

(f) The density could represent a mixture of males and females with normal distributions and different means. Another possibility is that the density is a mixture of juveniles and adults.

4. $\sigma = 1.5$ mm. So $P(-1 < Z < 1) = 0.6826$.

6. (*a*) $f(0) = \frac{e^{-3}(3)^0}{0!} = 0.0498$ or use Table C.2 where $f(0) = F(0) = 0.0498$.

    (*b*) $1 - f(0) = 0.9502$.

8. (*a*) $P(X < 10) = F(9)$. From Table C.1 with $p = 0.60$ and $n = 20$, $F(9) = 0.1275$.

    (*b*) Again $p = 0.60$ but $n = 13$. $f(8) = \frac{13!}{8!5!}(0.6)^8(0.4)^5 = 0.2214$. From Table C.1 $F(8) - F(7) = 0.2214$.

10. (*a*) $P(X = 3) = f(3) = F(3) - F(2) = 0.6472 - 0.4232 = 0.2240$.

    (*b*) 0.2240

    (*c*) $P(X > 3) = 1 - F(3) = 1 - 0.6472 = 0.3528$.

12. (*a*) $P(X = 9) = P(8.5 \leq X < 9.5) = P(X < 9.5) - P(X < 8.5) = P\left(Z < \frac{9.5 - 8.4}{2.4}\right) - P\left(Z < \frac{8.5 - 8.4}{2.4}\right) = P(Z < 0.4583) - P(Z < 0.042) \approx 0.6772 - 0.5160 = 0.1612$.

    (*b*) From Table C.3, $P(Z < 1.28) = 0.90$. So we must find the score x such that $P(X < x) = 0.90$. But $P(X < x) = P\left(Z < \frac{x - 8.4}{2.4}\right) = 0.90 \iff \frac{x - 8.4}{2.4} = 1.28 \iff x = 8.4 + 1.28(2.4) = 11.472$. Because scores are reported as integers, an applicant would need a minimum score of 11 to be considered.

14. (*a*) $P(X \geq 80) = P\left(Z \geq \frac{80 - 70}{4}\right) = P(Z \geq 2.5) = 0.0062$ or 0.62%.

    (*b*) $P(X \geq 77) = P\left(Z \geq \frac{77 - 70}{4}\right) = P(Z \geq 1.75) = 0.0401$ or 4.01%.

16. $X = \mu + 1.28\sigma = 70 + 1.28(10) = 82.8$.

18. Use a normal approximation of the binomial with $\mu = np = 67(0.25) = 16.75$ and $\sigma = \sqrt{np(1-p)} = \sqrt{67(0.25)(0.75)} = 3.54$.

    (*a*) $F_B(12) = F_N(12.5) = P\left(Z < \frac{12.5 - 16.75}{3.54}\right) = 0.1151$.

    (*b*) $1 - F_B(11) = 1 - F_N(11.5) = P\left(Z < \frac{11.5 - 16.75}{3.54}\right) = 1 - 0.0694 = 0.9306$.

20. (*a*) The expected value is $E(X) = \mu = np = 100(0.60) = 60$.

    (*b*) First, $\sigma = 4.90$. So $P(65 \leq X \leq 70) = F_B(70) - F_B(64) \approx F_N(70.5) - F_N(64.5) = 0.9838 - 0.8212 = 0.1626$.

22. (*a*) $E\left(\frac{X-\mu}{\sigma}\right) = \frac{1}{\sigma}E(X - \mu) = \frac{1}{\sigma}[E(X) - E(\mu)] = \mu - \mu = 0$.

    (*b*) $\mathrm{Var}(4X - 3) = \mathrm{Var}(4X) - \mathrm{Var}(3) = 4^2\,\mathrm{Var}(X) - 0 = 16(100) = 1600$.

24. Use the binomial distribution with $n = 19$ and $p = 0.3$ in Table C.1.

    (*a*) $P(X = 0) = F(0) = 0.0011$.

    (*b*) $P(X \geq 7) = 1 - F(6) = 1 - 0.6655 = 0.3345$.

    (*c*) $P(4 \leq X \leq 7) = F(7) - F(3) = 0.8180 - 0.1332 = 0.6848$.

    (*d*) $P(X \leq 12) = F(12) = 0.9994$.

    (*e*) The probability of at least 11 people having this condition in a random sample is $P(X \geq 11) = 1 - F(10) = 1 - 0.9895 = 0.0105$. Thus, it is very unlikely. However, this was not a random sample because many of the people at the picnic are related and the condition is inherited.

26. Assuming the patient is normal, $P(X \geq 225) = P(Z \geq 3.0) = 1 - F(3) = 0.0013$. If the patient has FH, $P(X \leq 225) = P(Z \leq -2.5) = 0.0062$.

28. $P(X \leq 0.5) = P\left(Z \leq \frac{0.5 - 0.4}{0.15}\right) = P(Z \leq 0.6667) = 0.7475$. No, we can expect only three-quarters of the samples to have measurements below 0.5 NTU.

30. This is a binomial problem with $p = 0.1$. For the first question, we are trying to determine the sample size n so that the probability of finding **no** infected crows is less than or equal to 50%. So we want $F(0) \leq 0.5$. Find the appropriate n from Table C.1. The smallest possible sample size is $n = 7$. To be 90% certain, we need $F(0) \leq 0.1$. Unfortunately, the largest sample size in Table C.1 is $n = 20$ and $F(0) = 0.1216$ is not small enough. So let's do the math. We need to find n so that

$$0.1 < F(0) = (0.1)^0 (1 - 0.1)^n = (0.9)^n.$$

For the mathematically inclined, take logs

$$\log 0.1 < \log(0.9^n) = n \log 0.9 \Rightarrow n > \frac{\log 0.1}{\log 0.9} = 21.85.$$

So $n = 22$ is the required sample size. If logs scare you, use your calculator. You already know that $(0.9)^{20}$ is not small enough. Try $(0.9)^{21}$. Nope, still too big. But $(0.9)^{22} = 0.0985$ works.

32. (a) $P(X < 140) = P\left(Z < \frac{140-120}{15}\right) = P(Z < 1.33) = 0.9082$.

(b) $P(100 < X < 140) = P\left(\frac{100-120}{15} < Z < \frac{140-120}{15}\right) = P(-1.33 < Z < 1.33) = 0.9082 - 0.0918 = 0.8164$.

(c) $P(X > 150) = P\left(Z > \frac{150-120}{15}\right) = P(Z > 2.0) = 1 - 0.9772 = 0.0228$.

(d) $0.0228(365) = 8.322$ days per year would be expected to be over the federal limit.

34. (a) $P(X \leq 120) = P\left(Z \leq \frac{120-110}{8}\right) = P(Z \leq 1.25) = 0.8944$ or 89.44%.

(b) $P(X \geq 140) = 1 - P\left(Z \leq \frac{140-110}{8}\right) = 1 - F(3.75) = 1 - 0.9999$ or 0.01%.

(c) $P(120 \leq X \leq 139) = P\left(Z \leq \frac{139-110}{8}\right) - P\left(Z \leq \frac{120-110}{8}\right) = F(3.625) - 0.8944 = 0.9999 - 0.8944 = 0.1055$ or 10.55%.

36. $z_{0.95} = 1.645$. So $\overline{X} + 1.645s = 81.3 + 1.645(6.7) = 92.32$ kg.

38. (a) $P(X \leq 3) = F(3) = 0.0818$.

(b) $P(X = 4) = F(4) - F(3) = 0.1730 - 0.0818 = 0.0912$.

(c) $P(X > 5) = 1 - F(5) = 1 - 0.3007 = 0.6993$.

(d) $P(X < 4) = F(3) = 0.0818$. So the probability that two randomly chosen hectares each have less than 4 specimens is $(0.0818)^2 = 0.0067$.

40. (a) $E(X) = \mu = np = 0.3 \times 20 = 6$.

(b) Use Table C.1. $P(X < 5) = F(4) = 0.2375$.

(c) $P(X \leq 5) = F(5) = 0.4164$.

(d) $P(X = 5) = F(5) - F(4) = 0.4164 - 0.2375 = 0.1789$.

(e) $P(X > 8) = 1 - F(8) = 1 - 0.8867 = 0.1133$.

(f) $P(X \geq 12) = 1 - F(11) = 1 - 0.9949 = 0.0051$. Yes. It's very unlikely that one would get a sample with 12 scarred in 20, if the average was 6 (30%).

42. (a) True, see page 66. (b) True, see page 72. (c) True, see page 74. (d) False, see page 76. (e) False, see page 77. (f) True, see page 80. (g) True, see page 82. (h) True, see page 86.

Answers for Chapter 4

2. For μ, from Formula 4.1,

$$L_1 = \overline{X} - t_{(1-\frac{\alpha}{2},\nu)}\frac{s}{\sqrt{n}} = 400 - 2.306\left(\frac{30}{\sqrt{9}}\right) = 376.9$$

$$L_2 = \overline{X} + t_{(1-\frac{\alpha}{2},\nu)}\frac{s}{\sqrt{n}} = 400 + 2.306\left(\frac{30}{\sqrt{9}}\right) = 423.1$$

or $C(376.9 \text{ hr} \le \mu \le 423.1 \text{ hr}) = 0.95$. For σ^2, from Formula 4.2,

$$L_1 = \frac{(n-1)s^2}{\chi^2_{1-\frac{\alpha}{2}}} = \frac{(8)(30)^2}{17.5} = 411.43 \quad \text{and} \quad L_2 = \frac{(n-1)s^2}{\chi^2_{\frac{\alpha}{2}}} = \frac{(8)(30)^2}{2.18} = 3302.75.$$

Taking square roots, we obtain $C(20.3 \text{ hr} \le \sigma \le 57.5 \text{ hr}) = 0.95$.

4. $C(5.58 \ \mu g/l \le \mu \le 6.82 \ \mu g/l) = 0.95$ and $C[1.37 \ (\mu g/l^2) \le \sigma^2 \le 4.35 \ (\mu g/l^2)] = 0.95$.

6. (a) $P(X \le 155) = P(Z \le -0.5) = 0.3085$.

(b) $P(125 < X < 215) = P(-1.5 < Z < 1.5) = 0.9332 - 0.0668 = 0.8664$.

(c) $n = 25$, the standard error is $\frac{\sigma}{\sqrt{n}} = 6$ mg/dl. So $P(\overline{X} < 182) = P(Z < \frac{182-170}{6}) = P(Z < 2.0) = 0.9772$.

(d) $n = 100$, $\frac{\sigma}{\sqrt{n}} = 3$ mg/dl, $P(\overline{X} < 164) = P(Z < \frac{164-170}{3}) = P(Z < -2.0) = 0.0228$.

8. $n = 10$, $\overline{X} = 14.75$ yr, $s = 0.84$ yr, so $C(14.15 \text{ yr} \le \mu \le 15.35 \text{ yr}) = 0.95$. The 95% confidence interval does not include 12.5, indicating that the mean for the swimmers is significantly higher.

10. $C(14.85 \text{ mg/100 ml} \le \mu \le 21.15 \text{ mg/100 ml}) = 0.95$ and $C(69.88 \ [\text{mg/100 ml}]^2 \le \sigma^2 \le 258.02 \ [\text{mg/100 ml}]^2) = 0.95$.

12. With $\hat{p} = \frac{29}{2000} = 0.0145$

$$L_1 = \hat{p} - z_{1-\frac{\alpha}{2}}\sqrt{\frac{\hat{p}(1-\hat{p})}{n}} = 0.0145 - 2.33\sqrt{\frac{0.0145(0.9855)}{2000}} = 0.00827$$

$$L_2 = \hat{p} + z_{1-\frac{\alpha}{2}}\sqrt{\frac{\hat{p}(1-\hat{p})}{n}} = 0.02072.$$

This is the same as an incident rate of 827 to 2072 per 100,000 people in the 40–49 age group.

14. Using the information in the table below, $\overline{X} = 15.4$ yr and $s = 2.2$ yr. The reading age appears to be too high because 35 of the 46 sites have reading ages above 14 years.

X_i	f_1	$f_i X_i$	$f_i X_i^2$
8	1	8	64
9	0	0	0
10	1	10	100
11	2	22	242
12	1	12	144
13	4	52	676
14	2	28	392
15	7	105	1575
16	4	64	1024
17	24	408	6936
Sum	46	709	11153

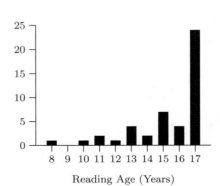

Reading Age (Years)

In the problem, $n = 46$, $\overline{X} = 15.4$ yr, and $s = 2.2$ yr. Since $df = n-1 = 45$, then from Table C.4 $t_0 = 2.014$. Therefore, using Formula 4.1 the confidence interval endpoints are

$$L_1 = 15.4 - 2.014\left(\frac{2.2}{\sqrt{46}}\right) = 14.7 \text{ yr}$$

and

$$L_2 = 15.4 + 2.014\left(\frac{2.2}{\sqrt{46}}\right) = 16.1 \text{ yr}.$$

The confidence interval lies completely above 14.0 yr, strongly suggesting that the reading age is not at or below 14.

16. (a) $\overline{X} = 232.9$ picograms/kg and $s = 39.9$ picograms/kg.

(b) Using Formula 4.1, the 95% confidence interval limits are

$$L_1 = \overline{X} - 2.145\frac{s}{\sqrt{n}} = 232.9 - 2.145\frac{39.9}{\sqrt{15}} = 232.9 - 22.1 = 210.8 \text{ picograms/kg}$$

and

$$L_2 = \overline{X} + 2.145\frac{s}{\sqrt{n}} = 232.9 + 22.1 = 255.0 \text{ picograms/kg}.$$

For a 95% confidence interval for σ^2, use Formula 4.2. From Table C.5, $\chi^2_{0.975(14)} = 26.1$ and $\chi^2_{0.025(14)} = 5.63$. So the limits for the confidence interval are

$$L_1 = \frac{(n-1)s^2}{\chi^2_{1-0.025}} = \frac{(15-1)(39.9)^2}{26.1} = 853.95 \text{ (picograms/kg)}^2$$

and

$$L_2 = \frac{(n-1)s^2}{\chi^2_{0.025}} = \frac{(15-1)(39.9)^2}{5.63} = 3958.82 \text{ (picograms/kg)}^2.$$

(c) Yes, 200 (picograms/kg)^2 lies well below the 95% confidence interval for the mean for the recyclers that was calculated above.

18. 99% confidence interval for μ:

$$L_1 = 144 - 2.977\left(\frac{42}{\sqrt{15}}\right) = 144 - 32.28 = 111.7 \text{ min}; \quad L_2 = 144 + 32.28 = 176.3 \text{ min}.$$

99% confidence interval for σ^2:

$$L_1 = \frac{(15-1)1764}{31.3} = 789.0 \text{ min}^2 \quad \text{and} \quad L_2 = \frac{(15-1)1764}{4.07} = 6067.8 \text{ min}^2.$$

99% confidence interval for σ: $L_1 = 28.1$ min and $L_2 = 77.9$ min.

20. For Method 1 use Formula 4.3 with $\hat{p} = 0.74$. Note that $n\hat{p} = 100 \times 0.74 = 74 > 5$ and $n(1 - \hat{p}) = 100 \times 0.26 = 26 > 5$. For a 95% confidence interval, the endpoints are

$$L_1 = \hat{p} - 1.960\sqrt{\frac{\hat{p}(1-\hat{p})}{n}} \quad \text{and} \quad L_2 = \hat{p} + 1.960\sqrt{\frac{\hat{p}(1-\hat{p})}{n}}.$$

Thus,

$$L_1 = 0.74 - 1.960\sqrt{\frac{(0.74)(0.26)}{100}} = 0.74 - 0.086 = 0.654 \text{ and } L_2 = 0.74 + 0.086 = 0.826.$$

For a 99% confidence interval, the endpoints are

$$L_1 = 0.74 - 2.58\sqrt{\frac{(0.74)(0.26)}{100}} = 0.74 - 0.113 = 0.627 \text{ and } L_2 = 0.74 + 0.113 = 0.853.$$

22. (*a*) The interval endpoints are

$$L_1 = \overline{X} - t\left(\frac{s}{\sqrt{n}}\right) = 1.8 - 2.131\left(\frac{0.5}{\sqrt{16}}\right) = 1.8 - 0.27 = 1.53 \text{ m}$$

$$L_2 = \overline{X} + t\left(\frac{s}{\sqrt{n}}\right) = 1.8 + 0.27 = 2.07 \text{ m.}$$

(*b*) The interval endpoints are

$$L_1 = \frac{(n-1)s^2}{\chi^2_{0.975}} = \frac{(16-1)(0.25)}{27.5} = 0.136 \text{ m}^2$$

$$L_2 = \frac{(n-1)s^2}{\chi^2_{0.025}} = \frac{(16-1)(0.25)}{6.26} = 0.599 \text{ m}^2.$$

(*c*) Since 2.0 m is within the 95% confidence interval for μ, the data support the claim about μ. Since 0.09 m^2 lies below the lower endpoint of the 95% confidence interval for σ^2, we have evidence that the horns are more variable than claimed.

24. (*a*) True, see page 107. (*b*) False, see page 117. (*c*) True, see page 111. (*d*) True, see page 115. (*e*) True, see page 115.

Answers for Chapter 5

2. The Central Limit Theorem describes the sampling distribution of $\overline{X}$ as normal or approximately normal with a standard deviation equal to the standard error of the mean.

4. In standard English, to *prove* something means to establish conclusively its truth or validity. With tests of hypothesis we always have some probability or possibility of being incorrect. Accepting an H_0 could lead to a Type II error (accepting a false H_0) and has a probability of β. Rejecting an H_0 could lead to a Type I error (rejecting a true H_0) and has a probability of α. While α and β are sometimes quite small each is never 0. We say H_0 is supported or not supported by the statistical test, never that the test proves H_0 to be true or false. See Table 5.1.

6. (*a*) $C(65.3 \text{ in} < \mu < 67.7 \text{ in}) = 0.95$.

(*b*) $C(5.48 \text{ in}^2 < \sigma^2 < 17.42 \text{ in}^2) = 0.95$.

(*c*) H_0: $\mu = 64.0$ in versus H_a: $\mu \neq 64.0$ in. Yes, 64.0 in is not included in a 95% confidence interval.

(*d*) H_0: $\sigma^2 = 4.0$ in^2, H_a: $\sigma^2 \neq 4.0$ in^2. Yes, 4.0 in^2 is not included in a 95% confidence interval.

(*e*) Confidence intervals serve as two-tailed tests of hypothesis. L_1 and L_2 are the critical values for $\overline{X}$ or s^2. If the sample statistic falls outside the confidence limits, H_a is supported.

8. (*a*) H_0: $\mu \leq 12.5$ yr versus H_a: $\mu > 12.5$ yr.

(*b*) H_0: $\mu \leq 24$ hr versus H_a: $\mu > 24$ hr.

(*c*) H_0: $\mu \leq 14$ yr versus H_a: $\mu > 14$ yr.

(*d*) H_0: $\mu \leq 200$ pg/kg versus H_a: $\mu > 200$pg/kg.

10. (*a*) True, see page 132. (*b*) True, see page 131. (*c*) True, see Section 5.2. (*d*) True, see page 134. (*e*) False, see Section 5.3. (*f*) False, see page 134. (*g*) True, see page 138.

Answers for Chapter 6

2. (*a*) $\overline{X} = 11.00$ cm, $s^2 = 2.644$ cm^2, and $n = 16$.

(*b*) $C(10.13 \text{ cm} \leq \mu \leq 11.87 \text{ cm}) = 0.95$. I am 95% confident that the parametric mean μ for arm length lies between 10.13 cm and 11.87 cm.

(*c*) The hypotheses are H_0: $\mu = 12$ cm versus H_a: $\mu \neq 12$ cm. Check that $t = -2.46$ and the c.v. $= \pm 2.131$. Since $-2.46 < -2.131$, reject H_0. The confidence interval in (*b*) uses $\pm 2.131 \frac{s}{\sqrt{n}}$ to generate limits. Algebraically, the two-tailed test of hypothesis at α and the confidence interval at $1 - \alpha$ are equivalent. The confidence interval does not contain 12 indicating a significant difference.

(*d*) $C(1.442 \text{ cm}^2 \leq \sigma^2 \leq 6.335 \text{ cm}^2) = 0.95$. I am 95% confident that the parametric variance, σ^2, for arm length lies between 1.442 cm^2 and 6.335 cm^2.

(*e*) The hypotheses are H_0: $\sigma^2 = 5.0$ cm^2 versus H_a: $\sigma^2 \neq 5.0$ cm^2. Calculate $\chi^2 = 7.932$. The c.v.'s $= 6.26, 27.5$. Since $6.26 < 7.932 < 27.5$, accept H_0. The variance cannot be shown to be significantly different from 5.0 cm^2. Again algebraically, the two-tailed test of hypothesis at α and the confidence interval at $1 - \alpha$ are equivalent. The sample variance of 5.0 cm^2 lies within the confidence limits at 95%.

4. The hypotheses are H_0: $\mu = 15$ lb versus H_a: $\mu \neq 15$ lb. Use $\alpha = 0.05$ with c.v.'s $= \pm 2.064$. Calculate $t = 1.5$ (and $0.10 < P$ value < 0.20). Since $-2.064 < 1.5 < 2.064$, accept H_0. The sample mean is not sufficiently different from the claim of 15 lb to reject H_0. By accepting H_0 we may be making a Type II error—accepting a false null hypothesis. The probability of this type of error is unknowable.

6. The hypotheses are H_0: $\mu \geq 1700$ mg versus H_a: $\mu < 1700$ mg. For a t test with $\alpha = 0.05$, the c.v. $= -1.895$. Calculate that $t = -2.44$. Since $-2.44 < -1.895$, reject H_0. Exposure to DEHP significantly decreases seminal vesicle weight.

8. (*a*) The hypotheses are H_0: $M = 0.618034$ versus H_a: $M \neq 0.618034$. Use the sign test. $S_- = 9$, $S_+ = 11$, so the test statistic is $S_- = 9$ for the two-tailed test. $F(9) = 0.4119$ from Table C.1. Since $0.4119 > 0.025$ accept H_0. Shoshoni Indians appear to be using the golden rectangle in their bead work.

(*b*) A 95% confidence interval is $[0.606, 0.672]$.

10. (*a*) The hypotheses are H_0: $M = 28.12°$ versus H_a: $M \neq 28.12°$. Test with $\alpha = 0.05$. $S_- = 9$, $S_+ = 1$, so the test statistic for a two-tailed test is $S_+ = 1$. $F(1) = 0.0107$ from Table C.1. Since $0.0107 < 0.025$, reject H_0. The webs don't appear to be made by *I. cicatricosa*.

(*b*) A 95% confidence interval is $[12, 26]$ which does not contain 28.12°.

12. The hypotheses are H_0: $\mu \leq 24$ hr versus H_a: $\mu > 24$ hr. For a t test with $\alpha = 0.05$. Calculate that $\overline{X} = 25.87$ hr, $n = 15$, and $s^2 = 0.849$ hr^2. The c.v. $= 1.761$.

$$t = \frac{\overline{X} - \mu}{\frac{s}{\sqrt{n}}} = \frac{25.87 - 24}{\frac{0.92}{\sqrt{15}}} = \frac{1.87}{0.24} = 7.79.$$

Since $7.79 > 1.761$, reject H_0. The average day for bunkered people is significantly longer than 24 hours.

14. The hypotheses are H_0: $M = 8$ m versus H_a: $M \neq 8$ m. Notice that $n = 15$ since one of the observations equals 8.0 m. The test statistic is the minimum of $S_+ = 6$ and $S_- = 9$. $F(6) = 0.3036$ and since this is a two-tailed test, the P value $= 0.6072$. So we accept H_0. There does not appear to be a change in the median Secchi depth from 1997.

16. The density and cumulative distribution function for W_+ for $n = 4$ are

W	0	1	2	3	4	5	6	7	8	9	10
$P(W = w)$	$\frac{1}{16}$	$\frac{1}{16}$	$\frac{1}{16}$	$\frac{2}{16}$	$\frac{2}{16}$	$\frac{2}{16}$	$\frac{2}{16}$	$\frac{2}{16}$	$\frac{1}{16}$	$\frac{1}{16}$	$\frac{1}{16}$
$P(W \leq w)$	$\frac{1}{16}$	$\frac{2}{16}$	$\frac{3}{16}$	$\frac{5}{16}$	$\frac{7}{16}$	$\frac{9}{16}$	$\frac{11}{16}$	$\frac{13}{16}$	$\frac{14}{16}$	$\frac{15}{16}$	$\frac{16}{16}$

which are derived from $4^2 = 16$ possible signed rankings:

Ranking				W_+
-1	-2	-3	-4	0
$+1$	-2	-3	-4	1
-1	$+2$	-3	-4	2
-1	-2	$+3$	-4	3
-1	-2	-3	$+4$	4
$+1$	$+2$	-3	-4	3
$+1$	-2	$+3$	-4	4
$+1$	-2	-3	$+4$	5
-1	$+2$	$+3$	-4	5
-1	$+2$	-3	$+4$	6
-1	-2	$+3$	$+4$	7
$+1$	$+2$	$+3$	-4	6
$+1$	$+2$	-3	$+4$	7
$+1$	-2	$+3$	$+4$	8
-1	$+2$	$+3$	$+4$	9
$+1$	$+2$	$+3$	$+4$	10

18. (*a*) From Table 6.1 with $n = 16$, we find $\left| \frac{\sigma}{\mu_0 - \mu} \right| = \left| \frac{5}{5} \right| = 1$.

    (*b*) $\left| \frac{\sigma}{\mu_0 - \mu} \right| = \left| \frac{5}{2} \right| = 2.5$, so $n = 99$. A 1 cm/msec decrease would require somewhat more than 253 observations according to Table 6.1. The 253 value would be appropriate for a 1.25 cm/msec decrease.

20. The hypotheses are H_0: $M \geq 165$ mmHg versus H_a: $M < 165$ mmHg. Use the Wilcoxon signed-rank test with $n = 14$. The test statistic is $W_+ = 46$ (see table below). $F(46) > 0.3349$, so accept H_0. We cannot conclude the mean systolic blood pressure is significantly less than 165 mmHg.

$X_i - 165$	18	-13	13	-8	29	-2	-21
Signed Rank	$+9$	-3.5	$+7$	-5	$+11.5$	-1.5	-10
$X_i - 165$	29	-2	-51	13	-13	-47	-7
Signed Rank	$+11.5$	-1.5	-14	$+7$	-7	-13	-3.5

22. The hypotheses are H_0: $\mu \leq 0.97$ mm versus H_a: $\mu > 0.97$ mm. Use a t test with $\alpha = 0.05$. The sample mean is $\overline{X} = 1.037$ mm and the standard deviation is $s = 0.031$ mm and $n = 10$. The c.v. = 1.833.

$$t = \frac{\overline{X} - \mu}{\frac{s}{\sqrt{n}}} = \frac{1.037 - 0.97}{\frac{0.031}{\sqrt{10}}} = 6.835.$$

Since $6.835 > 1.833$, reject H_0. 2m female anogenital distances are significantly longer than 0.97 mm.

24. Think of this problem as a sign test question. We toss out the two "no change" responses and consider the new medication as effective ($P(X \leq 2) = 0.0193$).

26. Use a t test and since you are skeptical about the company's claim it should be one-tailed with H_0: $\mu \leq 14$ min and H_a: $\mu > 14$ min. Since

$$t = \frac{\overline{X} - \mu}{\frac{s}{\sqrt{n}}} = \frac{19 - 14}{\frac{7}{\sqrt{25}}} = 3.57,$$

and its corresponding P value is less than 0.0025, we reject H_0. There is evidence that the time for relief is significantly longer than 14 minutes.

28. (*a*) Use a t test. The hypotheses are H_0: $\mu \leq 74.2$ yr versus H_a: $\mu > 74.2$ yr. Use $n = 16$ and $\nu = 15$. With $\alpha = 0.05$, the c.v. = 1.75. The test statistic is

$$t = \frac{\overline{X} - \mu}{\frac{s}{\sqrt{n}}} = \frac{81.2 - 74.2}{\frac{8.0}{\sqrt{16}}} = \frac{7.0}{2} = 3.5.$$

Reject H_0. The life span of people with the CETP variant gene was significantly longer.

(*b*) Use a chi-square test. The hypotheses are H_0: $\sigma^2 \geq 100$ (yr)^2 versus H_a: $\sigma^2 < 100$ (yr)^2. With $\alpha = 0.05$ and $\nu = 15$, the c.v. = 7.26. The test statistic is

$$\chi^2 = \frac{(n-1)s^2}{\sigma^2} = \frac{(16-1)(8.0)^2}{100} = 9.60.$$

Since $7.26 < 9.60$, the chi-square statistic is not significantly less than expected under H_0. Accept H_0. People with the CETP variant gene are not significantly less variable in their life spans.

30. The summary statistics are $\overline{X} = 24.0$ days and $s = 2.9$ days. Use a t test with H_0: $\mu \leq 20$ days versus H_a: $\mu > 20$ days. Calculate that

$$t = \frac{24.0 - 20}{\frac{2.9}{\sqrt{16}}} = \frac{4.0}{0.725} = 5.517.$$

From Table C.4 $P < 0.0005$. The data indicate that the modified *C. elegans* live significantly longer than unmodified ones.

32. We are anticipating a deviation from the mean in a particular direction, that is, an increase in NE levels. Use the Wilcoxon signed-rank test with $n = 9$. H_0: $M \leq 530$ ng/gm and H_a: $M > 530$ ng/gm. The test statistic is $W_- = 9$ (see table below). $F(9) = 0.0645 > 0.05$, so accept H_0.

$X_i - 530$	13	-7	-99	105	34	50	70	80	20
Signed Rank	$+2$	-1	-8	$+9$	$+4$	$+5$	$+6$	$+7$	$+3$

34. (*a*) False, see Section 6.1. (*b*) True, see page 165. (*c*) True, see page 165. (*d*) True, see Section 6.2. (*e*) True, see page 155. (*f*) False, see page 160. (*g*) False, see page 158.

Answers for Chapter 7

2. The hypotheses are H_0: $\mu_1 \leq \mu_2$ versus H_a: $\mu_1 > \mu_2$. A preliminary F test with $\alpha = 0.05$ and c.v. $= 0.35$ and c.v. $= 2.86$ is required. Test H_0: $\sigma_1^2 = \sigma_2^2$ versus H_a: $\sigma_2^2 \neq \sigma_2^2$. We find that $F = 1.56$. Since $0.35 < 1.56 < 2.86$, accept H_0. A pooled t test is appropriate. With $\alpha = 0.05$ and $\nu = 30$, the c.v. $= 1.697$.

$$s_p^2 = \frac{15(6.25) + 15(4.00)}{30} = 5.125 \quad \text{so} \quad t = \frac{12.0 - 8.5}{\sqrt{5.125 \left(\frac{1}{16} + \frac{1}{16} \right)}} = 4.375.$$

Since $4.375 > 1.697$, reject H_0. We can demonstrate a significant difference in cloacal gland widths with these data.

4. A paired t test is appropriate. The hypotheses are H_0: $\mu_d \leq 0$ versus H_a: $\mu_d > 0$, where $d =$ While Taking $-$ Before. Using $\alpha = 0.05$ with $\nu = 9$, the c.v. $= 1.833$. $\overline{X}_d = 4.0$ and $s_d^2 = 12.22$. Calculate that

$$t = \frac{4.0 - 0}{\sqrt{\frac{12.22}{10}}} = 3.618.$$

Since $3.618 > 1.833$, reject H_0. Using oral contraceptives significantly increased the systolic blood pressure. We could have made a Type I error here—rejecting a true null hypothesis. This is very unlikely since $0.0025 < P$ value < 0.005.

6. (*a*) The hypotheses are H_0: $\mu_U \leq \mu_E$ versus H_a: $\mu_U > \mu_E$. A preliminary F test is required. Test H_0: $\sigma_U^2 = \sigma_E^2$ versus H_a: $\sigma_U^2 \neq \sigma_E^2$. If $\alpha = 0.05$, then the c.v.'s $= 0.28, 3.25$. Calculate that $F = 1.39$. Since $0.28 < 1.39 < 3.25$, accept H_0. A pooled t test is appropriate. Use $\alpha = 0.05$ and $\nu = 23$. The c.v. $= 1.714$. Calculate that $t = 5.53$. Since $5.53 > 1.714$, reject H_0. Exposure to copper sulfate significantly reduces the number of mucus cells in the gill filaments in largemouth bass.

(*b*) A confidence interval for $\mu_U - \mu_E$ has endpoints

$$(\overline{X}_U - \overline{X}_E) \pm t_0 \sqrt{s_p^2 \left(\frac{1}{n_U} + \frac{1}{n_E} \right)} = (15.0 - 10.0) \pm 2.807 \sqrt{5.0435 \left(\frac{1}{11} + \frac{1}{14} \right)}$$

$$= 5.0 \pm 2.54.$$

A 99% confidence interval for $\mu_U - \mu_E$ is $[2.46, 7.54]$. We are 99% confident that exposure to copper sulfate reduces the number of mucus cells per square micron between 2.46 and 7.54 in largemouth bass.

8. (*a*) The hypotheses are H_0: $M_{X-Y} \geq 0$ versus H_a: $M_{X-Y} < 0$. Use the sign test with paired data. From the differences below, $S_+ = 1$. With $n = 7$, $F(1) = 0.0625$. Since the P value is greater than $\alpha = 0.05$, accept H_0. With this limited data set, we cannot demonstrate that the model of *P. baibaiye* is attacked less frequently than the colorless model.

$$X - Y \quad -8 \quad -4 \quad 2 \quad -3 \quad -8 \quad -4 \quad -2$$

(*b*) The small sample size here requires that *all* differences be negative in order to reject H_0. While the data show a clear trend, it cannot be deemed statistically significant due to the sample size. One should consider sample size in the design of any study.

10. (*a*) The hypotheses are H_0: $\mu_5 = \mu_8$ versus H_a: $\mu_5 \neq \mu_8$. A preliminary F test is required. Test H_0: $\sigma_5^2 = \sigma_8^2$ versus H_a: $\sigma_5^2 \neq \sigma_8^2$ at the $\alpha = 0.05$ level. $s_5^2 = 1.5888$ and $s_8^2 = 2.117$. $F = 0.750$ is between $F_{0.025(4,4)} = 0.104$ and $F_{0.975(4,4)} = 9.60$, accept H_0. A pooled t test should be used. Calculate that $t = -3.346$ and the corresponding P value is less than 0.02. The means are significantly different at the two pH's used. In fact, the mean at pH $= 5$ is significantly lower.

(*b*) A confidence interval for $\mu_5 - \mu_8$:

$$(\overline{X}_5 - \overline{X}_8) \pm t_0 \sqrt{s_p^2 \left(\frac{1}{n_5} + \frac{1}{n_8} \right)} = (11.24 - 14.12) \pm 2.306 \sqrt{1.8525 \left(\frac{1}{5} + \frac{1}{5} \right)}$$

$$= -2.88 \pm 1.985.$$

A 95% confidence interval for $\mu_5 - \mu_8$ is $[-4.865, -0.895]$.

12. (*a*) Begin with a preliminary F test with hypotheses H_0: $\sigma_M^2 = \sigma_F^2$ versus H_a: $\sigma_M^2 \neq \sigma_F^2$.

$$F = \frac{s_M^2}{s_F^2} = \frac{(11.72)^2}{(9.92)^2} = 1.40.$$

The critical value from Table C.7 is $F_{0.025[30,30]} = 2.07$. Since $1.40 < 2.07$, accept H_0. A pooled t test is now appropriate with H_0: $\mu_M = \mu_F$ versus H_a: $\mu_M \neq \mu_F$.

$$t_s = \frac{\overline{X}_M - \overline{X}_F}{\sqrt{s_p^2 \left(\frac{1}{n_M} + \frac{1}{n_F} \right)}},$$

where

$$s_p^2 = \frac{(34 - 1)(11.72)^2 + (38 - 1)(9.92)^2}{34 + 38 - 2} = 116.77.$$

Therefore,

$$t_s = \frac{94.6 - 96.4}{\sqrt{116.77 \left(\frac{1}{34} + \frac{1}{38} \right)}} = -0.706.$$

From Table C.4, $t_{0.025[70]} = 1.993$. Since $-1.993 < -0.706 < 1.993$, accept H_0. There is no evidence for a difference in adult lengths for males and females.

(*b*) For the length in neonates:

$$F = \frac{s_M^2}{s_F^2} = \frac{(1.76)^2}{(1.69)^2} = 1.08.$$

The critical value from Table C.7 is $F_{0.025[25,28]} = 2.16$. Since $1.08 < 2.16$, accept H_0. A pooled t test is appropriate with H_0: $\mu_M = \mu_F$ versus H_a: $\mu_M \neq \mu_F$.

$$s_p^2 = \frac{(30 - 1)(1.76)^2 + (29 - 1)(1.69)^2}{30 + 29 - 2} = 2.98.$$

Therefore,

$$t_s = \frac{\overline{X}_M - \overline{X}_F}{\sqrt{s_p^2 \left(\frac{1}{n_M} + \frac{1}{n_F} \right)}} = \frac{21.53 - 21.47}{\sqrt{2.98 \left(\frac{1}{30} + \frac{1}{29} \right)}} = 0.133.$$

From Table C.4, $t_{0.025[57]} = 2.002$. Since $-2.002 < 0.133 < 2.002$, accept H_0. There is no evidence for a difference in lengths for neonate males and females.
    For the weights of neonates:

$$F = \frac{s_F^2}{s_M^2} = \frac{(1.42)^2}{(1.33)^2} = 1.14.$$

(Why did we switch the numerator and denominator from the earlier parts?) The critical value from Table C.7 is $F_{0.025[25,29]} = 2.14$. Since $1.14 < 2.14$, accept H_0. A pooled t test is appropriate with H_0: $\mu_M = \mu_F$ versus H_a: $\mu_M \neq \mu_F$.

$$s_p^2 = \frac{(30-1)(1.33)^2 + (29-1)(1.42)^2}{30 + 29 - 2} = 1.89.$$

Therefore,

$$t_s = \frac{7.55 - 7.94}{\sqrt{1.89 \left(\frac{1}{30} + \frac{1}{29} \right)}} = -1.089.$$

From Table C.4, $t_{0.025[57]} = 2.002$. Since $-2.002 < -1.089 < 2.002$, accept H_0. There is no evidence for a difference in weights of neonate males and females.

14. The hypotheses are H_0: $M_I = M_{II}$ versus H_a: $M_I \neq M_{II}$. Using the Wilcoxon rank-sum test with $\alpha = 0.05$, the c.v.'s are 49 and 87.

Ranks I	4	13	5.5	8.5	8.5	14	12	1.5
Ranks II	3	7	11	16	1.5	15	5.5	10

$W_I = 67$. Since $49 < W_I < 87$, accept H_0. The repellants have the same efficacy.

16. (a) Notice that we do not have any numerical measure of the differences or levels of odor in this problem. All we have is the sign of the difference in odor for each subject. In this case a right-tailed sign test for paired data is appropriate. If O represents the original formulation and N the new, then the alternative hypothesis is H_a: $M_{O-N} > 0$ because researchers are testing whether the new formulation (N) lowers odor levels when compared with the original formulation (O)). The null hypothesis is H_0: $M_{O-N} \leq 0$; the odor levels with the old formulation are no greater than those with the new.

Sixteen of the differences ($O_i - N_i$) were positive, two were negative, and in two cases there was no perceived difference. Therefore, the sample size when adjusted for ties is $n = 18$. For a right-tailed test, the test statistic is S_-, the number of negative differences, which is 2. From Table C.1, the corresponding P value is 0.0007. We reject the null hypothesis; there is significant evidence that the new deodorant formulation lowers odor levels.

(b) The hypotheses remain the same. The sample size is $n = 60$ and the test statistic is $S_- = 19$. The large sample size requires using the normal approximation to binomial.

$$F(19) = P(S \leq 19) = P\left(Z \leq \frac{19.5 - \frac{60}{2}}{\frac{\sqrt{60}}{2}} \right) = P(Z \leq -2.71) = 0.0034.$$

Again, reject the null hypothesis; there is significant evidence that the new deodorant formulation lowers odor levels.

18. The hypotheses are H_0: $M_I \leq M_Q$ (the median concentration of dioxin in breast milk in Inuit women is no greater than that in women from southern Quebec) versus H_a: $M_I > M_Q$ (the median concentration of dioxin in breast milk in Inuit women is greater than that in women from southern Quebec).

The test statistic is the sum of the ranks for southern Quebec women (the smaller sample).

$$W_Q = 1 + 2 + 3 + 4 + 5.5 + 7 + 8 + 9 = 39.5.$$

From Table C.8, with $m = 8$ and $n = 10$, the P value is less than 0.0217, so reject H_0. There is evidence that the median concentration of dioxin in breast milk in Inuit women is signficantly greater than that in women from southern Quebec.

20. The hypotheses are H_0: $M_N \leq M_{HS}$ versus H_a: $M_N > M_{HS}$.

Nonsmoker ranks	8.5	8.5	7	13	17	15	20	16	19	18
Heavy Smoker ranks	10	1	6	2	4	14	11	3	5	12

The rank sums are $W_N = 142$ and $W_{HS} = 68$. For $\alpha = 0.05$ in Table C.8 with $m = n = 10$, we find the c.v.'s = 82, 128. Since $W_N = 142 > 128$, reject H_0. (Alternatively, $W_{HS} = 68 < 82$, reject H_0.) Data support the hypothesis that heavy smokers have reduced serum-testosterone levels.

22. (a) $E(X^2) = n - 1$

(b) $E(S_-) = E(S_+) = \frac{n}{2}$

(c) $E(W_-) = E(W_+) = \frac{n(n-1)}{4}$

(d) $E(t) = 0$

(e) $E(F) = 1$

(f) $E(t) = 0$

24. Use a sign test for paired data. The hypotheses are H_0: $M_{B-A} \geq 0$ versus H_a: $M_{B-A} < 0$. The test statistic is $S_+ = 0$. $F(S_+ = 0) = 0.0078 < 0.05$. Reject H_0. Data support the hypothesis that secondary cigarette smoke exposure increases cotinine levels.

26. The difference data are provided below.

Subject	Before	After	$A - B = d$	Signed Rank
1	10	7	3	7.5
2	8	4	4	10
3	7	5	2	4.5
4	9	7	2	4.5
5	8	3	5	11.5
6	10	5	5	11.5
7	6	6	0	—
8	7	4	3	7.5
9	4	5	−1	−2
10	10	7	3	7.5
11	10	7	3	7.5
12	9	3	6	13.5
13	5	6	−1	−2
14	7	6	1	2
15	10	4	6	13.5

If you use a paired sign test with $\alpha = 0.05$, then $n = 14$ because of the tied values. The test statistics are $S_- = 2$ and $S_+ = 12$. The P value for S_- is $0.0065 \ll 0.05$. Reject H_0. The nasal spray significantly decreased hunger levels.

If you use the Wilcoxon signed-rank test for paired data, the test statistics are $W_- = 4$ and $W_+ = 101$. The P value for W_- is $0.0004 \ll 0.05$. Reject H_0. The nasal spray significantly decreased hunger levels.

28. The hypotheses are H_0: $\mu_A = \mu_B$ versus H_a: $\mu_A \neq \mu_B$. A preliminary F test with $\alpha = 0.05$ and c.v.'s $= 0.14$ and 7.15 is required. Test H_0: $\sigma_A^2 = \sigma_B^2$ versus H_a: $\sigma_A^2 \neq \sigma_B^2$. We find that

$$F = \frac{s_A^2}{s_B^2} = \frac{434.17}{114.17} = 3.80.$$

Since $0.135 < 3.80 < 7.39$, accept H_0. A pooled t test is appropriate. With $\alpha = 0.05$ and $\nu = 10$, the c.v. $= 2.228$. $s_p^2 = 274.17$ and

$$t = \frac{274.2 - 235.8}{\sqrt{274.17 \left(\frac{1}{6} + \frac{1}{6}\right)}} = 4.016.$$

Since $4.016 > 2.228$, reject H_0. The brands of green tea have significantly different mean concentrations of EGCG.

30. (a) Assuming the test scores are normally distributed, a paired t test is the most appropriate analysis with H_0: $\mu_{A-B} \leq 0$ and H_a: $\mu_{A-B} > 0$. With $n = 15$, calculate that $\overline{X}_d = 5.1$, $s_d = 4.2$, and $t = 4.703$. From Table C.4 the P value is less than 0.0005; reject H_0. The IQ scores were significantly higher.

(b) Assuming the test scores are symmetrically but not normally distributed, a Wilcoxon signed-rank test for paired data is appropriate with H_0: $M_{A-B} \leq 0$ and H_a: $M_{A-B} > 0$. Calculate that $W_- = 7$ and $W_+ = 113$. From Table C.6 $P = 0.0006$, so reject H_0. The IQ scores were significantly higher.

(c) Making no assumptions about the distribution of IQ scores, the sign test for paired data is an appropriate analysis using H_0: $M_{A-B} \leq 0$ and H_a: $M_{A-B} > 0$. Check that $S_- = 2$ and $S_+ = 13$. From Table C.1 $P = 0.0037$, so reject H_0. Again conclude the IQ scores were significantly higher.

32. Use a t test to analyze the difference in the means of two *independent* samples. Perform an F test to see if the populations have similar variances. The hypotheses are H_0: $\sigma_U^2 = \sigma_K^2$ and H_a: $\sigma_U^2 \neq \sigma_K^2$. Then

$$F = \frac{s_U^2}{s_K^2} = \frac{(0.55)^2}{(0.47)^2} = 1.37.$$

The c.v. for $\alpha = 0.05$, $\nu_U = 19$ and $\nu_K = 23$ from Table C.7 is 2.39. Since $1.37 < 2.39$, the variances are assumed to not be significantly different.

The question about mean root weight can be addressed. This is a right-tailed test because we anticipate a deviation from equality in a particular direction. H_0: $\mu_U \geq \mu_K$ versus H_a: $\mu_U < \mu_K$. The pooled sample variance is

$$s_p^2 = \frac{(20-1)(0.55)^2 + (24-1)(0.47)^2}{20 + 24 - 2} = 0.258.$$

So

$$t = \frac{(\overline{X}_U - \overline{X}_K) - 0}{\sqrt{s_p^2 \left(\frac{1}{n_U} + \frac{1}{n_K}\right)}} = \frac{(7.22 - 6.25) - 0}{\sqrt{0.258 \left(\frac{1}{20} + \frac{1}{24}\right)}} = 6.307.$$

From Table C.4 the c.v. is 1.682 for a right-tailed test. Since $6.307 > 1.682$, reject H_0. Unrelated plants appear to have significantly greater root weights.

34. Note that these are paired data as the controls were matched to alcoholics. A paired t test is appropriate. The differences d_i (alcoholics − controls) are reported below.

Pair	d_i
1	−2
2	−2
3	−1
4	−5
5	−3
6	0
7	−4
8	+1

Since we are testing whether brain densities are diminished in alcoholics, the test is left-tailed. The hypotheses are H_0: $\mu_d \geq 0$ and H_a: $\mu_d < 0$. Choose $\alpha = 0.05$. The test statistic for the paired sample t test is

$$t = \frac{\overline{X}_d - \mu_d}{\frac{s_d}{\sqrt{n_d}}} = \frac{-2.0 - 0}{\frac{2}{\sqrt{8}}} = -\sqrt{8} = -2.828,$$

with $\nu = n - 1 = 7$. The critical value for this left-tailed test from Table C.4 is $t_{0.05(7)} = -1.895$. Since $-2.828 < -1.895$, reject H_0. There is evidence that brain densities are diminished significantly in alcoholics.

36. Use a left-tailed paired sign test for paired data with H_0: $M_{I-NI} \geq 0$ versus H_a: $M_{I-NI} < 0$. The differences are given below.

Pair	1	2	3	4	5	6	7	8	9	10	11	12
d_{I-NI}	−0.5	0.0	−0.5	0.5	−0.5	−0.5	−1.5	−1.5	−1.0	−0.5	0.5	−0.5

The test statistic is $S_+ = 2$ with $n = 11$. The corresponding P value from Table C.1 is 0.0327, which is smaller than 0.05. Reject H_0. There is evidence that instruction improves oral hygiene.

38. (a) Use a t test to analyze the difference in the means of two *independent* samples. Perform an F test to see if the populations have similar variances. The hypotheses are H_0: $\sigma_V^2 = \sigma_S^2$ and H_a: $\sigma_V^2 \neq \sigma_S^2$. Then

$$F = \frac{s_S^2}{s_V^2} = \frac{40}{25} = 1.60.$$

The c.v. for $\alpha = 0.05$, $\nu_S = 24$ and $\nu_V = 24$ from Table C.7 is 2.27. Since $1.60 < 2.27$, the variances are assumed to not be significantly different.

The question about mean restart can be addressed. This is a left-tailed test because we anticipate a deviation from equality in a particular direction. H_0: $\mu_V \geq \mu_S$ versus H_a: $\mu_V < \mu_S$. Since we believe the populations have equal variance, we calculate a pooled sample variance (which is the average of the two variances since the sample sizes are equal):

$$s_p^2 = \frac{25 + 40}{2} = 32.5.$$

So

$$t = \frac{(\overline{X}_V - \overline{X}_S) - 0}{\sqrt{s_p^2 \left(\frac{1}{n_V} + \frac{1}{n_S}\right)}} = \frac{(6 - 12) - 0}{\sqrt{32.5 \left(\frac{1}{25} + \frac{1}{25}\right)}} = -3.721.$$

From Table C.4 the c.v. is -1.677 for a left-tailed test. Since $-3.721 < -1.677$, reject H_0. Sildenafil (Viagra) significantly shortened reset time.

(b) Again, perform an F test to see if the populations have similar variances. The hypotheses are H_0: $\sigma_V^2 = \sigma_S^2$ and H_a: $\sigma_V^2 \neq \sigma_S^2$. Then

$$F = \frac{s_V^2}{s_S^2} = \frac{36}{25} = 1.44.$$

The c.v. for $\alpha = 0.05$, $\nu_V = \nu_S = 19$ from Table C.7 is 2.60. Since $1.44 < 2.60$, the variances are assumed to not be significantly different.

The question about restart can be addressed. H_0: $\mu_V \geq \mu_S$ versus H_a: $\mu_V < \mu_S$. This time

$$s_p^2 = \frac{36 + 25}{2} = 30.5.$$

So

$$t = \frac{(\overline{X}_V - \overline{X}_S) - 0}{\sqrt{s_p^2 \left(\frac{1}{n_V} + \frac{1}{n_S}\right)}} = \frac{(10 - 12.5) - 0}{\sqrt{30.5 \left(\frac{1}{20} + \frac{1}{20}\right)}} = -1.431.$$

From Table C.4 the c.v. is -1.686 for a left-tailed test. Since $-1.431 > -1.686$, we cannot reject H_0. There is no significant evidence that sildenafil (Viagra) shortened westbound reset time.

40. The hypotheses are H_0: $\mu_{Cr} \leq \mu_{Un}$ versus H_a: $\mu_{Cr} > \mu_{Un}$. A preliminary F test with $\alpha = 0.05$ and c.v.'s $= 0.35$ and 2.86 is required. Test H_0: $\sigma_{Cr}^2 = \sigma_{Un}^2$ versus H_a: $\sigma_{Cr}^2 \neq \sigma_{Un}^2$. Calculate $F = \frac{16.0}{25.0} = 0.64$ and since $0.35 < 0.64 < 2.86$, accept H_0. A pooled t test is appropriate, where $s_p^2 = 20.5$ is the average of the two variances since the sample sizes are equal.

With $\alpha = 0.05$ and $\nu = 30$, the c.v. $= 1.697$. Since

$$t = \frac{(20 - 14) - 0}{\sqrt{20.5 \left(\frac{1}{16} + \frac{1}{16}\right)}} = 3.75$$

and $3.75 > 1.697$, reject H_0 ($P < 0.0005$). There is evidence to support that crowding causes the crinoids to travel significantly farther.

42. (a) True, see Section 7.2. (b) False, see page 180. (c) True, see Section 7.3. (d) True, see Section 7.4. (e) False, see Section 7.5. (f) True, see page 192.

Answers for Chapter 8

2. The hypotheses are H_0: $\mu_N = \mu_D = \mu_T$ versus H_a: At least one pair of μ_i's is different. From the table below, since $16.03 \gg 4.26$, reject H_0. There are significant differences in hemoglobin levels among the groups.

Source of variation	Sum of squares	df	MS	F	c.v.
Sickle cell status	78.87	2	39.44	16.03	4.26
Error	66.33	27	2.46		
Total	145.20	29			

Use Bonferroni t tests to separate means. The table below summarizes the results.

	Normal − Disease	Normal − Trait	Disease − Trait
Bonferroni t	5.42	1.28	−4.13
P value	< 0.001	> 0.20	< 0.001

The normal and sickle cell trait subjects are not different, but both of these groups differ from the sickle cell disease subjects in hemoglobin level.

Sickle cell disease	Sickle cell trait	Normal subjects
9.8^a	12.7^b	13.6^b

4. The hypotheses are $H_0: \mu_1 = \mu_2 = \mu_3$ versus H_a: At least one pair of μ_i's is different. The summary data and ANOVA table are given below.

	Treatment 1	Treatment 2	Treatment 3
$\overline{X}_{i.}$	29.0	32.0	28.0
n_i	4	5	3

$$N = 12, \qquad \sum_i \sum_j X_{ij} = 360, \qquad \sum_i \sum_j X_{ij}^2 = 11,046$$

Source of variation	Sum of squares	df	MS	F	c.v.
Treatments	36	2	18.0	0.77	4.26
Error	210	9	23.3		
Total	246	11			

Since $0.77 < 4.26$, accept H_0. The treatments are not significantly different in their ability to increase growth in sunflowers. Further analysis is not warranted.

6. (a) H_0: All three populations have the same median versus H_a: At least one of the population medians is different from the others. Use the Kruskal-Wallis test. The rank sums and rank averages are given in the table that follows. df $= k - 1 = 2$. If $\alpha = 0.05$, then the c.v. $= 5.99$. Using the table that follows,

$$
\begin{aligned}
H &= \frac{12}{N(N+1)} \left(\sum_{i=1}^{k} \frac{R_i^2}{n_i} \right) - 3(N+1) \\
&= \frac{12}{40(40+1)} \left[\frac{(152.5)^2}{16} + \frac{(220.5)^2}{10} + \frac{(447)^2}{14} \right] - 3(40+1) \\
&= \frac{12}{1640} (20{,}587.6) - 123 \\
&= 27.64.
\end{aligned}
$$

Since $H = 27.64 > 5.99$, reject H_0. There is evidence that at least one population has a significantly different median.

HB SS	Rank	HB S/-thalassemia	Rank	HB SC	Rank
7.2	1	8.1	4.5	10.7	22
7.7	2	9.2	16	11.3	25
8.0	3	10.0	17	11.5	26
8.1	4.5	10.4	20	11.6	27
8.3	6	10.6	21	11.7	28
8.4	7.5	10.9	23	11.8	29
8.4	9	11.1	24	12.0	31.5
8.5	10	11.9	30	12.1	33.5
8.6	11	12.0	31.5	12.3	35
8.7	13	12.1	33.5	12.6	36
9.1	13			13.3	37.5
9.1	13			13.3	37.5
9.1	13			13.8	39
9.8	15			13.9	40
10.1	18				
10.3	19				
R_1	152.5	R_2	220.5	R_3	447
Ave. Rk	9.53	Ave. Rk	22.05	Ave. Rk	31.93

(b) Comparisons are required since H_0 was rejected. $n_1 = 16$, $n_2 = 10$, $n_3 = 14$, and $N = 40$. The average ranks are listed in the table above. $k = 3$. Let $\alpha' = 0.05$, so $\alpha = 0.05/\binom{3}{2} = 0.0167$. So $1 - \frac{\alpha}{2} = 0.9917$. Thus, for all comparisons the c.v. = 2.40. For all three comparisons the hypotheses are H_0: The medians of the ith and jth populations are the same versus H_a: The medians of the ith and jth populations are different.

Comparing HB SS and HB S/-thalassemia:

$$Z_{12} = \frac{\left| \frac{R_1}{n_1} - \frac{R_2}{n_2} \right|}{\sqrt{\frac{N(N+1)}{12} \left(\frac{1}{n_1} + \frac{1}{n_2} \right)}} = \frac{|9.53 - 22.05|}{\sqrt{\frac{40(41)}{12} \left(\frac{1}{16} + \frac{1}{10} \right)}} = \frac{12.52}{4.71} = 2.65.$$

$Z_{12} > 2.40$, so reject H_0. There is evidence that the medians for HB SS and HB S/-thalassemia are different. Comparing HB SS and HB SC:

$$Z_{13} = \frac{|9.53 - 31.93|}{\sqrt{\frac{40(41)}{12} \left(\frac{1}{16} + \frac{1}{14} \right)}} = \frac{22.40}{4.28} = 5.23.$$

$Z_{13} > 2.40$, so reject H_0. There is evidence that the medians for HB SS and HB SC are different. Comparing HB S/-thalassemia and HB SC:

$$Z_{23} = \frac{|22.05 - 31.93|}{\sqrt{\frac{40(41)}{12} \left(\frac{1}{10} + \frac{1}{14} \right)}} = \frac{9.88}{4.84} = 2.04.$$

$Z_{23} < 2.40$, so accept H_0. The medians for HB S/-thalassemia and HB SC are not different.

8. The ANOVA hypotheses are H_0: There is no significant difference in mean scores among the various treatments versus H_a: At least one pair of μ_i's is different. Using the summary data, $N = \sum_i n_i = 252$ and $T_{..} = \sum T_{i.} = 1301$. Therefore, by Formula 8.3,

$$SS_{\text{Total}} = \sum_i \sum_j X_{ij}^2 - \frac{T_{..}^2}{N} = 10,720 - \frac{(1301)^2}{252} = 4003.33.$$

Using Formula 8.4,

$$SS_{\text{Treat}} = \left(\frac{(428)^2}{84} + \frac{(448)^2}{83} + \frac{(425)^2}{85} \right) - \frac{(1301)^2}{252} = 7.21.$$

Therefore

$$SS_{\text{Error}} = SS_{\text{Total}} - SS_{\text{Treat}} = 4003.33 - 7.21 = 3996.12.$$

The ANOVA table is

Source of variation	Sum of squares	df	MS	F	c.v.
Treatments	7.21	2	3.605	0.225	3.06
Error	3996.12	249	16.049		
Total	4003.33	251			

Since $0.225 \ll 3.06$, we fail to reject H_0. There is no evidence of a significant difference in the mean number of hot flashes per day for the different treatments.

10. The hypotheses are H_0: $\mu_A = \mu_B = \mu_C$ versus H_a: At least one pair of μ_i's is different.

$$SS_{\text{Error}} = \sum_i CSS_i = 11.63.$$

$$SS_{\text{Treat}} = 15.49 - 11.63 = 3.86.$$

Source of variation	Sum of squares	df	MS	F	c.v.
Fertilizer mixtures	3.86	2	1.93	4.49	3.35
Error	11.63	27	0.43		
Total	15.49	29			

Since $4.49 > 3.35$, reject H_0. At least one treatment mean is significantly different from the others. Use SNK to locate the differences.

$$SE = \sqrt{\frac{MS_E}{n}} = \sqrt{\frac{0.43}{10}} = 0.21.$$

From Table C.9 with $\alpha = 0.05$ and $\nu = 27$, we obtain

k	2	3
q c.v.	2.919	3.532

Comparison	q statistic	k	c.v.	Conclusion
B to A	3.762	3	3.532	Reject H_0
B to C	0.286	2	2.919	Accept H_0
C to A	3.476	2	2.919	Reject H_0

Treatment A is significantly different from treatments B and C.

$$\begin{array}{ccc} \mathbf{B} & \mathbf{C} & \mathbf{A} \\ 2.22^a & 2.28^a & 3.01^b \end{array}$$

12. The hypotheses are H_0: $\sigma^2_{\text{sites}} = 0$ versus H_0: $\sigma^2_{\text{sites}} > 0$.

Source of variation	Sum of squares	df	MS	F	c.v.
Sites	69.75	3	23.25	83.78	3.49
Error	3.33	12	0.2775		
Total	73.08	15			

Since $83.78 \gg 3.49$, we can conclude that the manufacturing sites produce significantly variable margarine products. Because this is a Model II ANOVA, mean separation techniques are not applicable here.

14. The hypotheses are H_0: There are no significant differences in mean increases in bench press totals among the four dietary supplements versus H_a: There are significant differences in mean increases in bench press totals between at least two of the four dietary supplements. The summary data and ANOVA table are given below.

Supplement	n	Mean	Variance
Carbohydrate	7	4.0	5.33
Whey	7	4.7	1.57
Creatine and carbohydrate	7	7.1	4.48
Creatine and whey	7	10.1	9.48

Source of variation	Sum of squares	df	MS	F	c.v.
Supplement	161.86	3	53.95	10.36	3.01
Error	125.14	24	5.21		
Total	287.00	27			

Since $10.36 > 3.01$, reject H_0. At least two supplements produce significant differences in mean increases in bench press totals. Use Bonferroni t tests to separate the means. There are $\binom{4}{2} = 6$ comparisons to make. To compare the first two supplements, the hypotheses would be H_0: $\mu_C = \mu_W$ versus H_a: $\mu_C \neq \mu_W$. The others are similar. The test statistic is

$$\text{Bonferroni } t = \frac{\overline{X}_{i.} - \overline{X}_{j.}}{\sqrt{\text{MS}_{\text{E}}\left(\frac{1}{n_i} + \frac{1}{n_j}\right)}} = \frac{\overline{X}_{i.} - \overline{X}_{j.}}{\sqrt{5.21\left(\frac{2}{7}\right)}}$$

with $\nu = N - k = 28 - 4 = 24$. Use $\alpha = \frac{0.05}{6}$.

	C – W	C – CC	C – CW	W – CC	W – CW	CC–CW
Bonferroni t	-0.57	-2.54	-5.00	-1.97	-4.43	-2.46
P value	> 0.50	> 0.01	< 0.001	> 0.05	< 0.001	> 0.02

The combination of creatine and whey is significantly better than a supplement of carbohydrate alone or whey alone.

C	W	CC	CW
4.0^a	4.7^a	7.1^{ab}	10.1^b

16. The hypotheses are H_0: All three repellants have the same median time before first bite. H_a: At least one repellant has a different median time before first bite. See the table below for the ranked data.

Citronella	Rank	DEET 20%	Rank	Avon Skin So Soft	Rank
5	1	12	10	6	2.5
6	2.5	16	14	7	4
8	5.5	25	16	9	7
8	5.5	27	17	10	8
14	12	28	18	11	9
15	13	31	19	13	11
				17	15
Sum	39.5	Sum	94	Sum	56.5
Ave. rank	6.58	Ave. rank	15.67	Ave. rank	8.07

Note that $N = 19$ and $\nu = \text{df} = k - 1 = 2$. If $\alpha = 0.05$, then the c.v. $= 5.99$. From the table above,

$$H = \frac{12}{N(N+1)} \sum_{i=1}^{k} n_i \left(\frac{R_i}{n_i} - \frac{N+1}{2} \right)^2$$

$$= \frac{12}{19(20)} \left[6(6.58 - 10)^2 + 6(15.67 - 10)^2 + 7(8.07 - 10)^2 \right]$$

$$= 9.13.$$

Since $H > 5.99$, reject H_0. There is evidence that the median time before first bite for at least one repellant is different from the others.

Comparisons are required since H_0 was rejected. $n_1 = n_2 = 6$, $n_3 = 7$, and $N = 19$. The average ranks are listed in the table above. $k = 3$. Let $\alpha' = 0.05$, so $\alpha = \frac{0.05}{\binom{3}{2}} = 0.0167$. So $1 - \frac{\alpha}{2} = 0.9917$. Thus, for all comparisons the c.v. $= 2.40$. For all three comparisons the hypotheses are H_0: The median times before first bite for the ith and jth repellants are the same versus H_a: The median times before first bite for the ith and jth repellants are different.

Comparing Treatments I and II:

$$Z_{12} = \frac{\left| \frac{R_1}{n_1} - \frac{R_2}{n_2} \right|}{\sqrt{\frac{N(N+1)}{12} \left(\frac{1}{n_1} + \frac{1}{n_2} \right)}} = \frac{|6.58 - 15.67|}{\sqrt{\frac{19(20)}{12} \left(\frac{1}{6} + \frac{1}{6} \right)}} = \frac{9.09}{\sqrt{10.556}} = 2.798.$$

$Z_{12} > 2.40$, so reject H_0: There is evidence that the median times before first bite are different for citronella and DEET. Comparing Treatments I and III:

$$Z_{13} = \frac{|6.58 - 8.07|}{\sqrt{\frac{19(20)}{12}\left(\frac{1}{6} + \frac{1}{7}\right)}} = \frac{1.49}{\sqrt{9.802}} = 0.476.$$

$Z_{13} < 2.40$, so accept H_0: There is no evidence that the median times before first bite are different for citronella and Skin So Soft. Comparing Treatments II and III:

$$Z_{23} = \frac{|15.67 - 8.07|}{\sqrt{\frac{19(20)}{12}\left(\frac{1}{6} + \frac{1}{7}\right)}} = \frac{7.60}{\sqrt{9.802}} = 2.427.$$

$Z_{23} > 2.40$, so reject H_0: There is evidence that the median times before first bite are different for DEET and Skin So Soft.

18. (a) The hypotheses are H_0: $\mu_S = \mu_F = \mu_E$, there are no significant differences in mean androstenedione levels among the three streams. H_a: At least one pair of μ_i's is different. The summary data and ANOVA table are given below.

Location	n	Sum	Mean	Variance
Spring Creek	8	43	5.4	13.98
Fenholloway River	8	135	16.9	3.26
Elevenmile Creek	8	150	18.8	5.64

Source of variation	Sum of squares	df	MS	F	c.v.
Location	839.08	2	419.54	54.99	3.47
Error	160.25	21	7.63		
Total	999.33	23			

Since $54.99 \gg 3.47$, reject H_0. At least two streams have significant differences in mean androstenedione levels. Use Bonferroni t tests to separate the means. There are $\binom{3}{2} = 3$ comparisons to make. To compare the first two streams, the hypotheses would be H_0: $\mu_S = \mu_F$ versus H_a: $\mu_S \neq \mu_F$. The others are similar. The test statistic is

$$\text{Bonferroni } t = \frac{\overline{X}_{i.} - \overline{X}_{j.}}{\sqrt{\text{MS}_E\left(\frac{1}{n_i} + \frac{1}{n_j}\right)}}$$

with $\nu = N - k = 24 - 3 = 21$.

	S − F	S − E	F − E
Bonferroni t	−8.33	−9.68	−1.36
P value	< 0.001	< 0.001	> 0.10

Spring Creek has a significantly different mean androstenedione level from the other two streams.

SC	FR	EC
5.4[a]	16.9[b]	18.8[b]

(b) No, but they do *support* the claim.

20. The hypotheses are H_0: $\mu_A = \mu_O = \mu_B = \mu_H$. There are no significant differences in mean tau protein to beta amyloid peptide ratio among those with various brain disorders and H_a: At least one pair of μ_i's is different. The basic ANOVA table is

Source of variation	Sum of squares	df	MS	F	c.v.
Groups	143,880	3	47,960	429.17	2.74
Error	8,493	76	111.75		
Total	152,373	79			

Reject H_0. Some of the means are significantly different. Use an SNK test. $MS_E = 111.75$ and $n = 20$, so

$$SE = \sqrt{\frac{MS_E}{n}} = \sqrt{\frac{111.75}{20}} = 2.36.$$

From Table C.9 with $\alpha = 0.05$ and $\nu = 76 \approx 60$, we find

k	2	3	4
q c.v.	2.829	3.399	3.737

Comparing HI to BD, we find that

$$q = \frac{48 - 39}{2.36} = 3.814.$$

Since HI to BD is the smallest difference and its q statistic 3.814 is greater than the critical value of 2.829, all means are significantly different.

HI	BD	OD	AP
39^a	48^b	74^c	147^d

The ratio of tau protein to beta-amyloid peptide appears to discriminate among these groups very well and is, therefore, a good candidate for a new diagnostic tool.

22. Use a Kruskal-Wallis test. The hypotheses are H_0: Suits from all three time periods have the same median height. H_a: At least one time period has a different median height. See the table below for the ordered and ranked data.

10th & 11th C.	Rank	12th & 13th C.	Rank	After 1400 A.D.	Rank
149	1.5	149	1.5	153	4.5
151	3	158	9	155	7.5
153	4.5	159	10.5	163	12.5
154	6	159	10.5	165	15
155	7.5	165	15	165	15
163	12.5			166	17
				170	18
				171	19
Sum	35	Sum	46.5	Sum	108.5
Ave. rank	5.83	Ave. rank	9.30	Ave. rank	13.56

Note that $N = 19$ and $\nu = k - 1 = 2$. If $\alpha = 0.05$, then the c.v. = 5.99. From the table above,

$$H = \frac{12}{N(N+1)} \sum_{i=1}^{k} n_i \left(\frac{R_i}{n_i} - \frac{N+1}{2} \right)^2$$

$$= \frac{12}{19(20)} \left[6(5.83 - 10)^2 + 5(9.30 - 10)^2 + 8(13.56 - 10)^2 \right]$$

$$= 6.57.$$

Since $H > 5.99$, reject H_0. There is evidence that at least one time period has a different median height from another.

Comparisons are required since H_0 was rejected. Use $n_1 = 6$, $n_2 = 5$, and $n_3 = 8$. The average ranks are listed in the table above. Let $\alpha = \frac{0.05}{\binom{3}{2}} = 0.0167$. So $1 - \frac{\alpha}{2} = 0.9917$. Thus, for all comparisons the c.v. = 2.40. For all 3 comparisons the hypotheses are H_0: The median heights for the ith and jth periods are the same versus H_a: The median heights for the ith and jth periods are different.

Comparing periods I and II:

$$Z_{12} = \frac{\left| \frac{R_1}{n_1} - \frac{R_2}{n_2} \right|}{\sqrt{\frac{N(N+1)}{12} \left(\frac{1}{n_1} + \frac{1}{n_2} \right)}} = \frac{|5.83 - 9.30|}{\sqrt{\frac{19(20)}{12} \left(\frac{1}{6} + \frac{1}{5} \right)}} = \frac{3.47}{\sqrt{11.611}} = 1.018.$$

$Z_{12} < 2.40$, so accept H_0: There is no evidence that the median heights for the first two time periods differ. Comparing periods I and III:

$$Z_{13} = \frac{|5.83 - 13.56|}{\sqrt{\frac{19(20)}{12} \left(\frac{1}{6} + \frac{1}{8} \right)}} = \frac{7.73}{\sqrt{9.802}} = 2.544.$$

$Z_{13} > 2.40$, so reject H_0: There is evidence that the median from the first and last periods differ. Comparing periods II and III:

$$Z_{23} = \frac{|9.30 - 13.56|}{\sqrt{\frac{19(20)}{12} \left(\frac{1}{5} + \frac{1}{8} \right)}} = \frac{4.26}{\sqrt{10.292}} = 1.328.$$

$Z_{23} < 2.40$, so accept H_0: There is no evidence that the median heights for the last two time periods differ.

24. (a) True, see page 219. (b) False, see page 213. (c) True, see page 228. (d) True, see page 221. (e) True, see Section 8.2. (f) False, see page 223. (g) False, see page 228.

Answers for Chapter 9

2. The hypotheses are H_0: $\mu_I = \mu_{II} = \mu_{III}$ versus H_a: At least one pair of μ_i's is different. $N = 18$ and the uncorrected sum of squares is $\sum_i \sum_j X_{ij}^2 = 3571$. The sums of squares

can now be calculated using the computational formulas.

$$SS_{Total} = \sum_i \sum_j X_{ij}^2 - \frac{T_{..}^2}{N} = 3571 - \frac{(249)^2}{18} = 126.50$$

$$SS_{Treat} = \sum_i \frac{T_{i.}^2}{b} - \frac{T_{..}^2}{N} = \left(\frac{95^2}{6} + \frac{78^2}{6} + \frac{76^2}{6}\right) - \frac{(249)^2}{18} = 36.33$$

$$SS_{Blocks} = \sum_j \frac{T_{.j}^2}{b} - \frac{T_{..}^2}{N}$$

$$= \left(\frac{47^2}{3} + \frac{42^2}{3} + \frac{45^2}{3} + \frac{48^2}{3} + \frac{34^2}{3} + \frac{33^2}{3}\right) - \frac{(249)^2}{18} = 71.17$$

$$SS_{Error} = SS_{Total} - (SS_{Treat} + SS_{Blocks}) = 126.50 - (36.33 + 71.17) = 19.00.$$

For the global F test, use $\alpha = 0.05$. The ANOVA table is

Source of variation	Sum of squares	df	MS	F	c.v.
Diets	36.33	2	18.17	9.56	4.10
Lines	71.17	5	14.23		
Error	19.00	10	1.90		
Total	126.50	17			

Since $9.56 > 4.10$, reject H_0. There are significant differences among diets. Blocking by line was useful here because a large portion of the SS_{Total} was accounted for by the sum of squares lines (71.17).

4. The hypotheses are $H_0: \mu_N = \mu_S$ versus $H_a: \mu_N \neq \mu_S$ and $H_0: \mu_J = \mu_M = \mu_S$ versus H_a: At least one pair of μ_i's is different.

Location	Month			
	January	May	September	
North	$T_{11.} = 35.8$	$T_{21.} = 35.2$	$T_{31.} = 22.8$	$T_{.1.} = 93.8$
South	$T_{12.} = 40.9$	$T_{22.} = 39.5$	$T_{32.} = 22.4$	$T_{.2.} = 102.8$
	$T_{1..} = 76.7$	$T_{2..} = 74.7$	$T_{3..} = 45.2$	

$$\sum_i \sum_j \sum_k X_{ijk}^2 = 1371.2 \qquad \sum_i \sum_j \sum_k X_{ijk} = T_{...} = 196.6$$

Next the sums of squares can be calculated using Formulas 9.4.

$$SS_{Total} = 1371.2 - \frac{(196.6)^2}{30} = 82.815$$

$$SS_{Cells} = \frac{(35.8)^2}{5} + \frac{(35.2)^2}{5} + \cdots + \frac{(22.4)^2}{5} - \frac{(196.6)^2}{30} = 66.683$$

$$SS_{Error} = SS_{Total} - SS_{Cells} = 82.815 - 66.683 = 16.132.$$

Months:

$$SS_A = \frac{(76.7)^2}{10} + \frac{(74.7)^2}{10} + \frac{(45.2)^2}{10} - \frac{(196.6)^2}{30} = 62.217.$$

Locations :

$$SS_B = \frac{(93.8)^2}{15} + \frac{(102.8)^2}{15} - \frac{(196.6)^2}{30} = 2.700.$$

Interactions:

$$SS_{A \times B} = SS_{Cells} - (SS_A + SS_B) = 66.683 - (62.217 + 2.700) = 1.766.$$

Now we organize the information in an ANOVA table.

Source of variation	Sum of squares	df	MS	F	c.v.
Cells	66.683	5			
Locations	2.700	1	2.700	4.018	4.26
Months	62.217	2	31.108	46.292	3.40
Interaction	1.766	2	0.883	1.314	3.40
Error	16.132	24	0.672		
Total	82.815	29			

There is no interaction ($1.314 < 3.40$) between the two main effects—months and locations. Locations are not significantly different ($4.018 < 4.26$). In the future the limnologist could sample one site at different times of the year to garner the same information.

6. The hypotheses are $H_0: \mu_{CC} = \mu_{IPM} = \mu_{MIPM}$ versus H_a: At least one pair of μ_i's is different. The data are summarized below.

Farm	Agricultural method			$T_{\cdot j}$	$\overline{X}_{\cdot j}$
	Chemical control	IPM	Modified IPM		
I	71	73	77	221	73.7
II	90	90	92	272	90.7
III	59	70	80	209	69.7
IV	75	80	82	237	79.0
V	65	60	67	192	64.0
VI	82	86	85	253	84.3
$T_{i\cdot}$	442	459	483	$T_{\cdot\cdot} = 1384$	
$\overline{X}_{i\cdot}$	73.7	76.5	80.5	$\overline{X}_{\cdot\cdot} = 76.9$	

$N = 18$ and the uncorrected sum of squares is $\sum_i \sum_j X_{ij}^2 = 108,152$. The sums of squares can now be calculated using the computational formulas.

$$SS_{Total} = \sum_i \sum_j X_{ij}^2 - \frac{T_{\cdot\cdot}^2}{N} = 108,152 - \frac{(1384)^2}{18} = 1737.78$$

$$SS_{Treat} = \sum_i \frac{T_{i\cdot}^2}{b} - \frac{T_{\cdot\cdot}^2}{N} = \left(\frac{442^2}{6} + \frac{459^2}{6} + \frac{483^2}{6}\right) - \frac{(1384)^2}{18} = 141.45$$

$$SS_{Blocks} = \sum_j \frac{T_{\cdot j}^2}{b} - \frac{T_{\cdot\cdot}^2}{N}$$

$$= \left(\frac{221^2}{3} + \frac{272^2}{3} + \frac{209^2}{3} + \frac{237^2}{3} + \frac{192^2}{3} + \frac{253^2}{3}\right) - \frac{(1384)^2}{18} = 1435.11$$

$$SS_{Error} = SS_{Total} - (SS_{Treat} + SS_{Blocks}) = 1737.78 - (141.45 + 1435.11) = 161.22.$$

For the global F test, use $\alpha = 0.05$. The ANOVA table is

Source of variation	Sum of squares	df	MS	F	c.v.
Agricultural method	141.45	2	70.725	4.39	4.10
Farm	1435.11	5	287.022		
Error	161.22	10	16.122		
Total	1737.78	17			

Since $4.39 > 4.10$, reject H_0. Analyze further with an SNK test. The linearly ordered means are

CC	IPM	MIPM
73.7	76.5	80.5

$MS_E = 16.122$ and $b = 6$, so

$$SE = \sqrt{\frac{MS_E}{b}} = \sqrt{\frac{16.122}{6}} = 1.639.$$

Using Table C.9 with $\alpha = 0.05$ and $\nu = 10$, we find

k	2	3
q c.v.	3.151	3.877

This leads to the following conclusions:

Comparison	q statistic	k	c.v.	Conclusion
CC to MIPM	4.149	3	3.877	Reject H_0
CC to IPM	1.708	2	3.151	Accept H_0
IPM to MIPM	2.441	2	3.151	Accept H_0

Modified IPM is significantly different than the chemical control, but there are no other significant differences.

CC	IPM	MIPM
73.7^a	76.5^{ab}	80.5^b

Blocking explains a large portion of the SS_{Total} and is, therefore, useful.

8. H_0: All four light treatments produce identical growth rates versus H_a: At least one of the treatments produces a different growth rate. Let $\alpha = 0.05$. Since $k = 4$, then df = 3 and the c.v. = 7.81. The expected rank sum is $\frac{b(k+1)}{2} = \frac{10(4+1)}{2} = 25$ and the actual rank sums are in the table below.

$$
\begin{aligned}
T &= \frac{12}{bk(k+1)} \sum_{j=1}^{4} \left(R_j - \frac{b(k+1)}{2} \right)^2 \\
&= \frac{12}{10(4)(5)} [(37-25)^2 + (25-25)^2 + (24-25)^2 + (14-25)^2] \\
&= 12 \left(\frac{144 + 0 + 1 + 121}{200} \right) \\
&= 15.96.
\end{aligned}
$$

Since $T = 15.96 > 7.81$, reject H_0. There is evidence that at least one treatment produces a different growth rate.

Species	400 nm	Rank	500 nm	Rank	600 nm	Rank	700 nm	Rank
A	960	4	800	1	950	3	910	2
B	780	4	750	3	700	2	630	1
C	1130	4	1040	2	1050	3	1000	1
D	670	4	740	4	630	2	610	1
E	790	3	770	3	700	1	720	2
F	960	4	720	1	810	2	820	3
G	1110	4	1000	2	1040	3	980	1
H	930	3	970	4	910	2	860	1
I	920	4	850	3	840	2	820	1
J	880	3	860	2	920	4	830	1
Rank sum		37		25		24		14

To carry out paired comparisons, let $\alpha' = 0.05$ and $\alpha = \frac{\alpha'}{\binom{4}{2}} = \frac{0.05}{6} = 0.0083$. Then the c.v. $= 2.64$. H_0: The effects of the ith and jth treatments are the same versus H_a: The effects of the ith and jth treatments are different. We reject H_0 if and only if

$$|R_k - R_j| \geq \text{c.v.} \sqrt{\frac{bk(k+1)}{6}} = 2.64\sqrt{\frac{10(4)(5)}{6}} = 15.24.$$

From the table above, only wavelengths 400 nm and 700 nm have rank sums differing by at least 15.24. So there is evidence that wavelengths 400 nm and 700 nm produce different growth rates.

10. First test H_0: $(\alpha\beta)_{ij} = 0$ for all i, j versus H_0: $(\alpha\beta)_{ij} \neq 0$ for some i, j. Since $F = 1.964$ has a P value > 0.05, accept H_0. There are no significant interactions between commercial fertilizers and planting densities.

Source of variation	Sum of squares	df	MS	F	P value
Cells	498.278	17			
Commercial fertilizer	192.278	5	38.456	14.940	< 0.01
Planting density	255.444	2	127.722	49.620	< 0.01
Interaction	50.556	10	5.056	1.964	> 0.05
Error	139.000	54	2.574		
Total	637.278	71			

Now test H_0: $\mu_I = \mu_{II} = \mu_{III} = \mu_{IV} = \mu_V = \mu_{VI}$ versus H_a: At least one pair of μ_i's is different. Since $F = 14.940$ has a P value < 0.0001, reject H_0. Some commercial fertilizers are significantly different.

Also test H_0: $\mu_A = \mu_B = \mu_C$ versus H_a: At least one pair of μ_j's is different. $F = 49.620$ has a P value < 0.0001, reject H_0. Some planting densities are significantly different.

Now separate main effects with an SNK test. For the fertilizers use

$$\text{SE} = \sqrt{\frac{\text{MS}_\text{E}}{b \times n}} = \sqrt{\frac{2.574}{3 \times 4}} = 0.463.$$

From Table C.9 with $\alpha = 0.05$ and $\nu = 54 \approx 40$, we obtain

k	2	3	4	5	6
q c.v.	2.858	3.442	3.791	4.039	4.232

The 15 comparisons yield the following results. Commercial fertilizers IV and III are the best.

V	II	I	VI	IV	III
27.7^a	29.0^{ab}	30.0^{bc}	30.8^{cd}	32.0^d	32.3^d

For the planting densities use

$$\text{SE} = \sqrt{\frac{\text{MS}_\text{E}}{a \times n}} = \sqrt{\frac{2.574}{6 \times 4}} = 0.327.$$

From Table C.9 with $\alpha = 0.05$ and $\nu = 54 \approx 40$, we obtain

k	2	3
q c.v.	2.858	3.442

Planting density C is significantly better than A or B:

A	B	C
28.2^a	30.0^b	32.8^c

Finally: Recommend that growers use commercial fertilizer III or IV with planting density C.

12. First test H_0: $(\alpha\beta)_{ij} = 0$ for all i, j versus H_a: $(\alpha\beta)_{ij} \neq 0$ for some i, j. Since $F = 0.26$ is less than the critical F value, accept H_0. There are no significant interactions.

Source of variation	Sum of squares	df	MS	F	c.v.
Cells	7553.13	3			
Environment	3626.04	1	3626.04	18.50	4.35
Food	3876.04	1	3876.04	19.77	4.35
Interaction	51.04	1	51.04	0.26	4.35
Error	3920.83	20	196.04		
Total	11473.96	23			

Now test for differences in brain development due to the environment. H_0: $\mu_S = \mu_C$ versus H_a: $\mu_S \neq \mu_C$. Since $F = 18.50$ exceeds the critical F value; reject H_0. There is a significant difference in brain development between the two environments. Also test for differences in brain development due to diet. H_0: $\mu_E = \mu_C$ versus H_a: $\mu_E \neq \mu_C$. Again $F = 19.77$ exceeds the critical F value; reject H_0. There is a significant difference in brain development between the two diets. Mean separation techniques are not required because there is only one degree of freedom in each main effect analysis. Clearly the rats with the enriched diet in the stimulating environment had the largest cortices.

	Mean cortex weight (mg)		
	Stimulating	Control	Total
Enriched	715.0	693.3	704.2
Control	692.5	665.0	678.8
Total	703.8	679.2	691.5

14. Use a Friedman test. H_0: All four treatments produce the same effect. H_a: At least one treatment produces a different effect. The ranked data are summarized below.

Treatment	I	II	III	IV
Rank sum R_i	10	19	7	14

$\nu = \mathrm{df} = k - 1 = 3$. If $\alpha = 0.05$, then the c.v. $= 7.81$. From the given information,

$$T = \frac{12}{bk(k+1)}\left(\sum_{i=1}^{4} R_i^2\right) - 3b(k+1) = \frac{12}{5 \cdot 4 \cdot 5}(100 + 361 + 49 + 196) - 3 \cdot 5 \cdot 5 = 9.72.$$

Since $H = 9.72 > 7.81$, reject H_0. There is evidence at least one treatment is different.

To locate the differences, use H_0: The ith and jth treatments are the same versus H_a: The ith and jth treatments are different. Set $\alpha = \frac{0.05}{6} = 0.0083$. So $1 - \frac{\alpha}{2} = 1 - \frac{0.0083}{2} = 0.9958$. The critical value from Table C.3 is $z_{1-\frac{\alpha}{2}} = z_{0.9958} = 2.64$. The ith and jth treatments are different if and only if

$$|R_i - R_j| \geq z_{1-\frac{\alpha}{2}}\sqrt{\frac{bk(k+1)}{6}} = 2.64\sqrt{\frac{5 \cdot 4 \cdot 5}{6}} = 10.778.$$

Treatments II and III differ, with treatment III being significantly more effective. Treatment III cannot be shown to be more effective than treatments I and IV.

16. The hypotheses are H_0: $\mu_A = \mu_B = \cdots = \mu_F$ versus H_a: At least one pair of means is different.

Source of variation	Sum of squares	df	MS	F	c.v.
Varieties	1218.587	5	243.717	5.05	2.71
Blocks	685.955	4	171.489		
Error	964.713	20	48.236		
Total	2869.255	29			

Since $5.05 > 2.71$, reject H_0. There are significant differences in yield among the six new varieties. Separate the means using an SNK test. The linearly ordered means are

A	B	E	F	C	D
39.69	41.30	46.14	49.52	52.08	58.36

$\mathrm{MS_E} = 48.236$ and $b = 5$, so

$$\mathrm{SE} = \sqrt{\frac{\mathrm{MS_E}}{b}} = \sqrt{\frac{48.236}{5}} = 3.106.$$

Using Table C.9 with $\alpha = 0.05$ and $\nu = 20$ we find

k	2	3	4	5	6
q c.v.	2.950	3.578	3.958	4.232	4.445

The differences in means are indicated below. Note the overlapping ranges.

A	B	E	F	C	D
39.69[a]	41.30[a]	46.14[ab]	49.52[ab]	52.08[ab]	58.36[b]

18. (*a*) True, see page 253. (*b*) True, see pages 250 and 260. (*c*) False, see page 255. (*d*) True, see page 261. (*e*) True, see page 255. (*f*) False, see page 253. (*g*) True, see page 264.

Answers for Chapter 10

2. (*a*) There appears to be a linear relationship between the partial pressure of CO_2 and respiration rate.

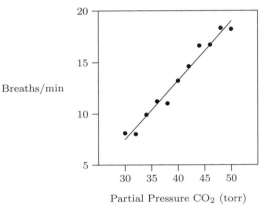

(*b*) The regression coefficient or slope is

$$b = \frac{SS_{XY}}{SS_X} = \frac{\sum XY - \frac{(\sum X)(\sum Y)}{n}}{\sum X^2 - \frac{(\sum X)^2}{n}} = \frac{253}{440} = 0.575.$$

$\hat{Y}_i = \overline{Y} + b(X_i - \overline{X})$, where $\overline{X} = 40.0$ and $\overline{Y} = 13.25$. So $\hat{Y}_i = 13.25 + 0.575(X_i - 40.0)$.

(*c*) The hypotheses are $H_0\colon \beta = 0$ versus $H_a\colon \beta \neq 0$.

Source of variation	Sum of squares	df	MS	F	P value
Regression	145.48	1	145.48	323.3	$\ll 0.01$
Residual	4.05	9	0.45		
Total	149.53	10			

Since $F = 323.3$ has a very small P value ($\ll 0.01$), reject H_0. There is a significant linear relationship between CO_2 and respiration rate. The expected value of MS_R is $\sigma_Y^2 + \beta^2 SS_R$, while the expected value of MS_E is σ_Y^2. When MS_R is significantly larger than MS_E, β is deemed significantly different from 0.

(*d*) For a 95% confidence interval for β:

$$s_b = \sqrt{\frac{MS_E}{SS_X}} = \sqrt{\frac{0.45}{440}} = 0.0320$$

so $C(b - 2.262s_b \leq \beta \leq b + 2.262s_b) = 0.95$. The 95% confidence interval for β is $[0.503, 0.647]$.

(*e*) For a 95% confidence interval for $\hat{\mu}_{Y|48}$: First $\hat{\mu}_{Y|48} = 13.25 + 0.575(48 - 40.0) = 17.85$. Next

$$s_{\hat{Y}|48} = \sqrt{\frac{SS_E}{n-2}\left(\frac{1}{n} + \frac{(X_i - \overline{X})^2}{\sum(X_i - \overline{X})^2}\right)} = \sqrt{0.45\left(\frac{1}{11} + \frac{64}{440}\right)} = 0.326.$$

Therefore, the confidence limits for $\hat{\mu}_{Y|48}$ are

$$L_1 = 17.85 - 2.262(0.326) = 17.11$$
$$L_2 = 17.85 + 2.262(0.326) = 18.59.$$

(f) For a 95% confidence interval for $\hat{\mu}_{Y|38}$, use $\hat{\mu}_{Y|40} = 13.25 + 0.575(38 - 40.0) = 12.10$ and $s_{\hat{Y}|38} = 0.212$. Then the confidence limits are

$$L_1 = 12.10 - 2.262(0.212) = 11.62$$
$$L_2 = 12.10 + 2.262(0.212) = 12.58.$$

(g) The confidence interval for a partial pressure of 38 is shorter than the confidence interval for 48 because 38 is closer to the mean or $\overline{X}$ and because β is not known exactly. Values closer to the mean will predict $\hat{\mu}_{Y|X}$ more accurately. See Figure 10.14.

4. A check shows
$$\tau = \frac{2(C - D)}{n(n - 1)} = \frac{2(105 - 15)}{16 \cdot 15} = 0.750.$$

We can test H_0: $\tau = 0$ versus H_a: $\tau \neq 0$. The test statistic is the smaller of $C = 105$ and $D = 15$. From Table C.12 with $n = 16$, the P value for a test statistic of 15 is less than 0.0001. Reject H_0. There is a (very) significant association between mean annual temperature and mortality index. That is, they covary in a similar fashion.

6. Since both variables are continuous, it is appropriate to use the Pearson product-moment correlation coefficient to measure the association between them. Preliminary calculations yield $\sum X = 1842$; $\sum X^2 = 340{,}196$; $\sum Y = 1867$; $\sum Y^2 = 349{,}283$; and $\sum XY = 344{,}648$. So $r = 0.932$.

To confirm the significance of the correlation we test H_0: $\rho = 0$ versus H_a: $\rho \neq 0$. Since $n = 10$, the test statistic is

$$t = \frac{r}{\sqrt{\frac{1-r^2}{n-2}}} = \frac{0.932}{\sqrt{\frac{1-0.932^2}{10-2}}} = 7.27.$$

The critical values from Table C.4 for $\nu = n - 2 = 8$ with $\alpha = 0.05$ are ± 2.306. Since $7.27 \gg 2.306$, we find a *strong linear correlation* between height and arm span.

8. The six rankings and the cumulative distribution for C are:

Y-ranks			**C**
1	2	3	3
1	3	2	2
2	1	3	2
2	3	1	1
3	1	2	1
3	2	1	0

c	0	1	2	3
$P(C \leq c)$	$\frac{1}{6}$	$\frac{3}{6}$	$\frac{5}{6}$	1

10. There appears to be a negative relationship between age of bitch and litter size.

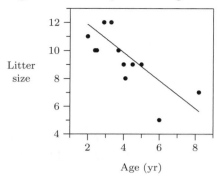

The regression coefficient or slope is

$$b = \frac{\text{SS}_{XY}}{\text{SS}_X} = \frac{\sum XY - \frac{(\sum X)(\sum Y)}{n}}{\sum X^2 - \frac{(\sum X)^2}{n}} = -\frac{29.485}{33.472} = -0.881.$$

$\hat{Y} = \overline{Y} + b(X - \overline{X})$, where $\overline{X} = 4.05$ and $\overline{Y} = 9.3$. So

$$\hat{Y} = 9.3 - 0.881(X - 4.05).$$

The hypotheses to test are H_0: $\beta = 0$ versus H_a: $\beta \neq 0$.

Source of variation	Sum of squares	df	MS	F	P value
Regression	25.972	1	25.972	15.20	≪ 0.01
Error	18.797	11	1.709		
Total	44.769	12			

Since $F = 15.19$ has a small P value (< 0.01), reject H_0: There is a significant linear relationship between age of the bitch and litter size. As the age increases, the litter size decreases. The regression line, shown in the figure, may be used to predict litter size for various ages.

12. (*a*) Use the fact that by additive and multiplicative coding $\overline{Y} = \alpha + \beta\overline{X}$. Therefore,

$$\sum(X_i - \overline{X})(Y_i - \overline{Y}) = \sum(X_i - \overline{X})[\alpha + \beta X_i - (\alpha + \beta\overline{X})]$$
$$= \sum(X_i - \overline{X})(\beta X_i - \beta\overline{X})$$
$$= \beta \sum(X_i - \overline{X})^2.$$

(*b*) Again from additive and multiplicative coding, $s_Y^2 = \beta^2 s_X^2$ so $s_Y = \sqrt{\beta^2 s_X^2} = |\beta| s_X$. The absolute value is necessary because β may be negative and we require the positive square root.

(*c*) Using the previous parts,

$$r = \frac{\sum(X_i - \overline{X})(Y_i - \overline{Y})}{(n-1) \cdot s_X \cdot s_Y} = \frac{\beta \sum(X_i - \overline{X})^2}{(n-1) \cdot s_X \cdot |\beta| s_X} = \frac{\beta s_X^2}{|\beta| s_X^2} = \frac{\beta}{|\beta|} = \pm 1,$$

depending on whether β is positive or negative.

14. (*a*) Using the table below, check that $C = 12$ and $D = 66$.

Drug	Cost	% Error	Rank X_i	Rank Y_i	Concordancies	Discordancies
Lanoxin	0.02	1450	1	13	0	12
Normison	0.06	283	2	11	1	10
Frumil	0.18	267	3	10	1	9
Ventolin	0.35	206	4	9	1	8
Capoten	0.40	313	5	12	0	8
Ponstan	0.48	8	6	2	6	1
Voltarol	0.63	71	7	8	0	6
Atrovent	0.80	55	8	6	1	4
Zantac	1.06	41	9	4	2	2
Augmentin	1.10	56	10	7	0	3
Losec	1.66	49	11	5	0	2
Zinnat	2.02	26	12	3	0	1
Claforan	13.50	−49	13	1	0	0

Kendall's correlation coefficient is

$$\tau = \frac{2(C - d)}{n(n-1)} = \frac{2(12 - 66)}{13(13 - 1)} = -0.692.$$

It appears that there is a moderately strong negative correlation between actual price and the percent difference of hospital doctors' price estimates. The strength of this correlation now can be tested. The hypotheses are H_0: $\tau = 0$ versus H_a: $\tau \neq 0$ with $n = 13$. The test statistic is $C = 12$ and $F(12) = 0.0003 \ll 0.05$. The negative correlation between actual price and the percent difference of hospital doctors' price estimates is significant. Reject H_0 and accept H_a.

(*b*) This time the data show that $C = 41$ while $D = 37$ and $\tau = 0.051$, which would indicate that the size of the error in the cost estimates is independent of the actual price of the medication. A hypothesis test confirms this: $F(37) = 0.4289 \gg 0.05$.

Drug	Cost	Difference	Rank X_i	Rank Y_i	Concordancies	Discordancies
Lanoxin	0.02	0.29	1	4	9	3
Normison	0.06	0.17	2	3	9	2
Frumil	0.18	0.48	3	8	5	5
Ventolin	0.35	0.72	4	11	2	7
Capoten	0.40	1.25	5	13	0	8
Ponstan	0.48	0.04	6	2	6	1
Voltarol	0.63	0.45	7	7	3	3
Atrovent	0.80	0.44	8	6	3	2
Zantac	1.06	0.43	9	5	3	1
Augmentin	1.10	0.62	10	10	1	2
Losec	1.66	0.82	11	12	0	2
Zinnat	2.02	0.52	12	9	0	1
Claforan	13.50	−6.65	13	1	0	0

16. (*a*) Carry out the preliminary calculations for the regression analysis.

$$n = 12 \qquad \sum X = 1980 \qquad \sum Y = 3620 \qquad \overline{X} = 165.0 \qquad \overline{Y} = 301.7$$

$$\sum X^2 = 424{,}400 \qquad \sum Y^2 = 1{,}321{,}400 \qquad \sum XY = 449{,}050$$

The regression coefficient or slope is

$$b = \frac{\text{SS}_{XY}}{\text{SS}_X} = \frac{\sum XY - \frac{(\sum X)(\sum Y)}{n}}{\sum X^2 - \frac{(\sum X)^2}{n}} = \frac{449,050 - \frac{(1980)3620}{12}}{424,400 - \frac{(1980)^2}{12}} = -1.517.$$

The least squares best fit equation for the data is

$$\widehat{Y} = \overline{Y} + b(X - \overline{X}) = 301.7 - 1.517(X - 165.0) = -1.517X + 552.005.$$

The hypotheses are $H_0\colon \beta = 0$ versus $H_a\colon \beta \neq 0$.

Source of variation	Sum of squares	df	MS	F	P value
Regression	224,835.935	1	224,835.935	496.247	< 0.01
Residual	4,530.732	10	453.073		
Total	229,366.667	11			

Since $F = 509.9$ has a small P value ($\ll 0.01$), reject H_0. There is a significant linear relationship between distance from the lake edge and 2,4-D level. The regression line is shown in the figure below.

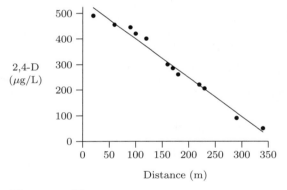

$$\text{Distance (m)}$$

(b) $\widehat{Y} = \mu_{Y|250} = \overline{Y} + b(X - \overline{X}) = 301.7 - 1.517(250 - 165.0) = 172.755$ μg/L. To find a 95% confidence interval for $\mu_{Y|250}$, first determine $\widehat{Y}_i = \widehat{\mu}_{Y|250}$.

$$s_{\widehat{Y}_i} = \sqrt{\text{MS}_\text{E}\left[\frac{1}{n} + \frac{(X_i - \overline{X})^2}{\sum(X_i - \bar{X})^2}\right]} = \sqrt{453.073\left[\frac{1}{12} + \frac{(250 - 165.0)^2}{97,700}\right]} = 8.44.$$

From Formula 10.6, the 95% confidence interval limits are

$$L_1 = \widehat{Y}_i - t_{1-\frac{\alpha}{2}}s_{\widehat{Y}_i} = 172.755 - 2.228(8.44) = 153.95 \ \mu\text{g/L}$$
$$L_2 = \widehat{Y}_i + t_{1-\frac{\alpha}{2}}s_{\widehat{Y}_i} = 172.755 + 2.228(8.44) = 191.56 \ \mu\text{g/L}.$$

(c) No, 85 μg/L does not lie in the 95% confidence for 2,4-D levels at 250 m.

18. (*a*) The transformed data are plotted below.

cm	log(bp)
1.30	4.3642
1.65	3.9739
1.90	3.8167
2.20	3.6396
2.90	3.3659
3.05	3.3069
4.30	2.7513

(*b*) The slope of the regression line is

$$b = \frac{\mathrm{SS}_{XY}}{\mathrm{SS}_X} = \frac{\sum XY - \frac{(\sum X)(\sum Y)}{n}}{\sum X^2 - \frac{(\sum X)^2}{n}} = -0.501.$$

The equation of the regression line is

$$\hat{Y}_i = \overline{Y} + b(X_i - \overline{X}) = 3.60264 - 0.501(X_i - 2.471).$$

(*c*) Test whether the slope is significantly different from 0. H_0: $\beta = 0$ versus H_a: $\beta \neq 0$.

Source of variation	Sum of squares	df	MS	F	P value
Regression	1.581	1	1.581	158.1	< 0.01
Error	0.052	5	0.010		
Total	1.633	6			

Since the P value is very small, reject H_0. There is a significant linear relationship between distance traveled by the segment and the log of the length of the segment.

(*d*) In this case

$$\hat{Y}_i = \overline{Y} + b(X_i - \overline{X}) = 3.60264 - 0.501(2.60 - 2.471) = 3.53801,$$

where $\hat{Y}_i$ is the log of the number of base pairs. Therefore, the predicted length is $10^{3.53801} = 3,452$ bp.

(*e*) First, evaluate the standard error

$$s_{\hat{Y}_i} = \sqrt{\mathrm{MS}_\mathrm{E}\left(\frac{1}{n} + \frac{(X_i - \overline{X})^2}{\sum(X_i - \bar{X})^2}\right)} = \sqrt{0.010\left(\frac{1}{7} + \frac{(2.6 - 2.471)^2}{6.309}\right)} = 0.0381.$$

So from Formula 10.6, the confidence interval endpoints are

$$L_1 = \hat{Y}_i - t_{1-\frac{\alpha}{2}}s_{\hat{Y}_i} = 3.53801 - 2.571(0.0381) = 3.4401$$
$$L_2 = \hat{Y}_i + t_{1-\frac{\alpha}{2}}s_{\hat{Y}_i} = 3.53801 + 2.571(0.0381) = 3.6360.$$

Transforming to base pairs, we get $10^{3.4401} = 2,755$ bp and $10^{3.6360} = 4,325$ bp.

20. Calculate the regression slope:

$$b = \frac{SS_{XY}}{SS_X} = \frac{\sum XY - \frac{(\sum X)(\sum Y)}{n}}{\sum X^2 - \frac{(\sum X)^2}{n}} = \frac{610 - \frac{(25)(247)}{6}}{155 - \frac{(25)^2}{6}} = -\frac{419.17}{50.83} = -8.25.$$

The least squares equation is

$$\hat{Y}_i = \hat{\mu}_{Y|X} = \overline{Y} + b(X_i - \overline{X}) = 41.167 - 8.25(X_i - 4.17).$$

Test the equation's significance with ANOVA: H_0: $\beta = 0$ and H_a: $\beta \neq 0$. Next

$$SS_{Total} = \sum Y^2 - \frac{(\sum Y)^2}{n} = 16,039 - \frac{(247)^2}{6} = 5870.83,$$
$$SS_R = b^2 SS_X = (-8.25)^2(50.83) = 3459.62,$$
$$SS_E = SS_{Total} - SS_R = 5870.83 - 3459.62 = 2411.21.$$

We summarize this information in an ANOVA table:

Source of variation	Sum of squares	df	MS	F	c.v.
Regression	3459.62	1	3459.62	5.74	7.71
Error	2411.21	4	602.80		
Total	5870.83	5			

Since $5.74 < 7.71$, the slope is not significantly different from 0. (Perhaps if the researchers had collected more data they might have been able to obtain a significant result.)

To calculate the 95% confidence interval for the slope, β begin by finding the standard error of b:

$$s_b^2 = \frac{MS_E}{SS_X} = \frac{602.80}{50.83} = 11.859$$

so $s_b = 3.44$. Therefore, the 95% confidence interval for β has endpoints

$$L_1 = -8.25 - 2.776(3.44) = -17.80$$
$$L_2 = -8.25 + 2.776(3.44) = 1.30.$$

While our estimate of the slope is $b = -8.25$, we are 95% confident that the true slope β is between -17.80 and 1.30. As expected from the ANOVA table, 0 is included in this interval, so we cannot claim that the slope is significantly different from 0.

22. (*a*) Yes. The slope of the regression line is

$$b = \frac{\sum XY - \frac{(\sum X)(\sum Y)}{n}}{\sum X^2 - \frac{(\sum X)^2}{n}} = \frac{2548 - 1620.5}{11287.5} = 0.08217.$$

The least squares best fit equation for the data is The regression equation is $\hat{V} = \overline{V} + b(T - \overline{T}) = 23.15 + 0.08217(T - 8.75)$ or

$$\hat{V} = 0.08217T + 22.4310.$$

The hypotheses to test are H_0: $\beta = 0$ versus H_a: $\beta \neq 0$.

Source of variation	Sum of squares	df	MS	F	P value
Regression	76.213	1	76.213	65,139	$\ll 0.01$
Residual	0.007	6	0.00117		
Total	76.220	7			

Since the P value is extremely small ($\ll 0.01$), reject H_0. There is a significant linear relationship between volume and temperature.

(b) $T = -272.98°\text{C}$, pretty close!

24. (a) True, see page 279. (b) False, see page 294. (c) True, see page 289. (d) False, see page 292. (e) True, see page 290. (f) True, see page 296. (g) True, see page 306.

Answers for Chapter 11

2. The hypotheses are H_0: $p = 0.25$ and H_a: $p \neq 0.25$. Use the binomial test with $n = 67$. Let $\alpha = 0.05$. The test statistic is $X = 12$. Use a normal approximation to the binomial with $\mu = np_0 = 67(0.25) = 16.75$ and $\sigma^2 = np_0(1 - p_0) = 67(0.25)(0.75) = 12.56$ so $\sigma = 3.54$. Then

$$P(X \leq 12) = P\left(Z \leq \frac{12.5 - 16.75}{3.54}\right) = P(Z \leq -1.20) = 0.1151.$$

Since $0.1151 > 0.025$, accept H_0. The data do not contradict the stated assumptions.

4. Use an extrinsic χ^2 test. The hypotheses are H_0: Light gradient had no effect; 1:1:1 ratio and H_a: The light gradient had an effect; not a 1:1:1 ratio. Use a chi-square test with $\alpha = 0.05$ and df $= 2$. The c.v. $= 5.99$. The expected values for all categories are the same: $E_i = 11$. So

$$\chi^2 = \frac{(15 - 11)^2}{11} + \frac{(12 - 11)^2}{11} + \frac{(6 - 11)^2}{11} = 3.818.$$

Since $3.818 < 5.99$, accept H_0. With this sample size, it is difficult to demonstrate that the light gradient had an effect on the distribution of these gastropods.

6. Use an extrinsic χ^2 test. The hypotheses are H_0: The distribution of fruit shape is 9 disc : 6 sphere : 1 long versus H_a: The fruit shapes do not follow a 9:6:1 distribution. Let $\alpha = 0.05$. Then df $= 3 - 1 = 2$ and the c.v. $= 5.99$. The expected values are given in the table below.

$$\chi^2 = \frac{81}{126} + \frac{36}{84} + \frac{9}{14} = 1.71.$$

Accept H_0. The data support the 9:6:1 ratio.

	O_i	E_i
A_B_	117	126
A_bb or aaB_	90	84
aabb	17	14
Total	224	224

8. Use a chi-square test (extrinsic) with H_0: The observed frequency distribution of exposure directions is random versus H_a: The distribution is not random.

Direction	Observed	Expected
North	3	10.5
East	8	10.5
South	24	10.5
West	7	10.5
Total	42	42

$$\chi^2 = \frac{(3 - 10.5)^2 + (8 - 10.5)^2 + (24 - 10.5)^2 + (7 - 10.5)^2}{10.5} = \frac{257}{10.5} = 24.5.$$

Since df $= 4 - 1 = 3$, the P value for the test statistic is less than 0.005. Reject H_0. There is evidence that the distribution of exposures is not random. The article suggests that ants prefer a southern exposure to maximize the solar warming of the nest.

10. Use an extrinsic χ^2 test. H_0: Accidents are independent of day of week (uniformly distributed) versus H_a: Accidents are dependent on day of week. Let $\alpha = 0.05$. This is an extrinsic model. Then df $= (6 - 1) = 5$. So the c.v. $= 11.07$. The expected values are given in the table below. $\chi^2 = 6.52 < 11.07$, so accept H_0, the data support accidents occurring independently of the day of week.

Day	Mon	Tues	Wed	Thurs	Fri	Sat	Total
Observed	19	34	33	36	35	29	186
Expected	31	31	31	31	31	31	186
$\frac{(O_i - E_i)^2}{E_i}$	4.65	0.29	0.13	0.81	0.52	0.13	6.52

12. Test H_0: Crab species and shell type are independent versus H_a: Crab species and shell type are dependent. The expected values are

Species	*Clusterwink*	*Cirithid*	Total
Crab A	133.0	62.0	195.0
Crab B	28.0	13.0	41.0
Total	161.0	75.0	236.0

Let $\alpha = 0.05$. Then df $= (2 - 1)(2 - 1) = 1$. So the c.v. $= 3.84$. Calculate that $\chi^2 = 81.53$. Since $\chi^2 \gg 3.84$, reject H_0. Different crab species select different shell types.

14. H_0: The distribution of amphipods and isopods is independent of the distance from shore versus H_a: The distribution of amphipods and isopods is dependent of the distance from shore. Let $\alpha = 0.05$. Then df $= (3 - 1)(2 - 1) = 2$. So the c.v. $= 5.99$. The expected values are given in the table below.

$$\chi^2 = \frac{(2 - 8.44)^2}{8.44} + \frac{(12 - 5.56)^2}{5.56} + \frac{(27 - 25.94)^2}{25.94} + \frac{(16 - 17.06)^2}{17.06}$$
$$+ \frac{(47 - 41.62)^2}{41.62} + \frac{(22 - 27.38)^2}{27.38}$$
$$= 14.23.$$

Since $\chi^2 > 5.99$, reject H_0. Amphipods and isopods are distributed differently at different distances.

Distance	Amphipods	Isopods
5 meters	8.44	5.56
10 meters	25.94	17.06
15 meters	41.62	27.38

16. (*a*) The hypotheses are H_0: Morphs and fates are independent versus H_a: An association between morphs and fates exists (one morph experiences more predation than the other in these circumstances). df $= (r-1)(k-1) = (2-1)(2-1) = 1$. The test statistic is

$$\chi^2 = \frac{80\left(|32 \times 22 - 12 \times 14| - \frac{80}{2}\right)^2}{44 \times 36 \times 46 \times 34} = 7.94.$$

From Table C.5 the critical value is 3.84 for $\alpha = 0.05$ and df $= 1$. Since $7.94 > 3.84$, we reject the H_0 that morph and fate are independent. It appears that the dark morphs are suffering significantly higher predation rates than the light morphs in this experimental protocol.

(*b*) Let population 1 denote the light morph. A two-tailed, two-sample proportions test at the $\alpha = 0.05$ level is appropriate. The hypotheses are H_0: $p_1 = p_2$ and H_a: $p_1 \neq p_2$, where p_1 and p_2 represent the proportions not eaten. We are given that $\hat{p}_1 = \frac{32}{46} = 0.6957$ and $\hat{p}_2 = \frac{12}{34} = 0.3529$. Therefore,

$$p_c = \frac{32 + 12}{80} = 0.5500.$$

The test statistic is

$$z = \frac{\hat{p}_1 - \hat{p}_2}{\sqrt{p_c(1 - p_c)\left(\frac{1}{n_1} + \frac{1}{n_2}\right)}} = \frac{0.6957 - 0.3529}{\sqrt{0.55(0.45)\left(\frac{1}{46} + \frac{1}{34}\right)}} = 3.05.$$

Using Table C.3 the probability of lying in the right tail is $1 - P(Z \leq 3.05) = 0.0011$, which is less than $\frac{\alpha}{2} = 0.025$. Reject H_0; there is evidence for a significant difference in predation rate for the two morphs.

18. The hypotheses are H_0: Data follows a binomial distribution with $p = 0.5$ versus H_a: Data are significantly different from a binomial with $p = 0.5$. The expected values are determined by

$$f_i = 402 \binom{8}{i}(0.5)^i(0.5)^{8-i} = 402 \binom{8}{i}(0.5)^8$$

and are listed in the table below.

X_i	f_i	$E(X_i)$
0	1	1.57
1	8	12.56
2	37	43.97
3	81	87.94
4	162	109.92
5	77	87.94
6	30	43.97
7	5	12.56
8	1	1.57
	402	402

Notice that categories 0 and 1 and categories 7 and 8 must be combined to produce expected values of at least 5. In each case the expected number in the combined category is 14.15. With these 7 categories, $\nu = k - 1 = 7 - 1 = 6$. At $\alpha = 0.05$, the c.v. $= 12.6$. The calculated value of $\chi^2 = 38.75$. Since $38.75 > 12.6$, reject H_0. The data do not fit a binomial distribution with $p = 0.5$.

20. The hypotheses are H_0: Species are independent of each other versus H_a: The species show a significant association. Analyze using a 2×2 contingency chi-square with $\alpha = 0.05$ and $\nu = 1$. The c.v. $= 3.84$.

$$\chi^2 = \frac{n_{..}(|n_{11}n_{22} - n_{12}n_{21}| - \frac{n_{..}}{2})^2}{n_{1.}n_{2.}n_{.1}n_{.2}} = \frac{100(|(50 \times 15) - (10 \times 25)| - \frac{100}{2})^2}{60 \times 40 \times 75 \times 25} = 4.5.$$

Since $4.5 > 3.84$, reject H_0. There is a positive association between the two species, i.e., they tend to appear together.

22. (a) The hypotheses are H_0: $p \leq 0.5$ versus H_a: $p > 0.5$. Use a binomial test with $\alpha = 0.05$ and $n = 20$. The test statistic is $X = 12$. $P(X \geq 12) = 1 - P(X \leq 11) = 1 - F(11) = 1 - 0.7483 = 0.2517$. Since $0.2517 > 0.05$, accept H_0. Data do not provide evidence that people prefer Coke.

(b) Now $n = 100$ and the test statistic is $X = 60$, so use a normal approximation. $\mu = np = 50$ and $\sigma^2 = np(1 - p) = 25$, so $\sigma = 5$. Then

$$P(X \geq 60) = 1 - P\left(Z < \frac{59.5 - 50}{5}\right) = 1 - P(Z < 1.9) = 1 - 0.9713 = 0.0287.$$

Since $0.0287 < 0.05$, reject H_0. With the larger sample size, the data now support the claim that people prefer Coke, *even though the ratio of those preferring Coke has stayed the same.*

24. Use a binomial test with $n = 20$. The hypotheses are H_0: There is no difference in taste between beer stored in light-tight containers or light exposed containers ($p = 0.5$) versus H_a: There is a perceptible difference in taste between beer stored in light-tight containers or light exposed containers ($p \neq 0.5$). The test statistic is $X = 16$ and, from Table C.1, $P(X \leq 16) = 0.9987$ and $P(X \geq 16) = 1 - P(X \leq 15) = 1 - 0.9941 = 0.0059$. Because the probability of lying in the right tail is very small (less than 0.025), reject H_0. The data indicate there is a significant difference in taste in beer from light-tight containers rather than light exposed containers.

26. The hypotheses are H_0: The presence of the APOE4 allele and the presence of Alzheimer's disease are independent versus H_a: The presence of the APOE4 allele and the presence of Alzheimer's disease are dependent. The expected values are as given in the following table.

Subjects	APOE4/APOE4	APOE4/+	+/+
Normal	138.19	321.05	1162.76
Alzheimer's	59.81	138.95	503.24

$$\chi^2 = \sum_{ij} \frac{(O_{ij} - E_{ij})^2}{E_{ij}} = 516.9.$$

df $= (3-1) \times (2-1) = 2$, so the critical value for the test is 5.99. Since $\chi^2 = 516.9 \gg 5.99$, the null hypothesis is rejected. Based on these data, the distribution of the APOE4 allele is not the same in normal subjects and Alzheimer's patients. There is an association between the presence of the APOE4 allele and Alzheimer's disease.

28. (*a*) The expected ratios for the offspring phenotypes assuming equal fitness and independent assortment are:

6/16	polled, roan
3/16	polled, red
3/16	polled, white
2/16	horned, roan
1/16	horned, red
1/16	horned, white

(*b*) The hypotheses are H_0: There is independent assortment in the genes versus H_a: There is not independent assortment. Let $\alpha = 0.05$. df $= \nu = 6 - 1 = 5$ and the c.v. $= 11.1$. Using the expected ratios above and a sample size of 160,

Phenotype	Observed	Expected
polled, roan	55	60
polled, red	27	30
polled, white	31	30
horned, roan	23	20
horned, red	11	10
horned, white	13	10

$$\chi^2 = \sum_{i=1}^{6} \frac{(O_i - E_i)^2}{E_i} = \frac{25}{60} + \frac{9}{30} + \frac{1}{30} + \frac{9}{20} + \frac{1}{10} + \frac{9}{10} = \frac{132}{60} = 2.2.$$

Since $2.2 < 11.1$, there is support for the claim of independent assortment.

30. Use a binomial test. The hypotheses are H_0: $P = 0.75$ versus H_a: $P \neq 0.75$. Since $P(X \leq 12) = 0.1018$ and $P(X \geq 12) = 1 - 0.0409 = 0.9591$ are both greater than 0.025; accept H_0. The preliminary test supports the claim that 75% will experience improvement in GVHD skin lesions.

32. Let the circumcised males represent the sample from the first population. Then a left-tailed two-sample proportions test at the $\alpha = 0.05$ level is appropriate. The hypotheses are H_0: $p_1 \geq p_2$ and H_a: $p_1 < p_2$. The test statistic is

$$z = \frac{\hat{p}_1 - \hat{p}_2}{\sqrt{p_c(1 - p_c)\left(\frac{1}{n_1} + \frac{1}{n_2}\right)}}.$$

The table below provides the details of the analysis.

Location	n_1, n_2	X_1	X_2	$\hat{p}_1$	$\hat{p}_2$	$\hat{p}_c$	z	P value
Kenya	1392	22	47	0.01580	0.03376	0.02478	-3.05	0.0011
Uganda	2498	22	43	0.00881	0.01721	0.01301	-2.62	0.0044

Because the P values were so small, the evidence was overwhelming that circumcision was effective in reducing HIV infection. Therefore, it was immoral to put the uncircumcised population at further risk.

34. A Kolmogorov-Smirnov test is appropriate since you want to know whether the sample is from a pre-specified normal population. Use $\alpha = 0.05$.

x_i	$S(x_i)$	$F(x_i)$	$|S(x_i) - F(x_i)|$	$|S(x_{i-1}) - F(x_i)|$
21.8	0.056	0.005	0.051	0.005
22.8	0.111	0.011	0.100	0.045
22.9	0.167	0.012	0.155	0.099
23.0	0.222	0.013	0.209	0.154
23.2	0.278	0.016	0.262	0.206
23.7	0.333	0.023	0.310	0.255
23.9	0.389	0.027	0.362	0.306
24.1	0.444	0.031	0.413	0.358
24.6	0.500	0.044	0.456	0.400
25.0	0.556	0.057	0.499	0.443
25.4	0.611	0.073	0.538	0.483
26.1	0.667	0.109	**0.558**	0.502
27.1	0.722	0.180	0.542	0.487
28.1	0.778	0.274	0.504	0.448
30.3	0.833	0.538	0.295	0.240
30.8	0.889	0.600	0.289	0.233
30.9	0.944	0.612	0.332	0.277
31.2	1.000	0.648	0.352	0.296

The test statistic is $K = 0.558$. In Table C.14 with $\alpha = 0.05$ level and $n = 18$, the c.v. $= 0.3094$, so H_0 is rejected. These froglets do not appear to be from a population with normally distributed lengths with $\mu = 30$ mm and $\sigma^2 = 10$ mm^2. The results do not support the hypothesis that these are clicking froglets.

36. A Lilliefors test is appropriate since the mean and variance were estimated from the data. The hypotheses are H_0: The sample is drawn from a normal distribution and H_a: The distribution is not normal. We carry out the test with $\alpha = 0.05$. In the following table, $F(x)$ denotes the normal distribution with $\mu = 136.9$ cm and $\sigma^2 = 394.55$.

x_i	$S(x_i)$	$F(x_i)$	$|S(x_i) - F(x_i)|$	$|S(x_{i-1}) - F(x_i)|$
95	0.0667	0.0174	0.0493	0.0174
110	0.1333	0.0885	0.0448	0.0218
116	0.2000	0.1469	0.0531	0.0136
125	0.2667	0.2743	0.0076	0.0743
128	0.3333	0.3264	0.0069	0.0597
135	0.4000	0.4602	0.0602	0.1269
140	0.4667	0.5636	0.0969	**0.1636**
140	0.5333	0.5636	0.0303	0.0969
141	0.6000	0.5832	0.0168	0.0499
145	0.6667	0.6591	0.0076	0.0591
146	0.7333	0.6772	0.0561	0.0105
150	0.8000	0.7454	0.0546	0.0121
150	0.8667	0.7454	0.1213	0.0546
157	0.9333	0.8438	0.0895	0.0229
175	1.0000	0.9726	0.0274	0.0393

The test statistic is $L = 0.1636$, which is smaller than the critical value of 0.220 from Table C.15. Accept H_0. We may continue to assume the leaf lengths come from a normally distributed population.

38. A Lilliefors test is appropriate since the mean and variance must be estimated from the sample. Use $\overline{X} = 76.06$ mm and $s = 6.14$ mm. The null hypothesis is H_0: Lengths of the gobies are distributed normally.

x_i	$S(x_i)$	$F(x_i)$	$|S(x_i) - F(x_i)|$	$|S(x_{i-1}) - F(x_i)|$
63.0	0.1	0.0166	0.0834	0.0166
69.5	0.2	0.1423	0.0577	0.0423
72.4	0.3	0.2743	0.0257	0.0743
75.4	0.4	0.4562	0.0562	0.1562
77.9	0.5	0.6179	0.1179	**0.2179**
77.9	0.6	0.6179	0.0179	0.1179
80.0	0.7	0.7389	0.0389	0.1389
80.6	0.8	0.7704	0.0296	0.0704
81.8	0.9	0.8238	0.0762	0.0238
82.1	1.0	0.8365	0.1635	0.0635

$L = 0.2179$. From Table C.15 the c.v. $= 0.258$ at the $\alpha = 0.05$ level for a sample size of $n = 10$. H_0 is accepted. The researchers may assume that the goby lengths are from a normally distributed population (with $\mu = 76.06$ mm and $\sigma = 6.14$ mm.)

40. Let population 1 denote those with limited literacy. A right-tailed, two-sample proportions test at the $\alpha = 0.05$ level is appropriate. The hypotheses are H_0: $p_1 \leq p_2$ and H_a: $p_1 > p_2$. Since there were $n_1 = 603$ participants with limited literacy, there were $n_2 = 2512 - 603 = 1909$ other participants. We are given that $\hat{p}_1 = 0.1973$, so $X_1 = 0.1973(603) = 119$. Similarly, $X_2 = 0.1058(1909) = 202$. Therefore,

$$p_c = \frac{119 + 202}{2512} = 0.1278.$$

The test statistic is

$$z = \frac{\hat{p}_1 - \hat{p}_2}{\sqrt{p_c(1 - p_c)\left(\frac{1}{n_1} + \frac{1}{n_2}\right)}} = \frac{0.1973 - 0.1058}{\sqrt{0.1278(0.8722)\left(\frac{1}{603} + \frac{1}{1909}\right)}} = 5.87.$$

Using Table C.3 the probability of lying in the right tail is $1 - P(Z \leq 5.87) = 0$. Reject H_0; there is evidence for a significant increase in death rate among the elderly with limited literacy.

42. Let Makemo represent the first population. A two-tailed, two-sample proportions test at the $\alpha = 0.05$ level is appropriate. The hypotheses are H_0: $p_1 = p_2$ and H_a: $p_1 \neq p_2$. Then

$$\hat{p}_1 = \frac{X_1}{n_1} = \frac{41}{64} = 0.6406 \qquad \text{and} \qquad \hat{p}_2 = \frac{X_2}{n_2} = \frac{81}{143} = 0.5664$$

so

$$p_c = \frac{41 + 81}{64 + 143} = 0.5894.$$

The test statistic is

$$z = \frac{\hat{p}_1 - \hat{p}_2}{\sqrt{p_c(1 - p_c)\left(\frac{1}{n_1} + \frac{1}{n_2}\right)}} = \frac{0.6406 - 0.5664}{\sqrt{0.5894(0.4106)\left(\frac{1}{64} + \frac{1}{143}\right)}} = 1.00.$$

Using Table C.3 the probability of lying in the left tail is $P(Z \leq 1.00) = 0.8413$ and the probability of lying in the right tail is $1 - P(Z \leq 1.00) = 0.1587$. We cannot reject H_0; there is no evidence of a significant difference in proportions of the white color morph at the two atolls.

44. Let population 1 denotes those exposed to dogs. Then

$$\hat{p}_1 = \frac{6}{184} = 0.0326 \qquad \text{and} \qquad \hat{p}_2 = \frac{19}{220} = 0.0864.$$

So $\hat{D} = 0.0326 - 0.0864 = -0.0538$. From Table C.3, $z_{0.975} = 1.96$.

$$\text{SE}_{D_u} = \sqrt{\frac{0.0326(0.9674)}{184} + \frac{0.0864(0.9136)}{220}} = 0.0230.$$

So

$$L_1 = -0.0538 - 1.96(0.0230) = -0.0989$$
$$L_2 = -0.0538 + 1.96(0.0230) = -0.0087.$$

Since the confidence interval does not contain 0 (just barely), there is a significant difference in the proportion of positive skin tests between those exposed to dogs before age one and those who were not. (Those exposed to dogs had a significantly lower proportion of positive skin tests.)

46. $\hat{p}_F = \frac{124}{418} = 0.2967$ and $\hat{p}_M = \frac{34}{182} = 0.1868$. So $\hat{D} = 0.2967 - 0.1868 = 0.1099$. From Table C.3, $z_{0.975} = 1.96$.

$$\text{SE}_{D_u} = \sqrt{\frac{0.2967(0.7033)}{418} + \frac{0.1868(0.8132)}{182}} = 0.0365.$$

So

$$L_1 = 0.1099 - 1.96(0.0365) = 0.0384$$
$$L_2 = 0.1099 + 1.96(0.0365) = 0.1814.$$

Since the confidence interval does not contains 0, there is a significant difference between the proportions of women and men who have adverse reactions to placebo.

48. (*a*) True, see page 320. (*b*) True, see page 327. (*c*) False, see page 328. (*d*) True, see page 332. (*e*) True, see page 335. (*f*) False, see Section 11.3. (*g*) True, see page 327. (*h*) False, see page 344.

Tables of Distributions and Critical Values

1. Cumulative binomial distribution

2. Cumulative Poisson distribution

3. Cumulative standard normal distribution

4. Student's t distribution

5. Cumulative chi-square distribution

6. Wilcoxon signed-rank test cumulative distribution

7. Cumulative F distribution

8. Critical values for the Wilcoxon rank-sum test

9. Critical values of the q statistic for the Student-Newman-Keuls test

10. Fisher's Z transformation of correlation coefficient, r

11. Correlation coefficient r corresponding to Fisher's Z transformation

12. Cumulative distribution for Kendall's test (τ)

13. Critical values for the Spearman rank correlation coefficient, r_s

14. Critical values for the Kolmogorov-Smirnov test

15. Critical values for the Lilliefors test

C.1   Cumulative Binomial Distribution

Table C.1 is a tabulation of the cumulative distribution function (F) for selected binomial distributions where

$$F(d) = P(X \le d) = \sum_{x=0}^{d} \binom{n}{x} p^x (1-p)^{n-x}.$$

If n is between 5 and 20 inclusive and p is found at the top of the table, one can find various pdfs and CDFs directly from this table. For example, if $n = 8$ and $p = 0.3$, the probability of 4 or fewer successes in 8 trials $P(X \le 4) = F(4) = 0.9420$ is found as the value at the intersection of the row "4" and column "0.3" in the $n = 8$ tabulation. The probability of exactly 4 successes in 8 trials is $P(X = 4) = f(4) = F(4) - F(3) = 0.9420 - 0.8059 = 0.1361$ from the $n = 8$ tabulation. For further examples, see Chapter 3.

| | | | | | | | p | | | | | |
n	d	0.1	0.2	0.25	0.3	0.4	0.5	0.6	0.7	0.75	0.8	0.9
5	0	0.5905	0.3277	0.2373	0.1681	0.0778	0.0313	0.0102	0.0024	0.0010	0.0003	0.0000
	1	0.9185	0.7373	0.6328	0.5282	0.3370	0.1875	0.0870	0.0308	0.0156	0.0067	0.0005
	2	0.9914	0.9421	0.8965	0.8369	0.6826	0.5000	0.3174	0.1631	0.1035	0.0579	0.0086
	3	0.9995	0.9933	0.9844	0.9692	0.9130	0.8125	0.6630	0.4718	0.3672	0.2627	0.0815
	4	1.0000	0.9997	0.9990	0.9976	0.9898	0.9688	0.9222	0.8319	0.7627	0.6723	0.4095
	5	1.0000	1.0000	1.0000	1.0000	1.0000	1.0000	1.0000	1.0000	1.0000	1.0000	1.0000
6	0	0.5314	0.2621	0.1780	0.1176	0.0467	0.0156	0.0041	0.0007	0.0002	0.0001	0.0000
	1	0.8857	0.6554	0.5339	0.4202	0.2333	0.1094	0.0410	0.0109	0.0046	0.0016	0.0001
	2	0.9842	0.9011	0.8306	0.7443	0.5443	0.3438	0.1792	0.0705	0.0376	0.0170	0.0013
	3	0.9987	0.9830	0.9624	0.9295	0.8208	0.6563	0.4557	0.2557	0.1694	0.0989	0.0159
	4	0.9999	0.9984	0.9954	0.9891	0.9590	0.8906	0.7667	0.5798	0.4661	0.3446	0.1143
	5	1.0000	0.9999	0.9998	0.9993	0.9959	0.9844	0.9533	0.8824	0.8220	0.7379	0.4686
	6	1.0000	1.0000	1.0000	1.0000	1.0000	1.0000	1.0000	1.0000	1.0000	1.0000	1.0000
7	0	0.4783	0.2097	0.1335	0.0824	0.0280	0.0078	0.0016	0.0002	0.0001	0.0000	0.0000
	1	0.8503	0.5767	0.4449	0.3294	0.1586	0.0625	0.0188	0.0038	0.0013	0.0004	0.0000
	2	0.9743	0.8520	0.7564	0.6471	0.4199	0.2266	0.0963	0.0288	0.0129	0.0047	0.0002
	3	0.9973	0.9667	0.9294	0.8740	0.7102	0.5000	0.2898	0.1260	0.0706	0.0333	0.0027
	4	0.9998	0.9953	0.9871	0.9712	0.9037	0.7734	0.5801	0.3529	0.2436	0.1480	0.0257
	5	1.0000	0.9996	0.9987	0.9962	0.9812	0.9375	0.8414	0.6706	0.5551	0.4233	0.1497
	6	1.0000	1.0000	0.9999	0.9998	0.9984	0.9922	0.9720	0.9176	0.8665	0.7903	0.5217
	7	1.0000	1.0000	1.0000	1.0000	1.0000	1.0000	1.0000	1.0000	1.0000	1.0000	1.0000
8	0	0.4305	0.1678	0.1001	0.0576	0.0168	0.0039	0.0007	0.0001	0.0000	0.0000	0.0000
	1	0.8131	0.5033	0.3671	0.2553	0.1064	0.0352	0.0085	0.0013	0.0004	0.0001	0.0000
	2	0.9619	0.7969	0.6785	0.5518	0.3154	0.1445	0.0498	0.0113	0.0042	0.0012	0.0000
	3	0.9950	0.9437	0.8862	0.8059	0.5941	0.3633	0.1737	0.0580	0.0273	0.0104	0.0004
	4	0.9996	0.9896	0.9727	0.9420	0.8263	0.6367	0.4059	0.1941	0.1138	0.0563	0.0050
	5	1.0000	0.9988	0.9958	0.9887	0.9502	0.8555	0.6846	0.4482	0.3215	0.2031	0.0381
	6	1.0000	0.9999	0.9996	0.9987	0.9915	0.9648	0.8936	0.7447	0.6329	0.4967	0.1869
	7	1.0000	1.0000	1.0000	0.9999	0.9993	0.9961	0.9832	0.9424	0.8999	0.8322	0.5695
	8	1.0000	1.0000	1.0000	1.0000	1.0000	1.0000	1.0000	1.0000	1.0000	1.0000	1.0000

							p					
n	d	0.1	0.2	0.25	0.3	0.4	0.5	0.6	0.7	0.75	0.8	0.9
9	0	0.3874	0.1342	0.0751	0.0404	0.0101	0.0020	0.0003	0.0000	0.0000	0.0000	0.0000
	1	0.7748	0.4362	0.3003	0.1960	0.0705	0.0195	0.0038	0.0004	0.0001	0.0000	0.0000
	2	0.9470	0.7382	0.6007	0.4628	0.2318	0.0898	0.0250	0.0043	0.0013	0.0003	0.0000
	3	0.9917	0.9144	0.8343	0.7297	0.4826	0.2539	0.0994	0.0253	0.0100	0.0031	0.0001
	4	0.9991	0.9804	0.9511	0.9012	0.7334	0.5000	0.2666	0.0988	0.0489	0.0196	0.0009
	5	0.9999	0.9969	0.9900	0.9747	0.9006	0.7461	0.5174	0.2703	0.1657	0.0856	0.0083
	6	1.0000	0.9997	0.9987	0.9957	0.9750	0.9102	0.7682	0.5372	0.3993	0.2618	0.0530
	7	1.0000	1.0000	0.9999	0.9996	0.9962	0.9805	0.9295	0.8040	0.6997	0.5638	0.2252
	8	1.0000	1.0000	1.0000	1.0000	0.9997	0.9980	0.9899	0.9596	0.9249	0.8658	0.6126
	9	1.0000	1.0000	1.0000	1.0000	1.0000	1.0000	1.0000	1.0000	1.0000	1.0000	1.0000
10	0	0.3487	0.1074	0.0563	0.0282	0.0060	0.0010	0.0001	0.0000	0.0000	0.0000	0.0000
	1	0.7361	0.3758	0.2440	0.1493	0.0464	0.0107	0.0017	0.0001	0.0000	0.0000	0.0000
	2	0.9298	0.6778	0.5256	0.3828	0.1673	0.0547	0.0123	0.0016	0.0004	0.0001	0.0000
	3	0.9872	0.8791	0.7759	0.6496	0.3823	0.1719	0.0548	0.0106	0.0035	0.0009	0.0000
	4	0.9984	0.9672	0.9219	0.8497	0.6331	0.3770	0.1662	0.0473	0.0197	0.0064	0.0001
	5	0.9999	0.9936	0.9803	0.9527	0.8338	0.6230	0.3669	0.1503	0.0781	0.0328	0.0016
	6	1.0000	0.9991	0.9965	0.9894	0.9452	0.8281	0.6177	0.3504	0.2241	0.1209	0.0128
	7	1.0000	0.9999	0.9996	0.9984	0.9877	0.9453	0.8327	0.6172	0.4744	0.3222	0.0702
	8	1.0000	1.0000	1.0000	0.9999	0.9983	0.9893	0.9536	0.8507	0.7560	0.6242	0.2639
	9	1.0000	1.0000	1.0000	1.0000	0.9999	0.9990	0.9940	0.9718	0.9437	0.8926	0.6513
	10	1.0000	1.0000	1.0000	1.0000	1.0000	1.0000	1.0000	1.0000	1.0000	1.0000	1.0000
11	0	0.3138	0.0859	0.0422	0.0198	0.0036	0.0005	0.0000	0.0000	0.0000	0.0000	0.0000
	1	0.6974	0.3221	0.1971	0.1130	0.0302	0.0059	0.0007	0.0000	0.0000	0.0000	0.0000
	2	0.9104	0.6174	0.4552	0.3127	0.1189	0.0327	0.0059	0.0006	0.0001	0.0000	0.0000
	3	0.9815	0.8389	0.7133	0.5696	0.2963	0.1133	0.0293	0.0043	0.0012	0.0002	0.0000
	4	0.9972	0.9496	0.8854	0.7897	0.5328	0.2744	0.0994	0.0216	0.0076	0.0020	0.0000
	5	0.9997	0.9883	0.9657	0.9218	0.7535	0.5000	0.2465	0.0782	0.0343	0.0117	0.0003
	6	1.0000	0.9980	0.9924	0.9784	0.9006	0.7256	0.4672	0.2103	0.1146	0.0504	0.0028
	7	1.0000	0.9998	0.9988	0.9957	0.9707	0.8867	0.7037	0.4304	0.2867	0.1611	0.0185
	8	1.0000	1.0000	0.9999	0.9994	0.9941	0.9673	0.8811	0.6873	0.5448	0.3826	0.0896
	9	1.0000	1.0000	1.0000	1.0000	0.9993	0.9941	0.9698	0.8870	0.8029	0.6779	0.3026
	10	1.0000	1.0000	1.0000	1.0000	1.0000	0.9995	0.9964	0.9802	0.9578	0.9141	0.6862
	11	1.0000	1.0000	1.0000	1.0000	1.0000	1.0000	1.0000	1.0000	1.0000	1.0000	1.0000
12	0	0.2824	0.0687	0.0317	0.0138	0.0022	0.0002	0.0000	0.0000	0.0000	0.0000	0.0000
	1	0.6590	0.2749	0.1584	0.0850	0.0196	0.0032	0.0003	0.0000	0.0000	0.0000	0.0000
	2	0.8891	0.5583	0.3907	0.2528	0.0834	0.0193	0.0028	0.0002	0.0000	0.0000	0.0000
	3	0.9744	0.7946	0.6488	0.4925	0.2253	0.0730	0.0153	0.0017	0.0004	0.0001	0.0000
	4	0.9957	0.9274	0.8424	0.7237	0.4382	0.1938	0.0573	0.0095	0.0028	0.0006	0.0000
	5	0.9995	0.9806	0.9456	0.8822	0.6652	0.3872	0.1582	0.0386	0.0143	0.0039	0.0001
	6	0.9999	0.9961	0.9857	0.9614	0.8418	0.6128	0.3348	0.1178	0.0544	0.0194	0.0005
	7	1.0000	0.9994	0.9972	0.9905	0.9427	0.8062	0.5618	0.2763	0.1576	0.0726	0.0043
	8	1.0000	0.9999	0.9996	0.9983	0.9847	0.9270	0.7747	0.5075	0.3512	0.2054	0.0256
	9	1.0000	1.0000	1.0000	0.9998	0.9972	0.9807	0.9166	0.7472	0.6093	0.4417	0.1109
	10	1.0000	1.0000	1.0000	1.0000	0.9997	0.9968	0.9804	0.9150	0.8416	0.7251	0.3410
	11	1.0000	1.0000	1.0000	1.0000	1.0000	0.9998	0.9978	0.9862	0.9683	0.9313	0.7176
	12	1.0000	1.0000	1.0000	1.0000	1.0000	1.0000	1.0000	1.0000	1.0000	1.0000	1.0000

							p					
n	d	0.1	0.2	0.25	0.3	0.4	0.5	0.6	0.7	0.75	0.8	0.9
13	0	0.2542	0.0550	0.0238	0.0097	0.0013	0.0001	0.0000	0.0000	0.0000	0.0000	0.0000
	1	0.6213	0.2336	0.1267	0.0637	0.0126	0.0017	0.0001	0.0000	0.0000	0.0000	0.0000
	2	0.8661	0.5017	0.3326	0.2025	0.0579	0.0112	0.0013	0.0001	0.0000	0.0000	0.0000
	3	0.9658	0.7473	0.5843	0.4206	0.1686	0.0461	0.0078	0.0007	0.0001	0.0000	0.0000
	4	0.9935	0.9009	0.7940	0.6543	0.3530	0.1334	0.0321	0.0040	0.0010	0.0002	0.0000
	5	0.9991	0.9700	0.9198	0.8346	0.5744	0.2905	0.0977	0.0182	0.0056	0.0012	0.0000
	6	0.9999	0.9930	0.9757	0.9376	0.7712	0.5000	0.2288	0.0624	0.0243	0.0070	0.0001
	7	1.0000	0.9988	0.9944	0.9818	0.9023	0.7095	0.4256	0.1654	0.0802	0.0300	0.0009
	8	1.0000	0.9998	0.9990	0.9960	0.9679	0.8666	0.6470	0.3457	0.2060	0.0991	0.0065
	9	1.0000	1.0000	0.9999	0.9993	0.9922	0.9539	0.8314	0.5794	0.4157	0.2527	0.0342
	10	1.0000	1.0000	1.0000	0.9999	0.9987	0.9888	0.9421	0.7975	0.6674	0.4983	0.1339
	11	1.0000	1.0000	1.0000	1.0000	0.9999	0.9983	0.9874	0.9363	0.8733	0.7664	0.3787
	12	1.0000	1.0000	1.0000	1.0000	1.0000	0.9999	0.9987	0.9903	0.9762	0.9450	0.7458
	13	1.0000	1.0000	1.0000	1.0000	1.0000	1.0000	1.0000	1.0000	1.0000	1.0000	1.0000
14	0	0.2288	0.0440	0.0178	0.0068	0.0008	0.0001	0.0000	0.0000	0.0000	0.0000	0.0000
	1	0.5846	0.1979	0.1010	0.0475	0.0081	0.0009	0.0001	0.0000	0.0000	0.0000	0.0000
	2	0.8416	0.4481	0.2811	0.1608	0.0398	0.0065	0.0006	0.0000	0.0000	0.0000	0.0000
	3	0.9559	0.6982	0.5213	0.3552	0.1243	0.0287	0.0039	0.0002	0.0000	0.0000	0.0000
	4	0.9908	0.8702	0.7415	0.5842	0.2793	0.0898	0.0175	0.0017	0.0003	0.0000	0.0000
	5	0.9985	0.9561	0.8883	0.7805	0.4859	0.2120	0.0583	0.0083	0.0022	0.0004	0.0000
	6	0.9998	0.9884	0.9617	0.9067	0.6925	0.3953	0.1501	0.0315	0.0103	0.0024	0.0000
	7	1.0000	0.9976	0.9897	0.9685	0.8499	0.6047	0.3075	0.0933	0.0383	0.0116	0.0002
	8	1.0000	0.9996	0.9978	0.9917	0.9417	0.7880	0.5141	0.2195	0.1117	0.0439	0.0015
	9	1.0000	1.0000	0.9997	0.9983	0.9825	0.9102	0.7207	0.4158	0.2585	0.1298	0.0092
	10	1.0000	1.0000	1.0000	0.9998	0.9961	0.9713	0.8757	0.6448	0.4787	0.3018	0.0441
	11	1.0000	1.0000	1.0000	1.0000	0.9994	0.9935	0.9602	0.8392	0.7189	0.5519	0.1584
	12	1.0000	1.0000	1.0000	1.0000	0.9999	0.9991	0.9919	0.9525	0.8990	0.8021	0.4154
	13	1.0000	1.0000	1.0000	1.0000	1.0000	0.9999	0.9992	0.9932	0.9822	0.9560	0.7712
	14	1.0000	1.0000	1.0000	1.0000	1.0000	1.0000	1.0000	1.0000	1.0000	1.0000	1.0000
15	0	0.2059	0.0352	0.0134	0.0047	0.0005	0.0000	0.0000	0.0000	0.0000	0.0000	0.0000
	1	0.5490	0.1671	0.0802	0.0353	0.0052	0.0005	0.0000	0.0000	0.0000	0.0000	0.0000
	2	0.8159	0.3980	0.2361	0.1268	0.0271	0.0037	0.0003	0.0000	0.0000	0.0000	0.0000
	3	0.9444	0.6482	0.4613	0.2969	0.0905	0.0176	0.0019	0.0001	0.0000	0.0000	0.0000
	4	0.9873	0.8358	0.6865	0.5155	0.2173	0.0592	0.0093	0.0007	0.0001	0.0000	0.0000
	5	0.9978	0.9389	0.8516	0.7216	0.4032	0.1509	0.0338	0.0037	0.0008	0.0001	0.0000
	6	0.9997	0.9819	0.9434	0.8689	0.6098	0.3036	0.0950	0.0152	0.0042	0.0008	0.0000
	7	1.0000	0.9958	0.9827	0.9500	0.7869	0.5000	0.2131	0.0500	0.0173	0.0042	0.0000
	8	1.0000	0.9992	0.9958	0.9848	0.9050	0.6964	0.3902	0.1311	0.0566	0.0181	0.0003
	9	1.0000	0.9999	0.9992	0.9963	0.9662	0.8491	0.5968	0.2784	0.1484	0.0611	0.0022
	10	1.0000	1.0000	0.9999	0.9993	0.9907	0.9408	0.7827	0.4845	0.3135	0.1642	0.0127
	11	1.0000	1.0000	1.0000	0.9999	0.9981	0.9824	0.9095	0.8392	0.7189	0.5519	0.1584
	12	1.0000	1.0000	1.0000	1.0000	0.9999	0.9991	0.9919	0.9525	0.8990	0.8021	0.4154
	13	1.0000	1.0000	1.0000	1.0000	1.0000	0.9999	0.9992	0.9932	0.9822	0.9560	0.7712
	14	1.0000	1.0000	1.0000	1.0000	1.0000	1.0000	0.9995	0.9953	0.9866	0.9648	0.7941
	15	1.0000	1.0000	1.0000	1.0000	1.0000	1.0000	1.0000	1.0000	1.0000	1.0000	1.0000

n	d	0.1	0.2	0.25	0.3	0.4	0.5	0.6	0.7	0.75	0.8	0.9
							p					
16	0	0.1853	0.0281	0.0100	0.0033	0.0003	0.0000	0.0000	0.0000	0.0000	0.0000	0.0000
	1	0.5147	0.1407	0.0635	0.0261	0.0033	0.0003	0.0000	0.0000	0.0000	0.0000	0.0000
	2	0.7892	0.3518	0.1971	0.0994	0.0183	0.0021	0.0001	0.0000	0.0000	0.0000	0.0000
	3	0.9316	0.5981	0.4050	0.2459	0.0651	0.0106	0.0009	0.0000	0.0000	0.0000	0.0000
	4	0.9830	0.7982	0.6302	0.4499	0.1666	0.0384	0.0049	0.0003	0.0000	0.0000	0.0000
	5	0.9967	0.9183	0.8103	0.6598	0.3288	0.1051	0.0191	0.0016	0.0003	0.0000	0.0000
	6	0.9995	0.9733	0.9204	0.8247	0.5272	0.2272	0.0583	0.0071	0.0016	0.0002	0.0000
	7	0.9999	0.9930	0.9729	0.9256	0.7161	0.4018	0.1423	0.0257	0.0075	0.0015	0.0000
	8	1.0000	0.9985	0.9925	0.9743	0.8577	0.5982	0.2839	0.0744	0.0271	0.0070	0.0001
	9	1.0000	0.9998	0.9984	0.9929	0.9417	0.7728	0.4728	0.1753	0.0796	0.0267	0.0005
	10	1.0000	1.0000	0.9997	0.9984	0.9809	0.8949	0.6712	0.3402	0.1897	0.0817	0.0033
	11	1.0000	1.0000	1.0000	0.9997	0.9951	0.9616	0.8334	0.5501	0.3698	0.2018	0.0170
	12	1.0000	1.0000	1.0000	1.0000	0.9991	0.9894	0.9349	0.7541	0.5950	0.4019	0.0684
	13	1.0000	1.0000	1.0000	1.0000	0.9999	0.9979	0.9817	0.9006	0.8029	0.6482	0.2108
	14	1.0000	1.0000	1.0000	1.0000	1.0000	0.9997	0.9967	0.9739	0.9365	0.8593	0.4853
	15	1.0000	1.0000	1.0000	1.0000	1.0000	1.0000	0.9997	0.9967	0.9900	0.9719	0.8147
	16	1.0000	1.0000	1.0000	1.0000	1.0000	1.0000	1.0000	1.0000	1.0000	1.0000	1.0000
17	0	0.1668	0.0225	0.0075	0.0023	0.0002	0.0000	0.0000	0.0000	0.0000	0.0000	0.0000
	1	0.4818	0.1182	0.0501	0.0193	0.0021	0.0001	0.0000	0.0000	0.0000	0.0000	0.0000
	2	0.7618	0.3096	0.1637	0.0774	0.0123	0.0012	0.0001	0.0000	0.0000	0.0000	0.0000
	3	0.9174	0.5489	0.3530	0.2019	0.0464	0.0064	0.0005	0.0000	0.0000	0.0000	0.0000
	4	0.9779	0.7582	0.5739	0.3887	0.1260	0.0245	0.0025	0.0001	0.0000	0.0000	0.0000
	5	0.9953	0.8943	0.7653	0.5968	0.2639	0.0717	0.0106	0.0007	0.0001	0.0000	0.0000
	6	0.9992	0.9623	0.8929	0.7752	0.4478	0.1662	0.0348	0.0032	0.0006	0.0001	0.0000
	7	0.9999	0.9891	0.9598	0.8954	0.6405	0.3145	0.0919	0.0127	0.0031	0.0005	0.0000
	8	1.0000	0.9974	0.9876	0.9597	0.8011	0.5000	0.1989	0.0403	0.0124	0.0026	0.0000
	9	1.0000	0.9995	0.9969	0.9873	0.9081	0.6855	0.3595	0.1046	0.0402	0.0109	0.0001
	10	1.0000	0.9999	0.9994	0.9968	0.9652	0.8338	0.5522	0.2248	0.1071	0.0377	0.0008
	11	1.0000	1.0000	0.9999	0.9993	0.9894	0.9283	0.7361	0.4032	0.2347	0.1057	0.0047
	12	1.0000	1.0000	1.0000	0.9999	0.9975	0.9755	0.8740	0.6113	0.4261	0.2418	0.0221
	13	1.0000	1.0000	1.0000	1.0000	0.9995	0.9936	0.9536	0.7981	0.6470	0.4511	0.0826
	14	1.0000	1.0000	1.0000	1.0000	0.9999	0.9988	0.9877	0.9226	0.8363	0.6904	0.2382
	15	1.0000	1.0000	1.0000	1.0000	1.0000	0.9999	0.9979	0.9807	0.9499	0.8818	0.5182
	16	1.0000	1.0000	1.0000	1.0000	1.0000	1.0000	0.9998	0.9977	0.9925	0.9775	0.8332
	17	1.0000	1.0000	1.0000	1.0000	1.0000	1.0000	1.0000	1.0000	1.0000	1.0000	1.0000

							p					
n	d	0.1	0.2	0.25	0.3	0.4	0.5	0.6	0.7	0.75	0.8	0.9
18	0	0.1501	0.0180	0.0056	0.0016	0.0001	0.0000	0.0000	0.0000	0.0000	0.0000	0.0000
	1	0.4503	0.0991	0.0395	0.0142	0.0013	0.0001	0.0000	0.0000	0.0000	0.0000	0.0000
	2	0.7338	0.2713	0.1353	0.0600	0.0082	0.0007	0.0000	0.0000	0.0000	0.0000	0.0000
	3	0.9018	0.5010	0.3057	0.1646	0.0328	0.0038	0.0002	0.0000	0.0000	0.0000	0.0000
	4	0.9718	0.7164	0.5187	0.3327	0.0942	0.0154	0.0013	0.0000	0.0000	0.0000	0.0000
	5	0.9936	0.8671	0.7175	0.5344	0.2088	0.0481	0.0058	0.0003	0.0000	0.0000	0.0000
	6	0.9988	0.9487	0.8610	0.7217	0.3743	0.1189	0.0203	0.0014	0.0002	0.0000	0.0000
	7	0.9998	0.9837	0.9431	0.8593	0.5634	0.2403	0.0576	0.0061	0.0012	0.0002	0.0000
	8	1.0000	0.9957	0.9807	0.9404	0.7368	0.4073	0.1347	0.0210	0.0054	0.0009	0.0000
	9	1.0000	0.9991	0.9946	0.9790	0.8653	0.5927	0.2632	0.0596	0.0193	0.0043	0.0000
	10	1.0000	0.9998	0.9988	0.9939	0.9424	0.7597	0.4366	0.1407	0.0569	0.0163	0.0002
	11	1.0000	1.0000	0.9998	0.9986	0.9797	0.8811	0.6257	0.2783	0.1390	0.0513	0.0012
	12	1.0000	1.0000	1.0000	0.9997	0.9942	0.9519	0.7912	0.4656	0.2825	0.1329	0.0064
	13	1.0000	1.0000	1.0000	1.0000	0.9987	0.9846	0.9058	0.6673	0.4813	0.2836	0.0282
	14	1.0000	1.0000	1.0000	1.0000	0.9998	0.9962	0.9672	0.8354	0.6943	0.4990	0.0982
	15	1.0000	1.0000	1.0000	1.0000	1.0000	0.9993	0.9918	0.9400	0.8647	0.7287	0.2662
	16	1.0000	1.0000	1.0000	1.0000	1.0000	0.9999	0.9987	0.9858	0.9605	0.9009	0.5497
	17	1.0000	1.0000	1.0000	1.0000	1.0000	1.0000	0.9999	0.9984	0.9944	0.9820	0.8499
	18	1.0000	1.0000	1.0000	1.0000	1.0000	1.0000	1.0000	1.0000	1.0000	1.0000	1.0000
19	0	0.1351	0.0144	0.0042	0.0011	0.0001	0.0000	0.0000	0.0000	0.0000	0.0000	0.0000
	1	0.4203	0.0829	0.0310	0.0104	0.0008	0.0000	0.0000	0.0000	0.0000	0.0000	0.0000
	2	0.7054	0.2369	0.1113	0.0462	0.0055	0.0004	0.0000	0.0000	0.0000	0.0000	0.0000
	3	0.8850	0.4551	0.2631	0.1332	0.0230	0.0022	0.0001	0.0000	0.0000	0.0000	0.0000
	4	0.9648	0.6733	0.4654	0.2822	0.0696	0.0096	0.0006	0.0000	0.0000	0.0000	0.0000
	5	0.9914	0.8369	0.6678	0.4739	0.1629	0.0318	0.0031	0.0001	0.0000	0.0000	0.0000
	6	0.9983	0.9324	0.8251	0.6655	0.3081	0.0835	0.0116	0.0006	0.0001	0.0000	0.0000
	7	0.9997	0.9767	0.9225	0.8180	0.4878	0.1796	0.0352	0.0028	0.0005	0.0000	0.0000
	8	1.0000	0.9933	0.9713	0.9161	0.6675	0.3238	0.0885	0.0105	0.0023	0.0003	0.0000
	9	1.0000	0.9984	0.9911	0.9674	0.8139	0.5000	0.1861	0.0326	0.0089	0.0016	0.0000
	10	1.0000	0.9997	0.9977	0.9895	0.9115	0.6762	0.3325	0.0839	0.0287	0.0067	0.0000
	11	1.0000	1.0000	0.9995	0.9972	0.9648	0.8204	0.5122	0.1820	0.0775	0.0233	0.0003
	12	1.0000	1.0000	0.9999	0.9994	0.9884	0.9165	0.6919	0.3345	0.1749	0.0676	0.0017
	13	1.0000	1.0000	1.0000	0.9999	0.9969	0.9682	0.8371	0.5261	0.3322	0.1631	0.0086
	14	1.0000	1.0000	1.0000	1.0000	0.9994	0.9904	0.9304	0.7178	0.5346	0.3267	0.0352
	15	1.0000	1.0000	1.0000	1.0000	0.9999	0.9978	0.9770	0.8668	0.7369	0.5449	0.1150
	16	1.0000	1.0000	1.0000	1.0000	1.0000	0.9996	0.9945	0.9538	0.8887	0.7631	0.2946
	17	1.0000	1.0000	1.0000	1.0000	1.0000	1.0000	0.9992	0.9896	0.9690	0.9171	0.5797
	18	1.0000	1.0000	1.0000	1.0000	1.0000	1.0000	0.9999	0.9989	0.9958	0.9856	0.8649
	19	1.0000	1.0000	1.0000	1.0000	1.0000	1.0000	1.0000	1.0000	1.0000	1.0000	1.0000

							p					
n	d	0.1	0.2	0.25	0.3	0.4	0.5	0.6	0.7	0.75	0.8	0.9
20	0	0.1216	0.0115	0.0032	0.0008	0.0000	0.0000	0.0000	0.0000	0.0000	0.0000	0.0000
	1	0.3917	0.0692	0.0243	0.0076	0.0005	0.0000	0.0000	0.0000	0.0000	0.0000	0.0000
	2	0.6769	0.2061	0.0913	0.0355	0.0036	0.0002	0.0000	0.0000	0.0000	0.0000	0.0000
	3	0.8670	0.4114	0.2252	0.1071	0.0160	0.0013	0.0000	0.0000	0.0000	0.0000	0.0000
	4	0.9568	0.6296	0.4148	0.2375	0.0510	0.0059	0.0003	0.0000	0.0000	0.0000	0.0000
	5	0.9887	0.8042	0.6172	0.4164	0.1256	0.0207	0.0016	0.0000	0.0000	0.0000	0.0000
	6	0.9976	0.9133	0.7858	0.6080	0.2500	0.0577	0.0065	0.0003	0.0000	0.0000	0.0000
	7	0.9996	0.9679	0.8982	0.7723	0.4159	0.1316	0.0210	0.0013	0.0002	0.0000	0.0000
	8	0.9999	0.9900	0.9591	0.8867	0.5956	0.2517	0.0565	0.0051	0.0009	0.0001	0.0000
	9	1.0000	0.9974	0.9861	0.9520	0.7553	0.4119	0.1275	0.0171	0.0039	0.0006	0.0000
	10	1.0000	0.9994	0.9961	0.9829	0.8725	0.5881	0.2447	0.0480	0.0139	0.0026	0.0000
	11	1.0000	0.9999	0.9991	0.9949	0.9435	0.7483	0.4044	0.1133	0.0409	0.0100	0.0001
	12	1.0000	1.0000	0.9998	0.9987	0.9790	0.8684	0.5841	0.2277	0.1018	0.0321	0.0004
	13	1.0000	1.0000	1.0000	0.9997	0.9935	0.9423	0.7500	0.3920	0.2142	0.0867	0.0024
	14	1.0000	1.0000	1.0000	1.0000	0.9984	0.9793	0.8744	0.5836	0.3828	0.1958	0.0113
	15	1.0000	1.0000	1.0000	1.0000	0.9997	0.9941	0.9490	0.7625	0.5852	0.3704	0.0432
	16	1.0000	1.0000	1.0000	1.0000	1.0000	0.9987	0.9840	0.8929	0.7748	0.5886	0.1330
	17	1.0000	1.0000	1.0000	1.0000	1.0000	0.9998	0.9964	0.9645	0.9087	0.7939	0.3231
	18	1.0000	1.0000	1.0000	1.0000	1.0000	1.0000	0.9995	0.9924	0.9757	0.9308	0.6083
	19	1.0000	1.0000	1.0000	1.0000	1.0000	1.0000	1.0000	0.9992	0.9968	0.9885	0.8784
	20	1.0000	1.0000	1.0000	1.0000	1.0000	1.0000	1.0000	1.0000	1.0000	1.0000	1.0000

C.2   Cumulative Poisson Distribution

Table C.2 is a tabulation of the cumulative distribution function (F) for Poisson distributions with means equal to 0.5 and the integers 1 through 15 where

$$F(d) = P(X \le d) = \sum_{x=0}^{d} \frac{e^{-\mu}\mu^x}{x!}.$$

One can find various pdfs and CDFs directly from this table for Poisson distributions with these means. For example, if $\mu = 4$ the probability of 3 or fewer events $P(X \le 3) = F(3) = 0.4335$ is found as the value at the intersection of the "3" row and "4.0" column. The probability of exactly 3 events $P(X = 3) = f(3) = F(3) - F(2) = 0.4335 - 0.2381 = 0.1954$ is found by subtracting consecutive entries in the "4.0" column. For further examples, see Chapter 3.

					μ					
d	0.5	1.0	2.0	3.0	4.0	5.0	6.0	7.0	8.0	9.0
0	0.6065	0.3679	0.1353	0.0498	0.0183	0.0067	0.0025	0.0009	0.0003	0.0001
1	0.9098	0.7358	0.4060	0.1991	0.0916	0.0404	0.0174	0.0073	0.0030	0.0012
2	0.9856	0.9197	0.6767	0.4232	0.2381	0.1247	0.0620	0.0296	0.0138	0.0062
3	0.9982	0.9810	0.8571	0.6472	0.4335	0.2650	0.1512	0.0818	0.0424	0.0212
4	0.9998	0.9963	0.9473	0.8153	0.6288	0.4405	0.2851	0.1730	0.0996	0.0550
5	1.0000	0.9994	0.9834	0.9161	0.7851	0.6160	0.4457	0.3007	0.1912	0.1157
6	1.0000	0.9999	0.9955	0.9665	0.8893	0.7622	0.6063	0.4497	0.3134	0.2068
7	1.0000	1.0000	0.9989	0.9881	0.9489	0.8666	0.7440	0.5987	0.4530	0.3239
8	1.0000	1.0000	0.9998	0.9962	0.9786	0.9319	0.8472	0.7291	0.5925	0.4557
9	1.0000	1.0000	1.0000	0.9989	0.9919	0.9682	0.9161	0.8305	0.7166	0.5874
10	1.0000	1.0000	1.0000	0.9997	0.9972	0.9863	0.9574	0.9015	0.8159	0.7060
11	1.0000	1.0000	1.0000	0.9999	0.9991	0.9945	0.9799	0.9467	0.8881	0.8030
12	1.0000	1.0000	1.0000	1.0000	0.9997	0.9980	0.9912	0.9730	0.9362	0.8758
13	1.0000	1.0000	1.0000	1.0000	0.9999	0.9993	0.9964	0.9872	0.9658	0.9261
14	1.0000	1.0000	1.0000	1.0000	1.0000	0.9998	0.9986	0.9943	0.9827	0.9585
15	1.0000	1.0000	1.0000	1.0000	1.0000	0.9999	0.9995	0.9976	0.9918	0.9780
16	1.0000	1.0000	1.0000	1.0000	1.0000	1.0000	0.9998	0.9990	0.9963	0.9889
17	1.0000	1.0000	1.0000	1.0000	1.0000	1.0000	0.9999	0.9996	0.9984	0.9947
18	1.0000	1.0000	1.0000	1.0000	1.0000	1.0000	1.0000	0.9999	0.9993	0.9976
19	1.0000	1.0000	1.0000	1.0000	1.0000	1.0000	1.0000	1.0000	0.9997	0.9989
20	1.0000	1.0000	1.0000	1.0000	1.0000	1.0000	1.0000	1.0000	0.9999	0.9996

	μ					
t	10	11	12	13	14	15
0	0.0000	0.0000	0.0000	0.0000	0.0000	0.0000
1	0.0005	0.0002	0.0001	0.0000	0.0000	0.0000
2	0.0028	0.0012	0.0005	0.0002	0.0001	0.0000
3	0.0103	0.0049	0.0023	0.0011	0.0005	0.0002
4	0.0293	0.0151	0.0076	0.0037	0.0018	0.0009
5	0.0671	0.0375	0.0203	0.0107	0.0055	0.0028
6	0.1301	0.0786	0.0458	0.0259	0.0142	0.0076
7	0.2202	0.1432	0.0895	0.0540	0.0316	0.0180
8	0.3328	0.2320	0.1550	0.0998	0.0621	0.0374
9	0.4579	0.3405	0.2424	0.1658	0.1094	0.0699
10	0.5830	0.4599	0.3472	0.2517	0.1757	0.1185
11	0.6968	0.5793	0.4616	0.3532	0.2600	0.1848
12	0.7916	0.6887	0.5760	0.4631	0.3585	0.2676
13	0.8645	0.7813	0.6815	0.5730	0.4644	0.3632
14	0.9165	0.8540	0.7720	0.6751	0.5704	0.4657
15	0.9513	0.9074	0.8444	0.7636	0.6694	0.5681
16	0.9730	0.9441	0.8987	0.8355	0.7559	0.6641
17	0.9857	0.9678	0.9370	0.8905	0.8272	0.7489
18	0.9928	0.9823	0.9626	0.9302	0.8826	0.8195
19	0.9965	0.9907	0.9787	0.9573	0.9235	0.8752
20	0.9984	0.9953	0.9884	0.9750	0.9521	0.9170
21	0.9993	0.9977	0.9939	0.9859	0.9712	0.9469
22	0.9997	0.9990	0.9970	0.9924	0.9833	0.9673
23	0.9999	0.9995	0.9985	0.9960	0.9907	0.9805
24	1.0000	0.9998	0.9993	0.9980	0.9950	0.9888
25	1.0000	0.9999	0.9997	0.9990	0.9974	0.9938
26	1.0000	1.0000	0.9999	0.9995	0.9987	0.9967
27	1.0000	1.0000	0.9999	0.9998	0.9994	0.9983
28	1.0000	1.0000	1.0000	0.9999	0.9997	0.9991
29	1.0000	1.0000	1.0000	1.0000	0.9999	0.9996

C.3   Cumulative Standard Normal Distribution

Table C.3 is a tabulation of the cumulative standard normal distribution (Z) for values from -3.99 to 3.99. This table can be used to calculate probabilities for various CDFs of normal distributions that have been standardized by the $\frac{X-\mu}{\sigma}$ transformation. While normal distributions can have various values for the mean and standard deviation, the Z transformation has a mean of $\mu = 0$ and a standard deviation of $\sigma = 1$. To use this table, transform the normal variate X to a Z value. For example, suppose X is distributed as normal with $\mu = 100$ and $\sigma = 6$, then

$$P(X \leq 93) = P\left(Z = \frac{93 - 100}{6}\right) = P(Z \leq -1.17).$$

Next find the value at the intersection of the "-1.1" row and the "0.07" column, 0.1210. This value is the probability of X less than or equal to 93 in the original distribution. For more examples, see Chapter 3.

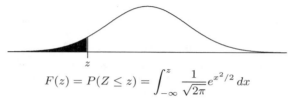

$$F(z) = P(Z \leq z) = \int_{-\infty}^{z} \frac{1}{\sqrt{2\pi}} e^{x^2/2} \, dx$$

z	0.00	0.01	0.02	0.03	0.04	0.05	0.06	0.07	0.08	0.09
−3.9	0.0000	0.0000	0.0000	0.0000	0.0000	0.0000	0.0000	0.0000	0.0000	0.0000
−3.8	0.0001	0.0001	0.0001	0.0001	0.0001	0.0001	0.0001	0.0001	0.0001	0.0001
−3.7	0.0001	0.0001	0.0001	0.0001	0.0001	0.0001	0.0001	0.0001	0.0001	0.0001
−3.6	0.0002	0.0002	0.0001	0.0001	0.0001	0.0001	0.0001	0.0001	0.0001	0.0001
−3.5	0.0002	0.0002	0.0002	0.0002	0.0002	0.0002	0.0002	0.0002	0.0002	0.0002
−3.4	0.0003	0.0003	0.0003	0.0003	0.0003	0.0003	0.0003	0.0003	0.0003	0.0002
−3.3	0.0005	0.0005	0.0005	0.0004	0.0004	0.0004	0.0004	0.0004	0.0004	0.0003
−3.2	0.0007	0.0007	0.0006	0.0006	0.0006	0.0006	0.0006	0.0005	0.0005	0.0005
−3.1	0.0010	0.0009	0.0009	0.0009	0.0008	0.0008	0.0008	0.0008	0.0007	0.0007
−3.0	0.0013	0.0013	0.0013	0.0012	0.0012	0.0011	0.0011	0.0011	0.0010	0.0010
−2.9	0.0019	0.0018	0.0018	0.0017	0.0016	0.0016	0.0015	0.0015	0.0014	0.0014
−2.8	0.0026	0.0025	0.0024	0.0023	0.0023	0.0022	0.0021	0.0021	0.0020	0.0019
−2.7	0.0035	0.0034	0.0033	0.0032	0.0031	0.0030	0.0029	0.0028	0.0027	0.0026
−2.6	0.0047	0.0045	0.0044	0.0043	0.0041	0.0040	0.0039	0.0038	0.0037	0.0036
−2.5	0.0062	0.0060	0.0059	0.0057	0.0055	0.0054	0.0052	0.0051	0.0049	0.0048
−2.4	0.0082	0.0080	0.0078	0.0075	0.0073	0.0071	0.0069	0.0068	0.0066	0.0064
−2.3	0.0107	0.0104	0.0102	0.0099	0.0096	0.0094	0.0091	0.0089	0.0087	0.0084
−2.2	0.0139	0.0136	0.0132	0.0129	0.0125	0.0122	0.0119	0.0116	0.0113	0.0110
−2.1	0.0179	0.0174	0.0170	0.0166	0.0162	0.0158	0.0154	0.0150	0.0146	0.0143
−2.0	0.0228	0.0222	0.0217	0.0212	0.0207	0.0202	0.0197	0.0192	0.0188	0.0183
−1.9	0.0287	0.0281	0.0274	0.0268	0.0262	0.0256	0.0250	0.0244	0.0239	0.0233
−1.8	0.0359	0.0351	0.0344	0.0336	0.0329	0.0322	0.0314	0.0307	0.0301	0.0294
−1.7	0.0446	0.0436	0.0427	0.0418	0.0409	0.0401	0.0392	0.0384	0.0375	0.0367
−1.6	0.0548	0.0537	0.0526	0.0516	0.0505	0.0495	0.0485	0.0475	0.0465	0.0455
−1.5	0.0668	0.0655	0.0643	0.0630	0.0618	0.0606	0.0594	0.0582	0.0571	0.0559
−1.4	0.0808	0.0793	0.0778	0.0764	0.0749	0.0735	0.0721	0.0708	0.0694	0.0681
−1.3	0.0968	0.0951	0.0934	0.0918	0.0901	0.0885	0.0869	0.0853	0.0838	0.0823
−1.2	0.1151	0.1131	0.1112	0.1093	0.1075	0.1056	0.1038	0.1020	0.1003	0.0985
−1.1	0.1357	0.1335	0.1314	0.1292	0.1271	0.1251	0.1230	0.1210	0.1190	0.1170
−1.0	0.1587	0.1562	0.1539	0.1515	0.1492	0.1469	0.1446	0.1423	0.1401	0.1379

z	0.00	0.01	0.02	0.03	0.04	0.05	0.06	0.07	0.08	0.09
−0.9	0.1841	0.1814	0.1788	0.1762	0.1736	0.1711	0.1685	0.1660	0.1635	0.1611
−0.8	0.2119	0.2090	0.2061	0.2033	0.2005	0.1977	0.1949	0.1922	0.1894	0.1867
−0.7	0.2420	0.2389	0.2358	0.2327	0.2296	0.2266	0.2236	0.2206	0.2177	0.2148
−0.6	0.2743	0.2709	0.2676	0.2643	0.2611	0.2578	0.2546	0.2514	0.2483	0.2451
−0.5	0.3085	0.3050	0.3015	0.2981	0.2946	0.2912	0.2877	0.2843	0.2810	0.2776
−0.4	0.3446	0.3409	0.3372	0.3336	0.3300	0.3264	0.3228	0.3192	0.3156	0.3121
−0.3	0.3821	0.3783	0.3745	0.3707	0.3669	0.3632	0.3594	0.3557	0.3520	0.3483
−0.2	0.4207	0.4168	0.4129	0.4090	0.4052	0.4013	0.3974	0.3936	0.3897	0.3859
−0.1	0.4602	0.4562	0.4522	0.4483	0.4443	0.4404	0.4364	0.4325	0.4286	0.4247
−0.0	0.5000	0.4960	0.4920	0.4880	0.4840	0.4801	0.4761	0.4721	0.4681	0.4641
0.0	0.5000	0.5040	0.5080	0.5120	0.5160	0.5199	0.5239	0.5279	0.5319	0.5359
0.1	0.5398	0.5438	0.5478	0.5517	0.5557	0.5596	0.5636	0.5675	0.5714	0.5753
0.2	0.5793	0.5832	0.5871	0.5910	0.5948	0.5987	0.6026	0.6064	0.6103	0.6141
0.3	0.6179	0.6217	0.6255	0.6293	0.6331	0.6368	0.6406	0.6443	0.6480	0.6517
0.4	0.6554	0.6591	0.6628	0.6664	0.6700	0.6736	0.6772	0.6808	0.6844	0.6879
0.5	0.6915	0.6950	0.6985	0.7019	0.7054	0.7088	0.7123	0.7157	0.7190	0.7224
0.6	0.7257	0.7291	0.7324	0.7357	0.7389	0.7422	0.7454	0.7486	0.7517	0.7549
0.7	0.7580	0.7611	0.7642	0.7673	0.7704	0.7734	0.7764	0.7794	0.7823	0.7852
0.8	0.7881	0.7910	0.7939	0.7967	0.7995	0.8023	0.8051	0.8078	0.8106	0.8133
0.9	0.8159	0.8186	0.8212	0.8238	0.8264	0.8289	0.8315	0.8340	0.8365	0.8389
1.0	0.8413	0.8438	0.8461	0.8485	0.8508	0.8531	0.8554	0.8577	0.8599	0.8621
1.1	0.8643	0.8665	0.8686	0.8708	0.8729	0.8749	0.8770	0.8790	0.8810	0.8830
1.2	0.8849	0.8869	0.8888	0.8907	0.8925	0.8944	0.8962	0.8980	0.8997	0.9015
1.3	0.9032	0.9049	0.9066	0.9082	0.9099	0.9115	0.9131	0.9147	0.9162	0.9177
1.4	0.9192	0.9207	0.9222	0.9236	0.9251	0.9265	0.9279	0.9292	0.9306	0.9319
1.5	0.9332	0.9345	0.9357	0.9370	0.9382	0.9394	0.9406	0.9418	0.9429	0.9441
1.6	0.9452	0.9463	0.9474	0.9484	0.9495	0.9505	0.9515	0.9525	0.9535	0.9545
1.7	0.9554	0.9564	0.9573	0.9582	0.9591	0.9599	0.9608	0.9616	0.9625	0.9633
1.8	0.9641	0.9649	0.9656	0.9664	0.9671	0.9678	0.9686	0.9693	0.9699	0.9706
1.9	0.9713	0.9719	0.9726	0.9732	0.9738	0.9744	0.9750	0.9756	0.9761	0.9767
2.0	0.9772	0.9778	0.9783	0.9788	0.9793	0.9798	0.9803	0.9808	0.9812	0.9817
2.1	0.9821	0.9826	0.9830	0.9834	0.9838	0.9842	0.9846	0.9850	0.9854	0.9857
2.2	0.9861	0.9864	0.9868	0.9871	0.9875	0.9878	0.9881	0.9884	0.9887	0.9890
2.3	0.9893	0.9896	0.9898	0.9901	0.9904	0.9906	0.9909	0.9911	0.9913	0.9916
2.4	0.9918	0.9920	0.9922	0.9925	0.9927	0.9929	0.9931	0.9932	0.9934	0.9936
2.5	0.9938	0.9940	0.9941	0.9943	0.9945	0.9946	0.9948	0.9949	0.9951	0.9952
2.6	0.9953	0.9955	0.9956	0.9957	0.9959	0.9960	0.9961	0.9962	0.9963	0.9964
2.7	0.9965	0.9966	0.9967	0.9968	0.9969	0.9970	0.9971	0.9972	0.9973	0.9974
2.8	0.9974	0.9975	0.9976	0.9977	0.9977	0.9978	0.9979	0.9979	0.9980	0.9981
2.9	0.9981	0.9982	0.9982	0.9983	0.9984	0.9984	0.9985	0.9985	0.9986	0.9986
3.0	0.9987	0.9987	0.9987	0.9988	0.9988	0.9989	0.9989	0.9989	0.9990	0.9990
3.1	0.9990	0.9991	0.9991	0.9991	0.9992	0.9992	0.9992	0.9992	0.9993	0.9993
3.2	0.9993	0.9993	0.9994	0.9994	0.9994	0.9994	0.9994	0.9995	0.9995	0.9995
3.3	0.9995	0.9995	0.9995	0.9996	0.9996	0.9996	0.9996	0.9996	0.9996	0.9997
3.4	0.9997	0.9997	0.9997	0.9997	0.9997	0.9997	0.9997	0.9997	0.9997	0.9998
3.5	0.9998	0.9998	0.9998	0.9998	0.9998	0.9998	0.9998	0.9998	0.9998	0.9998
3.6	0.9998	0.9998	0.9999	0.9999	0.9999	0.9999	0.9999	0.9999	0.9999	0.9999
3.7	0.9999	0.9999	0.9999	0.9999	0.9999	0.9999	0.9999	0.9999	0.9999	0.9999
3.8	0.9999	0.9999	0.9999	0.9999	0.9999	0.9999	0.9999	0.9999	0.9999	0.9999
3.9	1.0000	1.0000	1.0000	1.0000	1.0000	1.0000	1.0000	1.0000	1.0000	1.0000

C.4 Student's t Distribution

Unlike Table C.3 that collapsed all normal distributions into a single standard normal distribution, Table C.4 represents a summary of partial CDFs for a number of t distributions. Each row of this table summarizes a different t distribution as indicated by its degrees of freedom $(n-1)$. The body of the table presents t_0 values that represent cutoff points for various probabilities ranging from 0.25 to 0.0005 for one tail and 0.50 to 0.001 for two tails and that are tabulated as various columns. For example, if $n = 20$, then df $= 19$ and for $P(-t_0 \leq t \leq t_0) = 0.95$, the table value t_0 would be 2.093. This value is found as the intersection of the row labeled "19" and the column headed "0.05" (2-tail). This means that -2.093 and 2.093 cut off 0.025 in each tail of the distribution. For $P(t \leq t_0) = 0.10$, the table value t_0 would be -1.328. Left tail values are negative. For $P(t \geq t_0) = 0.10$, the table value t_0 would be $+1.328$. For $P(t \leq t_0) = 0.90$, the table value t_0 would again be $+1.328$. For more examples see Chapters 4 and 6. Notice that the limit of the t distribution family (where df $= \infty$) is the standard normal distribution (Z). The values in the last row of Table C.4 can also be found using Table C.3.

| 1-tail | 0.25 | 0.10 | 0.05 | 0.025 | 0.01 | 0.005 | 0.0025 | 0.001 | 0.0005 | 1-tail |
2-tail	0.50	0.20	0.10	0.05	0.02	0.010	0.005	0.002	0.001	2-tail
df: 1	1.000	3.078	6.314	12.71	31.82	63.66	127.3	318.3	636.6	df: 1
2	0.816	1.886	2.920	4.303	6.965	9.925	14.09	22.33	31.60	2
3	0.765	1.638	2.353	3.182	4.541	5.841	7.453	10.21	12.92	3
4	0.741	1.533	2.132	2.776	3.747	4.604	5.598	7.173	8.610	4
5	0.727	1.476	2.015	2.571	3.365	4.032	4.773	5.894	6.869	5
6	0.718	1.440	1.943	2.447	3.143	3.707	4.317	5.208	5.959	6
7	0.711	1.415	1.895	2.365	2.998	3.499	4.029	4.785	5.408	7
8	0.706	1.397	1.860	2.306	2.896	3.355	3.833	4.501	5.041	8
9	0.703	1.383	1.833	2.262	2.821	3.250	3.690	4.297	4.781	9
10	0.700	1.372	1.812	2.228	2.764	3.169	3.581	4.144	4.587	10
11	0.697	1.363	1.796	2.201	2.718	3.106	3.497	4.025	4.437	11
12	0.695	1.356	1.782	2.179	2.681	3.055	3.428	3.930	4.318	12
13	0.694	1.350	1.771	2.160	2.650	3.012	3.372	3.852	4.221	13
14	0.692	1.345	1.761	2.145	2.624	2.977	3.326	3.787	4.140	14
15	0.691	1.341	1.753	2.131	2.602	2.947	3.286	3.733	4.073	15
16	0.690	1.337	1.746	2.120	2.583	2.921	3.252	3.686	4.015	16
17	0.689	1.333	1.740	2.110	2.567	2.898	3.222	3.646	3.965	17
18	0.688	1.330	1.734	2.101	2.552	2.878	3.197	3.610	3.922	18
19	0.688	1.328	1.729	2.093	2.539	2.861	3.174	3.579	3.883	19
20	0.687	1.325	1.725	2.086	2.528	2.845	3.153	3.552	3.850	20
21	0.686	1.323	1.721	2.080	2.518	2.831	3.135	3.527	3.819	21
22	0.686	1.321	1.717	2.074	2.508	2.819	3.119	3.505	3.792	22
23	0.685	1.319	1.714	2.069	2.500	2.807	3.104	3.485	3.768	23
24	0.685	1.318	1.711	2.064	2.492	2.797	3.091	3.467	3.745	24
25	0.684	1.316	1.708	2.060	2.485	2.787	3.078	3.450	3.725	25

1-tail 2-tail	0.25 0.50	0.10 0.20	0.05 0.10	0.025 0.05	0.01 0.02	0.005 0.010	0.0025 0.005	0.001 0.002	0.0005 0.001	1-tail 2-tail
df: 26	0.684	1.315	1.706	2.056	2.479	2.779	3.067	3.435	3.707	df: 26
27	0.684	1.314	1.703	2.052	2.473	2.771	3.057	3.421	3.689	27
28	0.683	1.313	1.701	2.048	2.467	2.763	3.047	3.408	3.674	28
29	0.683	1.311	1.699	2.045	2.462	2.756	3.038	3.396	3.660	29
30	0.683	1.310	1.697	2.042	2.457	2.750	3.030	3.385	3.646	30
31	0.682	1.309	1.696	2.040	2.453	2.744	3.022	3.375	3.633	31
32	0.682	1.309	1.694	2.037	2.449	2.738	3.015	3.365	3.622	32
33	0.682	1.308	1.692	2.035	2.445	2.733	3.008	3.356	3.611	33
34	0.682	1.307	1.691	2.032	2.441	2.728	3.002	3.348	3.601	34
35	0.682	1.306	1.690	2.030	2.438	2.724	2.996	3.340	3.591	35
36	0.681	1.306	1.688	2.028	2.434	2.719	2.990	3.333	3.582	36
37	0.681	1.305	1.687	2.026	2.431	2.715	2.985	3.326	3.574	37
38	0.681	1.304	1.686	2.024	2.429	2.712	2.980	3.319	3.566	38
39	0.681	1.304	1.685	2.023	2.426	2.708	2.976	3.313	3.558	39
40	0.681	1.303	1.684	2.021	2.423	2.704	2.971	3.307	3.551	40
41	0.681	1.303	1.683	2.020	2.421	2.701	2.967	3.301	3.544	41
42	0.680	1.302	1.682	2.018	2.418	2.698	2.963	3.296	3.538	42
43	0.680	1.302	1.681	2.017	2.416	2.695	2.959	3.291	3.532	43
44	0.680	1.301	1.680	2.015	2.414	2.692	2.956	3.286	3.526	44
45	0.680	1.301	1.679	2.014	2.412	2.690	2.952	3.281	3.520	45
46	0.680	1.300	1.679	2.013	2.410	2.687	2.949	3.277	3.515	46
47	0.680	1.300	1.678	2.012	2.408	2.685	2.946	3.273	3.510	47
48	0.680	1.299	1.677	2.011	2.407	2.682	2.943	3.269	3.505	48
49	0.680	1.299	1.677	2.010	2.405	2.680	2.940	3.265	3.500	49
50	0.679	1.299	1.676	2.009	2.403	2.678	2.937	3.261	3.496	50
51	0.679	1.298	1.675	2.008	2.402	2.676	2.934	3.258	3.492	51
52	0.679	1.298	1.675	2.007	2.400	2.674	2.932	3.255	3.488	52
53	0.679	1.298	1.674	2.006	2.399	2.672	2.929	3.251	3.484	53
54	0.679	1.297	1.674	2.005	2.397	2.670	2.927	3.248	3.480	54
55	0.679	1.297	1.673	2.004	2.396	2.668	2.925	3.245	3.476	55
56	0.679	1.297	1.673	2.003	2.395	2.667	2.923	3.242	3.473	56
57	0.679	1.297	1.672	2.002	2.394	2.665	2.920	3.239	3.469	57
58	0.679	1.296	1.672	2.002	2.392	2.663	2.918	3.237	3.466	58
59	0.679	1.296	1.671	2.001	2.391	2.662	2.916	3.234	3.463	59
60	0.679	1.296	1.671	2.000	2.390	2.660	2.915	3.232	3.460	60
61	0.679	1.296	1.670	2.000	2.389	2.659	2.913	3.229	3.457	61
62	0.678	1.295	1.670	1.999	2.388	2.657	2.911	3.227	3.454	62
63	0.678	1.295	1.669	1.998	2.387	2.656	2.909	3.225	3.452	63
64	0.678	1.295	1.669	1.998	2.386	2.655	2.908	3.223	3.449	64
65	0.678	1.295	1.669	1.997	2.385	2.654	2.906	3.220	3.447	65
66	0.678	1.295	1.668	1.997	2.384	2.652	2.904	3.218	3.444	66
67	0.678	1.294	1.668	1.996	2.383	2.651	2.903	3.216	3.442	67
68	0.678	1.294	1.668	1.995	2.382	2.650	2.902	3.214	3.439	68
69	0.678	1.294	1.667	1.995	2.382	2.649	2.900	3.213	3.437	69
70	0.678	1.294	1.667	1.994	2.381	2.648	2.899	3.211	3.435	70
71	0.678	1.294	1.667	1.994	2.380	2.647	2.897	3.209	3.433	71
72	0.678	1.293	1.666	1.993	2.379	2.646	2.896	3.207	3.431	72
73	0.678	1.293	1.666	1.993	2.379	2.645	2.895	3.206	3.429	73
74	0.678	1.293	1.666	1.993	2.378	2.644	2.894	3.204	3.427	74
75	0.678	1.293	1.665	1.992	2.377	2.643	2.892	3.202	3.425	75

1-tail 2-tail	0.25 0.50	0.10 0.20	0.05 0.10	0.025 0.05	0.01 0.02	0.005 0.010	0.0025 0.005	0.001 0.002	0.0005 0.001	1-tail 2-tail
df: 76	0.678	1.293	1.665	1.992	2.376	2.642	2.891	3.201	3.423	df: 76
77	0.678	1.293	1.665	1.991	2.376	2.641	2.890	3.199	3.421	77
78	0.678	1.292	1.665	1.991	2.375	2.640	2.889	3.198	3.420	78
79	0.678	1.292	1.664	1.990	2.374	2.639	2.888	3.197	3.418	79
80	0.678	1.292	1.664	1.990	2.374	2.639	2.887	3.195	3.416	80
81	0.678	1.292	1.664	1.990	2.373	2.638	2.886	3.194	3.415	81
82	0.677	1.292	1.664	1.989	2.373	2.637	2.885	3.193	3.413	82
83	0.677	1.292	1.663	1.989	2.372	2.636	2.884	3.191	3.412	83
84	0.677	1.292	1.663	1.989	2.372	2.636	2.883	3.190	3.410	84
85	0.677	1.292	1.663	1.988	2.371	2.635	2.882	3.189	3.409	85
86	0.677	1.291	1.663	1.988	2.370	2.634	2.881	3.188	3.407	86
87	0.677	1.291	1.663	1.988	2.370	2.634	2.880	3.187	3.406	87
88	0.677	1.291	1.662	1.987	2.369	2.633	2.880	3.185	3.405	88
89	0.677	1.291	1.662	1.987	2.369	2.632	2.879	3.184	3.403	89
90	0.677	1.291	1.662	1.987	2.368	2.632	2.878	3.183	3.402	90
91	0.677	1.291	1.662	1.986	2.368	2.631	2.877	3.182	3.401	91
92	0.677	1.291	1.662	1.986	2.368	2.630	2.876	3.181	3.399	92
93	0.677	1.291	1.661	1.986	2.367	2.630	2.876	3.180	3.398	93
94	0.677	1.291	1.661	1.986	2.367	2.629	2.875	3.179	3.397	94
95	0.677	1.291	1.661	1.985	2.366	2.629	2.874	3.178	3.396	95
96	0.677	1.290	1.661	1.985	2.366	2.628	2.873	3.177	3.395	96
97	0.677	1.290	1.661	1.985	2.365	2.627	2.873	3.176	3.394	97
98	0.677	1.290	1.661	1.984	2.365	2.627	2.872	3.176	3.393	98
99	0.677	1.290	1.660	1.984	2.365	2.626	2.871	3.175	3.391	99
100	0.677	1.290	1.660	1.984	2.364	2.626	2.871	3.174	3.390	100
110	0.677	1.289	1.659	1.982	2.361	2.621	2.865	3.166	3.381	110
120	0.677	1.289	1.658	1.980	2.358	2.617	2.860	3.160	3.373	120
130	0.676	1.288	1.657	1.978	2.355	2.614	2.856	3.154	3.367	130
140	0.676	1.288	1.656	1.977	2.353	2.611	2.852	3.149	3.361	140
150	0.676	1.287	1.655	1.976	2.351	2.609	2.849	3.145	3.357	150
160	0.676	1.287	1.654	1.975	2.350	2.607	2.847	3.142	3.352	160
170	0.676	1.287	1.654	1.974	2.348	2.605	2.844	3.139	3.349	170
180	0.676	1.286	1.653	1.973	2.347	2.603	2.842	3.136	3.345	180
190	0.676	1.286	1.653	1.973	2.346	2.602	2.840	3.134	3.342	190
200	0.676	1.286	1.653	1.972	2.345	2.601	2.838	3.131	3.340	200
250	0.675	1.285	1.651	1.969	2.341	2.596	2.832	3.123	3.330	250
300	0.675	1.284	1.650	1.968	2.339	2.592	2.828	3.118	3.323	300
350	0.675	1.284	1.649	1.967	2.337	2.590	2.825	3.114	3.319	350
400	0.675	1.284	1.649	1.966	2.336	2.588	2.823	3.111	3.315	400
450	0.675	1.283	1.648	1.965	2.335	2.587	2.821	3.108	3.312	450
500	0.675	1.283	1.648	1.965	2.334	2.586	2.820	3.107	3.310	500
550	0.675	1.283	1.648	1.964	2.333	2.585	2.818	3.105	3.308	550
600	0.675	1.283	1.647	1.964	2.333	2.584	2.817	3.104	3.307	600
650	0.675	1.283	1.647	1.964	2.332	2.583	2.817	3.103	3.306	650
700	0.675	1.283	1.647	1.963	2.332	2.583	2.816	3.102	3.304	700
750	0.675	1.283	1.647	1.963	2.331	2.582	2.815	3.101	3.304	750
800	0.675	1.283	1.647	1.963	2.331	2.582	2.815	3.100	3.303	800
900	0.675	1.282	1.647	1.963	2.330	2.581	2.814	3.099	3.301	900
1000	0.675	1.282	1.646	1.962	2.330	2.581	2.813	3.098	3.300	1000
∞	0.674	1.282	1.645	1.960	2.326	2.576	2.807	3.090	3.290	∞

C.5 Cumulative Chi-Square Distribution

Table C.5 summarizes partial CDFs for χ^2 distributions with degrees of freedom from 1 to 60, where

$$F(\chi^2) = P(X^2 \le \chi^2).$$

Each row represents a different χ^2 distribution and each column a different CDF from 0.005 to 0.995. The body of the table presents χ^2 values associated with these various probabilities. For example, if $n = 20$, then df $= 19$ and $\chi^2_{0.025} = 8.91$ is found as the intersection of row "19" and column "0.025." For a χ^2 distribution with 19 degrees of freedom $P(\chi^2 \le 8.91) = 0.025$. $\chi^2_{0.90} = 27.2$ found as the intersection of row "19" and column "0.90" implies $P(\chi^2 \le 27.2) = 0.90$. For more examples see Chapters 4 and 6.

df	0.005	0.01	0.025	0.05	0.10	0.90	0.95	0.975	0.99	0.995
1	0.0000393	0.000157	0.000982	0.00393	0.0158	2.71	3.84	5.02	6.63	7.88
2	0.0100	0.0201	0.0506	0.103	0.211	4.61	5.99	7.38	9.21	10.6
3	0.0717	0.115	0.216	0.352	0.584	6.25	7.81	9.35	11.3	12.8
4	0.207	0.297	0.484	0.711	1.06	7.78	9.49	11.1	13.3	14.9
5	0.412	0.554	0.831	1.15	1.61	9.24	11.1	12.8	15.1	16.7
6	0.676	0.872	1.24	1.64	2.20	10.6	12.6	14.4	16.8	18.5
7	0.989	1.24	1.69	2.17	2.83	12.0	14.1	16.0	18.5	20.3
8	1.34	1.65	2.18	2.73	3.49	13.4	15.5	17.5	20.1	22.0
9	1.73	2.09	2.70	3.33	4.17	14.7	16.9	19.0	21.7	23.6
10	2.16	2.56	3.25	3.94	4.87	16.0	18.3	20.5	23.2	25.2
11	2.60	3.05	3.82	4.57	5.58	17.3	19.7	21.9	24.7	26.8
12	3.07	3.57	4.40	5.23	6.30	18.5	21.0	23.3	26.2	28.3
13	3.57	4.11	5.01	5.89	7.04	19.8	22.4	24.7	27.7	29.8
14	4.07	4.66	5.63	6.57	7.79	21.1	23.7	26.1	29.1	31.3
15	4.60	5.23	6.26	7.26	8.55	22.3	25.0	27.5	30.6	32.8
16	5.14	5.81	6.91	7.96	9.31	23.5	26.3	28.8	32.0	34.3
17	5.70	6.41	7.56	8.67	10.1	24.8	27.6	30.2	33.4	35.7
18	6.26	7.01	8.23	9.39	10.9	26.0	28.9	31.5	34.8	37.2
19	6.84	7.63	8.91	10.1	11.7	27.2	30.1	32.9	36.2	38.6
20	7.43	8.26	9.59	10.9	12.4	28.4	31.4	34.2	37.6	40.0

df	0.005	0.01	0.025	0.05	0.10	0.90	0.95	0.975	0.99	0.995
21	8.03	8.90	10.3	11.6	13.2	29.6	32.7	35.5	38.9	41.4
22	8.64	9.54	11.0	12.3	14.0	30.8	33.9	36.8	40.3	42.8
23	9.26	10.2	11.7	13.1	14.8	32.0	35.2	38.1	41.6	44.2
24	9.89	10.9	12.4	13.8	15.7	33.2	36.4	39.4	43.0	45.6
25	10.5	11.5	13.1	14.6	16.5	34.4	37.7	40.6	44.3	46.9
26	11.2	12.2	13.8	15.4	17.3	35.6	38.9	41.9	45.6	48.3
27	11.8	12.9	14.6	16.2	18.1	36.7	40.1	43.2	47.0	49.6
28	12.5	13.6	15.3	16.9	18.9	37.9	41.3	44.5	48.3	51.0
29	13.1	14.3	16.0	17.7	19.8	39.1	42.6	45.7	49.6	52.3
30	13.8	15.0	16.8	18.5	20.6	40.3	43.8	47.0	50.9	53.7
31	14.5	15.7	17.5	19.3	21.4	41.4	45.0	48.2	52.2	55.0
32	15.1	16.4	18.3	20.1	22.3	42.6	46.2	49.5	53.5	56.3
33	15.8	17.1	19.0	20.9	23.1	43.7	47.4	50.7	54.8	57.6
34	16.5	17.8	19.8	21.7	24.0	44.9	48.6	52.0	56.1	59.0
35	17.2	18.5	20.6	22.5	24.8	46.1	49.8	53.2	57.3	60.3
36	17.9	19.2	21.3	23.3	25.6	47.2	51.0	54.4	58.6	61.6
37	18.6	20.0	22.1	24.1	26.5	48.4	52.2	55.7	59.9	62.9
38	19.3	20.7	22.9	24.9	27.3	49.5	53.4	56.9	61.2	64.2
39	20.0	21.4	23.7	25.7	28.2	50.7	54.6	58.1	62.4	65.5
40	20.7	22.2	24.4	26.5	29.1	51.8	55.8	59.3	63.7	66.8
41	21.4	22.9	25.2	27.3	29.9	52.9	56.9	60.6	64.9	68.1
42	22.1	23.7	26.0	28.1	30.8	54.1	58.1	61.8	66.2	69.3
43	22.9	24.4	26.8	29.0	31.6	55.2	59.3	63.0	67.5	70.6
44	23.6	25.1	27.6	29.8	32.5	56.4	60.5	64.2	68.7	71.9
45	24.3	25.9	28.4	30.6	33.4	57.5	61.7	65.4	70.0	73.2
46	25.0	26.7	29.2	31.4	34.2	58.6	62.8	66.6	71.2	74.4
47	25.8	27.4	30.0	32.3	35.1	59.8	64.0	67.8	72.4	75.7
48	26.5	28.2	30.8	33.1	35.9	60.9	65.2	69.0	73.7	77.0
49	27.2	28.9	31.6	33.9	36.8	62.0	66.3	70.2	74.9	78.2
50	28.0	29.7	32.4	34.8	37.7	63.2	67.5	71.4	76.2	79.5
51	28.7	30.5	33.2	35.6	38.6	64.3	68.7	72.6	77.4	80.7
52	29.5	31.2	34.0	36.4	39.4	65.4	69.8	73.8	78.6	82.0
53	30.2	32.0	34.8	37.3	40.3	66.5	71.0	75.0	79.8	83.3
54	31.0	32.8	35.6	38.1	41.2	67.7	72.2	76.2	81.1	84.5
55	31.7	33.6	36.4	39.0	42.1	68.8	73.3	77.4	82.3	85.7
56	32.5	34.3	37.2	39.8	42.9	69.9	74.5	78.6	83.5	87.0
57	33.2	35.1	38.0	40.6	43.8	71.0	75.6	79.8	84.7	88.2
58	34.0	35.9	38.8	41.5	44.7	72.2	76.8	80.9	86.0	89.5
59	34.8	36.7	39.7	42.3	45.6	73.3	77.9	82.1	87.2	90.7
60	35.5	37.5	40.5	43.2	46.5	74.4	79.1	83.3	88.4	92.0

C.6   Wilcoxon Signed-Rank Test Cumulative Distribution

Table C.6 gives the cumulative distribution function of the Wilcoxon signed-rank test for effective sample sizes from 5 to 25. To use this table locate the effective sample size n (the number of observations minus those equaling the claimed median) as a column heading. Find the value of the test statistic W as a row value under column "w." The intersection of the row and column yield a cumulative probability $P(W \leq w)$. If this probability is greater than the chosen α level accept the null hypothesis.

						n					
w	5	6	7	8	9	10	11	12	13	14	15
0	0.0313	0.0156	0.0078	0.0039	0.0020	0.0010	0.0005	0.0002	0.0001	0.0001	0.0000
1	0.0625	0.0313	0.0156	0.0078	0.0039	0.0020	0.0010	0.0005	0.0002	0.0001	0.0001
2	0.0938	0.0469	0.0234	0.0117	0.0059	0.0029	0.0015	0.0007	0.0004	0.0002	0.0001
3	0.1563	0.0781	0.0391	0.0195	0.0098	0.0049	0.0024	0.0012	0.0006	0.0003	0.0002
4	0.2188	0.1094	0.0547	0.0273	0.0137	0.0068	0.0034	0.0017	0.0009	0.0004	0.0002
5	0.3125	0.1563	0.0781	0.0391	0.0195	0.0098	0.0049	0.0024	0.0012	0.0006	0.0003
6	0.4063	0.2188	0.1094	0.0547	0.0273	0.0137	0.0068	0.0034	0.0017	0.0009	0.0004
7	0.5000	0.2813	0.1484	0.0742	0.0371	0.0186	0.0093	0.0046	0.0023	0.0012	0.0006
8	0.5938	0.3438	0.1875	0.0977	0.0488	0.0244	0.0122	0.0061	0.0031	0.0015	0.0008
9	0.6875	0.4219	0.2344	0.1250	0.0645	0.0322	0.0161	0.0081	0.0040	0.0020	0.0010
10	0.7813	0.5000	0.2891	0.1563	0.0820	0.0420	0.0210	0.0105	0.0052	0.0026	0.0013
11	0.8438	0.5781	0.3438	0.1914	0.1016	0.0527	0.0269	0.0134	0.0067	0.0034	0.0017
12	0.9063	0.6563	0.4063	0.2305	0.1250	0.0654	0.0337	0.0171	0.0085	0.0043	0.0021
13	0.9375	0.7188	0.4688	0.2734	0.1504	0.0801	0.0415	0.0212	0.0107	0.0054	0.0027
14	0.9688	0.7813	0.5313	0.3203	0.1797	0.0967	0.0508	0.0261	0.0133	0.0067	0.0034
15	1.0000	0.8438	0.5938	0.3711	0.2129	0.1162	0.0615	0.0320	0.0164	0.0083	0.0042
16		0.8906	0.6563	0.4219	0.2480	0.1377	0.0737	0.0386	0.0199	0.0101	0.0051
17		0.9219	0.7109	0.4727	0.2852	0.1611	0.0874	0.0461	0.0239	0.0123	0.0062
18		0.9531	0.7656	0.5273	0.3262	0.1875	0.1030	0.0549	0.0287	0.0148	0.0075
19		0.9688	0.8125	0.5781	0.3672	0.2158	0.1201	0.0647	0.0341	0.0176	0.0090
20		0.9844	0.8516	0.6289	0.4102	0.2461	0.1392	0.0757	0.0402	0.0209	0.0108
21		1.0000	0.8906	0.6797	0.4551	0.2783	0.1602	0.0881	0.0471	0.0247	0.0128
22			0.9219	0.7266	0.5000	0.3125	0.1826	0.1018	0.0549	0.0290	0.0151
23			0.9453	0.7695	0.5449	0.3477	0.2065	0.1167	0.0636	0.0338	0.0177
24			0.9609	0.8086	0.5898	0.3848	0.2324	0.1331	0.0732	0.0392	0.0206
25			0.9766	0.8438	0.6328	0.4229	0.2598	0.1506	0.0839	0.0453	0.0240
26			0.9844	0.8750	0.6738	0.4609	0.2886	0.1697	0.0955	0.0520	0.0277
27			0.9922	0.9023	0.7148	0.5000	0.3188	0.1902	0.1082	0.0594	0.0319
28			1.0000	0.9258	0.7520	0.5391	0.3501	0.2119	0.1219	0.0676	0.0365
29				0.9453	0.7871	0.5771	0.3823	0.2349	0.1367	0.0765	0.0416
30				0.9609	0.8203	0.6152	0.4155	0.2593	0.1527	0.0863	0.0473
31				0.9727	0.8496	0.6523	0.4492	0.2847	0.1698	0.0969	0.0535
32				0.9805	0.8750	0.6875	0.4829	0.3110	0.1879	0.1083	0.0603
33				0.9883	0.8984	0.7217	0.5171	0.3386	0.2072	0.1206	0.0677
34				0.9922	0.9180	0.7539	0.5508	0.3667	0.2274	0.1338	0.0757
35				0.9961	0.9355	0.7842	0.5845	0.3955	0.2487	0.1479	0.0844
36				1.0000	0.9512	0.8125	0.6177	0.4250	0.2709	0.1629	0.0938
37					0.9629	0.8389	0.6499	0.4548	0.2939	0.1788	0.1039
38					0.9727	0.8623	0.6812	0.4849	0.3177	0.1955	0.1147
39					0.9805	0.8838	0.7114	0.5151	0.3424	0.2131	0.1262

		n						n		
w	16	17	18	19	20	w	16	17	18	19	20
0	0.0000	0.0000	0.0000	0.0000	0.0000	50	0.1877	0.1123	0.0649	0.0364	0.0200
1	0.0000	0.0000	0.0000	0.0000	0.0000	51	0.2019	0.1217	0.0708	0.0399	0.0220
2	0.0000	0.0000	0.0000	0.0000	0.0000	52	0.2166	0.1317	0.0770	0.0437	0.0242
3	0.0001	0.0000	0.0000	0.0000	0.0000	53	0.2319	0.1421	0.0837	0.0478	0.0266
4	0.0001	0.0001	0.0000	0.0000	0.0000	54	0.2477	0.1530	0.0907	0.0521	0.0291
5	0.0002	0.0001	0.0000	0.0000	0.0000	55	0.2641	0.1645	0.0982	0.0567	0.0319
6	0.0002	0.0001	0.0001	0.0000	0.0000	56	0.2809	0.1764	0.1061	0.0616	0.0348
7	0.0003	0.0001	0.0001	0.0000	0.0000	57	0.2983	0.1889	0.1144	0.0668	0.0379
8	0.0004	0.0002	0.0001	0.0000	0.0000	58	0.3161	0.2019	0.1231	0.0723	0.0413
9	0.0005	0.0003	0.0001	0.0001	0.0000	59	0.3343	0.2153	0.1323	0.0782	0.0448
10	0.0007	0.0003	0.0002	0.0001	0.0000	60	0.3529	0.2293	0.1419	0.0844	0.0487
11	0.0008	0.0004	0.0002	0.0001	0.0001	61	0.3718	0.2437	0.1519	0.0909	0.0527
12	0.0011	0.0005	0.0003	0.0001	0.0001	62	0.3910	0.2585	0.1624	0.0978	0.0570
13	0.0013	0.0007	0.0003	0.0002	0.0001	63	0.4104	0.2738	0.1733	0.1051	0.0615
14	0.0017	0.0008	0.0004	0.0002	0.0001	64	0.4301	0.2895	0.1846	0.1127	0.0664
15	0.0021	0.0010	0.0005	0.0003	0.0001	65	0.4500	0.3056	0.1964	0.1206	0.0715
16	0.0026	0.0013	0.0006	0.0003	0.0002	66	0.4699	0.3221	0.2086	0.1290	0.0768
17	0.0031	0.0016	0.0008	0.0004	0.0002	67	0.4900	0.3389	0.2211	0.1377	0.0825
18	0.0038	0.0019	0.0010	0.0005	0.0002	68	0.5100	0.3559	0.2341	0.1467	0.0884
19	0.0046	0.0023	0.0012	0.0006	0.0003	69	0.5301	0.3733	0.2475	0.1562	0.0947
20	0.0055	0.0028	0.0014	0.0007	0.0004	70	0.5500	0.3910	0.2613	0.1660	0.1012
21	0.0065	0.0033	0.0017	0.0008	0.0004	71	0.5699	0.4088	0.2754	0.1762	0.1081
22	0.0078	0.0040	0.0020	0.0010	0.0005	72	0.5896	0.4268	0.2899	0.1868	0.1153
23	0.0091	0.0047	0.0024	0.0012	0.0006	73	0.6090	0.4450	0.3047	0.1977	0.1227
24	0.0107	0.0055	0.0028	0.0014	0.0007	74	0.6282	0.4633	0.3198	0.2090	0.1305
25	0.0125	0.0064	0.0033	0.0017	0.0008	75	0.6471	0.4816	0.3353	0.2207	0.1387
26	0.0145	0.0075	0.0038	0.0020	0.0010	76	0.6657	0.5000	0.3509	0.2327	0.1471
27	0.0168	0.0087	0.0045	0.0023	0.0012	77	0.6839	0.5184	0.3669	0.2450	0.1559
28	0.0193	0.0101	0.0052	0.0027	0.0014	78	0.7017	0.5367	0.3830	0.2576	0.1650
29	0.0222	0.0116	0.0060	0.0031	0.0016	79	0.7191	0.5550	0.3994	0.2706	0.1744
30	0.0253	0.0133	0.0069	0.0036	0.0018	80	0.7359	0.5732	0.4159	0.2839	0.1841
31	0.0288	0.0153	0.0080	0.0041	0.0021	81	0.7523	0.5912	0.4325	0.2974	0.1942
32	0.0327	0.0174	0.0091	0.0047	0.0024	82	0.7681	0.6090	0.4493	0.3113	0.2045
33	0.0370	0.0198	0.0104	0.0054	0.0028	83	0.7834	0.6267	0.4661	0.3254	0.2152
34	0.0416	0.0224	0.0118	0.0062	0.0032	84	0.7981	0.6441	0.4831	0.3397	0.2262
35	0.0467	0.0253	0.0134	0.0070	0.0036	85	0.8123	0.6611	0.5000	0.3543	0.2375
36	0.0523	0.0284	0.0152	0.0080	0.0042	86	0.8258	0.6779	0.5169	0.3690	0.2490
37	0.0583	0.0319	0.0171	0.0090	0.0047	87	0.8387	0.6944	0.5339	0.3840	0.2608
38	0.0649	0.0357	0.0192	0.0102	0.0053	88	0.8511	0.7105	0.5507	0.3991	0.2729
39	0.0719	0.0398	0.0216	0.0115	0.0060	89	0.8628	0.7262	0.5675	0.4144	0.2853
40	0.0795	0.0443	0.0241	0.0129	0.0068	90	0.8739	0.7415	0.5841	0.4298	0.2979
41	0.0877	0.0492	0.0269	0.0145	0.0077	91	0.8844	0.7563	0.6006	0.4453	0.3108
42	0.0964	0.0544	0.0300	0.0162	0.0086	92	0.8943	0.7707	0.6170	0.4609	0.3238
43	0.1057	0.0601	0.0333	0.0180	0.0096	93	0.9036	0.7847	0.6331	0.4765	0.3371
44	0.1156	0.0662	0.0368	0.0201	0.0107	94	0.9123	0.7981	0.6491	0.4922	0.3506
45	0.1261	0.0727	0.0407	0.0223	0.0120	95	0.9205	0.8111	0.6647	0.5078	0.3643
46	0.1372	0.0797	0.0449	0.0247	0.0133	96	0.9281	0.8236	0.6802	0.5235	0.3781
47	0.1489	0.0871	0.0494	0.0273	0.0148	97	0.9351	0.8355	0.6953	0.5391	0.3921
48	0.1613	0.0950	0.0542	0.0301	0.0164	98	0.9417	0.8470	0.7101	0.5547	0.4062
49	0.1742	0.1034	0.0594	0.0331	0.0181	99	0.9477	0.8579	0.7246	0.5702	0.4204

			n						n		
w	21	22	23	24	25	w	21	22	23	24	25
≤ 30	≤ 0.0009	≤ 0.0005	≤ 0.0002	≤ 0.0001	≤ 0.0001	80	0.1145	0.0687	0.0401	0.0228	0.0128
31	0.0011	0.0005	0.0003	0.0001	0.0001	81	0.1214	0.0733	0.0429	0.0245	0.0137
32	0.0012	0.0006	0.0003	0.0002	0.0001	82	0.1286	0.0780	0.0459	0.0263	0.0148
33	0.0014	0.0007	0.0004	0.0002	0.0001	83	0.1361	0.0829	0.0490	0.0282	0.0159
34	0.0016	0.0008	0.0004	0.0002	0.0001	84	0.1439	0.0881	0.0523	0.0302	0.0171
35	0.0019	0.0010	0.0005	0.0002	0.0001	85	0.1519	0.0935	0.0557	0.0323	0.0183
36	0.0021	0.0011	0.0006	0.0003	0.0001	86	0.1602	0.0991	0.0593	0.0346	0.0197
37	0.0024	0.0013	0.0006	0.0003	0.0002	87	0.1688	0.1050	0.0631	0.0369	0.0211
38	0.0028	0.0014	0.0007	0.0004	0.0002	88	0.1777	0.1111	0.0671	0.0394	0.0226
39	0.0031	0.0016	0.0008	0.0004	0.0002	89	0.1869	0.1174	0.0712	0.0420	0.0241
40	0.0036	0.0018	0.0009	0.0005	0.0002	90	0.1963	0.1239	0.0755	0.0447	0.0258
41	0.0040	0.0021	0.0011	0.0005	0.0003	91	0.2060	0.1308	0.0801	0.0475	0.0275
42	0.0045	0.0023	0.0012	0.0006	0.0003	92	0.2160	0.1378	0.0848	0.0505	0.0294
43	0.0051	0.0026	0.0014	0.0007	0.0004	93	0.2262	0.1451	0.0897	0.0537	0.0313
44	0.0057	0.0030	0.0015	0.0008	0.0004	94	0.2367	0.1527	0.0948	0.0570	0.0334
45	0.0063	0.0033	0.0017	0.0009	0.0005	95	0.2474	0.1604	0.1001	0.0604	0.0355
46	0.0071	0.0037	0.0019	0.0010	0.0005	96	0.2584	0.1685	0.1056	0.0640	0.0377
47	0.0079	0.0042	0.0022	0.0011	0.0006	97	0.2696	0.1767	0.1113	0.0678	0.0401
48	0.0088	0.0046	0.0024	0.0013	0.0006	98	0.2810	0.1853	0.1172	0.0717	0.0426
49	0.0097	0.0052	0.0027	0.0014	0.0007	99	0.2927	0.1940	0.1234	0.0758	0.0452
50	0.0108	0.0057	0.0030	0.0016	0.0008	100	0.3046	0.2030	0.1297	0.0800	0.0479
51	0.0119	0.0064	0.0034	0.0018	0.0009	101	0.3166	0.2122	0.1363	0.0844	0.0507
52	0.0132	0.0070	0.0037	0.0020	0.0010	102	0.3289	0.2217	0.1431	0.0890	0.0537
53	0.0145	0.0078	0.0041	0.0022	0.0011	103	0.3414	0.2314	0.1501	0.0938	0.0567
54	0.0160	0.0086	0.0046	0.0024	0.0013	104	0.3540	0.2413	0.1573	0.0987	0.0600
55	0.0175	0.0095	0.0051	0.0027	0.0014	105	0.3667	0.2514	0.1647	0.1038	0.0633
56	0.0192	0.0104	0.0056	0.0029	0.0015	106	0.3796	0.2618	0.1723	0.1091	0.0668
57	0.0210	0.0115	0.0061	0.0033	0.0017	107	0.3927	0.2723	0.1802	0.1146	0.0705
58	0.0230	0.0126	0.0068	0.0036	0.0019	108	0.4058	0.2830	0.1883	0.1203	0.0742
59	0.0251	0.0138	0.0074	0.0040	0.0021	109	0.4191	0.2940	0.1965	0.1261	0.0782
60	0.0273	0.0151	0.0082	0.0044	0.0023	110	0.4324	0.3051	0.2050	0.1322	0.0822
61	0.0298	0.0164	0.0089	0.0048	0.0025	111	0.4459	0.3164	0.2137	0.1384	0.0865
62	0.0323	0.0179	0.0098	0.0053	0.0028	112	0.4593	0.3278	0.2226	0.1448	0.0909
63	0.0351	0.0195	0.0107	0.0058	0.0031	113	0.4729	0.3394	0.2317	0.1514	0.0954
64	0.0380	0.0212	0.0117	0.0063	0.0034	114	0.4864	0.3512	0.2410	0.1583	0.1001
65	0.0411	0.0231	0.0127	0.0069	0.0037	115	0.5000	0.3631	0.2505	0.1653	0.1050
66	0.0444	0.0250	0.0138	0.0075	0.0040	116	0.5136	0.3751	0.2601	0.1724	0.1100
67	0.0479	0.0271	0.0150	0.0082	0.0044	117	0.5271	0.3873	0.2700	0.1798	0.1152
68	0.0516	0.0293	0.0163	0.0089	0.0048	118	0.5407	0.3995	0.2800	0.1874	0.1205
69	0.0555	0.0317	0.0177	0.0097	0.0053	119	0.5541	0.4119	0.2902	0.1951	0.1261
70	0.0597	0.0342	0.0192	0.0106	0.0057	120	0.5676	0.4243	0.3005	0.2031	0.1317
71	0.0640	0.0369	0.0208	0.0115	0.0062	121	0.5809	0.4368	0.3110	0.2112	0.1376
72	0.0686	0.0397	0.0224	0.0124	0.0068	122	0.5942	0.4494	0.3217	0.2195	0.1436
73	0.0735	0.0427	0.0242	0.0135	0.0074	123	0.6073	0.4620	0.3325	0.2279	0.1498
74	0.0786	0.0459	0.0261	0.0146	0.0080	124	0.6204	0.4746	0.3434	0.2366	0.1562
75	0.0839	0.0492	0.0281	0.0157	0.0087	125	0.6333	0.4873	0.3545	0.2454	0.1627
76	0.0895	0.0527	0.0303	0.0170	0.0094	126	0.6460	0.5000	0.3657	0.2544	0.1694
77	0.0953	0.0564	0.0325	0.0183	0.0101	127	0.6586	0.5127	0.3770	0.2635	0.1763
78	0.1015	0.0603	0.0349	0.0197	0.0110	128	0.6711	0.5254	0.3884	0.2728	0.1833
79	0.1078	0.0644	0.0374	0.0212	0.0118	129	0.6834	0.5380	0.3999	0.2823	0.1905

C.7 Cumulative F Distribution

Table C.7 is a tabulation of values for the 0.90, 0.95, 0.975, and 0.99 points in the cumulative distribution functions of selected F distributions. To locate a critical value, select the page with the appropriate probability. For example, to find $P(F_{12, 20}) \leq$ 0.975, use page 453. Find the critical value as the intersection of the column with the degrees of freedom for the numerator and the row with the degrees of freedom for the denominator. So $P(F_{12, 20}) \leq 0.975 = 2.68$. For degrees of freedom not found in the table always use the next *lower* value in the table. For example, $P(F_{14, 55}) \leq 0.90 \approx$ $P(F_{12, 50}) \leq 0.90 = 1.68$, use page 449. See Chapter 7 for other examples.

$P(F_{\nu_1,\nu_2}) \leq 0.90$

ν_1 ν_2	1	2	3	4	5	6	7	8	9	10	12	15
1	39.86	49.50	53.59	55.83	57.24	58.20	58.91	59.44	59.86	60.19	60.71	61.22
2	8.53	9.00	9.16	9.24	9.29	9.33	9.35	9.37	9.38	9.39	9.41	9.42
3	5.54	5.46	5.39	5.34	5.31	5.28	5.27	5.25	5.24	5.23	5.22	5.20
4	4.54	4.32	4.19	4.11	4.05	4.01	3.98	3.95	3.94	3.92	3.90	3.87
5	4.06	3.78	3.62	3.52	3.45	3.40	3.37	3.34	3.32	3.30	3.27	3.24
6	3.78	3.46	3.29	3.18	3.11	3.05	3.01	2.98	2.96	2.94	2.90	2.87
7	3.59	3.26	3.07	2.96	2.88	2.83	2.78	2.75	2.72	2.70	2.67	2.63
8	3.46	3.11	2.92	2.81	2.73	2.67	2.62	2.59	2.56	2.54	2.50	2.46
9	3.36	3.01	2.81	2.69	2.61	2.55	2.51	2.47	2.44	2.42	2.38	2.34
10	3.29	2.92	2.73	2.61	2.52	2.46	2.41	2.38	2.35	2.32	2.28	2.24
11	3.23	2.86	2.66	2.54	2.45	2.39	2.34	2.30	2.27	2.25	2.21	2.17
12	3.18	2.81	2.61	2.48	2.39	2.33	2.28	2.24	2.21	2.19	2.15	2.10
13	3.14	2.76	2.56	2.43	2.35	2.28	2.23	2.20	2.16	2.14	2.10	2.05
14	3.10	2.73	2.52	2.39	2.31	2.24	2.19	2.15	2.12	2.10	2.05	2.01
15	3.07	2.70	2.49	2.36	2.27	2.21	2.16	2.12	2.09	2.06	2.02	1.97
16	3.05	2.67	2.46	2.33	2.24	2.18	2.13	2.09	2.06	2.03	1.99	1.94
17	3.03	2.64	2.44	2.31	2.22	2.15	2.10	2.06	2.03	2.00	1.96	1.91
18	3.01	2.62	2.42	2.29	2.20	2.13	2.08	2.04	2.00	1.98	1.93	1.89
19	2.99	2.61	2.40	2.27	2.18	2.11	2.06	2.02	1.98	1.96	1.91	1.86
20	2.97	2.59	2.38	2.25	2.16	2.09	2.04	2.00	1.96	1.94	1.89	1.84
21	2.96	2.57	2.36	2.23	2.14	2.08	2.02	1.98	1.95	1.92	1.87	1.83
22	2.95	2.56	2.35	2.22	2.13	2.06	2.01	1.97	1.93	1.90	1.86	1.81
23	2.94	2.55	2.34	2.21	2.11	2.05	1.99	1.95	1.92	1.89	1.84	1.80
24	2.93	2.54	2.33	2.19	2.10	2.04	1.98	1.94	1.91	1.88	1.83	1.78
25	2.92	2.53	2.32	2.18	2.09	2.02	1.97	1.93	1.89	1.87	1.82	1.77
26	2.91	2.52	2.31	2.17	2.08	2.01	1.96	1.92	1.88	1.86	1.81	1.76
27	2.90	2.51	2.30	2.17	2.07	2.00	1.95	1.91	1.87	1.85	1.80	1.75
28	2.89	2.50	2.29	2.16	2.06	2.00	1.94	1.90	1.87	1.84	1.79	1.74
29	2.89	2.50	2.28	2.15	2.06	1.99	1.93	1.89	1.86	1.83	1.78	1.73
30	2.88	2.49	2.28	2.14	2.05	1.98	1.93	1.88	1.85	1.82	1.77	1.72
40	2.84	2.44	2.23	2.09	2.00	1.93	1.87	1.83	1.79	1.76	1.71	1.66
50	2.81	2.41	2.20	2.06	1.97	1.90	1.84	1.80	1.76	1.73	1.68	1.63
60	2.79	2.39	2.18	2.04	1.95	1.87	1.82	1.77	1.74	1.71	1.66	1.60
70	2.78	2.38	2.16	2.03	1.93	1.86	1.80	1.76	1.72	1.69	1.64	1.59
80	2.77	2.37	2.15	2.02	1.92	1.85	1.79	1.75	1.71	1.68	1.63	1.57
90	2.76	2.36	2.15	2.01	1.91	1.84	1.78	1.74	1.70	1.67	1.62	1.56
100	2.76	2.36	2.14	2.00	1.91	1.83	1.78	1.73	1.69	1.66	1.61	1.56
120	2.75	2.35	2.13	1.99	1.90	1.82	1.77	1.72	1.68	1.65	1.60	1.55
150	2.74	2.34	2.12	1.98	1.89	1.81	1.76	1.71	1.67	1.64	1.59	1.53
∞	2.71	2.30	2.08	1.94	1.85	1.77	1.72	1.67	1.63	1.60	1.55	1.49

$$P(F_{\nu_1,\nu_2}) \leq 0.90$$

ν_1 / ν_2	18	20	24	25	30	40	50	60	90	120	∞
1	61.57	61.74	62.00	62.05	62.26	62.53	62.69	62.79	62.97	63.06	63.33
2	9.44	9.44	9.45	9.45	9.46	9.47	9.47	9.47	9.48	9.48	9.49
3	5.19	5.18	5.18	5.17	5.17	5.16	5.15	5.15	5.15	5.14	5.13
4	3.85	3.84	3.83	3.83	3.82	3.80	3.80	3.79	3.78	3.78	3.76
5	3.22	3.21	3.19	3.19	3.17	3.16	3.15	3.14	3.13	3.12	3.11
6	2.85	2.84	2.82	2.81	2.80	2.78	2.77	2.76	2.75	2.74	2.72
7	2.61	2.59	2.58	2.57	2.56	2.54	2.52	2.51	2.50	2.49	2.47
8	2.44	2.42	2.40	2.40	2.38	2.36	2.35	2.34	2.32	2.32	2.29
9	2.31	2.30	2.28	2.27	2.25	2.23	2.22	2.21	2.19	2.18	2.16
10	2.22	2.20	2.18	2.17	2.16	2.13	2.12	2.11	2.09	2.08	2.06
11	2.14	2.12	2.10	2.10	2.08	2.05	2.04	2.03	2.01	2.00	1.97
12	2.08	2.06	2.04	2.03	2.01	1.99	1.97	1.96	1.94	1.93	1.90
13	2.02	2.01	1.98	1.98	1.96	1.93	1.92	1.90	1.89	1.88	1.85
14	1.98	1.96	1.94	1.93	1.91	1.89	1.87	1.86	1.84	1.83	1.80
15	1.94	1.92	1.90	1.89	1.87	1.85	1.83	1.82	1.80	1.79	1.76
16	1.91	1.89	1.87	1.86	1.84	1.81	1.79	1.78	1.76	1.75	1.72
17	1.88	1.86	1.84	1.83	1.81	1.78	1.76	1.75	1.73	1.72	1.69
18	1.85	1.84	1.81	1.80	1.78	1.75	1.74	1.72	1.70	1.69	1.66
19	1.83	1.81	1.79	1.78	1.76	1.73	1.71	1.70	1.68	1.67	1.63
20	1.81	1.79	1.77	1.76	1.74	1.71	1.69	1.68	1.65	1.64	1.61
21	1.79	1.78	1.75	1.74	1.72	1.69	1.67	1.66	1.63	1.62	1.59
22	1.78	1.76	1.73	1.73	1.70	1.67	1.65	1.64	1.62	1.60	1.57
23	1.76	1.74	1.72	1.71	1.69	1.66	1.64	1.62	1.60	1.59	1.55
24	1.75	1.73	1.70	1.70	1.67	1.64	1.62	1.61	1.58	1.57	1.53
25	1.74	1.72	1.69	1.68	1.66	1.63	1.61	1.59	1.57	1.56	1.52
26	1.72	1.71	1.68	1.67	1.65	1.61	1.59	1.58	1.56	1.54	1.50
27	1.71	1.70	1.67	1.66	1.64	1.60	1.58	1.57	1.54	1.53	1.49
28	1.70	1.69	1.66	1.65	1.63	1.59	1.57	1.56	1.53	1.52	1.48
29	1.69	1.68	1.65	1.64	1.62	1.58	1.56	1.55	1.52	1.51	1.47
30	1.69	1.67	1.64	1.63	1.61	1.57	1.55	1.54	1.51	1.50	1.46
40	1.62	1.61	1.57	1.57	1.54	1.51	1.48	1.47	1.44	1.42	1.38
50	1.59	1.57	1.54	1.53	1.50	1.46	1.44	1.42	1.39	1.38	1.33
60	1.56	1.54	1.51	1.50	1.48	1.44	1.41	1.40	1.36	1.35	1.29
70	1.55	1.53	1.49	1.49	1.46	1.42	1.39	1.37	1.34	1.32	1.27
80	1.53	1.51	1.48	1.47	1.44	1.40	1.38	1.36	1.33	1.31	1.24
90	1.52	1.50	1.47	1.46	1.43	1.39	1.36	1.35	1.31	1.29	1.23
100	1.52	1.49	1.46	1.45	1.42	1.38	1.35	1.34	1.30	1.28	1.21
120	1.50	1.48	1.45	1.44	1.41	1.37	1.34	1.32	1.28	1.26	1.19
150	1.49	1.47	1.43	1.43	1.40	1.35	1.33	1.30	1.27	1.25	1.17
∞	1.44	1.42	1.38	1.38	1.34	1.30	1.26	1.24	1.20	1.17	1.00

$$P(F_{\nu_1,\nu_2}) \leq 0.95$$

ν_1 \ ν_2	1	2	3	4	5	6	7	8	9	10	12	15
1	161.4	199.5	215.7	224.6	230.2	234.0	236.8	238.9	240.5	241.9	243.9	245.9
2	18.51	19.00	19.16	19.25	19.30	19.33	19.35	19.37	19.38	19.40	19.41	19.43
3	10.13	9.55	9.28	9.12	9.01	8.94	8.89	8.85	8.81	8.79	8.74	8.70
4	7.71	6.94	6.59	6.39	6.26	6.16	6.09	6.04	6.00	5.96	5.91	5.86
5	6.61	5.79	5.41	5.19	5.05	4.95	4.88	4.82	4.77	4.74	4.68	4.62
6	5.99	5.14	4.76	4.53	4.39	4.28	4.21	4.15	4.10	4.06	4.00	3.94
7	5.59	4.74	4.35	4.12	3.97	3.87	3.79	3.73	3.68	3.64	3.57	3.51
8	5.32	4.46	4.07	3.84	3.69	3.58	3.50	3.44	3.39	3.35	3.28	3.22
9	5.12	4.26	3.86	3.63	3.48	3.37	3.29	3.23	3.18	3.14	3.07	3.01
10	4.96	4.10	3.71	3.48	3.33	3.22	3.14	3.07	3.02	2.98	2.91	2.85
11	4.84	3.98	3.59	3.36	3.20	3.09	3.01	2.95	2.90	2.85	2.79	2.72
12	4.75	3.89	3.49	3.26	3.11	3.00	2.91	2.85	2.80	2.75	2.69	2.62
13	4.67	3.81	3.41	3.18	3.03	2.92	2.83	2.77	2.71	2.67	2.60	2.53
14	4.60	3.74	3.34	3.11	2.96	2.85	2.76	2.70	2.65	2.60	2.53	2.46
15	4.54	3.68	3.29	3.06	2.90	2.79	2.71	2.64	2.59	2.54	2.48	2.40
16	4.49	3.63	3.24	3.01	2.85	2.74	2.66	2.59	2.54	2.49	2.42	2.35
17	4.45	3.59	3.20	2.96	2.81	2.70	2.61	2.55	2.49	2.45	2.38	2.31
18	4.41	3.55	3.16	2.93	2.77	2.66	2.58	2.51	2.46	2.41	2.34	2.27
19	4.38	3.52	3.13	2.90	2.74	2.63	2.54	2.48	2.42	2.38	2.31	2.23
20	4.35	3.49	3.10	2.87	2.71	2.60	2.51	2.45	2.39	2.35	2.28	2.20
21	4.32	3.47	3.07	2.84	2.68	2.57	2.49	2.42	2.37	2.32	2.25	2.18
22	4.30	3.44	3.05	2.82	2.66	2.55	2.46	2.40	2.34	2.30	2.23	2.15
23	4.28	3.42	3.03	2.80	2.64	2.53	2.44	2.37	2.32	2.27	2.20	2.13
24	4.26	3.40	3.01	2.78	2.62	2.51	2.42	2.36	2.30	2.25	2.18	2.11
25	4.24	3.39	2.99	2.76	2.60	2.49	2.40	2.34	2.28	2.24	2.16	2.09
26	4.23	3.37	2.98	2.74	2.59	2.47	2.39	2.32	2.27	2.22	2.15	2.07
27	4.21	3.35	2.96	2.73	2.57	2.46	2.37	2.31	2.25	2.20	2.13	2.06
28	4.20	3.34	2.95	2.71	2.56	2.45	2.36	2.29	2.24	2.19	2.12	2.04
29	4.18	3.33	2.93	2.70	2.55	2.43	2.35	2.28	2.22	2.18	2.10	2.03
30	4.17	3.32	2.92	2.69	2.53	2.42	2.33	2.27	2.21	2.16	2.09	2.01
40	4.08	3.23	2.84	2.61	2.45	2.34	2.25	2.18	2.12	2.08	2.00	1.92
50	4.03	3.18	2.79	2.56	2.40	2.29	2.20	2.13	2.07	2.03	1.95	1.87
60	4.00	3.15	2.76	2.53	2.37	2.25	2.17	2.10	2.04	1.99	1.92	1.84
70	3.98	3.13	2.74	2.50	2.35	2.23	2.14	2.07	2.02	1.97	1.89	1.81
80	3.96	3.11	2.72	2.49	2.33	2.21	2.13	2.06	2.00	1.95	1.88	1.79
90	3.95	3.10	2.71	2.47	2.32	2.20	2.11	2.04	1.99	1.94	1.86	1.78
100	3.94	3.09	2.70	2.46	2.31	2.19	2.10	2.03	1.97	1.93	1.85	1.77
120	3.92	3.07	2.68	2.45	2.29	2.18	2.09	2.02	1.96	1.91	1.83	1.75
150	3.90	3.06	2.66	2.43	2.27	2.16	2.07	2.00	1.94	1.89	1.82	1.73
∞	3.84	3.00	2.60	2.37	2.21	2.10	2.01	1.94	1.88	1.83	1.75	1.67

$$P(F_{\nu_1, \nu_2}) \leq 0.95$$

ν_1 \ ν_2	18	20	24	25	30	40	50	60	90	120	∞
1	247.3	248.0	249.1	249.3	250.1	251.1	251.8	252.2	252.9	253.3	254.3
2	19.44	19.45	19.45	19.46	19.46	19.47	19.48	19.48	19.48	19.49	19.50
3	8.67	8.66	8.64	8.63	8.62	8.59	8.58	8.57	8.56	8.55	8.53
4	5.82	5.80	5.77	5.77	5.75	5.72	5.70	5.69	5.67	5.66	5.63
5	4.58	4.56	4.53	4.52	4.50	4.46	4.44	4.43	4.41	4.40	4.37
6	3.90	3.87	3.84	3.83	3.81	3.77	3.75	3.74	3.72	3.70	3.67
7	3.47	3.44	3.41	3.40	3.38	3.34	3.32	3.30	3.28	3.27	3.23
8	3.17	3.15	3.12	3.11	3.08	3.04	3.02	3.01	2.98	2.97	2.93
9	2.96	2.94	2.90	2.89	2.86	2.83	2.80	2.79	2.76	2.75	2.71
10	2.80	2.77	2.74	2.73	2.70	2.66	2.64	2.62	2.59	2.58	2.54
11	2.67	2.65	2.61	2.60	2.57	2.53	2.51	2.49	2.46	2.45	2.40
12	2.57	2.54	2.51	2.50	2.47	2.43	2.40	2.38	2.36	2.34	2.30
13	2.48	2.46	2.42	2.41	2.38	2.34	2.31	2.30	2.27	2.25	2.21
14	2.41	2.39	2.35	2.34	2.31	2.27	2.24	2.22	2.19	2.18	2.13
15	2.35	2.33	2.29	2.28	2.25	2.20	2.18	2.16	2.13	2.11	2.07
16	2.30	2.28	2.24	2.23	2.19	2.15	2.12	2.11	2.07	2.06	2.01
17	2.26	2.23	2.19	2.18	2.15	2.10	2.08	2.06	2.03	2.01	1.96
18	2.22	2.19	2.15	2.14	2.11	2.06	2.04	2.02	1.98	1.97	1.92
19	2.18	2.16	2.11	2.11	2.07	2.03	2.00	1.98	1.95	1.93	1.88
20	2.15	2.12	2.08	2.07	2.04	1.99	1.97	1.95	1.91	1.90	1.84
21	2.12	2.10	2.05	2.05	2.01	1.96	1.94	1.92	1.88	1.87	1.81
22	2.10	2.07	2.03	2.02	1.98	1.94	1.91	1.89	1.86	1.84	1.78
23	2.08	2.05	2.01	2.00	1.96	1.91	1.88	1.86	1.83	1.81	1.76
24	2.05	2.03	1.98	1.97	1.94	1.89	1.86	1.84	1.81	1.79	1.73
25	2.04	2.01	1.96	1.96	1.92	1.87	1.84	1.82	1.79	1.77	1.71
26	2.02	1.99	1.95	1.94	1.90	1.85	1.82	1.80	1.77	1.75	1.69
27	2.00	1.97	1.93	1.92	1.88	1.84	1.81	1.79	1.75	1.73	1.67
28	1.99	1.96	1.91	1.91	1.87	1.82	1.79	1.77	1.73	1.71	1.65
29	1.97	1.94	1.90	1.89	1.85	1.81	1.77	1.75	1.72	1.70	1.64
30	1.96	1.93	1.89	1.88	1.84	1.79	1.76	1.74	1.70	1.68	1.62
40	1.87	1.84	1.79	1.78	1.74	1.69	1.66	1.64	1.60	1.58	1.51
50	1.81	1.78	1.74	1.73	1.69	1.63	1.60	1.58	1.53	1.51	1.44
60	1.78	1.75	1.70	1.69	1.65	1.59	1.56	1.53	1.49	1.47	1.39
70	1.75	1.72	1.67	1.66	1.62	1.57	1.53	1.50	1.46	1.44	1.35
80	1.73	1.70	1.65	1.64	1.60	1.54	1.51	1.48	1.44	1.41	1.32
90	1.72	1.69	1.64	1.63	1.59	1.53	1.49	1.46	1.42	1.39	1.30
100	1.71	1.68	1.63	1.62	1.57	1.52	1.48	1.45	1.40	1.38	1.28
120	1.69	1.66	1.61	1.60	1.55	1.50	1.46	1.43	1.38	1.35	1.25
150	1.67	1.64	1.59	1.58	1.54	1.48	1.44	1.41	1.36	1.33	1.22
∞	1.60	1.57	1.52	1.51	1.46	1.39	1.35	1.32	1.26	1.22	1.00

$$P(F_{\nu_1,\nu_2}) \leq 0.975$$

ν_1 / ν_2	1	2	3	4	5	6	7	8	9	10	12	15
1	647.8	799.5	864.2	899.6	921.8	937.1	948.2	956.6	963.3	968.6	976.7	984.9
2	38.51	39.00	39.17	39.25	39.30	39.33	39.36	39.37	39.39	39.40	39.41	39.43
3	17.44	16.04	15.44	15.10	14.88	14.73	14.62	14.54	14.47	14.42	14.34	14.25
4	12.22	10.65	9.98	9.60	9.36	9.20	9.07	8.98	8.90	8.84	8.75	8.66
5	10.01	8.43	7.76	7.39	7.15	6.98	6.85	6.76	6.68	6.62	6.52	6.43
6	8.81	7.26	6.60	6.23	5.99	5.82	5.70	5.60	5.52	5.46	5.37	5.27
7	8.07	6.54	5.89	5.52	5.29	5.12	4.99	4.90	4.82	4.76	4.67	4.57
8	7.57	6.06	5.42	5.05	4.82	4.65	4.53	4.43	4.36	4.30	4.20	4.10
9	7.21	5.71	5.08	4.72	4.48	4.32	4.20	4.10	4.03	3.96	3.87	3.77
10	6.94	5.46	4.83	4.47	4.24	4.07	3.95	3.85	3.78	3.72	3.62	3.52
11	6.72	5.26	4.63	4.28	4.04	3.88	3.76	3.66	3.59	3.53	3.43	3.33
12	6.55	5.10	4.47	4.12	3.89	3.73	3.61	3.51	3.44	3.37	3.28	3.18
13	6.41	4.97	4.35	4.00	3.77	3.60	3.48	3.39	3.31	3.25	3.15	3.05
14	6.30	4.86	4.24	3.89	3.66	3.50	3.38	3.29	3.21	3.15	3.05	2.95
15	6.20	4.77	4.15	3.80	3.58	3.41	3.29	3.20	3.12	3.06	2.96	2.86
16	6.12	4.69	4.08	3.73	3.50	3.34	3.22	3.12	3.05	2.99	2.89	2.79
17	6.04	4.62	4.01	3.66	3.44	3.28	3.16	3.06	2.98	2.92	2.82	2.72
18	5.98	4.56	3.95	3.61	3.38	3.22	3.10	3.01	2.93	2.87	2.77	2.67
19	5.92	4.51	3.90	3.56	3.33	3.17	3.05	2.96	2.88	2.82	2.72	2.62
20	5.87	4.46	3.86	3.51	3.29	3.13	3.01	2.91	2.84	2.77	2.68	2.57
21	5.83	4.42	3.82	3.48	3.25	3.09	2.97	2.87	2.80	2.73	2.64	2.53
22	5.79	4.38	3.78	3.44	3.22	3.05	2.93	2.84	2.76	2.70	2.60	2.50
23	5.75	4.35	3.75	3.41	3.18	3.02	2.90	2.81	2.73	2.67	2.57	2.47
24	5.72	4.32	3.72	3.38	3.15	2.99	2.87	2.78	2.70	2.64	2.54	2.44
25	5.69	4.29	3.69	3.35	3.13	2.97	2.85	2.75	2.68	2.61	2.51	2.41
26	5.66	4.27	3.67	3.33	3.10	2.94	2.82	2.73	2.65	2.59	2.49	2.39
27	5.63	4.24	3.65	3.31	3.08	2.92	2.80	2.71	2.63	2.57	2.47	2.36
28	5.61	4.22	3.63	3.29	3.06	2.90	2.78	2.69	2.61	2.55	2.45	2.34
29	5.59	4.20	3.61	3.27	3.04	2.88	2.76	2.67	2.59	2.53	2.43	2.32
30	5.57	4.18	3.59	3.25	3.03	2.87	2.75	2.65	2.57	2.51	2.41	2.31
40	5.42	4.05	3.46	3.13	2.90	2.74	2.62	2.53	2.45	2.39	2.29	2.18
50	5.34	3.97	3.39	3.05	2.83	2.67	2.55	2.46	2.38	2.32	2.22	2.11
60	5.29	3.93	3.34	3.01	2.79	2.63	2.51	2.41	2.33	2.27	2.17	2.06
70	5.25	3.89	3.31	2.97	2.75	2.59	2.47	2.38	2.30	2.24	2.14	2.03
80	5.22	3.86	3.28	2.95	2.73	2.57	2.45	2.35	2.28	2.21	2.11	2.00
90	5.20	3.84	3.26	2.93	2.71	2.55	2.43	2.34	2.26	2.19	2.09	1.98
100	5.18	3.83	3.25	2.92	2.70	2.54	2.42	2.32	2.24	2.18	2.08	1.97
120	5.15	3.80	3.23	2.89	2.67	2.52	2.39	2.30	2.22	2.16	2.05	1.94
150	5.13	3.78	3.20	2.87	2.65	2.49	2.37	2.28	2.20	2.13	2.03	1.92
∞	5.02	3.69	3.12	2.79	2.57	2.41	2.29	2.19	2.11	2.05	1.94	1.83

$$P(F_{\nu_1,\nu_2}) \leq 0.975$$

ν_1 \ ν_2	18	20	24	25	30	40	50	60	90	120	∞
1	990.3	993.1	997.3	998.1	1001	1006	1008	1010	1013	1014	1018
2	39.44	39.45	39.46	39.46	39.46	39.47	39.48	39.48	39.49	39.49	39.50
3	14.20	14.17	14.12	14.12	14.08	14.04	14.01	13.99	13.96	13.95	13.90
4	8.59	8.56	8.51	8.50	8.46	8.41	8.38	8.36	8.33	8.31	8.26
5	6.36	6.33	6.28	6.27	6.23	6.18	6.14	6.12	6.09	6.07	6.02
6	5.20	5.17	5.12	5.11	5.07	5.01	4.98	4.96	4.92	4.90	4.85
7	4.50	4.47	4.41	4.40	4.36	4.31	4.28	4.25	4.22	4.20	4.14
8	4.03	4.00	3.95	3.94	3.89	3.84	3.81	3.78	3.75	3.73	3.67
9	3.70	3.67	3.61	3.60	3.56	3.51	3.47	3.45	3.41	3.39	3.33
10	3.45	3.42	3.37	3.35	3.31	3.26	3.22	3.20	3.16	3.14	3.08
11	3.26	3.23	3.17	3.16	3.12	3.06	3.03	3.00	2.96	2.94	2.88
12	3.11	3.07	3.02	3.01	2.96	2.91	2.87	2.85	2.81	2.79	2.72
13	2.98	2.95	2.89	2.88	2.84	2.78	2.74	2.72	2.68	2.66	2.60
14	2.88	2.84	2.79	2.78	2.73	2.67	2.64	2.61	2.57	2.55	2.49
15	2.79	2.76	2.70	2.69	2.64	2.59	2.55	2.52	2.48	2.46	2.40
16	2.72	2.68	2.63	2.61	2.57	2.51	2.47	2.45	2.40	2.38	2.32
17	2.65	2.62	2.56	2.55	2.50	2.44	2.41	2.38	2.34	2.32	2.25
18	2.60	2.56	2.50	2.49	2.44	2.38	2.35	2.32	2.28	2.26	2.19
19	2.55	2.51	2.45	2.44	2.39	2.33	2.30	2.27	2.23	2.20	2.13
20	2.50	2.46	2.41	2.40	2.35	2.29	2.25	2.22	2.18	2.16	2.09
21	2.46	2.42	2.37	2.36	2.31	2.25	2.21	2.18	2.14	2.11	2.04
22	2.43	2.39	2.33	2.32	2.27	2.21	2.17	2.14	2.10	2.08	2.00
23	2.39	2.36	2.30	2.29	2.24	2.18	2.14	2.11	2.07	2.04	1.97
24	2.36	2.33	2.27	2.26	2.21	2.15	2.11	2.08	2.03	2.01	1.94
25	2.34	2.30	2.24	2.23	2.18	2.12	2.08	2.05	2.01	1.98	1.91
26	2.31	2.28	2.22	2.21	2.16	2.09	2.05	2.03	1.98	1.95	1.88
27	2.29	2.25	2.19	2.18	2.13	2.07	2.03	2.00	1.95	1.93	1.85
28	2.27	2.23	2.17	2.16	2.11	2.05	2.01	1.98	1.93	1.91	1.83
29	2.25	2.21	2.15	2.14	2.09	2.03	1.99	1.96	1.91	1.89	1.81
30	2.23	2.20	2.14	2.12	2.07	2.01	1.97	1.94	1.89	1.87	1.79
40	2.11	2.07	2.01	1.99	1.94	1.88	1.83	1.80	1.75	1.72	1.64
50	2.03	1.99	1.93	1.92	1.87	1.80	1.75	1.72	1.67	1.64	1.55
60	1.98	1.94	1.88	1.87	1.82	1.74	1.70	1.67	1.61	1.58	1.48
70	1.95	1.91	1.85	1.83	1.78	1.71	1.66	1.63	1.57	1.54	1.44
80	1.92	1.88	1.82	1.81	1.75	1.68	1.63	1.60	1.54	1.51	1.40
90	1.91	1.86	1.80	1.79	1.73	1.66	1.61	1.58	1.52	1.48	1.37
100	1.89	1.85	1.78	1.77	1.71	1.64	1.59	1.56	1.50	1.46	1.35
120	1.87	1.82	1.76	1.75	1.69	1.61	1.56	1.53	1.47	1.43	1.31
150	1.84	1.80	1.74	1.72	1.67	1.59	1.54	1.50	1.44	1.40	1.27
∞	1.75	1.71	1.64	1.63	1.57	1.48	1.43	1.39	1.31	1.27	1.00

$$P(F_{\nu_1,\nu_2}) \leq 0.99$$

ν_1 / ν_2	1	2	3	4	5	6	7	8	9	10	12	15
1	4052	4999	5404	5624	5764	5859	5928	5981	6022	6056	6107	6157
2	98.50	99.00	99.16	99.25	99.30	99.33	99.36	99.38	99.39	99.40	99.42	99.43
3	34.12	30.82	29.46	28.71	28.24	27.91	27.67	27.49	27.34	27.23	27.05	26.87
4	21.20	18.00	16.69	15.98	15.52	15.21	14.98	14.80	14.66	14.55	14.37	14.20
5	16.26	13.27	12.06	11.39	10.97	10.67	10.46	10.29	10.16	10.05	9.89	9.72
6	13.75	10.92	9.78	9.15	8.75	8.47	8.26	8.10	7.98	7.87	7.72	7.56
7	12.25	9.55	8.45	7.85	7.46	7.19	6.99	6.84	6.72	6.62	6.47	6.31
8	11.26	8.65	7.59	7.01	6.63	6.37	6.18	6.03	5.91	5.81	5.67	5.52
9	10.56	8.02	6.99	6.42	6.06	5.80	5.61	5.47	5.35	5.26	5.11	4.96
10	10.04	7.56	6.55	5.99	5.64	5.39	5.20	5.06	4.94	4.85	4.71	4.56
11	9.65	7.21	6.22	5.67	5.32	5.07	4.89	4.74	4.63	4.54	4.40	4.25
12	9.33	6.93	5.95	5.41	5.06	4.82	4.64	4.50	4.39	4.30	4.16	4.01
13	9.07	6.70	5.74	5.21	4.86	4.62	4.44	4.30	4.19	4.10	3.96	3.82
14	8.86	6.51	5.56	5.04	4.69	4.46	4.28	4.14	4.03	3.94	3.80	3.66
15	8.68	6.36	5.42	4.89	4.56	4.32	4.14	4.00	3.89	3.80	3.67	3.52
16	8.53	6.23	5.29	4.77	4.44	4.20	4.03	3.89	3.78	3.69	3.55	3.41
17	8.40	6.11	5.19	4.67	4.34	4.10	3.93	3.79	3.68	3.59	3.46	3.31
18	8.29	6.01	5.09	4.58	4.25	4.01	3.84	3.71	3.60	3.51	3.37	3.23
19	8.18	5.93	5.01	4.50	4.17	3.94	3.77	3.63	3.52	3.43	3.30	3.15
20	8.10	5.85	4.94	4.43	4.10	3.87	3.70	3.56	3.46	3.37	3.23	3.09
21	8.02	5.78	4.87	4.37	4.04	3.81	3.64	3.51	3.40	3.31	3.17	3.03
22	7.95	5.72	4.82	4.31	3.99	3.76	3.59	3.45	3.35	3.26	3.12	2.98
23	7.88	5.66	4.76	4.26	3.94	3.71	3.54	3.41	3.30	3.21	3.07	2.93
24	7.82	5.61	4.72	4.22	3.90	3.67	3.50	3.36	3.26	3.17	3.03	2.89
25	7.77	5.57	4.68	4.18	3.85	3.63	3.46	3.32	3.22	3.13	2.99	2.85
26	7.72	5.53	4.64	4.14	3.82	3.59	3.42	3.29	3.18	3.09	2.96	2.81
27	7.68	5.49	4.60	4.11	3.78	3.56	3.39	3.26	3.15	3.06	2.93	2.78
28	7.64	5.45	4.57	4.07	3.75	3.53	3.36	3.23	3.12	3.03	2.90	2.75
29	7.60	5.42	4.54	4.04	3.73	3.50	3.33	3.20	3.09	3.00	2.87	2.73
30	7.56	5.39	4.51	4.02	3.70	3.47	3.30	3.17	3.07	2.98	2.84	2.70
40	7.31	5.18	4.31	3.83	3.51	3.29	3.12	2.99	2.89	2.80	2.66	2.52
50	7.17	5.06	4.20	3.72	3.41	3.19	3.02	2.89	2.78	2.70	2.56	2.42
60	7.08	4.98	4.13	3.65	3.34	3.12	2.95	2.82	2.72	2.63	2.50	2.35
70	7.01	4.92	4.07	3.60	3.29	3.07	2.91	2.78	2.67	2.59	2.45	2.31
80	6.96	4.88	4.04	3.56	3.26	3.04	2.87	2.74	2.64	2.55	2.42	2.27
90	6.93	4.85	4.01	3.53	3.23	3.01	2.84	2.72	2.61	2.52	2.39	2.24
100	6.90	4.82	3.98	3.51	3.21	2.99	2.82	2.69	2.59	2.50	2.37	2.22
120	6.85	4.79	3.95	3.48	3.17	2.96	2.79	2.66	2.56	2.47	2.34	2.19
150	6.81	4.75	3.91	3.45	3.14	2.92	2.76	2.63	2.53	2.44	2.31	2.16
∞	6.64	4.61	3.78	3.32	3.02	2.80	2.64	2.51	2.41	2.32	2.18	2.04

$$P(F_{\nu_1,\nu_2}) \leq 0.99$$

ν_2 \ ν_1	18	20	24	25	30	40	50	60	90	120	∞
1	6191	6209	6234	6240	6260	6286	6302	6313	6331	6340	6366
2	99.44	99.45	99.46	99.46	99.47	99.48	99.48	99.48	99.49	99.49	99.50
3	26.75	26.69	26.60	26.58	26.50	26.41	26.35	26.32	26.25	26.22	26.13
4	14.08	14.02	13.93	13.91	13.84	13.75	13.69	13.65	13.59	13.56	13.46
5	9.61	9.55	9.47	9.45	9.38	9.29	9.24	9.20	9.14	9.11	9.02
6	7.45	7.40	7.31	7.30	7.23	7.14	7.09	7.06	7.00	6.97	6.88
7	6.21	6.16	6.07	6.06	5.99	5.91	5.86	5.82	5.77	5.74	5.65
8	5.41	5.36	5.28	5.26	5.20	5.12	5.07	5.03	4.97	4.95	4.86
9	4.86	4.81	4.73	4.71	4.65	4.57	4.52	4.48	4.43	4.40	4.31
10	4.46	4.41	4.33	4.31	4.25	4.17	4.12	4.08	4.03	4.00	3.91
11	4.15	4.10	4.02	4.01	3.94	3.86	3.81	3.78	3.72	3.69	3.60
12	3.91	3.86	3.78	3.76	3.70	3.62	3.57	3.54	3.48	3.45	3.36
13	3.72	3.66	3.59	3.57	3.51	3.43	3.38	3.34	3.28	3.25	3.17
14	3.56	3.51	3.43	3.41	3.35	3.27	3.22	3.18	3.12	3.09	3.00
15	3.42	3.37	3.29	3.28	3.21	3.13	3.08	3.05	2.99	2.96	2.87
16	3.31	3.26	3.18	3.16	3.10	3.02	2.97	2.93	2.87	2.84	2.75
17	3.21	3.16	3.08	3.07	3.00	2.92	2.87	2.83	2.78	2.75	2.65
18	3.13	3.08	3.00	2.98	2.92	2.84	2.78	2.75	2.69	2.66	2.57
19	3.05	3.00	2.92	2.91	2.84	2.76	2.71	2.67	2.61	2.58	2.49
20	2.99	2.94	2.86	2.84	2.78	2.69	2.64	2.61	2.55	2.52	2.42
21	2.93	2.88	2.80	2.79	2.72	2.64	2.58	2.55	2.49	2.46	2.36
22	2.88	2.83	2.75	2.73	2.67	2.58	2.53	2.50	2.43	2.40	2.31
23	2.83	2.78	2.70	2.69	2.62	2.54	2.48	2.45	2.39	2.35	2.26
24	2.79	2.74	2.66	2.64	2.58	2.49	2.44	2.40	2.34	2.31	2.21
25	2.75	2.70	2.62	2.60	2.54	2.45	2.40	2.36	2.30	2.27	2.17
26	2.72	2.66	2.58	2.57	2.50	2.42	2.36	2.33	2.26	2.23	2.13
27	2.68	2.63	2.55	2.54	2.47	2.38	2.33	2.29	2.23	2.20	2.10
28	2.65	2.60	2.52	2.51	2.44	2.35	2.30	2.26	2.20	2.17	2.06
29	2.63	2.57	2.49	2.48	2.41	2.33	2.27	2.23	2.17	2.14	2.03
30	2.60	2.55	2.47	2.45	2.39	2.30	2.25	2.21	2.14	2.11	2.01
40	2.42	2.37	2.29	2.27	2.20	2.11	2.06	2.02	1.95	1.92	1.80
50	2.32	2.27	2.18	2.17	2.10	2.01	1.95	1.91	1.84	1.80	1.68
60	2.25	2.20	2.12	2.10	2.03	1.94	1.88	1.84	1.76	1.73	1.60
70	2.20	2.15	2.07	2.05	1.98	1.89	1.83	1.78	1.71	1.67	1.54
80	2.17	2.12	2.03	2.01	1.94	1.85	1.79	1.75	1.67	1.63	1.49
90	2.14	2.09	2.00	1.99	1.92	1.82	1.76	1.72	1.64	1.60	1.46
100	2.12	2.07	1.98	1.97	1.89	1.80	1.74	1.69	1.61	1.57	1.43
120	2.09	2.03	1.95	1.93	1.86	1.76	1.70	1.66	1.58	1.53	1.38
150	2.06	2.00	1.92	1.90	1.83	1.73	1.66	1.62	1.54	1.49	1.33
∞	1.93	1.88	1.79	1.77	1.70	1.59	1.52	1.47	1.38	1.32	1.00

C.8 Critical Values for the Wilcoxon Rank-Sum Test

Table C.8 provides critical values for the Wilcoxon rank-sum test for independent samples with sizes from 3 to 25. Column m is the sample size for the smaller sample and column n is the sample size for the larger sample. If the sample sizes are equal, either sample can be designated m. For each pair of sample sizes (m, n) there are two sets of critical values, one set for one-tail $\alpha = 0.025$ and two-tail $\alpha = 0.05$ and a second set for one-tail $\alpha = 0.05$ and two-tail $\alpha = 0.10$. Suppose for a two-tailed test at $\alpha = 0.05$ we have $m = 8$ and $n = 9$. In the appropriate row and column we find the following numbers 51, 93 0.0232. The 51 and 93 are the lower and upper critical values for W_X, the statistic testing H_0: $M_X = M_Y$. If $W_X \leq 51$ or $W_X \geq 93$, H_0 would be rejected. The value 0.0232 is the exact P value for the critical values of 51 or 93. For more examples see Chapter 7.

1-tail 2-tail		$\alpha = 0.025$ $\alpha = 0.05$		$\alpha = 0.05$ $\alpha = 0.10$		1-tail 2-tail		$\alpha = 0.025$ $\alpha = 0.05$		$\alpha = 0.05$ $\alpha = 0.10$	
n_1 / m	n_2 / n	W	P	W	P	n_1 / m	n_2 / n	W	P	W	P
3	3			6, 15	0.0500	5	10	23, 57	0.0200	26, 54	0.0496
3	4			6, 18	0.0286	5	11	24, 61	0.0190	27, 58	0.0449
3	5	6, 21	0.0179	7, 20	0.0357	5	12	26, 64	0.0242	28, 62	0.0409
3	6	7, 23	0.0238	8, 22	0.0476	5	13	27, 68	0.0230	30, 65	0.0473
3	7	7, 26	0.0167	8, 25	0.0333	5	14	28, 72	0.0218	31, 69	0.0435
3	8	8, 28	0.0242	9, 27	0.0424	5	15	29, 76	0.0209	33, 72	0.0491
3	9	8, 31	0.0182	10, 29	0.0500	5	16	30, 80	0.0201	34, 76	0.0455
3	10	9, 33	0.0245	10, 32	0.0385	5	17	32, 83	0.0238	35, 80	0.0425
3	11	9, 36	0.0192	11, 34	0.0440	5	18	33, 87	0.0229	37, 83	0.0472
3	12	10, 38	0.0242	11, 37	0.0352	5	19	34, 91	0.0220	38, 87	0.0442
3	13	10, 41	0.0196	12, 39	0.0411	5	20	35, 95	0.0212	40, 90	0.0485
3	14	11, 43	0.0235	13, 41	0.0456	5	21	37, 98	0.0243	41, 94	0.0457
3	15	11, 46	0.0196	13, 44	0.0380	5	22	38, 102	0.0234	43, 97	0.0496
3	16	12, 48	0.0237	14, 46	0.0423	5	23	39, 106	0.0226	44, 101	0.0469
3	17	12, 51	0.0202	15, 48	0.0465	5	24	40, 110	0.0219	45, 105	0.0445
3	18	13, 53	0.0233	15, 51	0.0398	5	25	42, 113	0.0246	47, 108	0.0480
3	19	13, 56	0.0201	16, 53	0.0435	6	6	26, 52	0.0206	28, 50	0.0465
3	20	14, 58	0.0232	17, 55	0.0469	6	7	27, 57	0.0175	29, 55	0.0367
3	21	14, 61	0.0203	17, 58	0.0410	6	8	29, 61	0.0213	31, 59	0.0406
3	22	15, 63	0.0230	18, 60	0.0443	6	9	31, 65	0.0248	33, 63	0.0440
3	23	15, 66	0.0204	19, 62	0.0473	6	10	32, 70	0.0210	35, 67	0.0467
3	24	16, 68	0.0229	19, 65	0.0421	6	11	34, 74	0.0238	37, 71	0.0491
3	25	16, 71	0.0205	20, 67	0.0449	6	12	35, 79	0.0207	38, 76	0.0415
4	4	10, 26	0.0143	11, 25	0.0286	6	13	37, 83	0.0231	40, 80	0.0437
4	5	11, 29	0.0159	12, 28	0.0317	6	14	38, 88	0.0204	42, 84	0.0457
4	6	12, 32	0.0190	13, 31	0.0333	6	15	40, 92	0.0224	44, 88	0.0474
4	7	13, 35	0.0212	14, 34	0.0364	6	16	42, 96	0.0244	46, 92	0.0490
4	8	14, 38	0.0242	15, 37	0.0364	6	17	43, 101	0.0219	47, 97	0.0433
4	9	14, 42	0.0168	16, 40	0.0378	6	18	45, 105	0.0236	49, 101	0.0448
4	10	15, 45	0.0180	17, 43	0.0380	6	19	46, 110	0.0214	51, 105	0.0462
4	11	16, 48	0.0198	18, 46	0.0388	6	20	48, 114	0.0229	53, 109	0.0475
4	12	17, 51	0.0209	19, 49	0.0390	6	21	50, 118	0.0244	55, 113	0.0487
4	13	18, 54	0.0223	20, 52	0.0395	6	22	51, 123	0.0224	57, 117	0.0498
4	14	19, 57	0.0232	21, 55	0.0395	6	23	53, 127	0.0237	58, 122	0.0452
4	15	20, 60	0.0243	22, 58	0.0400	6	24	54, 132	0.0219	60, 126	0.0463
4	16	21, 63	0.0250	24, 60	0.0497	6	25	56, 136	0.0231	62, 130	0.0473
4	17	21, 67	0.0202	25, 63	0.0493	7	7	36, 69	0.0189	39, 66	0.0487
4	18	22, 70	0.0212	26, 66	0.0491	7	8	38, 74	0.0200	41, 71	0.0469
4	19	23, 73	0.0219	27, 69	0.0487	7	9	40, 79	0.0209	43, 76	0.0454
4	20	24, 76	0.0227	28, 72	0.0485	7	10	42, 84	0.0215	45, 81	0.0439
4	21	25, 79	0.0233	29, 75	0.0481	7	11	44, 89	0.0221	47, 86	0.0427
4	22	26, 82	0.0240	30, 78	0.0480	7	12	46, 94	0.0225	49, 91	0.0416
4	23	27, 85	0.0246	31, 81	0.0477	7	13	48, 99	0.0228	52, 95	0.0484
4	24	27, 89	0.0211	32, 84	0.0475	7	14	50, 104	0.0230	54, 100	0.0469
4	25	28, 92	0.0217	33, 87	0.0473	7	15	52, 109	0.0233	56, 105	0.0455
5	5	17, 38	0.0159	19, 36	0.0476	7	16	54, 114	0.0234	58, 110	0.0443
5	6	18, 42	0.0152	20, 40	0.0411	7	17	56, 119	0.0236	61, 114	0.0497
5	7	20, 45	0.0240	21, 44	0.0366	7	18	58, 124	0.0237	63, 119	0.0484
5	8	21, 49	0.0225	23, 47	0.0466	7	19	60, 129	0.0238	65, 124	0.0471
5	9	22, 53	0.0210	24, 51	0.0415	7	20	62, 134	0.0239	67, 129	0.0460

1-tail	2-tail	$\alpha = 0.025$ $\alpha = 0.05$		$\alpha = 0.05$ $\alpha = 0.10$		1-tail	2-tail	$\alpha = 0.025$ $\alpha = 0.05$		$\alpha = 0.05$ $\alpha = 0.10$	
n_1 / m	n_2 / n	W	P	W	P	n_1 / m	n_2 / n	W	P	W	P
7	21	64, 139	0.0240	69, 134	0.0449	10	20	110, 200	0.0245	117, 193	0.0498
7	22	66, 144	0.0240	72, 138	0.0492	10	21	113, 207	0.0241	120, 200	0.0478
7	23	68, 149	0.0241	74, 143	0.0481	10	22	116, 214	0.0237	123, 207	0.0459
7	24	70, 154	0.0241	76, 148	0.0470	10	23	119, 221	0.0233	127, 213	0.0482
7	25	72, 159	0.0242	78, 153	0.0461	10	24	122, 228	0.0230	130, 220	0.0465
8	8	49, 87	0.0249	51, 85	0.0415	10	25	126, 234	0.0248	134, 226	0.0486
8	9	51, 93	0.0232	54, 90	0.0464	11	11	96, 157	0.0237	100, 153	0.0440
8	10	53, 99	0.0217	56, 96	0.0416	11	12	99, 165	0.0219	104, 160	0.0454
8	11	55, 105	0.0204	59, 101	0.0454	11	13	103, 172	0.0237	108, 167	0.0467
8	12	58, 110	0.0237	62, 106	0.0489	11	14	106, 180	0.0221	112, 174	0.0477
8	13	60, 116	0.0223	64, 112	0.0445	11	15	110, 187	0.0236	116, 181	0.0486
8	14	62, 122	0.0211	67, 117	0.0475	11	16	113, 195	0.0221	120, 188	0.0494
8	15	65, 127	0.0237	69, 123	0.0437	11	17	117, 202	0.0235	123, 196	0.0453
8	16	67, 133	0.0224	72, 128	0.0463	11	18	121, 209	0.0247	127, 203	0.0461
8	17	70, 138	0.0247	75, 133	0.0487	11	19	124, 217	0.0233	131, 210	0.0468
8	18	72, 144	0.0235	77, 139	0.0452	11	20	128, 224	0.0244	135, 217	0.0474
8	19	74, 150	0.0224	80, 144	0.0475	11	21	131, 232	0.0230	139, 224	0.0480
8	20	77, 155	0.0244	83, 149	0.0495	11	22	135, 239	0.0240	143, 231	0.0486
8	21	79, 161	0.0233	85, 155	0.0464	11	23	139, 246	0.0250	147, 238	0.0490
8	22	81, 167	0.0223	88, 160	0.0483	11	24	142, 254	0.0237	151, 245	0.0495
8	23	84, 172	0.0240	90, 166	0.0454	11	25	146, 261	0.0246	155, 252	0.0499
8	24	86, 178	0.0231	93, 171	0.0472	12	12	115, 185	0.0225	120, 180	0.0444
8	25	89, 183	0.0247	96, 176	0.0488	12	13	119, 193	0.0229	125, 187	0.0488
9	9	62, 109	0.0200	66, 105	0.0470	12	14	123, 201	0.0232	129, 195	0.0475
9	10	65, 115	0.0217	69, 111	0.0474	12	15	127, 209	0.0234	133, 203	0.0463
9	11	68, 121	0.0232	72, 117	0.0476	12	16	131, 217	0.0236	138, 210	0.0500
9	12	71, 127	0.0245	75, 123	0.0477	12	17	135, 225	0.0238	142, 218	0.0486
9	13	73, 134	0.0217	78, 129	0.0478	12	18	139, 233	0.0239	146, 226	0.0474
9	14	76, 140	0.0228	81, 135	0.0478	12	19	143, 241	0.0240	150, 234	0.0463
9	15	79, 146	0.0238	84, 141	0.0478	12	20	147, 249	0.0241	155, 241	0.0493
9	16	82, 152	0.0247	87, 147	0.0477	12	21	151, 257	0.0242	159, 249	0.0481
9	17	84, 159	0.0223	90, 153	0.0476	12	22	155, 265	0.0242	163, 257	0.0471
9	18	87, 165	0.0231	93, 159	0.0475	12	23	159, 273	0.0243	168, 264	0.0496
9	19	90, 171	0.0239	96, 165	0.0474	12	24	163, 281	0.0243	172, 272	0.0486
9	20	93, 177	0.0245	99, 171	0.0473	12	25	167, 289	0.0243	176, 280	0.0475
9	21	95, 184	0.0225	102, 177	0.0472	13	13	136, 215	0.0221	142, 209	0.0454
9	22	98, 190	0.0231	105, 183	0.0471	13	14	141, 223	0.0241	147, 217	0.0472
9	23	101, 196	0.0237	108, 189	0.0470	13	15	145, 232	0.0232	152, 225	0.0489
9	24	104, 202	0.0243	111, 195	0.0469	13	16	150, 240	0.0250	156, 234	0.0458
9	25	107, 208	0.0249	114, 201	0.0468	13	17	154, 249	0.0240	161, 242	0.0472
10	10	78, 132	0.0216	82, 128	0.0446	13	18	158, 258	0.0232	166, 250	0.0485
10	11	81, 139	0.0215	86, 134	0.0493	13	19	163, 266	0.0247	171, 258	0.0497
10	12	84, 146	0.0213	89, 141	0.0465	13	20	167, 275	0.0238	175, 267	0.0470
10	13	88, 152	0.0247	92, 148	0.0441	13	21	171, 284	0.0231	180, 275	0.0481
10	14	91, 159	0.0242	96, 154	0.0478	13	22	176, 292	0.0243	185, 283	0.0491
10	15	94, 166	0.0238	99, 161	0.0455	13	23	180, 301	0.0236	189, 292	0.0467
10	16	97, 173	0.0234	103, 167	0.0487	13	24	185, 309	0.0247	194, 300	0.0476
10	17	100, 180	0.0230	106, 174	0.0465	13	25	189, 318	0.0240	199, 308	0.0485
10	18	103, 187	0.0226	110, 180	0.0493	14	14	160, 246	0.0249	166, 240	0.0469
10	19	107, 193	0.0250	113, 187	0.0472	14	15	164, 256	0.0229	171, 249	0.0466

1-tail 2-tail		$\alpha = 0.025$ $\alpha = 0.05$		$\alpha = 0.05$ $\alpha = 0.10$	
n_1 / m	n_2 / n	W	P	W	P
14	16	169, 265	0.0236	176, 258	0.0463
14	17	174, 274	0.0242	182, 266	0.0500
14	18	179, 283	0.0247	187, 275	0.0495
14	19	183, 293	0.0230	192, 284	0.0489
14	20	188, 302	0.0235	197, 293	0.0484
14	21	193, 311	0.0239	202, 302	0.0480
14	22	198, 320	0.0243	207, 311	0.0475
14	23	203, 329	0.0247	212, 320	0.0471
14	24	207, 339	0.0233	218, 328	0.0498
14	25	212, 348	0.0236	223, 337	0.0492
15	15	184, 281	0.0227	192, 273	0.0488
15	16	190, 290	0.0247	197, 283	0.0466
15	17	195, 300	0.0243	203, 292	0.0485
15	18	200, 310	0.0239	208, 302	0.0465
15	19	205, 320	0.0235	214, 311	0.0482
15	20	210, 330	0.0232	220, 320	0.0497
15	21	216, 339	0.0247	225, 330	0.0478
15	22	221, 349	0.0243	231, 339	0.0492
15	23	226, 359	0.0239	236, 349	0.0474
15	24	231, 369	0.0235	242, 358	0.0486
15	25	237, 378	0.0248	248, 367	0.0499
16	16	211, 317	0.0234	219, 309	0.0469
16	17	217, 327	0.0243	225, 319	0.0471
16	18	222, 338	0.0231	231, 329	0.0473
16	19	228, 348	0.0239	237, 339	0.0474
16	20	234, 358	0.0247	243, 349	0.0475
16	21	239, 369	0.0235	249, 359	0.0475
16	22	245, 379	0.0242	255, 369	0.0476
16	23	251, 389	0.0248	261, 379	0.0476
16	24	256, 400	0.0238	267, 389	0.0476
16	25	262, 410	0.0243	273, 399	0.0476
17	17	240, 355	0.0243	249, 346	0.0493
17	18	246, 366	0.0243	255, 357	0.0479
17	19	252, 377	0.0243	262, 367	0.0499
17	20	258, 388	0.0242	268, 378	0.0485
17	21	264, 399	0.0242	274, 389	0.0473
17	22	270, 410	0.0241	281, 399	0.0490
17	23	276, 421	0.0240	287, 410	0.0477
17	24	282, 432	0.0239	294, 420	0.0492
17	25	288, 443	0.0238	300, 431	0.0480
18	18	270, 396	0.0235	280, 386	0.0485
18	19	277, 407	0.0246	287, 397	0.0490
18	20	283, 419	0.0238	294, 408	0.0495
18	21	290, 430	0.0247	301, 419	0.0499
18	22	296, 442	0.0240	307, 431	0.0474
18	23	303, 453	0.0248	314, 442	0.0478
18	24	309, 465	0.0240	321, 453	0.0481
18	25	316, 476	0.0248	328, 464	0.0484
19	19	303, 438	0.0248	313, 428	0.0482
19	20	309, 451	0.0234	320, 440	0.0474
19	21	316, 463	0.0236	328, 451	0.0494
19	22	323, 475	0.0238	335, 463	0.0486
19	23	330, 487	0.0240	342, 475	0.0478
19	24	337, 499	0.0241	350, 486	0.0496
19	25	344, 511	0.0243	357, 498	0.0488
20	20	337, 483	0.0245	348, 472	0.0482
20	21	344, 496	0.0241	356, 484	0.0490
20	22	351, 509	0.0236	364, 496	0.0497
20	23	359, 521	0.0246	371, 509	0.0478
20	24	366, 534	0.0242	379, 521	0.0484
20	25	373, 547	0.0237	387, 533	0.0490
21	21	373, 530	0.0245	385, 518	0.0486
21	22	381, 543	0.0249	393, 531	0.0482
21	23	388, 557	0.0238	401, 544	0.0478
21	24	396, 570	0.0242	410, 556	0.0497
21	25	404, 583	0.0245	418, 569	0.0492
22	22	411, 579	0.0247	424, 566	0.0491
22	23	419, 593	0.0244	432, 580	0.0477
22	24	427, 607	0.0242	441, 593	0.0486
22	25	435, 621	0.0240	450, 606	0.0494
23	23	451, 630	0.0249	465, 616	0.0499
23	24	459, 645	0.0242	474, 630	0.0497
23	25	468, 659	0.0246	483, 644	0.0495
24	24	492, 684	0.0241	507, 669	0.0486
24	25	501, 699	0.0241	517, 683	0.0496
25	25	536, 739	0.0247	552, 723	0.0497

C.9   Critical Values of the q Statistic for the Student-Newman-Keuls Test

Table C.9 is a tabulation of critical values for the Student-Newman-Keuls test used to separate means in a Model I ANOVA with a significant global F test. To find the critical values for the SNK q statistics choose either $\alpha = 0.05$ or $\alpha = 0.01$. Find the number of means (k) in rank order to be compared at the top of the table and the degrees of freedom (ν) for MS_E as a row value. As with earlier tables if the degrees of freedom are not found, be conservative, and use the closest lower value. For example, at $\alpha = 0.05$ with 35 degrees of freedom in MS_E and five means to separate, the critical values would be

k	2	3	4	5
q c.v.	2.888	3.486	3.845	4.102

To compare the first and the last of 5 ordered means use 4.102 as the critical value for the q statistic. To compare any two adjacent means use 2.888 as the critical value for the q statistic. For means separated by one other mean use 3.486 as the critical value for the q statistic. In general, k is the number of means in the range of interest. For further examples, see Chapter 8.

$\alpha = 0.05$

ν	\(k\) 2	3	4	5	6	7	8	9	10
1	17.97	26.98	32.82	37.08	40.41	43.12	45.40	47.36	49.07
2	6.085	8.331	9.798	10.88	11.74	12.44	13.03	13.54	13.99
3	4.501	5.910	6.825	7.502	8.037	8.478	8.853	9.177	9.462
4	3.927	5.040	5.757	6.287	6.707	7.053	7.347	7.602	7.826
5	3.635	4.602	5.218	5.673	6.033	6.330	6.582	6.802	6.995
6	3.461	4.339	4.896	5.305	5.628	5.895	6.122	6.319	6.493
7	3.344	4.165	4.681	5.060	5.359	5.606	5.815	5.998	6.158
8	3.261	4.041	4.529	4.886	5.167	5.399	5.597	5.767	5.918
9	3.199	3.949	4.415	4.756	5.024	5.244	5.432	5.595	5.739
10	3.151	3.877	4.327	4.654	4.912	5.124	5.305	5.461	5.599
11	3.113	3.820	4.256	4.574	4.823	5.028	5.202	5.353	5.487
12	3.082	3.773	4.199	4.508	4.751	4.950	5.119	5.265	5.395
13	3.055	3.735	4.151	4.453	4.690	4.885	5.049	5.192	5.318
14	3.033	3.702	4.111	4.407	4.639	4.829	4.990	5.131	5.254
15	3.014	3.674	4.076	4.367	4.595	4.782	4.940	5.077	5.198
16	2.998	3.649	4.046	4.333	4.557	4.741	4.897	5.031	5.150
17	2.984	3.628	4.020	4.303	4.524	4.705	4.858	4.991	5.108
18	2.971	3.609	3.997	4.277	4.495	4.673	4.824	4.956	5.071
19	2.960	3.593	3.977	4.253	4.469	4.645	4.794	4.924	5.038
20	2.950	3.578	3.958	4.232	4.445	4.620	4.768	4.896	5.008
24	2.919	3.532	3.901	4.166	4.373	4.541	4.684	4.807	4.915
30	2.888	3.486	3.845	4.102	4.302	4.464	4.602	4.720	4.824
40	2.858	3.442	3.791	4.039	4.232	4.389	4.521	4.635	4.735
60	2.829	3.399	3.737	3.977	4.163	4.314	4.441	4.550	4.646
120	2.800	3.356	3.685	3.917	4.096	4.241	4.363	4.468	4.560
∞	2.772	3.314	3.633	3.858	4.030	4.170	4.286	4.387	4.474

					$\alpha = 0.01$				
					k				
v	2	3	4	5	6	7	8	9	10
1	90.03	135.0	164.3	185.6	202.2	215.8	227.2	237.0	245.6
2	14.04	19.02	22.29	24.72	26.63	28.20	29.53	30.68	31.69
3	8.261	10.62	12.17	13.33	14.24	15.00	15.64	16.20	16.69
4	6.512	8.120	9.173	9.958	10.58	11.10	11.55	11.93	12.27
5	5.702	6.976	7.804	8.421	8.913	9.321	9.669	9.972	10.24
6	5.243	6.331	7.033	7.556	7.973	8.318	8.613	8.869	9.097
7	4.949	5.919	6.543	7.005	7.373	7.679	7.939	8.166	8.368
8	4.746	5.635	6.204	6.625	6.960	7.237	7.474	7.681	7.863
9	4.596	5.428	5.957	6.348	6.658	6.915	7.134	7.325	7.495
10	4.482	5.270	5.769	6.136	6.428	6.669	6.875	7.055	7.213
11	4.392	5.146	5.621	5.970	6.247	6.476	6.672	6.842	6.992
12	4.320	5.046	5.502	5.836	6.101	6.321	6.507	6.670	6.814
13	4.260	4.964	5.404	5.727	5.981	6.192	6.372	6.528	6.667
14	4.210	4.895	5.322	5.634	5.881	6.085	6.258	6.409	6.543
15	4.168	4.836	5.252	5.556	5.796	5.994	6.162	6.309	6.439
16	4.131	4.786	5.192	5.489	5.722	5.915	6.079	6.222	6.349
17	4.099	4.742	5.140	5.430	5.659	5.847	6.007	6.147	6.270
18	4.071	4.703	5.094	5.379	5.603	5.788	5.944	6.081	6.201
19	4.046	4.670	5.054	5.334	5.554	5.735	5.889	6.022	6.141
20	4.024	4.639	5.018	5.294	5.510	5.688	5.839	5.970	6.087
24	3.956	4.546	4.907	5.168	5.374	5.542	5.685	5.809	5.919
30	3.889	4.455	4.799	5.048	5.242	5.401	5.536	5.653	5.756
40	3.825	4.367	4.696	4.931	5.114	5.265	5.392	5.502	5.559
60	3.762	4.282	4.595	4.818	4.991	5.133	5.253	5.356	5.447
120	3.702	4.200	4.497	4.709	4.872	5.005	5.118	5.214	5.299
∞	3.643	4.120	4.403	4.603	4.757	4.882	4.987	5.078	5.157

H. Leon Harter. 1960. Tables of Range and Studentized Range. *The Annals of Mathematical Statistics.* **31** (4). 1132, 1136, and 1138. Used with permission.

C.10 Fisher's Z Transformation of Correlation Coefficient r

Table C.10 is used to transform correlation coefficients (r) into Fisher's z values. This transformation is used to generate confidence intervals for ρ when the r values have been deemed significantly different from 0 via a t test. The transformation is the inverse hyperbolic tangent of r and is utilized to normalize the distribution of r, so that values from the cumulative standard normal distribution (Table C.3) can be used in the confidence intervals. Simply look up the first two decimals of r along the side of the table and the third decimal across the top of the table. The intersection of the appropriate row and column is the z transformation. For example, if $r = 0.342$, then the z transformation would be 0.3564.

This transformation can also be done by calculator utilizing

$$z = \tanh^{-1} r = 0.5 \ln \left(\frac{1+r}{1-r} \right).$$

A 95% confidence interval for ζ (zeta), the transformed correlation coefficient, would have the following form

$$L_1 = z - 1.960 \sqrt{\frac{1}{n-3}}$$

$$L_2 = z + 1.960 \sqrt{\frac{1}{n-3}}$$

with the ±1.960 coming from Table C.3. For a 99% confidence interval use ±2.575 also from Table C.3. These limits for ζ are then transformed back using Table C.11 to generate confidence values for ρ (rho). For further discussion see Chapter 10.

r	0	1	2	3	4	5	6	7	8	9	r
0.00	0.0000	0.0010	0.0020	0.0030	0.0040	0.0050	0.0060	0.0070	0.0080	0.0090	0.00
0.01	0.0100	0.0110	0.0120	0.0130	0.0140	0.0150	0.0160	0.0170	0.0180	0.0190	0.01
0.02	0.0200	0.0210	0.0220	0.0230	0.0240	0.0250	0.0260	0.0270	0.0280	0.0290	0.02
0.03	0.0300	0.0310	0.0320	0.0330	0.0340	0.0350	0.0360	0.0370	0.0380	0.0390	0.03
0.04	0.0400	0.0410	0.0420	0.0430	0.0440	0.0450	0.0460	0.0470	0.0480	0.0490	0.04
0.05	0.0500	0.0510	0.0520	0.0530	0.0541	0.0551	0.0561	0.0571	0.0581	0.0591	0.05
0.06	0.0601	0.0611	0.0621	0.0631	0.0641	0.0651	0.0661	0.0671	0.0681	0.0691	0.06
0.07	0.0701	0.0711	0.0721	0.0731	0.0741	0.0751	0.0761	0.0772	0.0782	0.0792	0.07
0.08	0.0802	0.0812	0.0822	0.0832	0.0842	0.0852	0.0862	0.0872	0.0882	0.0892	0.08
0.09	0.0902	0.0913	0.0923	0.0933	0.0943	0.0953	0.0963	0.0973	0.0983	0.0993	0.09
0.10	0.1003	0.1013	0.1024	0.1034	0.1044	0.1054	0.1064	0.1074	0.1084	0.1094	0.10
0.11	0.1104	0.1115	0.1125	0.1135	0.1145	0.1155	0.1165	0.1175	0.1186	0.1196	0.11
0.12	0.1206	0.1216	0.1226	0.1236	0.1246	0.1257	0.1267	0.1277	0.1287	0.1297	0.12
0.13	0.1307	0.1318	0.1328	0.1338	0.1348	0.1358	0.1368	0.1379	0.1389	0.1399	0.13
0.14	0.1409	0.1419	0.1430	0.1440	0.1450	0.1460	0.1471	0.1481	0.1491	0.1501	0.14
0.15	0.1511	0.1522	0.1532	0.1542	0.1552	0.1563	0.1573	0.1583	0.1593	0.1604	0.15
0.16	0.1614	0.1624	0.1634	0.1645	0.1655	0.1665	0.1676	0.1686	0.1696	0.1706	0.16
0.17	0.1717	0.1727	0.1737	0.1748	0.1758	0.1768	0.1779	0.1789	0.1799	0.1809	0.17
0.18	0.1820	0.1830	0.1841	0.1851	0.1861	0.1872	0.1882	0.1892	0.1903	0.1913	0.18
0.19	0.1923	0.1934	0.1944	0.1955	0.1965	0.1975	0.1986	0.1996	0.2007	0.2017	0.19
0.20	0.2027	0.2038	0.2048	0.2059	0.2069	0.2079	0.2090	0.2100	0.2111	0.2121	0.20
0.21	0.2132	0.2142	0.2153	0.2163	0.2174	0.2184	0.2195	0.2205	0.2216	0.2226	0.21
0.22	0.2237	0.2247	0.2258	0.2268	0.2279	0.2289	0.2300	0.2310	0.2321	0.2331	0.22
0.23	0.2342	0.2352	0.2363	0.2374	0.2384	0.2395	0.2405	0.2416	0.2427	0.2437	0.23
0.24	0.2448	0.2458	0.2469	0.2480	0.2490	0.2501	0.2512	0.2522	0.2533	0.2543	0.24
0.25	0.2554	0.2565	0.2575	0.2586	0.2597	0.2608	0.2618	0.2629	0.2640	0.2650	0.25
0.26	0.2661	0.2672	0.2683	0.2693	0.2704	0.2715	0.2726	0.2736	0.2747	0.2758	0.26
0.27	0.2769	0.2779	0.2790	0.2801	0.2812	0.2823	0.2833	0.2844	0.2855	0.2866	0.27
0.28	0.2877	0.2888	0.2899	0.2909	0.2920	0.2931	0.2942	0.2953	0.2964	0.2975	0.28
0.29	0.2986	0.2997	0.3008	0.3018	0.3029	0.3040	0.3051	0.3062	0.3073	0.3084	0.29
0.30	0.3095	0.3106	0.3117	0.3128	0.3139	0.3150	0.3161	0.3172	0.3183	0.3194	0.30
0.31	0.3205	0.3217	0.3228	0.3239	0.3250	0.3261	0.3272	0.3283	0.3294	0.3305	0.31
0.32	0.3316	0.3328	0.3339	0.3350	0.3361	0.3372	0.3383	0.3395	0.3406	0.3417	0.32
0.33	0.3428	0.3440	0.3451	0.3462	0.3473	0.3484	0.3496	0.3507	0.3518	0.3530	0.33
0.34	0.3541	0.3552	0.3564	0.3575	0.3586	0.3598	0.3609	0.3620	0.3632	0.3643	0.34
0.35	0.3654	0.3666	0.3677	0.3689	0.3700	0.3712	0.3723	0.3734	0.3746	0.3757	0.35
0.36	0.3769	0.3780	0.3792	0.3803	0.3815	0.3826	0.3838	0.3850	0.3861	0.3873	0.36
0.37	0.3884	0.3896	0.3907	0.3919	0.3931	0.3942	0.3954	0.3966	0.3977	0.3989	0.37
0.38	0.4001	0.4012	0.4024	0.4036	0.4047	0.4059	0.4071	0.4083	0.4094	0.4106	0.38
0.39	0.4118	0.4130	0.4142	0.4153	0.4165	0.4177	0.4189	0.4201	0.4213	0.4225	0.39
0.40	0.4236	0.4248	0.4260	0.4272	0.4284	0.4296	0.4308	0.4320	0.4332	0.4344	0.40
0.41	0.4356	0.4368	0.4380	0.4392	0.4404	0.4416	0.4428	0.4441	0.4453	0.4465	0.41
0.42	0.4477	0.4489	0.4501	0.4513	0.4526	0.4538	0.4550	0.4562	0.4574	0.4587	0.42
0.43	0.4599	0.4611	0.4624	0.4636	0.4648	0.4660	0.4673	0.4685	0.4698	0.4710	0.43
0.44	0.4722	0.4735	0.4747	0.4760	0.4772	0.4784	0.4797	0.4809	0.4822	0.4834	0.44
0.45	0.4847	0.4860	0.4872	0.4885	0.4897	0.4910	0.4922	0.4935	0.4948	0.4960	0.45
0.46	0.4973	0.4986	0.4999	0.5011	0.5024	0.5037	0.5049	0.5062	0.5075	0.5088	0.46
0.47	0.5101	0.5114	0.5126	0.5139	0.5152	0.5165	0.5178	0.5191	0.5204	0.5217	0.47
0.48	0.5230	0.5243	0.5256	0.5269	0.5282	0.5295	0.5308	0.5321	0.5334	0.5347	0.48
0.49	0.5361	0.5374	0.5387	0.5400	0.5413	0.5427	0.5440	0.5453	0.5466	0.5480	0.49

r	0	1	2	3	4	5	6	7	8	9	r
0.50	0.5493	0.5506	0.5520	0.5533	0.5547	0.5560	0.5573	0.5587	0.5600	0.5614	**0.50**
0.51	0.5627	0.5641	0.5654	0.5668	0.5682	0.5695	0.5709	0.5722	0.5736	0.5750	**0.51**
0.52	0.5763	0.5777	0.5791	0.5805	0.5818	0.5832	0.5846	0.5860	0.5874	0.5888	**0.52**
0.53	0.5901	0.5915	0.5929	0.5943	0.5957	0.5971	0.5985	0.5999	0.6013	0.6027	**0.53**
0.54	0.6042	0.6056	0.6070	0.6084	0.6098	0.6112	0.6127	0.6141	0.6155	0.6169	**0.54**
0.55	0.6184	0.6198	0.6213	0.6227	0.6241	0.6256	0.6270	0.6285	0.6299	0.6314	**0.55**
0.56	0.6328	0.6343	0.6358	0.6372	0.6387	0.6401	0.6416	0.6431	0.6446	0.6460	**0.56**
0.57	0.6475	0.6490	0.6505	0.6520	0.6535	0.6550	0.6565	0.6580	0.6595	0.6610	**0.57**
0.58	0.6625	0.6640	0.6655	0.6670	0.6685	0.6700	0.6716	0.6731	0.6746	0.6761	**0.58**
0.59	0.6777	0.6792	0.6807	0.6823	0.6838	0.6854	0.6869	0.6885	0.6900	0.6916	**0.59**
0.60	0.6931	0.6947	0.6963	0.6978	0.6994	0.7010	0.7026	0.7042	0.7057	0.7073	**0.60**
0.61	0.7089	0.7105	0.7121	0.7137	0.7153	0.7169	0.7185	0.7201	0.7218	0.7234	**0.61**
0.62	0.7250	0.7266	0.7283	0.7299	0.7315	0.7332	0.7348	0.7365	0.7381	0.7398	**0.62**
0.63	0.7414	0.7431	0.7447	0.7464	0.7481	0.7498	0.7514	0.7531	0.7548	0.7565	**0.63**
0.64	0.7582	0.7599	0.7616	0.7633	0.7650	0.7667	0.7684	0.7701	0.7718	0.7736	**0.64**
0.65	0.7753	0.7770	0.7788	0.7805	0.7823	0.7840	0.7858	0.7875	0.7893	0.7910	**0.65**
0.66	0.7928	0.7946	0.7964	0.7981	0.7999	0.8017	0.8035	0.8053	0.8071	0.8089	**0.66**
0.67	0.8107	0.8126	0.8144	0.8162	0.8180	0.8199	0.8217	0.8236	0.8254	0.8273	**0.67**
0.68	0.8291	0.8310	0.8328	0.8347	0.8366	0.8385	0.8404	0.8423	0.8441	0.8460	**0.68**
0.69	0.8480	0.8499	0.8518	0.8537	0.8556	0.8576	0.8595	0.8614	0.8634	0.8653	**0.69**
0.70	0.8673	0.8693	0.8712	0.8732	0.8752	0.8772	0.8792	0.8812	0.8832	0.8852	**0.70**
0.71	0.8872	0.8892	0.8912	0.8933	0.8953	0.8973	0.8994	0.9014	0.9035	0.9056	**0.71**
0.72	0.9076	0.9097	0.9118	0.9139	0.9160	0.9181	0.9202	0.9223	0.9245	0.9266	**0.72**
0.73	0.9287	0.9309	0.9330	0.9352	0.9373	0.9395	0.9417	0.9439	0.9461	0.9483	**0.73**
0.74	0.9505	0.9527	0.9549	0.9571	0.9594	0.9616	0.9639	0.9661	0.9684	0.9707	**0.74**
0.75	0.9730	0.9752	0.9775	0.9798	0.9822	0.9845	0.9868	0.9892	0.9915	0.9939	**0.75**
0.76	0.9962	0.9986	1.0010	1.0034	1.0058	1.0082	1.0106	1.0130	1.0154	1.0179	**0.76**
0.77	1.0203	1.0228	1.0253	1.0277	1.0302	1.0327	1.0352	1.0378	1.0403	1.0428	**0.77**
0.78	1.0454	1.0479	1.0505	1.0531	1.0557	1.0583	1.0609	1.0635	1.0661	1.0688	**0.78**
0.79	1.0714	1.0741	1.0768	1.0795	1.0822	1.0849	1.0876	1.0903	1.0931	1.0958	**0.79**
0.80	1.0986	1.1014	1.1042	1.1070	1.1098	1.1127	1.1155	1.1184	1.1212	1.1241	**0.80**
0.81	1.1270	1.1299	1.1329	1.1358	1.1388	1.1417	1.1447	1.1477	1.1507	1.1538	**0.81**
0.82	1.1568	1.1599	1.1630	1.1660	1.1692	1.1723	1.1754	1.1786	1.1817	1.1849	**0.82**
0.83	1.1881	1.1914	1.1946	1.1979	1.2011	1.2044	1.2077	1.2111	1.2144	1.2178	**0.83**
0.84	1.2212	1.2246	1.2280	1.2315	1.2349	1.2384	1.2419	1.2454	1.2490	1.2526	**0.84**
0.85	1.2562	1.2598	1.2634	1.2671	1.2707	1.2745	1.2782	1.2819	1.2857	1.2895	**0.85**
0.86	1.2933	1.2972	1.3011	1.3050	1.3089	1.3129	1.3169	1.3209	1.3249	1.3290	**0.86**
0.87	1.3331	1.3372	1.3414	1.3456	1.3498	1.3540	1.3583	1.3626	1.3670	1.3714	**0.87**
0.88	1.3758	1.3802	1.3847	1.3892	1.3938	1.3984	1.4030	1.4077	1.4124	1.4171	**0.88**
0.89	1.4219	1.4268	1.4316	1.4365	1.4415	1.4465	1.4516	1.4566	1.4618	1.4670	**0.89**
0.90	1.4722	1.4775	1.4828	1.4882	1.4937	1.4992	1.5047	1.5103	1.5160	1.5217	**0.90**
0.91	1.5275	1.5334	1.5393	1.5453	1.5513	1.5574	1.5636	1.5698	1.5762	1.5826	**0.91**
0.92	1.5890	1.5956	1.6022	1.6089	1.6157	1.6226	1.6296	1.6366	1.6438	1.6510	**0.92**
0.93	1.6584	1.6658	1.6734	1.6811	1.6888	1.6967	1.7047	1.7129	1.7211	1.7295	**0.93**
0.94	1.7380	1.7467	1.7555	1.7645	1.7736	1.7828	1.7923	1.8019	1.8117	1.8216	**0.94**
0.95	1.8318	1.8421	1.8527	1.8635	1.8745	1.8857	1.8972	1.9090	1.9210	1.9333	**0.95**
0.96	1.9459	1.9588	1.9721	1.9857	1.9996	2.0139	2.0287	2.0439	2.0595	2.0756	**0.96**
0.97	2.0923	2.1095	2.1273	2.1457	2.1649	2.1847	2.2054	2.2269	2.2494	2.2729	**0.97**
0.98	2.2976	2.3235	2.3507	2.3796	2.4101	2.4427	2.4774	2.5147	2.5550	2.5987	**0.98**
0.99	2.6467	2.6996	2.7587	2.8257	2.9031	2.9945	3.1063	3.2504	3.4534	3.8002	**0.99**

C.11   Correlation Coefficient r Corresponding to Fisher's Z Transformation

Table C.11 reverses the transformation accomplished by Table C.10

$$r = \tanh z = \frac{e^{2z} - 1}{e^{2z} + 1}.$$

Find the first two significant figures of the z value in the appropriate row and the third significant figure in the appropriate column. Their intersection will generate the correlation coefficient (r) corresponding to this Fisher's z value. See Table C.10 and Chapter 10.

z	0	1	2	3	4	5	6	7	8	9	z
0.00	0.0000	0.0010	0.0020	0.0030	0.0040	0.0050	0.0060	0.0070	0.0080	0.0090	0.00
0.01	0.0100	0.0110	0.0120	0.0130	0.0140	0.0150	0.0160	0.0170	0.0180	0.0190	0.01
0.02	0.0200	0.0210	0.0220	0.0230	0.0240	0.0250	0.0260	0.0270	0.0280	0.0290	0.02
0.03	0.0300	0.0310	0.0320	0.0330	0.0340	0.0350	0.0360	0.0370	0.0380	0.0390	0.03
0.04	0.0400	0.0410	0.0420	0.0430	0.0440	0.0450	0.0460	0.0470	0.0480	0.0490	0.04
0.05	0.0500	0.0510	0.0520	0.0530	0.0539	0.0549	0.0559	0.0569	0.0579	0.0589	0.05
0.06	0.0599	0.0609	0.0619	0.0629	0.0639	0.0649	0.0659	0.0669	0.0679	0.0689	0.06
0.07	0.0699	0.0709	0.0719	0.0729	0.0739	0.0749	0.0759	0.0768	0.0778	0.0788	0.07
0.08	0.0798	0.0808	0.0818	0.0828	0.0838	0.0848	0.0858	0.0868	0.0878	0.0888	0.08
0.09	0.0898	0.0907	0.0917	0.0927	0.0937	0.0947	0.0957	0.0967	0.0977	0.0987	0.09
0.10	0.0997	0.1007	0.1016	0.1026	0.1036	0.1046	0.1056	0.1066	0.1076	0.1086	0.10
0.11	0.1096	0.1105	0.1115	0.1125	0.1135	0.1145	0.1155	0.1165	0.1175	0.1184	0.11
0.12	0.1194	0.1204	0.1214	0.1224	0.1234	0.1244	0.1253	0.1263	0.1273	0.1283	0.12
0.13	0.1293	0.1303	0.1312	0.1322	0.1332	0.1342	0.1352	0.1361	0.1371	0.1381	0.13
0.14	0.1391	0.1401	0.1411	0.1420	0.1430	0.1440	0.1450	0.1460	0.1469	0.1479	0.14
0.15	0.1489	0.1499	0.1508	0.1518	0.1528	0.1538	0.1547	0.1557	0.1567	0.1577	0.15
0.16	0.1586	0.1596	0.1606	0.1616	0.1625	0.1635	0.1645	0.1655	0.1664	0.1674	0.16
0.17	0.1684	0.1694	0.1703	0.1713	0.1723	0.1732	0.1742	0.1752	0.1761	0.1771	0.17
0.18	0.1781	0.1790	0.1800	0.1810	0.1820	0.1829	0.1839	0.1849	0.1858	0.1868	0.18
0.19	0.1877	0.1887	0.1897	0.1906	0.1916	0.1926	0.1935	0.1945	0.1955	0.1964	0.19
0.20	0.1974	0.1983	0.1993	0.2003	0.2012	0.2022	0.2031	0.2041	0.2051	0.2060	0.20
0.21	0.2070	0.2079	0.2089	0.2098	0.2108	0.2117	0.2127	0.2137	0.2146	0.2156	0.21
0.22	0.2165	0.2175	0.2184	0.2194	0.2203	0.2213	0.2222	0.2232	0.2241	0.2251	0.22
0.23	0.2260	0.2270	0.2279	0.2289	0.2298	0.2308	0.2317	0.2327	0.2336	0.2346	0.23
0.24	0.2355	0.2364	0.2374	0.2383	0.2393	0.2402	0.2412	0.2421	0.2430	0.2440	0.24
0.25	0.2449	0.2459	0.2468	0.2477	0.2487	0.2496	0.2506	0.2515	0.2524	0.2534	0.25
0.26	0.2543	0.2552	0.2562	0.2571	0.2580	0.2590	0.2599	0.2608	0.2618	0.2627	0.26
0.27	0.2636	0.2646	0.2655	0.2664	0.2673	0.2683	0.2692	0.2701	0.2711	0.2720	0.27
0.28	0.2729	0.2738	0.2748	0.2757	0.2766	0.2775	0.2784	0.2794	0.2803	0.2812	0.28
0.29	0.2821	0.2831	0.2840	0.2849	0.2858	0.2867	0.2876	0.2886	0.2895	0.2904	0.29
0.30	0.2913	0.2922	0.2931	0.2941	0.2950	0.2959	0.2968	0.2977	0.2986	0.2995	0.30
0.31	0.3004	0.3013	0.3023	0.3032	0.3041	0.3050	0.3059	0.3068	0.3077	0.3086	0.31
0.32	0.3095	0.3104	0.3113	0.3122	0.3131	0.3140	0.3149	0.3158	0.3167	0.3176	0.32
0.33	0.3185	0.3194	0.3203	0.3212	0.3221	0.3230	0.3239	0.3248	0.3257	0.3266	0.33
0.34	0.3275	0.3284	0.3293	0.3302	0.3310	0.3319	0.3328	0.3337	0.3346	0.3355	0.34
0.35	0.3364	0.3373	0.3381	0.3390	0.3399	0.3408	0.3417	0.3426	0.3435	0.3443	0.35
0.36	0.3452	0.3461	0.3470	0.3479	0.3487	0.3496	0.3505	0.3514	0.3522	0.3531	0.36
0.37	0.3540	0.3549	0.3557	0.3566	0.3575	0.3584	0.3592	0.3601	0.3610	0.3618	0.37
0.38	0.3627	0.3636	0.3644	0.3653	0.3662	0.3670	0.3679	0.3688	0.3696	0.3705	0.38
0.39	0.3714	0.3722	0.3731	0.3739	0.3748	0.3757	0.3765	0.3774	0.3782	0.3791	0.39

z	0	1	2	3	4	5	6	7	8	9	z
0.40	0.3799	0.3808	0.3817	0.3825	0.3834	0.3842	0.3851	0.3859	0.3868	0.3876	**0.40**
0.41	0.3885	0.3893	0.3902	0.3910	0.3919	0.3927	0.3936	0.3944	0.3952	0.3961	**0.41**
0.42	0.3969	0.3978	0.3986	0.3995	0.4003	0.4011	0.4020	0.4028	0.4036	0.4045	**0.42**
0.43	0.4053	0.4062	0.4070	0.4078	0.4087	0.4095	0.4103	0.4112	0.4120	0.4128	**0.43**
0.44	0.4136	0.4145	0.4153	0.4161	0.4170	0.4178	0.4186	0.4194	0.4203	0.4211	**0.44**
0.45	0.4219	0.4227	0.4235	0.4244	0.4252	0.4260	0.4268	0.4276	0.4285	0.4293	**0.45**
0.46	0.4301	0.4309	0.4317	0.4325	0.4333	0.4342	0.4350	0.4358	0.4366	0.4374	**0.46**
0.47	0.4382	0.4390	0.4398	0.4406	0.4414	0.4422	0.4430	0.4438	0.4446	0.4454	**0.47**
0.48	0.4462	0.4470	0.4478	0.4486	0.4494	0.4502	0.4510	0.4518	0.4526	0.4534	**0.48**
0.49	0.4542	0.4550	0.4558	0.4566	0.4574	0.4582	0.4590	0.4598	0.4605	0.4613	**0.49**
0.50	0.4621	0.4629	0.4637	0.4645	0.4653	0.4660	0.4668	0.4676	0.4684	0.4692	**0.50**
0.51	0.4699	0.4707	0.4715	0.4723	0.4731	0.4738	0.4746	0.4754	0.4762	0.4769	**0.51**
0.52	0.4777	0.4785	0.4792	0.4800	0.4808	0.4815	0.4823	0.4831	0.4839	0.4846	**0.52**
0.53	0.4854	0.4861	0.4869	0.4877	0.4884	0.4892	0.4900	0.4907	0.4915	0.4922	**0.53**
0.54	0.4930	0.4937	0.4945	0.4953	0.4960	0.4968	0.4975	0.4983	0.4990	0.4998	**0.54**
0.55	0.5005	0.5013	0.5020	0.5028	0.5035	0.5043	0.5050	0.5057	0.5065	0.5072	**0.55**
0.56	0.5080	0.5087	0.5095	0.5102	0.5109	0.5117	0.5124	0.5132	0.5139	0.5146	**0.56**
0.57	0.5154	0.5161	0.5168	0.5176	0.5183	0.5190	0.5198	0.5205	0.5212	0.5219	**0.57**
0.58	0.5227	0.5234	0.5241	0.5248	0.5256	0.5263	0.5270	0.5277	0.5285	0.5292	**0.58**
0.59	0.5299	0.5306	0.5313	0.5320	0.5328	0.5335	0.5342	0.5349	0.5356	0.5363	**0.59**
0.60	0.5370	0.5378	0.5385	0.5392	0.5399	0.5406	0.5413	0.5420	0.5427	0.5434	**0.60**
0.61	0.5441	0.5448	0.5455	0.5462	0.5469	0.5476	0.5483	0.5490	0.5497	0.5504	**0.61**
0.62	0.5511	0.5518	0.5525	0.5532	0.5539	0.5546	0.5553	0.5560	0.5567	0.5574	**0.62**
0.63	0.5581	0.5587	0.5594	0.5601	0.5608	0.5615	0.5622	0.5629	0.5635	0.5642	**0.63**
0.64	0.5649	0.5656	0.5663	0.5669	0.5676	0.5683	0.5690	0.5696	0.5703	0.5710	**0.64**
0.65	0.5717	0.5723	0.5730	0.5737	0.5744	0.5750	0.5757	0.5764	0.5770	0.5777	**0.65**
0.66	0.5784	0.5790	0.5797	0.5804	0.5810	0.5817	0.5823	0.5830	0.5837	0.5843	**0.66**
0.67	0.5850	0.5856	0.5863	0.5869	0.5876	0.5883	0.5889	0.5896	0.5902	0.5909	**0.67**
0.68	0.5915	0.5922	0.5928	0.5935	0.5941	0.5948	0.5954	0.5961	0.5967	0.5973	**0.68**
0.69	0.5980	0.5986	0.5993	0.5999	0.6005	0.6012	0.6018	0.6025	0.6031	0.6037	**0.69**
0.70	0.6044	0.6050	0.6056	0.6063	0.6069	0.6075	0.6082	0.6088	0.6094	0.6100	**0.70**
0.71	0.6107	0.6113	0.6119	0.6126	0.6132	0.6138	0.6144	0.6150	0.6157	0.6163	**0.71**
0.72	0.6169	0.6175	0.6181	0.6188	0.6194	0.6200	0.6206	0.6212	0.6218	0.6225	**0.72**
0.73	0.6231	0.6237	0.6243	0.6249	0.6255	0.6261	0.6267	0.6273	0.6279	0.6285	**0.73**
0.74	0.6291	0.6297	0.6304	0.6310	0.6316	0.6322	0.6328	0.6334	0.6340	0.6346	**0.74**
0.75	0.6351	0.6357	0.6363	0.6369	0.6375	0.6381	0.6387	0.6393	0.6399	0.6405	**0.75**
0.76	0.6411	0.6417	0.6423	0.6428	0.6434	0.6440	0.6446	0.6452	0.6458	0.6463	**0.76**
0.77	0.6469	0.6475	0.6481	0.6487	0.6492	0.6498	0.6504	0.6510	0.6516	0.6521	**0.77**
0.78	0.6527	0.6533	0.6539	0.6544	0.6550	0.6556	0.6561	0.6567	0.6573	0.6578	**0.78**
0.79	0.6584	0.6590	0.6595	0.6601	0.6607	0.6612	0.6618	0.6624	0.6629	0.6635	**0.79**
0.80	0.6640	0.6646	0.6652	0.6657	0.6663	0.6668	0.6674	0.6679	0.6685	0.6690	**0.80**
0.81	0.6696	0.6701	0.6707	0.6712	0.6718	0.6723	0.6729	0.6734	0.6740	0.6745	**0.81**
0.82	0.6751	0.6756	0.6762	0.6767	0.6772	0.6778	0.6783	0.6789	0.6794	0.6799	**0.82**
0.83	0.6805	0.6810	0.6815	0.6821	0.6826	0.6832	0.6837	0.6842	0.6847	0.6853	**0.83**
0.84	0.6858	0.6863	0.6869	0.6874	0.6879	0.6884	0.6890	0.6895	0.6900	0.6905	**0.84**
0.85	0.6911	0.6916	0.6921	0.6926	0.6932	0.6937	0.6942	0.6947	0.6952	0.6957	**0.85**
0.86	0.6963	0.6968	0.6973	0.6978	0.6983	0.6988	0.6993	0.6998	0.7004	0.7009	**0.86**
0.87	0.7014	0.7019	0.7024	0.7029	0.7034	0.7039	0.7044	0.7049	0.7054	0.7059	**0.87**
0.88	0.7064	0.7069	0.7074	0.7079	0.7084	0.7089	0.7094	0.7099	0.7104	0.7109	**0.88**
0.89	0.7114	0.7119	0.7124	0.7129	0.7134	0.7139	0.7143	0.7148	0.7153	0.7158	**0.89**

z	0	1	2	3	4	5	6	7	8	9	z
0.90	0.7163	0.7168	0.7173	0.7178	0.7182	0.7187	0.7192	0.7197	0.7202	0.7207	0.90
0.91	0.7211	0.7216	0.7221	0.7226	0.7230	0.7235	0.7240	0.7245	0.7249	0.7254	0.91
0.92	0.7259	0.7264	0.7268	0.7273	0.7278	0.7283	0.7287	0.7292	0.7297	0.7301	0.92
0.93	0.7306	0.7311	0.7315	0.7320	0.7325	0.7329	0.7334	0.7338	0.7343	0.7348	0.93
0.94	0.7352	0.7357	0.7361	0.7366	0.7371	0.7375	0.7380	0.7384	0.7389	0.7393	0.94
0.95	0.7398	0.7402	0.7407	0.7411	0.7416	0.7420	0.7425	0.7429	0.7434	0.7438	0.95
0.96	0.7443	0.7447	0.7452	0.7456	0.7461	0.7465	0.7469	0.7474	0.7478	0.7483	0.96
0.97	0.7487	0.7491	0.7496	0.7500	0.7505	0.7509	0.7513	0.7518	0.7522	0.7526	0.97
0.98	0.7531	0.7535	0.7539	0.7544	0.7548	0.7552	0.7557	0.7561	0.7565	0.7569	0.98
0.99	0.7574	0.7578	0.7582	0.7586	0.7591	0.7595	0.7599	0.7603	0.7608	0.7612	0.99
1.0	0.7616	0.7620	0.7624	0.7629	0.7633	0.7637	0.7641	0.7645	0.7649	0.7653	1.0
1.1	0.8005	0.8009	0.8012	0.8016	0.8019	0.8023	0.8026	0.8030	0.8034	0.8037	1.1
1.2	0.8337	0.8340	0.8343	0.8346	0.8349	0.8352	0.8355	0.8358	0.8361	0.8364	1.2
1.3	0.8617	0.8620	0.8622	0.8625	0.8627	0.8630	0.8633	0.8635	0.8638	0.8640	1.3
1.4	0.8854	0.8856	0.8858	0.8860	0.8862	0.8864	0.8866	0.8869	0.8871	0.8873	1.4
1.5	0.9051	0.9053	0.9055	0.9057	0.9059	0.9060	0.9062	0.9064	0.9066	0.9068	1.5
1.6	0.9217	0.9218	0.9220	0.9221	0.9223	0.9224	0.9226	0.9227	0.9229	0.9230	1.6
1.7	0.9354	0.9355	0.9357	0.9358	0.9359	0.9360	0.9362	0.9363	0.9364	0.9365	1.7
1.8	0.9468	0.9469	0.9470	0.9471	0.9472	0.9473	0.9474	0.9475	0.9476	0.9477	1.8
1.9	0.9562	0.9563	0.9564	0.9565	0.9566	0.9567	0.9567	0.9568	0.9569	0.9570	1.9
2.0	0.9640	0.9641	0.9642	0.9642	0.9643	0.9644	0.9644	0.9645	0.9646	0.9647	2.0
2.1	0.9705	0.9705	0.9706	0.9706	0.9707	0.9707	0.9708	0.9709	0.9709	0.9710	2.1
2.2	0.9757	0.9758	0.9758	0.9759	0.9759	0.9760	0.9760	0.9761	0.9761	0.9762	2.2
2.3	0.9801	0.9801	0.9802	0.9802	0.9803	0.9803	0.9803	0.9804	0.9804	0.9804	2.3
2.4	0.9837	0.9837	0.9837	0.9838	0.9838	0.9838	0.9839	0.9839	0.9839	0.9840	2.4
2.5	0.9866	0.9866	0.9867	0.9867	0.9867	0.9867	0.9868	0.9868	0.9868	0.9869	2.5
2.6	0.9890	0.9890	0.9891	0.9891	0.9891	0.9891	0.9892	0.9892	0.9892	0.9892	2.6
2.7	0.9910	0.9910	0.9910	0.9911	0.9911	0.9911	0.9911	0.9911	0.9911	0.9912	2.7
2.8	0.9926	0.9926	0.9927	0.9927	0.9927	0.9927	0.9927	0.9927	0.9927	0.9928	2.8
2.9	0.9940	0.9940	0.9940	0.9940	0.9940	0.9940	0.9940	0.9940	0.9941	0.9941	2.9
3.0	0.9951	0.9951	0.9951	0.9951	0.9951	0.9951	0.9951	0.9951	0.9951	0.9951	3.0
3.1	0.9959	0.9960	0.9960	0.9960	0.9960	0.9960	0.9960	0.9960	0.9960	0.9960	3.1
3.2	0.9967	0.9967	0.9967	0.9967	0.9967	0.9967	0.9967	0.9967	0.9967	0.9967	3.2
3.3	0.9973	0.9973	0.9973	0.9973	0.9973	0.9973	0.9973	0.9973	0.9973	0.9973	3.3
3.4	0.9978	0.9978	0.9978	0.9978	0.9978	0.9978	0.9978	0.9978	0.9978	0.9978	3.4
3.5	0.9982	0.9982	0.9982	0.9982	0.9982	0.9982	0.9982	0.9982	0.9982	0.9982	3.5
3.6	0.9985	0.9985	0.9985	0.9985	0.9985	0.9985	0.9985	0.9985	0.9985	0.9985	3.6
3.7	0.9988	0.9988	0.9988	0.9988	0.9988	0.9988	0.9988	0.9988	0.9988	0.9988	3.7
3.8	0.9990	0.9990	0.9990	0.9990	0.9990	0.9990	0.9990	0.9990	0.9990	0.9990	3.8
3.9	0.9992	0.9992	0.9992	0.9992	0.9992	0.9992	0.9992	0.9992	0.9992	0.9992	3.9
4.0	0.9993	0.9993	0.9993	0.9993	0.9993	0.9993	0.9993	0.9993	0.9993	0.9993	4.0
4.1	0.9995	0.9995	0.9995	0.9995	0.9995	0.9995	0.9995	0.9995	0.9995	0.9995	4.1
4.2	0.9996	0.9996	0.9996	0.9996	0.9996	0.9996	0.9996	0.9996	0.9996	0.9996	4.2
4.3	0.9996	0.9996	0.9996	0.9996	0.9996	0.9996	0.9996	0.9996	0.9996	0.9996	4.3
4.4	0.9997	0.9997	0.9997	0.9997	0.9997	0.9997	0.9997	0.9997	0.9997	0.9997	4.4
4.5	0.9998	0.9998	0.9998	0.9998	0.9998	0.9998	0.9998	0.9998	0.9998	0.9998	4.5
4.6	0.9998	0.9998	0.9998	0.9998	0.9998	0.9998	0.9998	0.9998	0.9998	0.9998	4.6
4.7	0.9998	0.9998	0.9998	0.9998	0.9998	0.9998	0.9998	0.9998	0.9998	0.9998	4.7
4.8	0.9999	0.9999	0.9999	0.9999	0.9999	0.9999	0.9999	0.9999	0.9999	0.9999	4.8
4.9	0.9999	0.9999	0.9999	0.9999	0.9999	0.9999	0.9999	0.9999	0.9999	0.9999	4.9

C.12 Cumulative Distribution for Kendall's Test (τ)

Table C.12 compiles the cumulative distribution function for Kendall's correlation coefficient, τ, for 4 to 25 pairs of observations. Find the sample size n across the top of the table. The smaller of the values $C + \frac{1}{2}E$ (the number of concordants plus half the ties) and $D + \frac{1}{2}E$ (the number of discordants plus half the ties) serves as the test statistic. Locate this value as a row heading for the appropriate column. The number found at the intersection of this row and column is the probability of the test statistic being less than or equal to the calculated value if there is no correlation between X and Y. In other words, it is the P value for a one-tailed test of hypothesis. For H_0: $\tau = 0$ versus H_a: $\tau \neq 0$, double this table value.

c	4	5	6	7	8	9	10	11	12	13	14	15
0	0.0417	0.0083	0.0014	0.0002	0.0000	0.0000	0.0000	0.0000	0.0000	0.0000	0.0000	0.0000
1	0.1667	0.0417	0.0083	0.0014	0.0002	0.0000	0.0000	0.0000	0.0000	0.0000	0.0000	0.0000
2	0.3750	0.1167	0.0278	0.0054	0.0009	0.0001	0.0000	0.0000	0.0000	0.0000	0.0000	0.0000
3	0.6250	0.2417	0.0681	0.0151	0.0028	0.0004	0.0001	0.0000	0.0000	0.0000	0.0000	0.0000
4	0.8333	0.4083	0.1361	0.0345	0.0071	0.0012	0.0002	0.0000	0.0000	0.0000	0.0000	0.0000
5	0.9583	0.5917	0.2347	0.0681	0.0156	0.0029	0.0005	0.0001	0.0000	0.0000	0.0000	0.0000
6	1.0000	0.7583	0.3597	0.1194	0.0305	0.0063	0.0011	0.0002	0.0000	0.0000	0.0000	0.0000
7		0.8833	0.5000	0.1907	0.0543	0.0124	0.0023	0.0004	0.0001	0.0000	0.0000	0.0000
8		0.9583	0.6403	0.2810	0.0894	0.0223	0.0046	0.0008	0.0001	0.0000	0.0000	0.0000
9		0.9917	0.7653	0.3863	0.1375	0.0376	0.0083	0.0016	0.0002	0.0000	0.0000	0.0000
10		1.0000	0.8639	0.5000	0.1994	0.0597	0.0143	0.0029	0.0005	0.0001	0.0000	0.0000
11			0.9319	0.6137	0.2742	0.0901	0.0233	0.0050	0.0009	0.0001	0.0000	0.0000
12			0.9722	0.7190	0.3598	0.1298	0.0363	0.0083	0.0016	0.0003	0.0000	0.0000
13			0.9917	0.8093	0.4524	0.1792	0.0542	0.0132	0.0027	0.0005	0.0001	0.0000
14			0.9986	0.8806	0.5476	0.2384	0.0779	0.0203	0.0044	0.0008	0.0001	0.0000
15			1.0000	0.9319	0.6402	0.3061	0.1082	0.0301	0.0069	0.0013	0.0002	0.0000
16				0.9655	0.7258	0.3807	0.1456	0.0433	0.0105	0.0021	0.0004	0.0001
17				0.9849	0.8006	0.4597	0.1904	0.0605	0.0155	0.0033	0.0006	0.0001
18				0.9946	0.8625	0.5403	0.2422	0.0823	0.0224	0.0051	0.0010	0.0002
19				0.9986	0.9106	0.6193	0.3003	0.1092	0.0314	0.0075	0.0015	0.0003
20				0.9998	0.9457	0.6939	0.3637	0.1415	0.0432	0.0108	0.0023	0.0004
21				1.0000	0.9695	0.7616	0.4309	0.1794	0.0580	0.0152	0.0034	0.0006
22					0.9844	0.8208	0.5000	0.2227	0.0763	0.0211	0.0049	0.0010
23					0.9929	0.8702	0.5691	0.2711	0.0985	0.0286	0.0069	0.0014
24					0.9972	0.9099	0.6363	0.3240	0.1248	0.0382	0.0096	0.0021
25					0.9991	0.9403	0.6997	0.3806	0.1554	0.0500	0.0132	0.0030
26					0.9998	0.9624	0.7578	0.4396	0.1904	0.0644	0.0178	0.0041
27					1.0000	0.9777	0.8096	0.5000	0.2295	0.0817	0.0236	0.0057
28					1.0000	0.9876	0.8544	0.5604	0.2726	0.1022	0.0308	0.0078
29						0.9937	0.8918	0.6194	0.3192	0.1259	0.0397	0.0104
30						0.9971	0.9221	0.6760	0.3687	0.1531	0.0505	0.0137
31						0.9988	0.9458	0.7289	0.4203	0.1837	0.0634	0.0179
32						0.9996	0.9637	0.7773	0.4733	0.2177	0.0786	0.0231
33						0.9999	0.9767	0.8206	0.5267	0.2549	0.0963	0.0295
34						1.0000	0.9857	0.8585	0.5797	0.2950	0.1166	0.0372
35						1.0000	0.9917	0.8908	0.6313	0.3377	0.1396	0.0463
36						1.0000	0.9954	0.9177	0.6808	0.3825	0.1654	0.0571
37							0.9977	0.9395	0.7274	0.4289	0.1940	0.0697
38							0.9989	0.9567	0.7705	0.4762	0.2253	0.0843
39							0.9995	0.9699	0.8096	0.5238	0.2591	0.1009

c	16	17	18	19	20	c	16	17	18	19	20
≤20	≤ 0.0001	0.0000	0.0000	0.0000	0.0000	70	0.8249	0.5803	0.3270	0.1492	0.0563
21	0.0001	0.0000	0.0000	0.0000	0.0000	71	0.8471	0.6118	0.3544	0.1660	0.0642
22	0.0002	0.0000	0.0000	0.0000	0.0000	72	0.8675	0.6425	0.3826	0.1840	0.0729
23	0.0003	0.0000	0.0000	0.0000	0.0000	73	0.8859	0.6723	0.4114	0.2031	0.0825
24	0.0004	0.0001	0.0000	0.0000	0.0000	74	0.9025	0.7012	0.4407	0.2234	0.0929
25	0.0006	0.0001	0.0000	0.0000	0.0000	75	0.9174	0.7288	0.4703	0.2447	0.1043
26	0.0008	0.0001	0.0000	0.0000	0.0000	76	0.9305	0.7552	0.5000	0.2670	0.1166
27	0.0012	0.0002	0.0000	0.0000	0.0000	77	0.9420	0.7802	0.5297	0.2903	0.1299
28	0.0017	0.0003	0.0001	0.0000	0.0000	78	0.9520	0.8036	0.5593	0.3144	0.1442
29	0.0023	0.0005	0.0001	0.0000	0.0000	79	0.9606	0.8256	0.5886	0.3394	0.1594
30	0.0032	0.0006	0.0001	0.0000	0.0000	80	0.9679	0.8459	0.6174	0.3650	0.1757
31	0.0043	0.0009	0.0002	0.0000	0.0000	81	0.9742	0.8647	0.6456	0.3913	0.1929
32	0.0057	0.0012	0.0002	0.0000	0.0000	82	0.9794	0.8819	0.6730	0.4180	0.2111
33	0.0076	0.0017	0.0003	0.0001	0.0000	83	0.9837	0.8976	0.6995	0.4451	0.2303
34	0.0099	0.0023	0.0004	0.0001	0.0000	84	0.9872	0.9117	0.7251	0.4725	0.2503
35	0.0128	0.0030	0.0006	0.0001	0.0000	85	0.9901	0.9243	0.7496	0.5000	0.2712
36	0.0163	0.0040	0.0008	0.0002	0.0000	86	0.9924	0.9356	0.7729	0.5275	0.2929
37	0.0206	0.0052	0.0011	0.0002	0.0000	87	0.9943	0.9456	0.7949	0.5549	0.3154
38	0.0258	0.0067	0.0015	0.0003	0.0001	88	0.9957	0.9543	0.8157	0.5820	0.3386
39	0.0321	0.0086	0.0020	0.0004	0.0001	89	0.9968	0.9619	0.8352	0.6087	0.3623
40	0.0394	0.0109	0.0026	0.0005	0.0001	90	0.9977	0.9685	0.8533	0.6350	0.3866
41	0.0480	0.0137	0.0033	0.0007	0.0001	91	0.9983	0.9741	0.8700	0.6606	0.4113
42	0.0580	0.0170	0.0043	0.0009	0.0002	92	0.9988	0.9789	0.8855	0.6856	0.4364
43	0.0695	0.0211	0.0054	0.0012	0.0002	93	0.9992	0.9830	0.8996	0.7097	0.4618
44	0.0826	0.0259	0.0069	0.0016	0.0003	94	0.9994	0.9863	0.9124	0.7330	0.4872
45	0.0975	0.0315	0.0086	0.0020	0.0004	95	0.9996	0.9891	0.9240	0.7553	0.5128
46	0.1141	0.0381	0.0107	0.0026	0.0005	96	0.9997	0.9914	0.9345	0.7766	0.5382
47	0.1325	0.0457	0.0132	0.0033	0.0007	97	0.9998	0.9933	0.9438	0.7969	0.5636
48	0.1529	0.0544	0.0162	0.0041	0.0009	98	0.9999	0.9948	0.9521	0.8160	0.5887
49	0.1751	0.0644	0.0197	0.0052	0.0012	99	0.9999	0.9960	0.9594	0.8340	0.6134
50	0.1992	0.0757	0.0239	0.0064	0.0015	100	1.0000	0.9970	0.9658	0.8508	0.6377
51	0.2251	0.0883	0.0287	0.0079	0.0019	101	1.0000	0.9977	0.9713	0.8666	0.6614
52	0.2528	0.1024	0.0342	0.0097	0.0024	102	1.0000	0.9983	0.9761	0.8811	0.6846
53	0.2821	0.1181	0.0406	0.0118	0.0030	103	1.0000	0.9988	0.9803	0.8945	0.7071
54	0.3129	0.1353	0.0479	0.0143	0.0037	104	1.0000	0.9991	0.9838	0.9069	0.7288
55	0.3450	0.1541	0.0562	0.0172	0.0045	105	1.0000	0.9994	0.9868	0.9181	0.7497
56	0.3783	0.1744	0.0655	0.0206	0.0056	106	1.0000	0.9995	0.9893	0.9284	0.7697
57	0.4124	0.1964	0.0760	0.0245	0.0068	107	1.0000	0.9997	0.9914	0.9376	0.7889
58	0.4472	0.2198	0.0876	0.0290	0.0082	108	1.0000	0.9998	0.9931	0.9459	0.8071
59	0.4823	0.2448	0.1004	0.0342	0.0099	109	1.0000	0.9999	0.9946	0.9534	0.8243
60	0.5177	0.2712	0.1145	0.0400	0.0119	110	1.0000	0.9999	0.9957	0.9600	0.8406
61	0.5528	0.2988	0.1300	0.0466	0.0142	111	1.0000	0.9999	0.9967	0.9658	0.8558
62	0.5876	0.3277	0.1467	0.0541	0.0168	112	1.0000	1.0000	0.9974	0.9710	0.8701
63	0.6217	0.3575	0.1648	0.0624	0.0199	113	1.0000	1.0000	0.9980	0.9755	0.8834
64	0.6550	0.3882	0.1843	0.0716	0.0234	114	1.0000	1.0000	0.9985	0.9794	0.8957
65	0.6871	0.4197	0.2051	0.0819	0.0274	115	1.0000	1.0000	0.9989	0.9828	0.9071
66	0.7179	0.4516	0.2271	0.0931	0.0319	116	1.0000	1.0000	0.9992	0.9857	0.9175
67	0.7472	0.4838	0.2504	0.1055	0.0370	117	1.0000	1.0000	0.9994	0.9882	0.9271
68	0.7749	0.5162	0.2749	0.1189	0.0428	118	1.0000	1.0000	0.9996	0.9903	0.9358
69	0.8008	0.5484	0.3005	0.1334	0.0492	119	1.0000	1.0000	0.9997	0.9921	0.9437

		n						n		
c	21	22	23	24	25	c	21	22	23	24	25
45	0.0001	0.0000	0.0000	0.0000	0.0000	95	0.2853	0.1314	0.0510	0.0170	0.0049
46	0.0001	0.0000	0.0000	0.0000	0.0000	96	0.3060	0.1438	0.0569	0.0193	0.0057
47	0.0001	0.0000	0.0000	0.0000	0.0000	97	0.3272	0.1570	0.0633	0.0218	0.0065
48	0.0002	0.0000	0.0000	0.0000	0.0000	98	0.3491	0.1709	0.0702	0.0246	0.0075
49	0.0002	0.0000	0.0000	0.0000	0.0000	99	0.3714	0.1856	0.0777	0.0277	0.0086
50	0.0003	0.0001	0.0000	0.0000	0.0000	100	0.3942	0.2010	0.0858	0.0312	0.0098
51	0.0004	0.0001	0.0000	0.0000	0.0000	101	0.4173	0.2172	0.0944	0.0349	0.0112
52	0.0005	0.0001	0.0000	0.0000	0.0000	102	0.4408	0.2340	0.1037	0.0390	0.0127
53	0.0007	0.0001	0.0000	0.0000	0.0000	103	0.4644	0.2515	0.1136	0.0435	0.0144
54	0.0008	0.0002	0.0000	0.0000	0.0000	104	0.4881	0.2697	0.1241	0.0484	0.0163
55	0.0010	0.0002	0.0000	0.0000	0.0000	105	0.5119	0.2885	0.1353	0.0537	0.0183
56	0.0013	0.0003	0.0001	0.0000	0.0000	106	0.5356	0.3079	0.1472	0.0594	0.0206
57	0.0016	0.0003	0.0001	0.0000	0.0000	107	0.5592	0.3278	0.1597	0.0656	0.0232
58	0.0020	0.0004	0.0001	0.0000	0.0000	108	0.5827	0.3482	0.1729	0.0723	0.0259
59	0.0025	0.0005	0.0001	0.0000	0.0000	109	0.6058	0.3690	0.1867	0.0795	0.0290
60	0.0030	0.0007	0.0001	0.0000	0.0000	110	0.6286	0.3903	0.2011	0.0872	0.0323
61	0.0037	0.0009	0.0002	0.0000	0.0000	111	0.6509	0.4118	0.2162	0.0954	0.0359
62	0.0045	0.0011	0.0002	0.0000	0.0000	112	0.6728	0.4336	0.2320	0.1041	0.0399
63	0.0054	0.0013	0.0003	0.0001	0.0000	113	0.6940	0.4556	0.2483	0.1134	0.0441
64	0.0066	0.0016	0.0003	0.0001	0.0000	114	0.7147	0.4778	0.2652	0.1233	0.0488
65	0.0078	0.0020	0.0004	0.0001	0.0000	115	0.7347	0.5000	0.2827	0.1338	0.0538
66	0.0093	0.0024	0.0005	0.0001	0.0000	116	0.7540	0.5222	0.3006	0.1448	0.0592
67	0.0111	0.0029	0.0007	0.0001	0.0000	117	0.7725	0.5444	0.3191	0.1564	0.0650
68	0.0131	0.0035	0.0008	0.0002	0.0000	118	0.7902	0.5664	0.3380	0.1686	0.0712
69	0.0154	0.0042	0.0010	0.0002	0.0000	119	0.8071	0.5882	0.3573	0.1813	0.0778
70	0.0180	0.0050	0.0012	0.0003	0.0001	120	0.8232	0.6097	0.3770	0.1947	0.0850
71	0.0210	0.0059	0.0015	0.0003	0.0001	121	0.8384	0.6310	0.3970	0.2086	0.0925
72	0.0244	0.0070	0.0018	0.0004	0.0001	122	0.8528	0.6518	0.4173	0.2230	0.1006
73	0.0282	0.0083	0.0021	0.0005	0.0001	123	0.8663	0.6722	0.4378	0.2380	0.1091
74	0.0325	0.0097	0.0025	0.0006	0.0001	124	0.8789	0.6921	0.4584	0.2536	0.1181
75	0.0373	0.0114	0.0030	0.0007	0.0002	125	0.8907	0.7115	0.4792	0.2696	0.1277
76	0.0426	0.0133	0.0036	0.0009	0.0002	126	0.9017	0.7303	0.5000	0.2861	0.1377
77	0.0485	0.0154	0.0043	0.0010	0.0002	127	0.9118	0.7485	0.5208	0.3031	0.1483
78	0.0551	0.0179	0.0050	0.0012	0.0003	128	0.9212	0.7660	0.5416	0.3205	0.1594
79	0.0623	0.0206	0.0059	0.0015	0.0003	129	0.9298	0.7828	0.5622	0.3383	0.1710
80	0.0702	0.0237	0.0069	0.0018	0.0004	130	0.9377	0.7990	0.5827	0.3564	0.1831
81	0.0788	0.0272	0.0081	0.0021	0.0005	131	0.9449	0.8144	0.6030	0.3749	0.1957
82	0.0882	0.0311	0.0094	0.0025	0.0006	132	0.9515	0.8291	0.6230	0.3936	0.2088
83	0.0983	0.0354	0.0109	0.0030	0.0007	133	0.9574	0.8430	0.6427	0.4126	0.2224
84	0.1093	0.0401	0.0126	0.0035	0.0008	134	0.9627	0.8562	0.6620	0.4318	0.2365
85	0.1211	0.0454	0.0146	0.0041	0.0010	135	0.9675	0.8686	0.6809	0.4512	0.2511
86	0.1337	0.0512	0.0167	0.0048	0.0012	136	0.9718	0.8803	0.6994	0.4707	0.2661
87	0.1472	0.0575	0.0192	0.0056	0.0014	137	0.9756	0.8913	0.7173	0.4902	0.2815
88	0.1616	0.0644	0.0219	0.0065	0.0017	138	0.9790	0.9015	0.7348	0.5098	0.2974
89	0.1768	0.0720	0.0249	0.0075	0.0020	139	0.9820	0.9110	0.7517	0.5293	0.3136
90	0.1929	0.0801	0.0283	0.0086	0.0023	140	0.9846	0.9199	0.7680	0.5488	0.3302
91	0.2098	0.0890	0.0320	0.0099	0.0027	141	0.9869	0.9280	0.7838	0.5682	0.3472
92	0.2275	0.0985	0.0361	0.0114	0.0032	142	0.9889	0.9356	0.7989	0.5874	0.3644
93	0.2460	0.1087	0.0406	0.0131	0.0037	143	0.9907	0.9425	0.8133	0.6064	0.3819
94	0.2653	0.1197	0.0456	0.0149	0.0043	144	0.9922	0.9488	0.8271	0.6251	0.3997

1-tail	0.10	0.05	0.025	0.01	0.005	1-tail	0.10	0.05	0.025	0.01	0.005
2-tail	0.20	0.10	0.050	0.02	0.010	2-tail	0.20	0.10	0.050	0.02	0.010
n: 1	0.9000	0.9500	0.9750	0.9900	0.9950	n: 51	0.1470	0.1680	0.1866	0.2086	0.2239
2	0.6838	0.7764	0.8419	0.9000	0.9293	52	0.1456	0.1664	0.1848	0.2067	0.2217
3	0.5648	0.6360	0.7076	0.7846	0.8290	53	0.1442	0.1648	0.1831	0.2048	0.2197
4	0.4927	0.5652	0.6239	0.6889	0.7342	54	0.1429	0.1633	0.1814	0.2029	0.2177
5	0.4470	0.5095	0.5633	0.6272	0.6685	55	0.1416	0.1619	0.1798	0.2011	0.2157
6	0.4104	0.4680	0.5193	0.5774	0.6166	56	0.1404	0.1604	0.1782	0.1993	0.2138
7	0.3815	0.4361	0.4834	0.5384	0.5758	57	0.1392	0.1591	0.1767	0.1976	0.2120
8	0.3583	0.4096	0.4543	0.5065	0.5418	58	0.1380	0.1577	0.1752	0.1959	0.2102
9	0.3391	0.3875	0.4300	0.4796	0.5133	59	0.1369	0.1564	0.1737	0.1943	0.2084
10	0.3226	0.3687	0.4092	0.4566	0.4889	60	0.1357	0.1551	0.1723	0.1927	0.2067
11	0.3083	0.3524	0.3912	0.4367	0.4677	61	0.1346	0.1539	0.1709	0.1911	0.2051
12	0.2958	0.3382	0.3754	0.4192	0.4490	62	0.1336	0.1526	0.1696	0.1896	0.2034
13	0.2847	0.3255	0.3614	0.4036	0.4325	63	0.1325	0.1514	0.1682	0.1881	0.2018
14	0.2748	0.3142	0.3489	0.3897	0.4176	64	0.1315	0.1503	0.1669	0.1867	0.2003
15	0.2659	0.3040	0.3376	0.3771	0.4042	65	0.1305	0.1491	0.1657	0.1853	0.1988
16	0.2578	0.2947	0.3273	0.3657	0.3920	66	0.1295	0.1480	0.1644	0.1839	0.1973
17	0.2504	0.2863	0.3180	0.3553	0.3809	67	0.1286	0.1469	0.1632	0.1825	0.1958
18	0.2436	0.2785	0.3094	0.3457	0.3706	68	0.1277	0.1459	0.1620	0.1812	0.1944
19	0.2373	0.2714	0.3014	0.3369	0.3612	69	0.1267	0.1448	0.1609	0.1799	0.1930
20	0.2316	0.2647	0.2941	0.3287	0.3524	70	0.1259	0.1438	0.1597	0.1786	0.1917
21	0.2262	0.2586	0.2872	0.3210	0.3443	71	0.1250	0.1428	0.1586	0.1774	0.1903
22	0.2212	0.2528	0.2809	0.3139	0.3367	72	0.1241	0.1418	0.1576	0.1762	0.1890
23	0.2165	0.2475	0.2749	0.3073	0.3295	73	0.1233	0.1409	0.1565	0.1750	0.1878
24	0.2120	0.2424	0.2693	0.3010	0.3229	74	0.1225	0.1399	0.1554	0.1738	0.1865
25	0.2079	0.2377	0.2640	0.2952	0.3166	75	0.1217	0.1390	0.1544	0.1727	0.1853
26	0.2040	0.2332	0.2591	0.2896	0.3106	76	0.1209	0.1381	0.1534	0.1716	0.1841
27	0.2003	0.2290	0.2544	0.2844	0.3050	77	0.1201	0.1372	0.1524	0.1705	0.1829
28	0.1968	0.2250	0.2499	0.2794	0.2997	78	0.1193	0.1364	0.1515	0.1694	0.1817
29	0.1935	0.2212	0.2457	0.2747	0.2947	79	0.1186	0.1355	0.1505	0.1683	0.1806
30	0.1903	0.2176	0.2417	0.2702	0.2899	80	0.1179	0.1347	0.1496	0.1673	0.1795
31	0.1873	0.2141	0.2379	0.2660	0.2853	81	0.1172	0.1339	0.1487	0.1663	0.1784
32	0.1844	0.2108	0.2342	0.2619	0.2809	82	0.1165	0.1330	0.1478	0.1653	0.1773
33	0.1817	0.2077	0.2308	0.2580	0.2768	83	0.1158	0.1323	0.1469	0.1643	0.1763
34	0.1791	0.2047	0.2274	0.2543	0.2728	84	0.1151	0.1315	0.1460	0.1633	0.1752
35	0.1766	0.2018	0.2242	0.2507	0.2690	85	0.1144	0.1307	0.1452	0.1624	0.1742
36	0.1742	0.1991	0.2212	0.2473	0.2653	86	0.1138	0.1300	0.1444	0.1614	0.1732
37	0.1719	0.1965	0.2183	0.2440	0.2618	87	0.1131	0.1292	0.1435	0.1605	0.1722
38	0.1697	0.1939	0.2154	0.2409	0.2584	88	0.1125	0.1285	0.1427	0.1596	0.1713
39	0.1675	0.1915	0.2127	0.2379	0.2552	89	0.1119	0.1278	0.1419	0.1587	0.1703
40	0.1655	0.1891	0.2101	0.2349	0.2521	90	0.1112	0.1271	0.1412	0.1579	0.1694
41	0.1635	0.1869	0.2076	0.2321	0.2490	91	0.1106	0.1264	0.1404	0.1570	0.1685
42	0.1616	0.1847	0.2052	0.2294	0.2461	92	0.1100	0.1257	0.1396	0.1562	0.1676
43	0.1597	0.1826	0.2028	0.2268	0.2433	93	0.1095	0.1251	0.1389	0.1553	0.1667
44	0.1580	0.1805	0.2006	0.2243	0.2406	94	0.1089	0.1244	0.1382	0.1545	0.1658
45	0.1562	0.1786	0.1984	0.2218	0.2380	95	0.1083	0.1238	0.1375	0.1537	0.1649
46	0.1546	0.1767	0.1963	0.2194	0.2354	96	0.1078	0.1231	0.1368	0.1529	0.1641
47	0.1530	0.1748	0.1942	0.2171	0.2330	97	0.1072	0.1225	0.1361	0.1521	0.1632
48	0.1514	0.1730	0.1922	0.2149	0.2306	98	0.1067	0.1219	0.1354	0.1514	0.1624
49	0.1499	0.1713	0.1903	0.2128	0.2283	99	0.1062	0.1213	0.1347	0.1506	0.1616
50	0.1484	0.1696	0.1884	0.2107	0.2260	100	0.1056	0.1207	0.1340	0.1499	0.1608
						$n > 100$	$\dfrac{1.0730}{\sqrt{n}}$	$\dfrac{1.2239}{\sqrt{n}}$	$\dfrac{1.3581}{\sqrt{n}}$	$\dfrac{1.5174}{\sqrt{n}}$	$\dfrac{1.6276}{\sqrt{n}}$

C.15   Critical Values for the Lilliefors Test

Table C.15 compiles the critical values for the Lilliefors test. The columns represent α levels of 0.10, 0.05, and 0.01. Sample sizes are indicated by row headings (4 through 30). The intersection of the appropriate column and row generates the critical value for the Lilliefors statistic

$$L = \max_{x_i} \left\{ |S(x_i) - F(x_i)|, \; |S(x_{i-1}) - F(x_i)| \right\}.$$

If L is greater than the table critical value, the null hyothesis H_0 that the sample is drawn from a normal distribution is rejected. For further discussion see Chapter 11.

n	0.10	0.05	0.01
4	0.352	0.381	0.417
5	0.315	0.337	0.405
6	0.294	0.319	0.364
7	0.276	0.300	0.348
8	0.261	0.285	0.331
9	0.249	0.271	0.311
10	0.239	0.258	0.294
11	0.230	0.249	0.284
12	0.223	0.242	0.275
13	0.214	0.234	0.268
14	0.207	0.227	0.261
15	0.201	0.220	0.257
16	0.195	0.213	0.250
17	0.189	0.206	0.245
18	0.184	0.200	0.239
19	0.179	0.195	0.235
20	0.174	0.190	0.231
25	0.158	0.173	0.200
30	0.144	0.161	0.187
$n > 30$	$\dfrac{0.805}{\sqrt{n}}$	$\dfrac{0.886}{\sqrt{n}}$	$\dfrac{1.031}{\sqrt{n}}$

Adapted from: Lilliefors. 1967. On the Kolmogorov-Smirnov test for normality with mean and variance unknown, *Journal of the American Statistical Association*, **62**: 400 and Lilliefors. 1969. Corrigenda: On the Kolmogorov-Smirnov Test for Normality with Mean and Variance Unknown, *Journal of the American Statistical Association*, **64**: 1702. Reprinted with permission from the *Journal of the American Statistical Association*. Copyright 1967 and 1969 by the American Statistical Association. All rights reserved.

REFERENCES

The following are sources of information for further study. Several of the data sets and problems in this text come from *A Handbook of Small Data Sets*. Such data sets are denoted *(HSDS, # n)*, which refers to data set # n in this book.

Chakraborti, S. and J. Gibbons. 2003. *Nonparametric Statistical Inference.* CRC Press. Boca Raton, Florida.

Daniel, W. W. 2005. *Biostatistics: A Foundation for Analysis in the Health Sciences.* 8th ed. John Wiley & Sons. New York, New York.

Gerstman, B. B. 2008. *Basic Biostatistics–Statistics for Public Health Practice.* Jones and Bartlett Publishers. Sudbury, Massachusetts.

Hand, D. J. *et al.*, editors. 1994. *A Handbook of Small Data Sets.* Chapman & Hall. London.

Rosner, B. 2006. *Fundamentals of Biostatistics.* 6th ed. Duxbury by Wadsworth Publishing Co. Belmont, California.

Samuels, M. L. and J. A. Witmer. 2003. *Statistics for the Life Sciences.* 3rd ed. Prentice Hall. Upper Saddle River, New Jersey.

Sokal, R. R. and F. J. Rohlf. 1995. *Biometry.* 3rd ed. W. H. Freeman and Company. New York.

Zar, J. H. 1999. *Biostatistical Analysis.* 4th ed. Prentice Hall. Upper Saddle River, New Jersey.

extrinsic model, 327
    intrinsic model, 327
    statistic, 151
chi-square test
    2×2 contingency table, 335–338
    $r \times k$ contingency table, 330–335
    for variance, 151–154
    goodness of fit, 326–330
coding, 14
    additive, 15
    multiplicative, 15
coefficient of determination, 296
combinations, 37
concordant, 300
confidence interval
    difference in means, 182, 183
    difference in proportions, 325–326
    for regression slope, 291
    population mean, 110, 113
    population proportion, 118
    population standard deviation, 118
    population variance, 117
    proportion, 120
    sign test, 158
contingency table, 320, 331
continuity correction
    2×2 contingency table, 335
    binomial, 89
cornflakes, 127
correlation
    linear, 294–300
    rank, 300–308
      Kendall's, 300–303
      Spearman's, 306–308
    test for independence, 303
correlation coefficient, 294
    characteristics, 295
    Kendall's τ, 301
    Pearson, 294
    Spearman's r_s, 306
critical value, 131
cross products
    corrected, 287

data
    categorical, 5, 319
    interval, 4
    type, 4

de Moivre, Abraham, 83
degrees of freedom, 113, 327
    ANOVA, 218
density function, 66
    binomial, 72
    continuous random variable, 81
    discrete random variable, 66
    normal, 84
    Poisson, 77
depth, 7
design
    completely random, 255
    factorial, 255
deviates, 10
diploid, 52
discordant, 300
distribution
    F, 174
    binomial, 71
    cumulative frequency, 16
    empirical cumulative, 339
    normal, 83
    Poisson, 77
    probability, 66
    relative cumulative frequency, 16
    sampling, 101
    standard normal, 86
    Student's t, 112
    uniform, 66, 104
distribution function
    cumulative
      continuous, 84
      discrete, 71

empirical cumulative distribution, 339
error mean squares, 218
estimate
    interval, 111
    point, 111
event, 34
    independent, 48
    mutually exclusive, 35, 41
    simple, 34
expected value
    discrete random variable, 68, 367
    generalized, 68
experiment, 33
extrinsic model, 327

F distribution, 174
F test, 173–176
factorial, 36
fence, 21
Fisher's Z transformation, 299
Fisher, Sir Ronald, 174, 213
five-number summary, 21
Friedmam, M., 266
Friedman k-sample test, 264–270
Friedman test
    statistic, 269

Galton, Francis, 83
Gauss, Karl Friedrich, 83
Geiger, Hans, 352
genotype, 52
goodness of fit, 319
    chi-square test, 326
    Kolmogorov-Smirnov test, 338
    Lilliefors test, 344
Gossett, W. S., 112
grand mean, 215

heterozygous, 52
histogram, 17
homozygous, 52
hypothesis
    alternative, 128
    null, 128
    research, 128

independence, 48, 332
independent, 48, 300
independent assortment, 54
index of association, 295
interaction, 258
interference, 255
interquartile range, 20
interval estimate, 111
intrinsic model, 328

Kendall's τ, 301
Kolmogorov-Smirnov test, 338–343
Kruskal-Wallis test, 228–232
    paired comparisons, 232–234

Laplace, Pierre Simon, 83
law of segregation, 52
least squares regression

line, 287
slope, 287
Lilliefors test, 344–347
limits
    sample range, 9

Mann-Whitney U test, 187–190
margin of error
    population proportion, 121
mean, 5
    coded, 15
    population, 5
    sample, 6
        grouped data, 13
mean saquares, 250
median, 7, 20
    population, 7
    sample, 7
Mendelian genetics, 51
mode, 7
Multiplication Principle, 38

normal approximation
    in sign test, 157
    to binomial, 89
normal distribution, 83
    standard, 86

one-way, 213
outlier, 8, 21

parameter, 3, 5
    population, 5
pdf, 66
Pearson correlation coefficient, 294
percentile, 20
permutations, 36
phenotype, 51
point estimate, 111
Poisson
    approximation to binomial, 80
    density function, 77
    random variable, 77
        expected value, 77
        variance, 77
Poisson, Siméon Denis, 77
pooled CSS, 218
population, 3
population proportion, 118

power, 134
precision, 23
probability, 33
    addition rule, 44
    classical approach, 34
    conditional, 46
    multiplication rule, 49
    relative frequency approach, 34
proportion
    population, 118
    sample, 119
proportions test, 323
    one-sample, 322–323
    two-sample, 324–325
Punnett, Reginald, 353

quartile, 20

random sample, 101
random variable, 3, 65
    continuous, 81
        density function, 81
    Poisson, 77
randomized response technique, 50
range, 9
rank correlation, *see* correlation, rank
regression
    linear, 282–294
    slope, 287
regression equation
    least squares, 287
Rutherford, Ernest, 352

sample, 3
    independent, 177
sample proportion, 119
sample space, 34, 40
sampling distribution, 101
sampling variability, 101
scientific method, 1
set, 40
    complement, 41
    conjoint, 40
    disjoint, 41
    empty set, 43
    universal, 40
sign test
    confidence interval, 158–159
    one-sample, 155–158

    paired data, 190–192
significance
    biological, 147
    economic, 147
significant, 139
    biologically, 3
    highly, 139
    statistically, 3, 129
    very highly, 139
Smith-Satterthwaite procedure, 180
SNK test, *see* Student-Newman-Keuls
        test
staged experiment, 38
standard deviation, 12
standard error, 31, 106
    sample, 112
statistic, 3, 5
Student's t, 112
Student-Newman-Keuls test, 223–226,
        253
sum of squares
    among treatments, 216
    corrected, 10, 11, 365
    in regression, 287, 288
    residual, 216
    total, 215
    two-factor analysis, 259
    uncorrected, 11
synergy, 255

t test
    one-sample, 143–147
    paired data, 184–187
    two-sample
        equal variances, 177–180
        unequal variances, 180–182
table
    contingency, *see* contingency table
    relative frequency, 16
thirty–three-hundred rule, 23
Tibbetts, Ian, 314
treatment mean squares, 218
two-sample tests
    summary, 194
Type I error, 132, 134
Type II error, 134

unbiased estimator, 6, 10, 370, 371

variable, 4
    categorical, 5
    continuous, 4
    discrete, 4
    measurement, 4
    ordinal, 5
    random, 65
      binomial, 72
      continuous, 65
      discrete, 65
    ranked, 5
variance, 9, 69
    population, 11, 371
      random variable, 69
    sample, 10, 371, 372
      coded, 15
      computational formula, 11
      grouped data, 13
Venn diagram, 40
Venn, John, 40

Welch's approximate t, 181
Wilcoxon rank-sum test, 187–190
Wilcoxon signed-rank test, 159–164, 192–
      193

Z transformation
    Fisher's, 299
Z-transformation, 86